Sheet Metal 2023

The 20th International Conference on Sheet Metal was hosted by the Institute of Manufacturing Technology at the Friedrich-Alexander-Universität Erlangen-Nürnberg, Germany.

It took place April 2nd - 5th, 2023.

https://www.shemet.org/2023

Editors

Marion Merklein, Hinnerk Hagenah, Joost R. Duflou, Livan Fratini, Paulo Martins, Gerson Meschut, Fabrizio Micari

AF721908

Peer review statement

All papers published in this volume of "Materials Research Proceedings" have been peer reviewed. The process of peer review was initiated and overseen by the above proceedings editors. All reviews were conducted by expert referees in accordance to Materials Research Forum LLC high standards.

Copyright © 2023 by authors

Content from this work may be used under the terms of the Creative Commons Attribution 3.0 license. Any further distribution of this work must maintain attribution to the author(s) and the title of the work, journal citation and DOI.

Published under License by **Materials Research Forum LLC**
Millersville, PA 17551, USA

Published as part of the proceedings series
Materials Research Proceedings
Volume 25 (2023)

ISSN 2474-3941 (Print)
ISSN 2474-395X (Online)

ISBN 978-1-64490-240-0 (Print)
ISBN 978-1-64490-241-7 (eBook)

This book contains information obtained from authentic and highly regarded sources. Reasonable efforts have been made to publish reliable data and information, but the author and publisher cannot assume responsibility for the validity of all materials or the consequences of their use. The authors and publishers have attempted to trace the copyright holders of all material reproduced in this publication and apologize to copyright holders if permission to publish in this form has not been obtained. If any copyright material has not been acknowledged please write and let us know so we may rectify in any future reprint.

Distributed worldwide by

Materials Research Forum LLC
105 Springdale Lane
Millersville, PA 17551
USA
https://www.mrforum.com

Manufactured in the United State of America
10 9 8 7 6 5 4 3 2 1

Table of Contents

Joining by forming

Material characterization

Modelling

Presses and press tools

Processes

Quality and reliability

Sustainability

Warm and hot forming

Committees

Organizing Committee

Prof. M. Merklein	Friedrich-Alexander-Universität Erlangen-Nürnberg, Germany
Prof. J. Duflou	Katholieke Universiteit Leuven, Belgium
Prof. L. Fratini	Università degli Studi di Palermo, Italy
Prof. H. Hagenah	Friedrich-Alexander-Universität Erlangen-Nürnberg, Germany
Prof. P. Martins	Universidade de Lisboa, Portugal
Prof. G. Meschut	Universität Paderborn, Germany
Prof. F. Micari	Università degli Studi di Palermo, Italy

Scientific Advisory Board

Prof. J. Allwood	University of Cambridge, Great Britain
Prof. M. Bambach	ETH Zürich, Switzerland
Prof. D. Banabic	Technical University from Cluj-Napoca, Romania
Prof. B.-A. Behrens	Leibniz Universität Hannover, Germany
Prof. T. Bergs	Rheinisch-Westfälische Technisch Hochschule Aachen, Germany
Prof. P. Boisse	Université de Lyon, France
Prof. A. Brosius	Technische Universität Dresden, Germany
Prof. S. Bruschi	Università degli Studi di Padova, Italy
Prof. G. Buffa	Università degli Studi di Palermo, Italy
Prof. A. H. van den Boogard	University of Twente, The Netherlands
Prof. J. Ciao	Nortwestern University, USA
Prof. E. Ceretti	Free University of Bozen-Bolzano, Italy
Prof. D. Drummer	Friedrich-Alexander-Universität Erlangen-Nürnberg, Germany
Prof. J. Duflou	Katholieke Universiteit Leuven, Belgium
Prof. L. Filice	Università della Calabria, Italy
Prof. L. Fratini	Università degli Studi di Palermo, Italy
Prof. A. Ghiotti	Università degli Studi di Padova, Italy
Prof. C. Giardini	Università degli Studi die Bergamo, Italy
Prof. P. Groche	Technische Universität Darmstadt, Germany
Prof. M. Gude	Technische Universität Dresden, Germany
Prof. H. Hagenah	Friedrich-Alexander-Universität Erlangen-Nürnber, Germany
Prof. G. Hirt	Rheinisch-Westfälische Technisch Hochschule Aachen, Germany
Prof. W. Homberg	Universität Paderborn, Germany
Prof. G. Ingarao	Univerità degli Studi di Palermo, Italy
Prof. V. Kräusel	Technische Universität Chemnitz, Germany
Prof. D. R. Kumar	IIT Delhi, India
Prof. M. Liewald	Universität Stuttgart, Germany
Prof. L. Madeij	AGH University of Science and Technology, Poland
Prof. P. Martins	Universidade de Lisboa, Portugal

Prof. M. Merklein	Friedrich-Alexander-Universität Erlangen-Nürnberg, Germany
Prof. G. Meschut	Universität Paderborn, Germany
Prof. F. Micari	Università degli Studi di Palermo, Italy
Prof. H. Palkowski	Technische Universität Clausthal, Germany
Prof. U. Prahl	Technische Universität Freiberg, Germany
Prof. B. Rolfe	Deakin University, Australia
Prof. M. Schmidt	Friedrich-Alexander-Universität Erlangen-Nürnberg, Germany
Prof. W. Volk	Technische Universität München, Germany
Prof. J. Yanagimoto	The University of Tokyo, Japan
Prof. Z. Zhao	Shanghai Jiao Tong University, PR China

Prof. M. Merklein	Friedrich-Alexander-Universität Erlangen-Nürnberg, Germany
Prof. G. Meschut	Universität Paderborn, Germany
Prof. F. Micari	Università degli Studi di Palermo, Italy
Prof. H. Palkowski	Technische Universität Clausthal, Germany
Prof. U. Prahl	Technische Universität Freiberg, Germany
Prof. B. Rolfe	Deakin University, Australia
Prof. M. Schmidt	Friedrich-Alexander-Universität Erlangen-Nürnberg, Germany
Prof. W. Volk	Technische Universität München, Germany
Prof. J. Yanagimoto	The University of Tokyo, Japan
Prof. Z. Zhao	Shanghai Jiao Tong University, PR China

High speed cutting and forming of sheet metals

Sheet Metal 2023 Materials Research Forum LLC
Materials Research Proceedings **25** (2023) 3-10 https://doi.org/10.21741/9781644902417-1

Temperature measurement during blanking with enhanced speeds

Agnes Schrepfer[1, a *], Markus Welm[1, b], Philipp Tröber[1, c], Florian Steinlehner[1, d] and Wolfram Volk[1, e]

[1]Chair of Metal Forming and Casting, Technical University of Munich, Walther-Meißner-Straße 4, 85748 Garching

[a]agnes.schrepfer@tum.de, [b]markus.welm@tum.de, [c]philipp.troeber@tum.de, [d]florian.steinlehner@tum.de, [e]wolfram.volk@tum.de

Keywords: Sheet Metal, Temperature, Enhanced-Speed Cutting

Abstract. Shear cutting is one of the most important manufacturing processes due to its high productivity and process stability. The advantages of shear cutting also obtain in high-speed cutting and cutting with enhanced speed. In addition, further advantages such as high dimensional accuracy and a predominant fracture zone accompany it. At the same time, according to literature, high cutting speeds lead to increased temperatures in the shear zone, which can entail tool damage and wear in the short or long term. Knowledge of the temperatures is therefore indispensable for forward planning and economical production processes. Therefore, measurement of temperature in the shear zone has already been approached by a wide variety of methods. In this paper, the temperatures are determined by recording the thermoelectric voltages occurring during shear cutting with enhanced speed up to 270 mm/s and converting these voltages into temperature values using knowledge of the Seebeck coefficients of the punch and sheet material.

Introduction

High-speed shear cutting can be characterized by a cutting speed of over 500 mm/s, since the fracture mechanism of the blanking process changes from this speed on [1]. Therefore, high-speed cutting has several advantages over the conventional shearing process. In addition to high dimensional accuracy and low deformation, the components have low burr height, cutting surfaces with a predominant fracture zone content and a particularly fine-grained surface at the same time. Furthermore, this process leads to a reduction of the deformation zone and less work hardening [2]. High-speed shearing consequently saves material as well as rework and eliminates the need for lubricants [3].

During the shear cutting process, the sheet metal is deformed both plastically and elastically until it is separated. For this, a defined amount of cutting work must be performed. A large part of the plastic deformation work, up to 95% according to Macdougall [4], is converted into heat. Only a small part is stored in the structure of the material [5].

If more forming heat is generated by increased forming speeds and, at the same time, only a limited amount can be dissipated by heat conduction, the originally isothermal process takes on an adiabatic character. Due to the high temperatures, softening processes occur in the forming zone. This adiabatic softening influences the fracture process during material separation and thus leads to an improvement in the quality of the blanking parts. [6]

In traditional blanking the punch speed is about 100 mm/s [3]. This paper deals with cutting with enhanced speed of up to 270 mm/s and investigates if the advantages for high-speed cutting also obtain here.

Although the high cutting speed leads to higher productivity and better part quality, it is also associated with the occurrence of high temperatures within the shear zone and in the contact area between the active element and the sheet.

Content from this work may be used under the terms of the Creative Commons Attribution 3.0 license. Any further distribution of this work must maintain attribution to the author(s) and the title of the work, journal citation and DOI. Published under license by Materials Research Forum LLC.

Knowing the values of these temperatures is of particular importance, since a significant influence of the temperature on the process can be assumed above a temperature of 200 °C. Especially increased tool wear due to local changes of the mechanical properties of the tool materials, thermally induced mechanical stresses in the tool surface [7] and increased adhesion behavior [6] result from an increased temperature. During cutting with a very small cutting clearance, thermal expansion of the active elements can also lead to tool damage [8]. Likewise, damage to the tool coating can occur due to the cyclic thermal stress [9] as well as a reduction in the lubricant effect [10]. The measurement of the resulting temperatures during cutting with high stroke rates under different process parameters is therefore indispensable to ensure an economical production process. However, profound knowledge of occurring temperatures while blanking is important to understand and improve the process. For example, wear phenomena such as adhesion formation, which is significantly influenced by temperature itself as well as side effects such as thermoelectricity [11].

In the past, several approaches to determine the temperature in the contact zone between sheet metal and active elements during shear cutting have been taken. These approaches can be divided into experimental, analytical and numerical. In experimental studies, thermocouples were often inserted into punches or cutting plates to determine temperatures [12]. Another method was the detection of thermal radiation from the punch surface [8]. Similarly, an examination of the microstructure of cut components can allow conclusions about temperatures. The temperatures determined range between 32 °C [12] and 600 °C [13].

Analytical calculations also led to values of around 500 °C in the forming zone [7]. In addition, investigations into the numerical modeling of the temperature rise have already been carried out several times. Gruner and Mauermann determined by far the highest values in cutting simulations with increased cutting speeds. Values from 450° to 500 °C were mentioned in this context [14].

In addition to these measured values, experts assume temperatures of up to 300 °C for conventional cutting speeds and up to 1000 °C for high-speed cutting when working with steel [15].

Demmel first pursued the approach of determining temperatures by means of thermoelectricity in shear cutting [16]. With his measurement method, he confirmed previous assumptions that the temperature in the shear zone increases with the cutting speed and reduces with die clearance. The sheet metal also has an influence on temperatures. When cutting steel sheet, temperatures are significantly higher compared to aluminum alloys [17]. Besides maximum temperatures, this measuring method also enables the possibility to determine the characteristic temperature profile for forming processes [16].

The thermoelectric phenomena, called the Seebeck effect, occurs when two different electrical conductors come into contact and are simultaneously subjected to a temperature gradient. During the blanking process, the punch and the sheet metal have electrical contact and therefore represent those two conductors. The temperature differences within the conductors induce the thermodiffusion of charge carriers in the punch as well as in the sheet. The resulting potential difference leads to an electric voltage measurable in an open circuit. As punch and sheet metal build the pair of conductors, the measuring principle is called tool-workpiece-thermocouple. With the use of this principle, temperature measurement is possible with minimal time delay, since the tool represents the sensor. Besides the temperature gradient, the difference in the material-specific Seebeck coefficients (SCs) of the contact partners determine the thermoelectric voltage. In turn, the SC is determined mainly by its chemical composition. However, the different crystal structures and grain sizes of the materials also change it. The SC describes the thermoelectric behavior of materials and is a measure of the increase in thermoelectric voltage with respect to temperature [18].

To infer temperatures from the measured thermoelectric voltages, it is necessary to know the Seebeck coefficients of both contact partners. These coefficients must be determined experimentally for the calibration of the measuring system.

Experimental Setup

Determination of the thermoelectric properties. An experimental determination of the SCs of the investigated tool and sheet materials was possible with a special test rig developed at the Chair of Metal Forming and Casting. For the measurement of the thermoelectric voltage, the two ends of the material samples are exposed to a defined temperature difference. This is achieved by cooling one end of the sample in an ice bath to 0 °C and heating the other end to 500 °C by contact with a copper block. The heated end is then cooled back down to 0 °C using dry ice. Due to the high inertness and temperature resistance, both sample ends were connected to a high precision voltmeter by platinum wires. This ensures a consistently high quality of the measurements. The maximum deviation of the voltage signals is 1.5 % in the measured temperature range. At the same time, platinum serves as a reference material for the thermoelectric characterization. Subsequently, the Seebeck coefficient is calculated from the measured thermoelectric voltage. [16]

Blanking tool and press. For the measurement of thermoelectric voltages during shear cutting and forming, a special tool was designed which, due to its four-track design, allows simultaneous investigation of different configurations. Both the punch and the sheet material are connected to the voltmeter by low resistance copper cables, as it can be seen in Fig. 1. The temperatures at the junctions were measured with high precision semiconductor temperature sensors. To minimize disturbances, the active elements and the sheet metal are electrically isolated from the rest of the tool. The single stroke tests were carried out on a BSTA 1600-181 high-performance stamping press, Bruderer, Frasnacht (Switzerland).

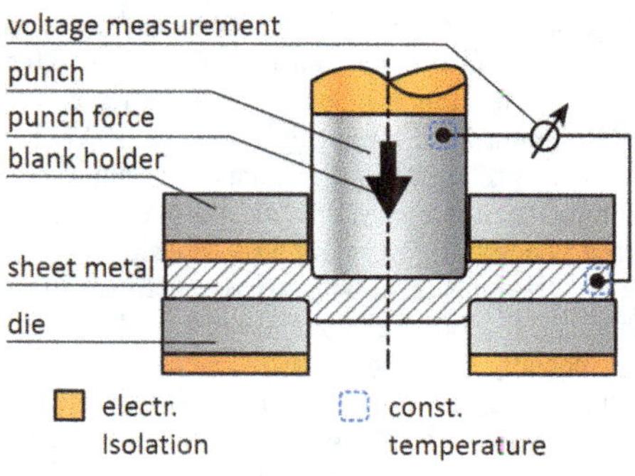

Fig. 1 Blanking tool [17]

Materials

In this publication, the hot-rolled fine-grain steel S355MC with a thickness of 4 mm serves as sheet material. It provides a tensile strength of 491 MPa and is a commonly used steel in cold forming and blanking operations. The punch material is a CF-H40S powder metallurgical cemented carbide. Due to its fine-grain structure, high homogeneity and negative Seebeck coefficient, the material ensures a high quality thermovoltage and thus temperature measurement. A hardness of 1400 HV 10 qualifies it for forming and blanking operations. Approximate temperature-dependent values of the relative Seebeck coefficient from a maximum of 17 µV/°C to a minimum of 10 µV/°C were determined for the sheet material and -10 µV/°C to

-3 μV/°C for the punch material over a temperature range of 0°C to 500°C. Tab. 1 shows the chemical composition of both materials, measured by an optical emission spectrometer.

Table 1: Mass fraction of the chemical composition of the investigated materials (in percent).

CF-H40S	**WC**		**Co**
	88.0		balance
S355MC	**C**	**Mn**	**Fe**
	0.1	0.45	balance

Results

During this investigation, the temperature and the respective punch force during shear cutting for three different speeds were observed. To examine a high velocity range, from normal shear cutting to cutting with enhanced speeds, the stroke rate was varied between 60 1/min and 300 1/min, corresponding to a punch impact speed of 50 mm/s respective 270 mm/s. For every configuration, three measurements were conducted. Fig. 2 shows the averaged punch force (a) and temperature (b) profiles. The x-axis represents the punch travel. While negative values indicate punch movement towards the bottom dead center at 0 mm, the return stroke afterwards is represented by positive values. All experiments were carried out at least three times in order to exclude random measurement errors. The highest standard deviations are 1 kN for all force measurements and 6 °C for the temperature range.

Punch force. The punch force profiles show a characteristic shape for blanking forces with five stages 1-5 but differ in detail. At the beginning (-8.0 mm) the punch comes into contact with the sheet metal and deforms it elastically (1), apparent in a rise in linear force that is the same for all velocities, since it represents the Young's Modulus. Afterwards, plastic deformation begins at -7.7 mm, as indicated by a change of curvature (2). For 60 1/min the force maximum is reached at -7.0 mm with 73 kN; for both higher velocities it amounts to 75 kN at -7.2 mm. Consequently, the force declines and the material is separated at -6.0 mm (3). For both lower speeds, the force profiles show a similar shape, with a strong decrease down to 18 kN for 60 1/min and 21 kN for 150 1/min. Both rise again for about 10 kN at -4.0 mm when the slug hits the slug from the previous stroke in the die channel. During cutting with a stroke rate of 300 1/min, after the maximum, the force stays higher at a plateau and decreases only to 36 kN. This difference can be traced back to the formation of adhesions in the die channel and on the lateral surface of the punch, resulting in varying frictional forces. Therefore, the punch force shows a slightly higher standard deviation of ± 2.5 kN in this phase. The force rebound because of the second slug is not especially sharp and amounts to only 4 kN. Towards the bottom dead center, a slight force decline can be observed for all profiles (4). The maximum return stroke forces amount to -21 kN for 60 1/min, -33 kN for 150 1/min and more than -50 kN for 300 1/min and exceed the maximum measurable tensile force (5).

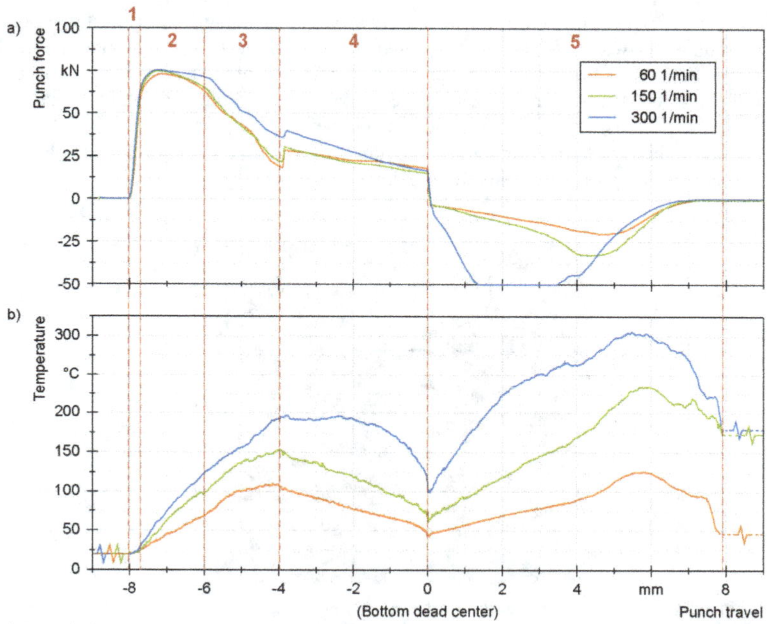

Fig. 2 Punch force (a) and temperature profiles (b) for three different cutting speeds and a die clearance of 1 %

Temperature Profiles. The five blanking stages can also be seen in the temperature profiles. While elastic deformation (1) does not entail heating, no temperature rise from the ambient temperature of 21 °C occurs in the beginning. This temperature corresponds to the volume temperature of the tool and remained constant over all experiments. When the sheet metal is plastically deformed (2), work dissipates and instantaneously triggers a temperature rise, more pronounced with higher cutting speeds. The temperatures increase during stages (2) and (3) and reach a local maximum at the final material separation at -4 mm, with 107 °C for 60 1/min, 156 °C for 150 1/min and 199 °C for 300 1/min, when the slug leaves the stamping grid. Both lower speeds show a slight temperature decrease when the slug comes into contact with the slug from the previous stroke, which is still in the die. Afterwards, the temperature declines toward the bottom dead center (4) to 42 °C, 65 °C and 97 °C. In the return stroke (5), high friction between stamping grid and punch occurs and causes the temperature to rise again, mainly due to frictional heating. All temperatures reach their respective global maximum of 124 °C, 237 °C and 292 °C at 5.7 mm.

Discussion

Regarding the force profiles, low-speed cutting (60 1/min and 150 1/min) shows a similar shape for the complete blanking process. Cutting with enhanced speed changes the slope in three ways. First, a force plateau occurs after the maximum force is reached; it is caused by a change in the material separation process, where the crack propagation is delayed and thus a higher area of clean cut occurs, which can clearly be seen in Fig. 3. Second, the force rise, when the slug from the previous stroke is hit, is not especially sharp and strong. This phenomenon can be traced back to the higher impact speed, thus the inertia force is higher and static friction is overcome faster. The third difference is the much higher return stroke force due to a clamping effect between

Materials Research Proceedings **25** (2023) 3-10 https://doi.org/10.21741/9781644902417-1

stamping grid and punch, caused by the larger contact area resulting from a greater area of clean cut in combination with adhesions at the lateral surface of the punch due to the higher temperature. Regarding the temperature profiles, it becomes obvious that higher speeds entail higher temperatures. The reason is less time for heat conduction and equalizing effects, together with the higher forces and thus dissipation. Hampered heat conduction can also be seen at -4 mm, where the slug of the previous stroke produces a small cooling effect, visible in a small dip in the temperature profile, for only the two lower velocities.

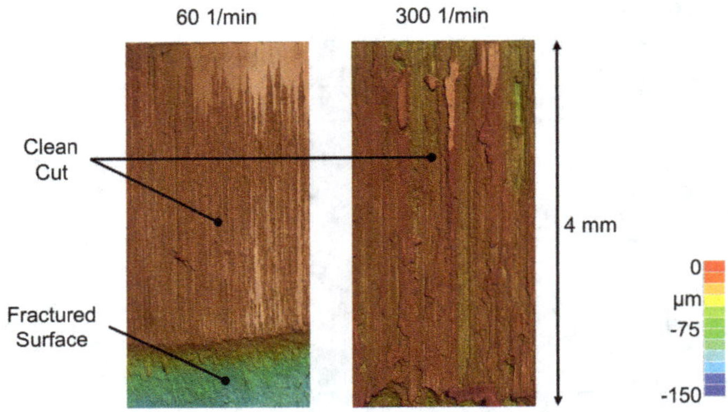

Fig. 3 Cutting surfaces of parts cut with 60 Strokes per minute (left) and 300 strokes per minute (right)

Conclusion

Cutting with enhanced speeds affects the blanking process in different ways. One is a change in the punch force slope, which entails a variation in the cutting result, apparent in a higher area of clean cut compared to components cut at slower speed. This shows that, although shear cutting leads to a very good cut surface quality due to the large clean cut area, the typical fracture behavior of a cut surface after shear cutting with high-speed is not yet evident here. Temperatures that lead to a loss of strength within the shear band and thus to the formation of a plane cut surface with a predominant fine-grained fracture zone are not reached with maximum temperatures of 292°C. The late failure point of the sheet material also leads to the disadvantage of high forces during cutting and the return stroke.

In addition, a clear difference compared to cutting with low speeds can be seen in an increase in temperatures during the whole cutting process. This difference is caused by the reduced time for heat equalization and the change in the higher forces. Currently, preparations are being made to accelerate the punching process into the range of high-speed blanking in order to measure temperatures at even higher punch velocities. In this way, on the one hand, the boundary between conventional and high-speed blanking can be precisely determined for the first time and, on the other hand, the separation mechanism in the adiabatic range can be investigated in more detail.

References

[1] F. P. Michalenko, Einfluss der Pressengeschwindigkeit und der Größe des Spiels auf die Kräfteverteilung und den Werkzeugverschleiß beim Lochen und Ausschneiden von Elektroblechen, Kuz.-Stamp-Proizvod, 11 (1969) 23-26.

[2] P. Huml, Der Einfluss der hohen Geschwindigkeit auf das Schneiden von Metallen, Annals of the CIRP, 23-1 (1974) 61-62. https://doi.org/10.1007/s11340-013-9738-1

[3] C. Gaudillière, N. Ranc, A. Larue, A. Maillard, High speed blanking: An experimental method to measure induced cutting forces, Experimental Mechanics, 53 (2013) 1117-1126. https://doi.org/10.1007/BF02327503

[3] D. Macdougall, Determination of the plastic work converted to heat using radiometry, Experimental Mechanics, 40-3 (2000) 298-306.

[4] A. T. Zehnder, A model for the heating due to plastic work, Mechanics Research Communications, 18-1 (1991) 23-28. https://doi.org/10.1016/0093-6413(91)90023-P

[5] R. Ebner, W. Ecker, S. Marsoner, S. Eck, P. Gruber, P. Raninger and M. Krobarth, Methodology for advanced tool load analysis and lifetime prediction of tools, Proceedings of the 9th International Tooling Conference (2012) 3-21.

[6] M. Schüßler, Hochgeschwindigkeits-Scherschneiden im geschlossenen Schnitt zur Verbesserung der Schnitteilequalität, Technical University Darmstadt, 1990.

[7] R. Dies, Die Vorgänge beim Lochen von Stahl und Nichteisenmetallen mit runden Stempeln, Der Maschinenmarkt, 69 (1960).

[8] D. Ugues, M. Pellizzari, A. Pisa, R. Trotti, V. Sisti, Thermal cracking and soldering on PVD coated steel with edges and different surface finishing, Proceedings of the 9th International Tooling Conference (2012) 571-580.

[9] M. Obst, Feinschneidöl – die Lösung für spezielle Prozesse, in: Blech in Form 4, Carl Hanser Verlag, Munich, (2007).

[10] M. Welm, P. Tröber, H.A. Weiss, P. Demmel, R. Golle and W. Volk, Thermoelectrically Based Approaches to Reduce Adhesive Wear During Blanking, The Journal of The Minerals, Metals & Materials Society, 72 (2020). https://doi.org/10.1007/s11837-020-04191-8

[11] C. Rentsch, Feinschneiden mit beschichteten Werkzeugen, Fortschritt-Berichte VDI, Reihe 2: Fertigungstechnik 413 (1996)

[12] Z. H. Chen, L. C. Chan, T. C. Lee, C. Y. Tang, An investigation on the formation and propagation of shear band in fine-blanking process, Journal of Materials Processing Technology 138 (2003) 610-614. https://doi.org/10.1016/S0924-0136(03)00141-9

[13] K. Lange, Umformtechnik, Handbuch für Industrie und Wissenschaft, Band 3: Blechbearbeitung, second ed., Springer-Verlag, Berlin, 1990.

[14] M. Gruner, R. Mauermann, Erweiterte Möglichkeiten in der Anwendung von Servospindelpressen, Kongress Stanztechnik, Dortmund, 2009.

[15] P. Demmel, P. Tröber, T. Kopp, R. Golle, W. Volk and H. Hoffmann, Characterization of the thermoelectric behavior of plastically deformed steels by means of relative Seebeck coefficient, Materials Science Forum, 755 (2013) 1-7. https://doi.org/10.4028/www.scientific.net/MSF.755.1

[16] P. Demmel, In-situ Temperaturmessung beim Scherschneiden, TU München (2014)

[17] P. Tröber, M. Welm, H. A. Weiss, P. Demmel, R. Golle and W. Volk, Temperature, thermoelectric current and adhesion formation during deep drawing, WEAR, 447 (2021). https://doi.org/10.1016/j.wear.2021.203839

[18] P. Tröber, M. Welm, H. A. Weiss, P. Demmel, R. Golle and W. Volk, The influence of process parameters and sheet material on the temperature development in the forming zone during cold forming, Manufacturing Review, 6 (2019). https://doi.org/10.1051/matecconf/201819014004, creative commons license 4.0 (CC BY 4.0)

Materials Research Proceedings **25** (2023) 11-18 https://doi.org/10.21741/9781644902417-2

Proof of concept for incremental sheet metal forming by means of electromagnetic and electrohydraulic high-speed forming

Maik Holzmüller[1, a*], Maik Linnemann[2, c], Werner Homberg[1, b], Verena Psyk[2, d], Verena Kräusel[2, e], Janika Kroos[1, f]

[1]Forming and Machining Technology, Paderborn University, Pohlweg 47-49, 33098 Paderborn, Germany

[2]Fraunhofer Institute for Machine Tools and Forming Technology IWU, Reichenhainer Strasse 88, 09126 Chemnitz, Germany

[a]mh@luf.uni-paderborn.de*, [b]maik.linnemann@iwu.fraunhofer.de, [c]wh@luf.uni-paderborn.de, [d]verena.psyk@iwu.fraunhofer.de, [e]verena.kraeusel@iwu.fraunhofer.de, [f]jkroos@mail.uni-paderborn.de

Keywords: Incremental Sheet Forming, Aluminium, High-Speed Forming

Abstract. The combination of incremental sheet metal forming and high-speed forming offers new possibilities for flexible forming processes in the production of large sheet metal components of increased complexity with relatively low forming energies. In this paper, the general feasibility and process differences between the pulse-driven high-speed forming technologies of electrohydraulic and electromagnetic forming were investigated. An example component made of EN AW-6016 aluminum sheet metal was thus formed incrementally by both processes and the forming result evaluated by an optical 3D measurement system. For this purpose, a forming strategy for electromagnetic incremental forming (EMIF) was developed, tested and adapted to the electrohydraulic incremental forming process (EHIF). The discharge energy, the tool displacement and the pressure field of the forming zone were determined as relevant parameters for the definition of an adequate tool path strategy. It was found that the EHIF process is less affected by larger distances between the tool and the blank, while this is a critical variable for force application to the component during EMIF. On the other hand, the more uniform pressure distribution of the EMIF process is advantageous for forming large steady component areas.

Introduction

The global trends towards individualized and regionally adapted products are leading to new challenges for production technologies due to the significantly higher number of variants and simultaneously decreasing batch sizes [1]. In addition, the reduction of greenhouse gas emissions and the sustainable use of resources are currently significant requirements for both manufacturing processes and products. Lightweight structures, for example, are a possible solution for addressing these challenges. The use of lightweight materials and more differentiated geometries leads to increasingly complex sheet metal components with the aim of saving materials and reducing CO_2 emissions. These demands on forming technology result in the need for flexible and cost-efficient production of complex components in small batch sizes. It is therefore necessary to develop innovative manufacturing processes that can meet these requirements in a good way.

Incremental sheet metal forming is one approach to the economical production of sheet metal components in small batches. In conventional incremental sheet metal forming a universal tool is used to create the desired workpiece shape. This universal tool (stylus) rotates and is moved by a CNC-machine or robot in paths along the desired geometry. The dimension of the mandrel is typically very small in comparison to the size of the workpiece and thus the part is formed gradually and locally. The small deformation zone of the tool makes it possible to dispense with

Content from this work may be used under the terms of the Creative Commons Attribution 3.0 license. Any further distribution of this work must maintain attribution to the author(s) and the title of the work, journal citation and DOI. Published under license by Materials Research Forum LLC.

at least one half of a forming tool such as those used in conventional forming processes like deep drawing. Incremental forming technologies therefore offer a high potential for increasing process flexibility and reducing tooling costs due to the decreased shape retention rate. A disadvantage of incremental sheet metal forming is that the production of large complex parts with fine details is only possible with long machining times. The process is also of limited accuracy for small radii and demanding geometries [2, 3].

High-speed processes, by contrast, offer a solution for manufacturing complex geometries with a high accuracy. These processes are characterized by high forming-speeds (e.g. 400 m/s) and strain rates (e.g. $10^4 s^{-1}$) which lead to increased formability for many materials. Sharp-edged shapes and small radii, in particular, can be produced more accurately than with quasi-static processes [4, 5]. Due to the short process times, inertia effects can be exploited in tool design. The required clamping forces are thus significantly reduced and the tools are more cost-effective [6]. Electromagnetic forming (EMF) is an active-energy-based high-speed forming process, which was first mentioned in the 1960s [7]. Electrohydraulic forming (EHF) is a working-media-based high-speed forming process. A limiting value for both methods is the size of the part to be formed, because they are usually only suitable for small or medium sized components or just for individual areas of a component due to equipment limitations. In the case of large components, very high capacitor charging energies are required, and the load on the tools increases significantly [8]. Other working principles for high-speed forming include the use of explosives and pneumomechanical compression [9].

In EMF the loads acting on the workpiece are generated using the energy density of pulsed magnetic fields. For this purpose, a transient current is imposed on the active tool (inductor), which generates a magnetic field (see Fig. 3 left). This in turn induces an eddy current in the workpiece, which flows in the opposite direction to the inductor current. Based on the fundamental principles of electrodynamics, so-called Lorentz forces are produced in this way. More simply, the acting loads can also be calculated as magnetic pressure according to Eq. 1 [10]. To obtain appropriate results, it is necessary to know the magnetic field intensity H_{gap} between the inductor and the workpiece, the penetrating magnetic field intensity H_{pen} on the other side of the workpiece and the permeability μ. As soon as the magnetic pressure reaches the yield strength of the material, the deformation of the workpiece begins. A detailed description of the process principles, equipment and applications is given in [11].

$$p = \frac{1}{2}\mu\left(H_{gap}^2 - H_{pen}^2\right) \tag{1}$$

In electrohydraulic forming, the workpiece is formed by a short but very high pressure pulse. This is achieved by applying a high voltage to an assembly of two electrodes inside a water-filled discharge chamber (see Fig. 3 left). The voltage is supplied by capacitors. For forming, these are short-circuited via a spark gap between the electrodes. As soon as the water resistance is overcome, a plasma channel is formed and expands at high speed in the working medium. The resulting shockwave accelerates the workpiece into the single-sided die. In conventional EHF setups, the blank is positioned between the die and the discharge chamber in direct contact with the working media. The insulation between the electrodes and the housing, and the sealing of the housing against high pressures are important for the stable operation of EHF tools [12].

The electromagnetic incremental forming process (EMIF), first proposed in [13], combines the advantages of high speed and incremental forming into a new forming process for larger sheet metal parts. The coil system is moved across the part and local deformation generated to shape the part incrementally. A process strategy for forming complex parts by EMIF was investigated in [6]. Discharge energy, tool displacement and the pressure field were found to be relevant parameters for defining a suitable tool path. In [14], a further approach to incremental forming technology was presented by combining it with the electrohydraulic forming process. The incremental application of the electrohydraulic effect for forming complex geometries with sheet metal has not, however, been proven as yet. This paper provides the proof of concept for the electrohydraulic incremental forming (EHIF) strategy and compares EMIF and EHIF on the basis of a specific demonstrator component. For this purpose, the EMIF forming strategy developed at the Fraunhofer Institute for Machine Tools and Forming Technology (IWU) in Chemnitz was transferred to the EHIF process by the Chair of Forming and Machining Technology (LUF) in Paderborn. The same geometry was considered for both technologies to allow a direct comparison.

Experimental setups

For EMIF, a flat spiral coil made of CuCrZr1 with a diameter of 120 mm is used as the active tool. It consists of 2.5 turns and each turn features a width of 10 mm and a height of 25 mm. The insulation distance between the turns is 6 mm. A pulsed-power generator of type Bluewave PS100/25 from PSTproducts, Alzenau, Germany, provides the electrical energy. The maximum capacitor charging energy is 100 kJ at a maximum voltage of 25 kV and a maximum capacitance of 330 μF, which can be changed incrementally. The sheet is positioned between the inductor and the tool via a clamping frame. The positioning is achieved by two orthogonally aligned linear axes. A more detailed description can be found in [6].

The tool used for the investigation of the EHIF process at the LUF consists of an electrode system with two CuCrZr12 electrodes arranged facing each other in a cylindrical discharge chamber. The discharge chamber has a diameter of 150 mm and a spherical reflection surface above the electrodes. The electrical energy is provided by an SSG-0620 pulsed-power generator from Poynting GmbH, Dortmund, Germany. The maximum capacitor charging energy is 6 kJ at a maximum voltage of 20 kV and a capacitance of 30 μF. In contrast to conventional EHF tools, the active medium (water) and the workpiece are separated here by an expandable membrane in high-grade NR-SBR (hardness = 40 Shore) with a thickness of 3 mm. This makes it possible to move and position the tool freely on the workpiece. A more detailed description of the process principles and the tooling can be found in [6, 14, 15].

The forming-tests were carried out using aluminum sheet made of the alloy EN AW-6016 with a thickness of 1 mm. In quasistatic tensile tests in the rolling direction, a yield strength $R_{p0.2}$ of 133 MPa and a tensile strength R_m of 241 MPa at a uniform elongation A_g of 16.1 % were determined.

Incremental manufacturing requires the development of a component-specific forming strategy. The pressure distribution in the forming zone must be taken into consideration here. Analysis of the membrane behavior is one possibility for achieving a qualitative description of the pressure distribution within the EHIF forming zone. Fig. 1 shows, by way of example, the evolution of free expansion of the membrane during a discharge with 3 kJ and 20 kV. The images were taken with a high-speed Fastcam SA1.1 camera from Photron at a frame rate of 5000 fps. Immediately after ignition, the membrane expands in a cone shape within 13.2 ms. The pressure maximum is at its center and the expansion attains a maximum of 134 mm. After the membrane has receded from the first expansion, the pressure waves reflected in the discharge chamber are superimposed and an annular expansion occurs. From this point onward, the membrane oscillates between conical and annular expansion with a rapid decrease in intensity. It can be assumed that, in the process of forming sheet metal, only the first two expansions contribute significantly to the forming, because the impulse is quickly absorbed by the damping of the membrane and energy dissipation during forming.

Materials Research Proceedings **25** (2023) 11-18 https://doi.org/10.21741/9781644902417-2

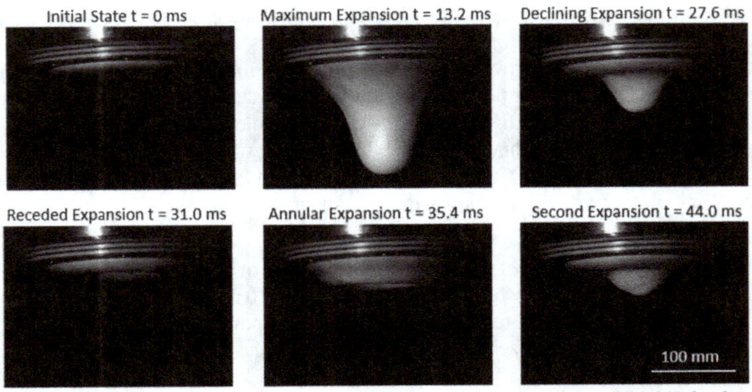

Fig. 1 High-speed camera images of the EHIF membrane at 3 kJ and 20 kV discharge parameters

Electromagnetic incremental forming of a chair seat

The concept for EMIF for large complex parts was already proven in [6]. For this, a small area of a chair seat, designed by elem design [16], was formed with different capacitor charging energies and distances between each step. Taking the experience gathered from these prior investigations on small sections of the part, a strategy for forming the full chair seat was developed in this work. It was found that the process parameters have to be adapted to the geometrical variations within the component. In detail this means that smaller relative displacements are necessary for fine structures. This in turn makes it possible to increase the displacement for large steady areas. In addition, as is familiar from single shot forming, sections with a higher depth of the die require higher capacitor charging energies. The forming strategy shown in Fig. 2 is the result of these findings.

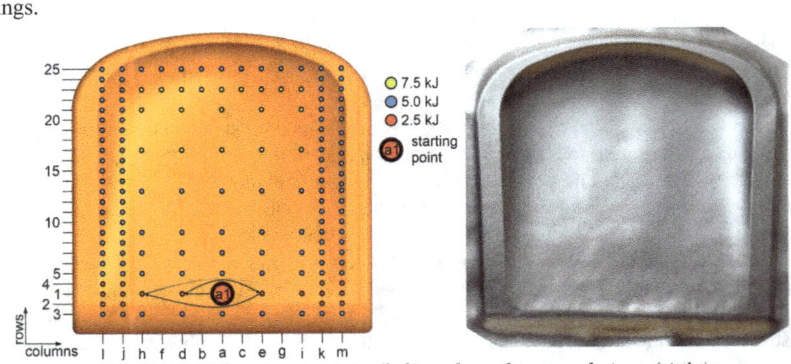

Fig. 2 EMIF forming strategy (left) and resulting workpiece (right)

A suitable starting point for the forming process is the center of the front area of the chair seat (a1). Here, only minimal forming is necessary and hence a low capacitor charging energy is required. In addition, the gradient of the forming depth below the effective area of the inductor is very low compared to all other areas. The value was therefore set to 2.5 kJ. For the area with the largest forming depth, on the other hand, a capacitor charging energy of 7.5 kJ was used. For all other areas of the chair seat surface, it was set at 5 kJ.

In terms of the geometric boundary conditions, the workpiece is well suited to a discontinuous trajectory. As shown in Fig. 2 left, the process was first performed in columns (i.e. in the x-direction), alternating from the inside to the outside (letters a to m). Then the rows (numbers 1 to 25) were formed. The distance from one line to the next was 15 mm. Between the columns there was 30 mm in each case. Since small edge radii run all around the u-shaped flat surface, small displacements (i.e. 15 mm) were chosen at these points. In the center of the workpiece, only a few geometric details are present. A displacement of 60 mm was therefore used in this area. All other areas were produced with a relative displacement of 30 mm. Based on this, the positioning sequences of the inductor were a1, d1, e1, h1, …, j2, k2, l2, m2, a3, d3, …, a5, d5, e5 and so on. With the selected path strategy, 158 capacitor discharges were necessary to form the chair seat. 15 discharges were performed using 7.5 kJ, 17 discharges performed using 2.5 kJ and all the other (i. e. 126) discharges were performed using 5 kJ.

Fig. 2 right shows the result. The u-shaped plane, including the surrounding radii, is well formed. The areas with the lowest forming depths show visible imprints of the inductor for the individual increments, but these are only visible on the side facing the inductor. These imprints should be prevented by homogenizing the magnetic pressure through adjusting the inductor geometry. Furthermore, the complete forming depth has not been reached in the deepest section of the die. This is because the inductor was only moved parallel to the sheet plane. In general, it was shown that workpieces of a good quality can be formed by EMIF.

Proof of concept for electrohydraulic incremental forming

Next, the feasibility of EHIF was demonstrated too. For this purpose, the capacitor charging energies had to be adapted to the different technology (EHIF) and equipment. In order to identify the corresponding energy level – i. e. energy levels leading to the same deformation – in EMF and EHF, free forming tests, similar to [14] were carried out, i. e. metal blanks (ø220 mm) were formed by EMF and EHF into a drawing ring without restricting the forming depth (Fig. 3).

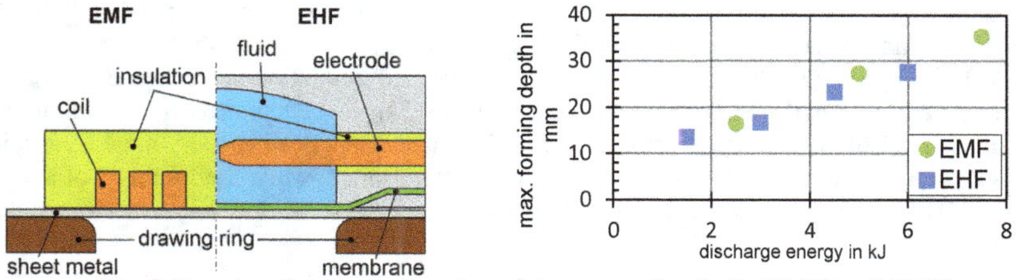

Fig. 3 Forming depth as a function of the energy levels for EMIF and EHIF

Achieving the same forming depth by EHF requires a slightly higher capacitor charging energy than EMF. However, the large number of increments required for forming at capacitor charging energies close to the power limit of the EHIF system at LUF significantly stresses the tooling system and hence the applied discharge energy was limited to 4.5 kJ. Fig. 4 shows the EHIF forming strategy. The starting point (a1) and the discontinuous trajectory were adopted from EMIF. The numbered row spacing in the y-direction (15 mm) and the alphanumeric column widths in the x-direction (30 mm) are identical to those for the EMIF forming strategy in Fig. 2 too.

Materials Research Proceedings **25** (2023) 11-18 https://doi.org/10.21741/9781644902417-2

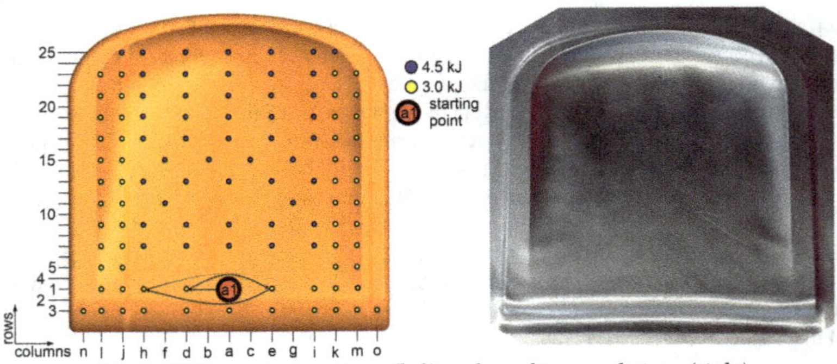

Fig. 4 EHIF forming strategy (left) and resulting workpiece (right)

In the same way as for the EMIF strategy, different capacitor charging energies were used for areas of different forming depths, but the energy level was adjusted to 3 kJ for the u-shaped flat area and 4.5 kJ for deeper sections of the die. While, for the EMIF forming strategy, a larger displacement of the tool could be chosen for successful deformation of steady areas without fine structures, the discharges in this area have to be distributed more evenly and with a smaller displacement for the EHIF forming strategy, because dents result in the seat surface if the displacements are chosen as large as those for EMIF. One possible cause is that the maximum pressure of the EHIF forming zone acts in its center, while in case of EMIF a ring-shaped area exists in the middle coil radius. In addition, discharges with high energies and large displacements caused already-formed areas to spring back or rebound. This effect was particularly pronounced for discharges in deep sections of the die. A smaller-step approach and staggered discharges in the middle section of the seat (see Fig. 4 columns 11 and 15) proved successful here. Overall, the displacement in y was changed from 15 mm to 30 mm. Previous forming tests showed that this did not cause any deterioration in the good forming of the radii for the u-shaped flat area. Two additional discharges (see Fig. 4 n3 and o3) were positioned in the bottom section of the seat to counteract wrinkling in the corner area. A total of 108 discharges were thus used for forming with this strategy, 60 of which were performed with 3 kJ and 48 performed with 4.5 kJ capacitor charging energy. The radii in the u-shaped area were well formed, while the front end of the seat is warped. This indicates subsequent deformation due to material being drawn into the deeper areas of the die.

Comparison of EHIF and EMIF

The forming result of the workpiece was measured using the Comet L3D optical measuring system from Steinbichler. Following the measurement, the workpiece and tool geometry were superimposed, and their deviation determined. Fig. 5 shows the forming results for EHIF (left) and EMIF (right). The deepest region of the die (area A) was better formed by EHIF than by EMIF although free forming tests showed that single-discharge EMF achieved higher forming depth for equivalent capacitor charging energies (Fig. 3). The reason for the inferior forming of EMIF is the longer distance between the inductor and the blank in this area, which is caused by pre-deformations resulting from preceding forming increments. It is well-known that increasing this distance significantly reduces the magnetic pressure and hence the deformation in that region. A possible solution is to shift the EMIF tool in the z-direction in order to reduce the distance again. The membrane, or the active medium of the EHIF process respectively, can easily overcome this distance as a moving mass with less energy dissipation.

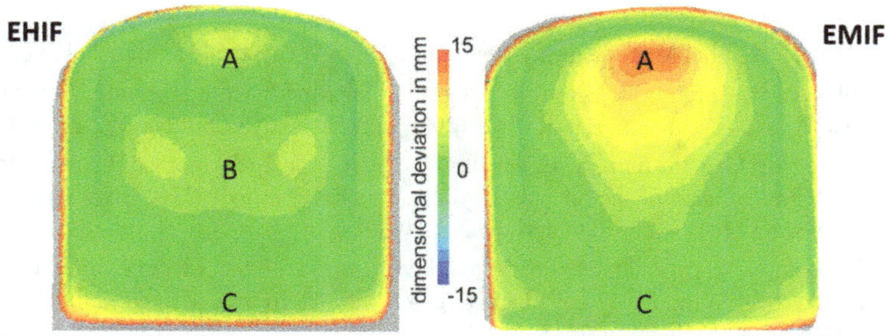

Fig. 5 Dimensional deviation between the workpiece and die for EHIF (left) and EMIF (right)

Furthermore, it is obvious that the middle section of the component produced by the EHIF process (area B) deviates from the target geometry. This might be due to the lever effect created by the point load on the sheet metal due to the pressure distribution of the EHIF forming zone. It may also be responsible for the dent formation in large steady areas of the component. The component produced by the EMIF process shows significantly less irregularities in this area due to the more uniform pressure distribution of the forming zone. The comparison of the seat edges (area C) in the lower part of the image shows, that the pressure in the outer area of the EHIF forming zone is not sufficient to achieve the set shape and therefore further discharges are necessary.

Based on these results, it is possible to make initial rough estimates of the economics of the process, especially with regard to the process time and the energy costs. As modern pulsed power generators typically allow a maximum discharging rate of four discharges per minute, the production of one full chair seat takes about 39.5 minutes in the case of EMIF and 27 minutes in the case of EHIF [17]. To estimate the energy cost for the forming process, the power consumption of the pulse generator was recorded at IWU as a function of the capacitor charging energy. For EMIF of the chair seat, the total energy consumption is 0.328 kWh. Taking an electricity price of € 0.15673/kWh, as averaged out over the last 10 years for industry in Germany, this energy consumption corresponds to electricity costs of € 0.0514 per component. Taking the same characteristic values, the electricity costs for the EHIF process are € 0.0265 per component.

Conclusion

This study has shown in broad terms that the incremental approach of electromagnetic and electrohydraulic forming can be used to form workpieces of a good quality. Nevertheless, there is still further potential for reducing the deviation between the set and the measured geometry. The process-related differences were mainly seen in the way the load is transferred to the blank. While in the EMIF process the distance between the die and the coil is a relevant factor for the forming depth, the active-media-based EHIF process can overcome larger distances. In terms of the evenness of the component surfaces produced, the more uniform pressure distribution of the EMIF process is advantageous. Further investigations should aim at improving the forming strategy. Adding a tool displacement in the z-direction in order to achieve greater forming depth as well as varying the number and distribution of discharges for a better fit of the desired geometries are possible approaches.

Acknowledgment

The authors would like to thank the German Research Foundation (DFG) for funding the research project "Incremental high-speed forming technologies" with reference numbers "HO 2356/17-1" and "PS 68/4-1". The research work conducted in this project forms the basis of the paper.

References

[1] Y. Koren, The Global Manufacturing Revolution, John Wiley & Sons, Inc., Hoboken, 2010.

[2] M. Callegari, D. Amodio, E. Ceretti, C. Giardini, Sheet Incremental Forming: Advantages of Robotised Cells vs. CNC Machines, ARS/plV (2006), pp. 702. https://doi.org/10.5772/4913

[3] W. C. Emmens, G. Sebastiani, A. H. van den Boogaard, The Technology of Incremental Sheet Forming — A brief review of the history, J. Mater. Process. Technol. 210, Germany, 2010, pp. 981–997. https://doi.org/10.1016/j.jmatprotec.2010.02.014

[4] W. Homberg, C. Beerwald, A. Pröbsting, Investigation of the Electrohydraulic Forming Process with Respect to the Design of Sharp Edged Contours. In Proceedings of the 4th ICHSF, Columbus, OH, USA, 2010, pp. 9–10.

[5] V. Psyk, C. Scheffler, A. Stalmann, M. Goede: Shaping of Sharp-Edged Design Elements by Electromagnetic Forming. In: Forming the Future. Miner. Met. Mater. Ser. Springer, Cham.

[6] M. Linnemann, V. Psyk, E. Djakow, R. Springer, W. Homberg, D. Landgrebe, High-Speed Incremental Forming—New Technologies for Flexible Production of Sheet Metal Parts. Procedia Manuf. 27, 2019, pp. 21–26. https://doi.org/10.1016/j.promfg.2018.12.038

[7] Harwey G W and Brower D F 1961 Metal Forming Device and Method US2976907 (Harwey G W and Brower D F)

[8] Z. Lai, Q. Cao, X. Han, N. Liu, X. Li, Y. Huang, M. Chen, H. Cai, G. Wang, L. Liu, W. Guo, Q. Chen, L. Li, A comprehensive electromagnetic forming approach for large sheet metal forming Procedia Eng. 207, 2017, pp. 54–9. https://doi.org/10.1016/j.proeng.2017.10.737

[9] V. Psyk, D. Risch, High-Velocity Forming, Sheet Metal Forming, ASM International, 2012, pp. 227-248. https://doi.org/10.31399/asm.tb.smfpa.t53500227

[10] Bühler H and Bauer D 1968 Ein Beitrag zur Magnetumformung rohrförmiger Werkstücke Werkstatt und Betrieb 9 513–6 (1968), S. 513–516.

[11] Psyk V, Risch D, Kinsey B L, Tekkaya A E and Kleiner M 2011 Electromagnetic Forming -- A Review, J. Mater. Process. Technol. 211 787–829. https://doi.org/10.1016/j.jmatprotec.2010.12.012

[12] S. Golovashchenko, Electrohydraulic Forming of Near-Net Shape Automotive Panels, Technical Report, U.S. Department of Energy Office of Scientific and Technical Information, Oak Ridge, TN, USA, 2013. https://doi.org/10.2172/1094831

[13] X. H. Cui, J. H. Mo, J. J. Li, J. Zhao, Y. Zhu, L. Huang, Z. W. Li, K. Zhong, Electromagnetic Incremental Forming (EMIF): A Novel Alumimum Alloy Sheet and Tube, Forming Technology J. Mater. Process. Technol. 214, 2014, pp. 409–27. https://doi.org/10.1016/j.jmatprotec.2013.05.024

[14] E. Djakow, R. Springer, W. Homberg, M. Piper, J. Tran, A. Zibart, E. Kenig, Incremental electrohydraulic forming - A new approach for the manufacture of structured multifunctional sheet metal blanks, AIP Conference Proceedings, 2017. https://doi.org/10.1063/1.5008083

[15] T. Heggemann, V. Psyk, A. Oesterwinter, M. Linnemann, V. Kräusel, W. Homberg, Comparative Analysis of Electrohydraulic and Electromagnetic Sheet Metal Forming against the Background of the Application as an Incremental Processing Technology, Metals 2022, 12, 660. https://doi.org/10.3390/met12040660

[16] Faber L 2018 KOM lightweight aluminum chair, Information on http://elem-id.de/#Kom

[17] M. Linnemann, Analyse und Automatisierung von inkrementellen elektromagnetischen Umformprozessen, Berichte aus dem IWU, Volume 128.

Materials Research Proceedings **25** (2023) 19-26 https://doi.org/10.21741/9781644902417-3

Development and testing of a flexible test bench for high-speed impact shear-cutting with linear motors

Pascal Krutz[1,a *], André Leonhardt[1,b], Alexander Graf[1,c], Sven Winter[2,d], Matthias Rehm[1,e], Verena Kräusel[2,f], Martin Dix[1,2,g]

[1]Institute for Machine Tools and Production Processes, Chemnitz University of Technology, Reichenhainer Str. 70, Building M, 09126 Chemnitz, Germany

[2]Fraunhofer Institute for Machine Tools and Forming Technology IWU, Reichenhainer Str. 88, 09126 Chemnitz, Germany

[a]pascal.krutz@mb.tu-chemnitz.de, [b]andre.leonhardt@mb.tu-chemnitz.de, [c]alexander.graf@mb.tu-chemnitz.de, [d]sven.winter@iwu.fraunhofer.de, [e]matthias.rehm@mb.tu-chemnitz.de, [f]verena.kraeusel@iwu.fraunhofer.de, [g]martin.dix@mb.tu-chemnitz.de

Keywords: Linear Motor, Blanking, High-Speed Impact Shear-Cutting

Abstract. Due to the use of high-strength steels to achieve lightweight construction goals, conventional shear cutting processes are reaching their limits. Therefore, so-called high-speed impact cutting (HSIC) is used to achieve the required cut surface qualities. A new machine concept consisting of linear motors and an impact mass is presented to investigate HSIC. It allows all relevant parameters to be flexibly adjusted and measured. The design and construction of the test bench are described. The validation was performed with HSIC of a mild deep-drawing steel sheet. The velocities as well as the cut surface were analysed.

Introduction

The manufacturing of complex components and assemblies, such as vehicle bodies, requires additional punching operations after the actual forming process in order to be able to realize the final component geometry. On the one hand, this includes the trimming of formed parts, and on the other hand, holes or forming contours are cut into the component to ensure further processing as well as the function of the component. Therefore, a primary challenge is to ensure the required cut quality. Due to the growing importance of lightweight construction, higher, high-strength and ultra-high-strength steels are increasingly being used in vehicle construction [1], which means that the application limits of conventional shear-cutting operations are being reached. One approach to optimize the shear-cutting of ultra-high-strength steels is to exploit high-speed effects in cutting processes. In recent years, the potential of high velocities ($v \geq 10$ m/s) in shear-cutting has been investigated, analysed and evaluated within the framework of various research projects [2–4]. In so-called high-speed impact cutting (HSIC), the early shear failure in the material is exploited by generating adiabatic shear bands [5]. Due to this effect, even ultra-high-strength steels can be cut with a particularly high quality of the cut surfaces in a resource- and energy-efficient way [6]. The current hydraulic HSIC presses can provide very high impact energies (up to 7000 J) and require a large installation space. However, recent studies show that significantly lower energies are required for cutting. Winter et al. [7] were able to show on the 3 mm thick and hardened steel 22MnB5 that already 250 J impact energy and velocities significantly below 10 m/s are sufficient to produce qualitatively very good cutting surfaces. Due to the fact that significantly lower energies and velocities are sufficient, this allows the use of other more flexible drive concepts with significantly smaller installation space. Conceivable here are electromagnetically accelerated drives, such as that of Linnemann et al. [8]. Therefore, an impact body is accelerated, which provides the impulse for the subsequent shear-cutting. However, the disadvantage is that the

Content from this work may be used under the terms of the Creative Commons Attribution 3.0 license. Any further distribution of this work must maintain attribution to the author(s) and the title of the work, journal citation and DOI. Published under license by Materials Research Forum LLC.

accelerated mass is usually below 5 kg, otherwise the coils and the currents are very large, consequently the energy provided is also very low. Pneumatic [9] or explosively [10] operated drives are also conceivable. If larger masses are to be accelerated, however, they quickly reach their limits. Likewise, the explosive drives are hardly transferable to larger industrial areas. So-called linear motors offer an interesting approach here, because in comparison with conventional drives they can reach higher dynamics due to the lack of mechanical coupling elements [11]. These are mainly used in the areas of handling systems, machine tool and special machine building as well as packaging and assembly systems [12]. Linear motors have a high degree of flexibility, are very compact and can easily accelerate masses significantly above 50 kg to the velocity range relevant for the HSIC (v > 2 m/s). Nevertheless, linear actuators are not currently used for HSIC. This study will be the first to use the advantages of a linear actuator for the HSIC.

Conception of the high-speed impact cutting process

The principle of a drop tower was adopted for the implementation of HSIC. A large impact mass (IM) is actively accelerated by a linear motor unit and transfers the impulse as an elastic impact to a smaller, float-mounted mass – the shear-cutting tool. The mass ratio and the physical relationship of the elastic impact result in a higher speed of the tool with mounted cutting punch, which should be in the adiabatic range. Fig. 1 describes the distribution of the different energies and their specific calculation. Initially, the kinetic energy of the IM is available, based on the velocity v_{IM0} of the IM before the impact. After the impact, the kinetic energy of the IM splits into a remaining kinetic energy of the IM and the kinetic energy of the upper part of the cutting tool. The latter is available for the blanking process and is divided into the required cutting energy and an available energy for cutting speed.

The quantity of this available energy finally determines the speed at which the cutting punch moves through the sheet metal strip in theory. The required masses of the IM and cutting tool as well as the required processing speeds are calculable inversely via the theoretically occurring cutting speed. This should be in adiabatic range for the generation of reproducible high-speed cutting effects on the workpiece.

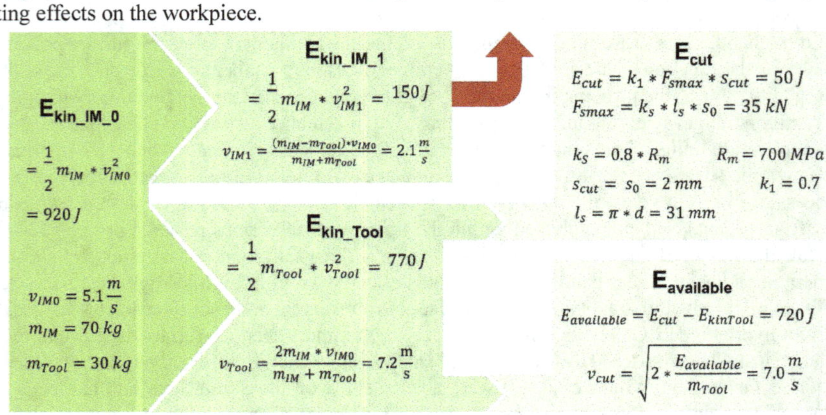

Fig. 1: Allocation of the present process energies for conceptual design and their calculation, assuming an uncoupling speed of the linear motors with 2 m/s

The calculation starts with an impact speed of IM. This is composed of the nominal speed of the linear motors (set to 2.0 m/s) and a residual drop height (assumed to 0.5 m). For the latter, a frictionless acceleration of 9.81 m/s² is assumed for simplification. With these assumptions, the impact velocity v_{IM0} amounts 5.1 m/s. To calculate the velocities of the IM v_{IM1} and the cutting tool v_{Tool} after the impact, the equations of elastic impact of two masses are used. They are included in an adapted form in Fig. 1. Since the IM has a significantly higher mass than the upper part of the cutting tool, the latter will move at a higher speed than the IM after the impact. In addition to the calculation of the kinetic energies, Fig. 1 also shows the calculation of the cutting energy required for blanking of a sheet metal strip (sheet thickness s_0) with a length of the cutting line l_s. The parameter k_1 is a scaling factor in the range 0.4 to 0.7 and k_s describes the shear resistance which can be approximated with 0.8 · tensile strength. After calculating the cutting energy and subtracting it from the kinetic energy of the cutting tool after the impact, the remaining amount is used to calculate the theoretical occurring cutting punch speed. In this calculation example, the final cutting speed is around 7 m/s.

Constructive realization of the novel test bench

For practical implementation, a linear motor test bench was considered, which consists of two separately controllable linear motors. They can perform highly dynamic motions up to an acceleration of 100 m/s² due to their design and are thus predestined for the acceleration of the IM.

For the integration of the shear-cutting tool into the linear motor test bench, a column guided plate frame was chosen into which the cutting tool was integrated. The plate frame includes a top and base plate, which hold together four cylindrical guide columns with a length of two meters. Two further plates are located between the top and base plates, which can move along the guide pillars with slide bearing sockets. One of these displaceable plates acts as the IM and has to be accelerated by the linear motors. The second movable plate is used for mounting the cutting tool.

The two vertically movable linear motors move synchronously in gantry mode and are connected via an adapter plate. The technical data for the two linear motors from SKF® are specified with maximum force of 3.1 kN each and a maximum feed rate with 7 m/s at nominal force and 2.3 m/s at maximum force. A linking mechanism was constructed on the adapter plate that enables the IM to be picked up. After lifting the IM to the upper end position, the linear motors accelerate downwards at an acceleration greater than the acceleration due to gravity. During this process, the IM is mechanically detached so that the motors decelerate until standstill after reaching the set speed. The detachment of the IM is necessary to protect the motors, as they must not collide with a stationary tool while the IM is linked. After detachment, the IM continues to fall downwards, gaining speed due to gravity.

The cutting tool is provided with an additional mass on the upper side to absorb an increased amount of kinetic energy from the IM during the elastic impact. After the impact of the cutting tool and the IM, the cutting punch immediately performs the shear-cut and the cutting tool subsequently closes by itself. Next, the IM continues to move downwards at a reduced speed and is cushioned by two buffers. Both the cutting tool and the buffers are mounted floating on the so-called assembly table. This table is connected to the frame of the test bench with a spring-damper system in order to transfer the dynamically occurring forces during acceleration and deceleration of the components to a large base area. In Fig. 2, the adapted design of the linear motor test stand is illustrated by comparing it with the concept sketch. Two high-speed cameras were used in combination with two high-power led spotlights to record the process speeds.

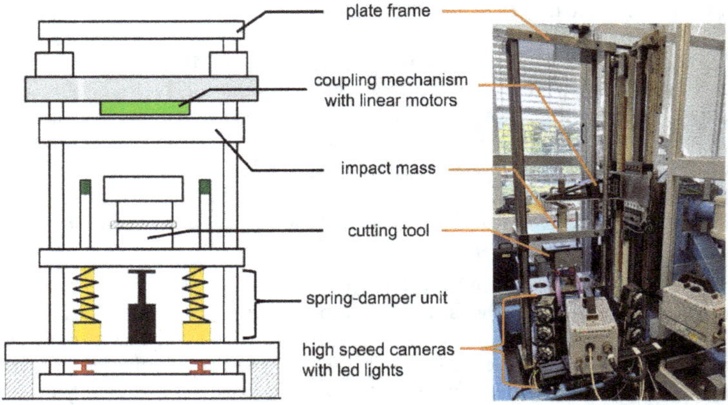

Fig. 2: HSIC test bench - comparison of concept sketch and constructive realisation

In accordance to Fig. 2, the most important requirements for the test stand, their constructive realisation and the resulting specific advantages are summarised in Tab. 1. The main advantage of the test bench is the high flexibility for examination of the HSIC processes, as the applied impact energies can be continuously adjusted and furthermore a good accessibility for the sensory recording of the shear-cutting process is guaranteed.

Tab. 1: Description of technological requirements, the corresponding constructive realisations and derivation of specific advantages

Technological requirement	**Constructive/technological implementation**
highly dynamic acceleration of the cutting punch	principle of elastic impact, acceleration of an impact mass with linear motors
central localized impact on cutting tool	selection of a frame with column guided plates, one plate is impact mass
variation of the cutting energy	flexible adjustment of the motor dynamics by setting control parameters, optional variation of impact mass and additional tool masses
controlled absorption of dynamic forces	use of a spring-damper system which absorbs impact energy and transfers it to the frame over a large area
linear motors must not collide with stationary masses	construction of a coupling mechanism which enables the automated pick-up and uncoupling of the impact mass
Technological advantages compared to other test benches:	
• Precise adjustment of impact energy by adjusting engine dynamics (speed, acceleration, jerk, setting of launch position) and variation of impact and cutting tool masses • High accessibility for sensory acquisition of the HSIC process • Recording of the cutting punch speed with high-speed cameras as no obscuring of the field of vision by enclosures etc. • Integrated motor measuring system enables feedback on the applied impact energy • Easy access to the cutting tool simplifies workpiece handling	

In the following, the implemented automation and the technological sequence, shown in Fig. 3, are described in more detail. At the beginning, the linear motors are in the start position. Here, the

IM is located on the cutting tool in the uncoupled state. First, the linear motors move down to the IM and pick it up via a fully mechanical coupling mechanism. Afterwards, the IM is lifted by a few centimetres, while the upper part of the cutting tool, including the blank holder, is also lifted. This motion opens the cutting tool and the sheet metal strip can be inserted. Subsequently, the cutting tool is closed and the IM is lifted to the upper end position of the linear motors. With release by the user, the IM is accelerated up to the parametrized uncoupling speed. After uncoupling of IM, it is further accelerated due to gravity until the elastic impact with the tool and the execution of the cutting process. After performing the shear-cutting operation and decelerating the moving masses with the spring-damper unit one cycle is completed.

A pulse-like change in speed of the IM and cutting tool occurs during the cutting process at the beginning of section C in Fig. 3. The speed of the cutting tool adjusts to the cutting speed, which results after the deduction of the required cutting energy. During the cutting process, the cutting punch moves faster through the metal strip than the IM falls at its residual speed. As a result, the cutting tool speed is temporarily at 0 m/s after the cutting process has been completed. As soon as the IM hits the buffers, both IM and cutting tool move downwards at the same speed and are decelerated by the spring-damper unit until they come to a standstill. The realized test bench is characterized in particular by its high flexibility on the part of the process parameters. In addition to the dynamic parameters like jerk, acceleration and velocity of the linear motors, the IM and tool masses can also be adjusted. This results in a fine granular setting of the inserted impact energy.

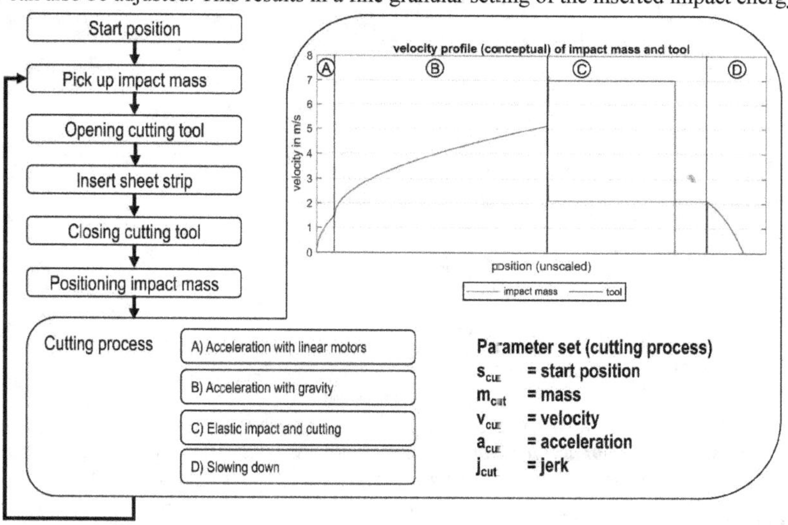

Fig. 3: Technological flowchart for the implementation of HSIC trials

Validation of test bench

After the successful commissioning of the test bench, the functionality of the conceptualised kinematics has to be proven. This includes in particular the effective coupling and decoupling of the IM for different process speeds. Two high-speed cameras were used to determine the process speeds of the IM and the tool. Fig. 4 shows the determined velocity curves of the IM for a motor velocity set to 2.0 m/s. Since the entire movement range cannot be captured with one measuring range of the camera, three relevant measuring ranges, each covering 130 mm in the vertical direction, were selected. The arrangement of the first two upper windows was selected in a way that, starting with the acceleration of IM from standstill, its speed was recorded approximately continuously over 260 mm. The set speed of the motors was already reached in the first window, with the result that only the acceleration due to gravity is shown in the second window. After passing out of the second window, IM continues to move downwards and is recorded with the third window subsequent at impact height. The intermediary speed curve is linearly interpolated according to the assumption of a uniform acceleration and constant friction and is shown with an interruption for better display of the subsequent camera window in Fig. 4. Window 3 shows a first maximum speed of the tool with approximately 3 m/s, indicating that the adiabatic cutting range has been reached. With 3.5 m/s, the impact velocity of IM is around 1.6 m/s below the theoretically calculated impact velocity v_{IM0} in Fig. 1. The main reasons for this are the frictional forces acting during the acceleration of IM. These were not considered in the conception. If the calculation of the cutting speed is repeated with the measured impact speed, the result is 4.5 m/s for v_{cut}. This implies that v_{cut} is still 1.5 m/s below the calculated speed. This deviation can be partly explained due to a not loss-free energy transfer during the elastic impact. In particular, the floating mounting of the tool contributes to the absorption of parts of the kinetic energy by IM from the spring-damper system. With a good approximation, the new calculated velocity v_{IM1} of 1.4 m/s agrees with the practically measured velocity of IM after the impact. As another cause for deviations, the multiple rebound of IM after the first impact can be mentioned, which means that the highly simplified assumptions for the calculation are only valid to a limited extent.

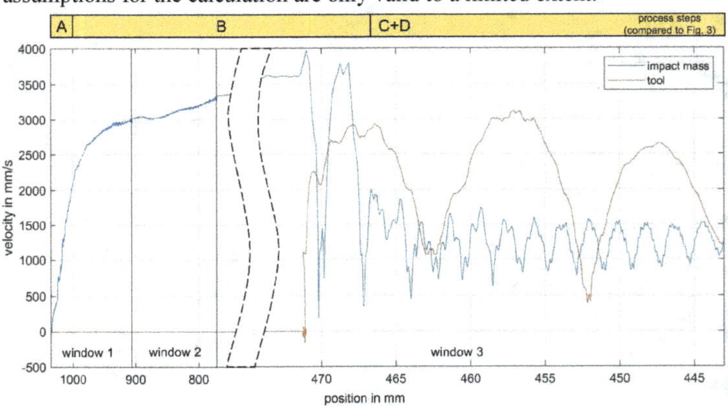

Fig. 4: Measured velocity profile of IM and tool, captured with three different recording windows, consisting of acceleration with linear motors (A), acceleration by free fall (B) and impact with the tool for a set uncoupling speed of 2 m/s (C+D)

Validation of the HSIC process

To validate the flexible test bench, circular blanks with a diameter of 10 mm were cut out of mild deep-drawing steel (1.0338). The fixed parameters for the test series were the sheet thickness (s = 2 mm), the related cutting gap (u_s = 2.5 %s) and the material mentioned above. The only variation during the validation was the cutting speed. These were three speeds for the uncoupling of IM (1.0, 1.5 and 2.0 m/s) and were supplemented by conventional cutting tests at 0.025 m/s. The first step was the evaluation of the cut surface quality according to VDI 2906 sheet 2 [13]. The summary of the results is shown in the Fig. 5. It can be clearly seen that the roll-over is reduced with increasing cutting speed. On the other hand, the proportion clean-cut zone is reduced and, consequently, the proportion of fracture zone increases with increasing cutting speed. Another positive aspect is the increase in the fracture surface angle when using HSIC. All results have a drastic change from v =

 Materials Research Forum LLC
https://doi.org/10.21741/9781644902417-3

0.025 m/s to v = 2.4 m/s in common, but a saturation or even a reversal of the results occurs with further increase of the cutting speed.

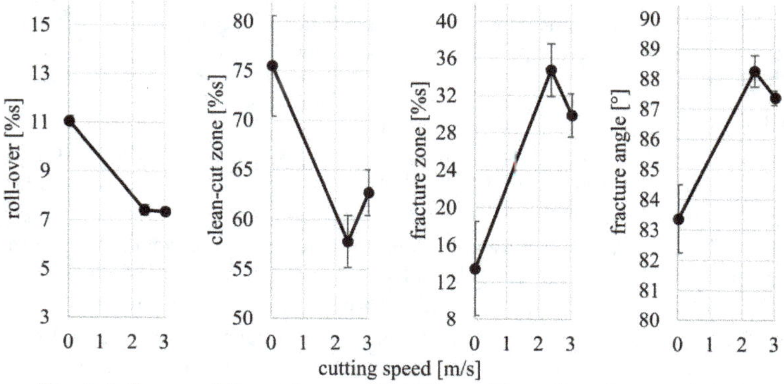

Fig. 5: Influence of the cutting speed on the cutting surface characteristics

With the characterisation of the microstructure by mapping the hardness, further influences of the punch velocity become apparent. At a punch speed of 0.025 m/s, an increasing hardening towards the cut edge is visible (Fig. 6 a). The hardening depth is up to 0.5 mm from the cutting edge. By increasing the punch speed to 3 m/s, the hardening depth decreases to 0.3 mm (Fig. 6 b). It is noticeable that there is a smaller increase in hardness compared to the punch speed of 0.025 m/s in the area from the roll-over up to the beginning of the fracture zone. Observing the microstructure (Fig. 6 c), it becomes clear that a large deformation has occurred in this area. This indicates an adiabatic shear band in this area.

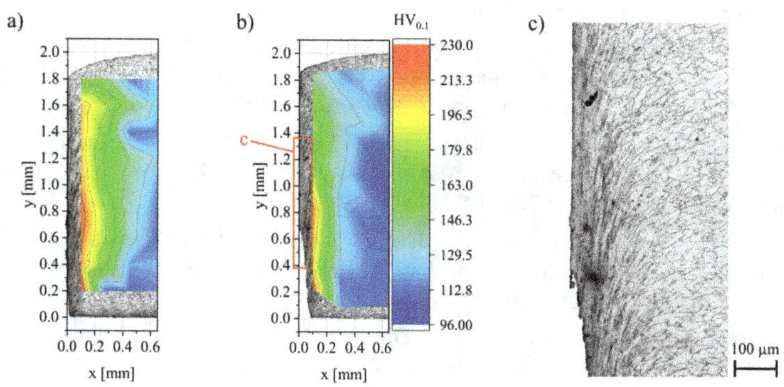

Fig. 6: Measured hardening profiles under variation of the punch speed: a) 0.025 m/s, b) 3 m/s and c) the enlargement of the microstructure in the area of large deformation at 3 m/s

Summary and outlook

HSIC tests were carried out with the presented test bench. Linear motors were used to generate the cutting energy. The construction, which is easily accessible for sensor technology, enabled the precise measuring of occurring tool velocities. Three different decoupling velocities (1.0, 1.5 and 2.0 m/s) were investigated for a mild deep-drawing steel. The determined IM and tool velocities were lower than the conceptually calculated velocities, which is mainly due to the neglect of frictional forces and energy losses during impact. In the verification of the cutting results, the high punch speed (up to 3 m/s) was proven. This is reflected in the influence on the quality of the cut surface and the microstructure. In the continuation of the presented investigations, the speed of the falling mass is to be further increased, with a main focus on the reduction of friction on guide columns. To further expand the process, it is also planned to cut other materials such as high-strength steels.

References

[1] Merklein, M.; Wieland, M.; Lechner, M.; Bruschi, S.; Ghiotti, A. Hot Stamping of Boron Steel Sheets with Tailored Properties: A Review. J. Mater. Process. Technol. 2016, 228, pp. 11–24. https://doi.org/10.1016/j.jmatprotec.2015.09.023

[2] Hu, D.; Chen, M.; Wang, L.; Wang, H. Microstructural Characterization of Blanked Surface of C5191 Phosphor Bronze Sheet under Ultra-High-Speed Blanking. Trans. Nonferrous Met. Soc. China 2021, 31, pp. 692–702. https://doi.org/10.1016/S1003-6326(21)65530-9

[3] Gaudillière, C.; Ranc, N.; Larue, A.; Lorong, P. Investigations in High Speed Blanking: Cutting Forces and Microscopic Observations. EPJ Web Conf. 2010, 6. https://doi.org/10.1051/epjconf/20100619003

[4] Winter, S.; Schmitz, F.; Clausmeyer, T.; Tekkaya, A.E.; Wagner, M.F.X. High Temperature and Dynamic Testing of AHSS for an Analytical Description of the Adiabatic Cutting Process. IOP Conf. Ser. Mater. Sci. Eng. 2017, 181. https://doi.org/10.1088/1757-899X/181/1/012026

[5] Rittel, D.; Wang, Z.G.; Merzer, M. Adiabatic Shear Failure and Dynamic Stored Energy of Cold Work. Phys. Rev. Lett. 2006, 96. https://doi.org/10.1103/PhysRevLett.96.075502

[6] Schmitz, F.; Winter, S.; Clausmeyer, T.; Wagner, M.F.-X.; Tekkaya, A.E. Adiabatic Blanking of Advanced High-Strength Steels. CIRP Ann. 2020, 69, pp. 269–272. https://doi.org/10.1016/j.cirp.2020.03.007

[7] Winter, S.; Nestler, M.; Galiev, E.; Hartmann, F.; Psyk, V.; Kräusel, V.; Dix, M. Adiabatic Blanking: Influence of Clearance, Impact Energy, and Velocity on the Blanked Surface. J. Manuf. Mater. Process. 2021, 5, 35. https://doi.org/10.3390/jmmp5020035

[8] Linnemann, M.; Scheffler, C.; Psyk, V. Numerically Assisted Design For Electromagnetically Driven Tools. Procedia Manuf. 2020, 47, pp. 1334–1338. https://doi.org/10.1016/j.promfg.2020.04.254

[9] Yaldız, S.; Sağlam, H.; Ünsaçar, F.; Işık, H. Design and Applications of a Pneumatic Accelerator for High Speed Punching. Mater. Des. 2007, 28, pp. 889–896. https://doi.org/10.1016/j.matdes.2005.10.009

[10] Mottram, A.R. High Energy Rate Forming Techniques. In Mechanical Engineer's Reference Book; Elsevier, 1973. https://doi.org/10.1016/B978-0-408-00083-3.50022-9

[11] Rehm, M.; Quellmalz, J.; Schlegel, H.; Neugebauer, R. Control Strategies for Mechanically-Coupled High Peed Linear Drives. In Proceedings of the Proceedings of the Ninth International Symposium on Linear Drives for Industry Applications; Hangzhou, 2013; pp. 273–278.

[12] Bernstein, H. Gleichstrom-, Schritt- Und Linearmotoren. In Elektrotechnik/Elektronik für Maschinenbauer; Springer Fachmedien Wiesbaden: Wiesbaden, 2018; pp. 289–329. https://doi.org/10.1007/978-3-658-20838-7_6

[13] VDI 2906 Blatt 2- Schnittflächenqualität beim Schneiden - Scherschneiden 1994.

Numerical and experimental study of high-speed blanking of DC06 steel

E. Galiev[1a*], S. Winter[1b], M. Linnemann[1c], L. Winter[2d], V. Psyk[1e], V. Kräusel[1f]

[1] Fraunhofer Institute for Machine Tools and Forming Technology IWU, Reichenhainer Strasse 88, 09126 Chemnitz, Germany

[2] Institute of Materials Science and Engineering, Chemnitz University of Technology Erfenschlager Strasse 73, 09125 Chemnitz, Germany

[a*] elmar.galiev@iwu.fraunhofer.de, [b] sven.winter@iwu.fraunhofer.de, [c] maik.linnemann@iwu.fraunhofer.de, [d] lisa.winter@mb.tu-chemnitz.de, [e] verena.psyk@iwu.fraunhofer.de, [f] verena.kraeusel@iwu.fraunhofer.de

Keywords: Simulation, Blanking, High-Speed Impact Cutting

Abstract. In high-speed forming processes, such as electromagnetic forming or adiabatic blanking, the yield stress of the material is influenced by two opposing effects: deformation hardening and thermal softening due to adiabatic heating. In most cases, it is difficult to determine the temperature in the deformation zone in the process, due to closed tools. In this study, the tool for high-speed blanking was modified to access the measuring technique to enable recording the temperature in the deformation zone of a 3 mm thick DC06 steel with a high-speed pyrometer. The experimental results validated numerical calculations in LS-DYNA using the strain rate-dependent plastic model and the GISSMO material damage model. The reached temperatures in the shear zone ranged from 225 °C in case of a cutting punch velocity of 6.15 m/s) to 345 °C in case of a cutting punch velocity of 9.64 m/s. Thus, the study shows that the used numerical model and the damage model have a great potential for the simulation of high-speed blanking processes.

Introduction

In the sheet metal processing industry, cutting processes play an important role due to their high cost-effectiveness [1]. Almost every structural component is subject to at least one cutting process during its production [2]. Due to the increasing complexity of part geometries and the use of high-strength materials [3] often traditional blanking techniques, are not able to cope with the quality requirements imposed for the resulting cut surface [4]. An alternative to conventional cutting by shearing is adiabatic blanking or high-speed impact cutting (HSIC) [5]. The adiabatic blanking process is performed using a cutting tool speed of more than 3 m/s and a local strain rate above 10^3 s^{-1} [6]. These parameters result in a local temperature increase in the shear zone during the process, since the heat resulting from the energy released cannot dissipate during the short process time. In general, the development and change of temperature in the fracture region of a material is extremely important [7] and can explain many phenomena occurring in the material, which are often misinterpreted as caused by the strain rate [8]. These two important effects, temperature dependence and strain rate dependence of a materials plasticity, damage and failure behavior, are fundamental for creating a material model that approximates the real process. In the high strain rate regime, experimental process analysis and process design are much more complicated, since the measuring equipment is limited with respect to measurement frequency, and the area where the local temperature maximum is reached shortly before material failure is relatively small. In addition, most high-speed forming processes are performed in closed working tools without the possibility of access to measuring equipment.

It is well known that having a high-quality FEM model is key to cost-effective industrial implementation of production processes. Modern process simulation requires constitutive material

Content from this work may be used under the terms of the Creative Commons Attribution 3.0 license. Any further distribution of this work must maintain attribution to the author(s) and the title of the work, journal citation and DOI. Published under license by Materials Research Forum LLC.

models to describe plasticity. These models must reflect the essential relationships between yield strength and stress state, strain, strain rate and temperature for the specific material [9]. In the particular case of high-speed blanking, strain rates in the range of 10^3 s^{-1} and higher require a wide range of yield curves and a suitable material damage model enabling the simulation of locally extremely high temperature changes in the fracture region. Here, the temperature in the shear zone is one of the dominant factors influencing the failure and further the quality of the cut surface. Thus, in the present study, for the first time, the temperature during HSIC will be measured directly in the shear zone using a specially modified tool and matched to a numerical model. The process validation will thus lead to a higher process understanding.

Materials and Methods

For the experimental HSIC tests and the subsequent numerical validation of the temperature, the steel DC06 was selected in a sheet metal thickness of 3 mm. The chemical composition of the used material received from the material supplier is shown in Tab. 1.

Tab. 1. Chemical compositions of the investigated steel DC06

	Composition (in wt-%)							
DC06	C	Si	Mn	S	Al	Ti	Nb	Fe
	0.007	0.01	0.12	0.00	0.06	0.06	0.01	bal.

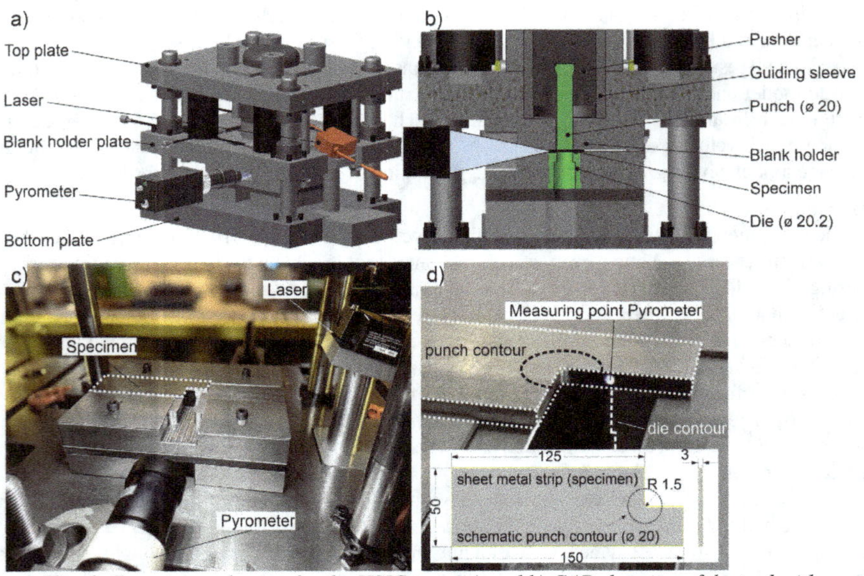

Fig. 1: Experimental setup for the HSIC tests. a) and b) CAD drawing of the tool with positioning of the measuring equipment and the relevant components. c) and d) Specimen geometry and position of the measuring point of the pyrometer.

The experimental tests were carried out on a high-speed ADIAflex press by MPM (Saint-Etienne, France). The tool shown in Fig. 1a was used for the cutting tests. The punch used had a diameter of 20 mm and the die had a diameter of 20.2 mm resulting in a mean circumferential cutting gap of 0.1 mm. For the used sheet thickness of 3 mm this cutting gap corresponds to a relative clearance of 3.33 %. Tests were performed at an energy level of 250, 500, 750, and 1000 J (adjustable machine parameter). The corresponding punch speed and the resulting temperature in the shear zone were determined using a tool-integrated LKH-157 laser displacement measurement system from Keyence (Neu-Isenburg, Germany) and an IGA 740 high-speed pyrometer from LumaSense Technologies GmbH (Raunheim, Germany). The pyrometer has a working range of 160 °C - 1000 °C and an extremely high sampling rate of 166 kHz. To allow for a temperature measurement directly in the cutting zone (Fig. 1b), the sheets under investigation were pre-cut before the experiments (Fig. 1c and d). Precisely, a square with 25 mm side length was cut out at the outer edge. Consequently, the pyrometer could obtain the temperature directly in the center of the shear zone during the cutting test. The measuring area has a diameter of 0.7 mm (Fig. 1d) and the output data of the pyrometer is an integral quantity representative for the entire measurement area. The punch velocity was used as input parameter for the numerical simulation and the temperature measurement was used to validate the numerical model. In addition, the cut surfaces were evaluated with regard to the rollover height, the burnish and the fracture surface.

FE-Simulation

FE Analysis was performed using the explicit thermal and mechanical solver in the commercial software from LS-DYNA (DYNAmore GmbH, Stuttgart, Germany). The plastic deformation behavior of the material was described using the classical *MAT 024 PIECEWISE LINEAR PLASTICITY model [10], using different strain rate dependent yield curves. Details of this model including the yield curves used are described in [11]. Strain rate dependant damage and failure was modeled via the GISSMO damage and failure model developed by Neukamm et al. [12]. This model combines the damage description, which is used to calculate crash simulations, and an incremental formulation for the description of the material instability and localization. This fracture model can be used as a complement to the plastic material model and is specified using the *MAT_ADD_EROSION file. More detailed parameters for this fracture model are presented in [13].

Due to the geometry of the precut sheet, the stress distribution in the shear zone is inhomogeneous. Therefore, it was necessary to simulate the full geometry of the setup without exploiting any simplifications due to symmetry. The simulations of the high-speed blanking of the precut sheet metal specimens were performed with fully integrated solid elements (ELFORM 1) with edge lengths of 0.1x0.1x0.1 mm for test specimen and LS-DYNA R12.0.0. Fig. 2 shows the model consisting of a punch, a blank holder, a die, and a specimen. All parts of the model except the direct sample were specified as rigid bodies. The blankholder force was set to 80 kN and the friction coefficient was set to 0.15. The punch speed obtained from the experiment was specified as input information for the simulation using the *BOUNDARY_PRESCRIBED_MOTION file.

The temperature distribution was calculated via the dissipated forming energy according to Eq. 1 based on the temperature increase $\Delta\vartheta$, the specific heat capacity c_ϑ, the Taylor-Quinney factor κ [14], mass density ρ and the forming work performed ΔW_{dis}. The Taylor-Quinney factor is used to define the fraction of plastic stress power that is not converted into heat [15]. For the thermomechanical calculations performed in LS-DYNA $\kappa = 0.9$ was set.

$$\Delta\vartheta = \frac{\kappa \cdot \Delta W_{\text{dis}}}{c_\vartheta \cdot \rho} \tag{1}$$

Materials Research Proceedings **25** (2023) 27-34 https://doi.org/10.21741/9781644902417-4

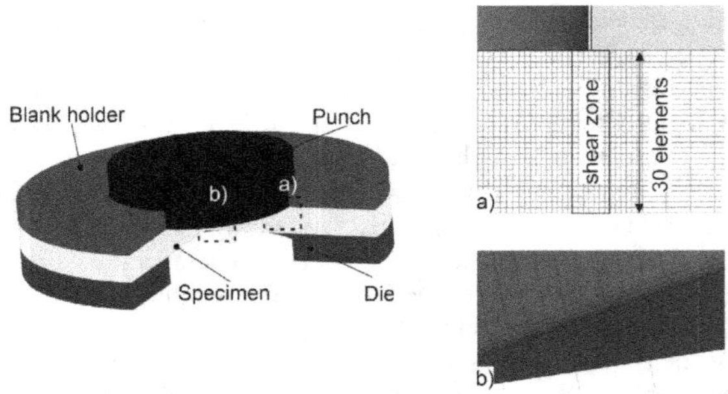

Fig. 2: Numerical model of the investigated HSIC process. a) Shear zone of the sheet metal specimen for the temperature calculation with fine mesh size. b) Punch with significantly larger mesh.

Results and Discussion

Fig. 3 summerizes results of the experimental determination of the punch velocity. The exemplary diagram shows three measured time-dependent displacement of the cutting punch as recorded by the laser displacement measurement system during tests at an energy level of 250 J together with the resulting mean value curve. The shape of the curve is representative for all measurements. The gradient of the curve in the region between Point 1 and Point 2 corresponds to the velocity during cutting. Subsequently, due to the impact, wave propagation occurs throughout the die and the punch consequently continues to move in a wave-like manner after the part is ejected, as documented by the curve after the straight initial line. For transfer to the numerical simulation, the gradient between Point 1 and 2 of the respective mean value curve was evaluated and transferred to the simulation as input variable for the punch velocity. The measured velocities for each energy level investigated here are shown in the table next to the curve (Fig. 3). The lowest energy reached a velocity of 6.15 m/s and the velocity increases with increasing energy as expected. At 1000 J the velocity was about 9.64 m/s.

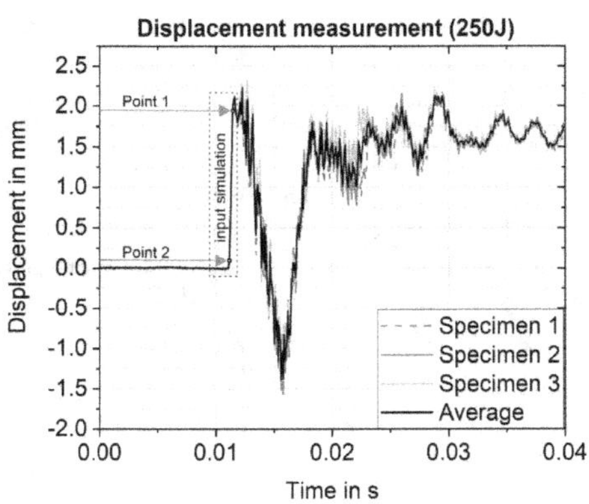

$$V_{Punch} = \frac{S_1 - S_2}{t_1 - t_2} = 6.147\ m/s$$

Energie	Velocity
250 J	6.147 m/s
500 J	8.244 m/s
750 J	8.978 m/s
1000 J	9.641 m/s

Fig. 3: Representative displacement-time measurements at an energy level of 250 J. The gradient of the curve between point 1 and 2 corresponds to the velocity and were used as input parameter for the numerical simulation. The measured velocities from 250 J to 1000 J are recorded in the table.

The HSIC tests at all four different energy levels led to complete separation of the parts in all cases (see Fig. 4). The optical microscopy images for the evaluation of the cut surface quality were all taken at maximum distance from the pre-cut area (Fig. 4, red marking). Regardless of the energy and the corresponding punch velocity, all specimens show excellent cut surfaces with very good quality. The parts have no burr and a rollover height of only 0.237–0.325 mm. The burnish is between 1.19–1.32 mm and the fracture surface between 1.41–1.51 mm. Significant differences or tendencies cannot be recognized, proving that even the lowest cutting speed or impact energy provides sufficient energy to cut the relatively soft steel DC06 (approx. 93 HV1) and to provide very good cut surface quality (no burr, very straight, high percentage of burnish).

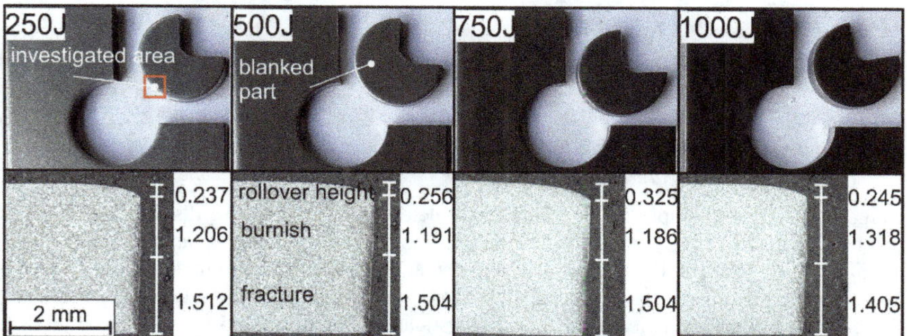

Fig. 4: Cut parts optical images of the corresponding cut surfaces produced using energy levels from 250 J to 1000 J. The rollover height, burnish and fracture surface fractions were measured in mm.

Fig. 5 shows the simulation results of the cutting process. Precisely, the temperature distribution in the circumferential shear zone at a penetration depth of the punch of 10 % of the sheet thickness is illustrated. The deformation and especially the temperature distribution shown here is highly inhomogeneous. In the area of the free surface where the temperature was measured in the experiment, the temperature maximum is 142 °C (see Fig. 5a). In the area of the longest distance to the measuring point, however, the temperature already reaches 242 °C (Fig. 5b). This can be explained by the free surface: Only on one side of this surface material can generate heat by deformation, while at the other side the material is cut away. In the experiment additional cooling effects due to convection can occur, but presumably these play a minor role and were therefore not considered in the simulation. Likewise, the slug is not moved homogeneously through the sheet, but is subject to bending stress. This proves that in order to simulate the process correctly, it is absolutely necessary to model the complete slug. Considering any symmetry in order to reduce calculation time will cause significant differences between experiment and simulation due to the inhomogeneous temperature distribution.

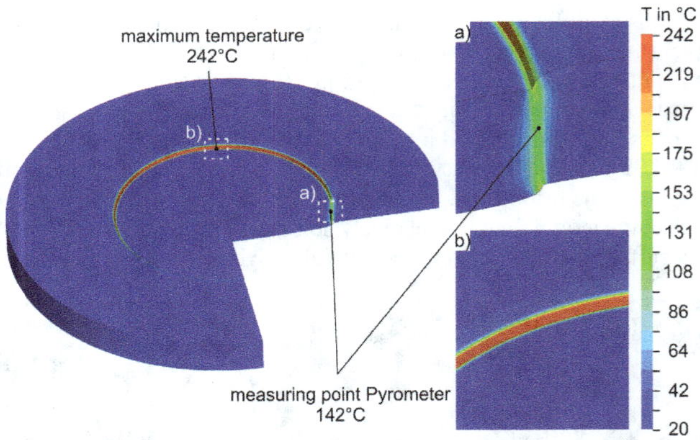

Fig. 5: Simulation of the pre-cut part at 9.64 m/s impact velocity with strongly inhomogeneous temperature distribution in the circumferential shear zone.

Fig. 6 shows an example of the numerically determined temperature distribution in the 2D section as a function of the punch penetration depth in the shear zone in the area of the measuring field of the pyrometer (cf. Fig. 5a) for a cutting operation with 9.641 m/s impact velocity (1000 J). The penetration depth is specified in relation to the sheet thickness. The qualitative temperature profile in the cutting test is identical for all investigated velocities. However, the maxima differ significantly (see Fig. 7). The maximum temperatures occur close to the contact surfaces between punch/sheet and sheet/die. The heat then propagates into the center of the shear zone. Due to the high velocity, the temperature increases only very locally and hardly spreads into the base material, so that the process can be considered quasi-adiabatic. The strong localization of heat leads to localized material softening, which counteracts the hardening during forming and can cause the formation of adiabatic shear bands in the shear zone [6]. The geometric properties (rollover height, burnish, fracture) of the cut surface formed at 25 % penetration depth are comparable to those of the real cut surface shown in Fig. 4. This demonstrates that the implemented damage model works well, even if it is only a qualitative comparison. The maximum temperature reached in Fig. 6 (at a penetration depth of 25 %) of 322 °C is not the maximum in the cutting process. This was reached just before failure and is discussed in Fig. 7.

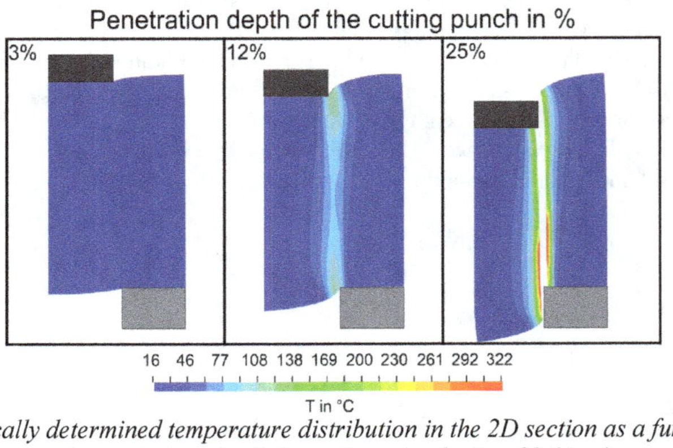

Fig. 6: Numerically determined temperature distribution in the 2D section as a function of the punch penetration depth for an impact velocity of 9.64 m/s.

The averaged results (3 samples for each velocity) of the temperature measurement by pyrometer and the numerically determined maximum temperatures are summarized in Fig. 7. For all velocity ranges from about 6.15 m/s to 9.64 m/s experimentally and numerically determined temperatures are in good agreement with each other. The temperatures in the experiment were between approx. 225 °C and 345 °C. In the simulation, the temperatures ranged between approx. 240 °C and 338 °C. Thus, the resulting standard deviation is 4.8% at 6.15 m/s, 5.4 % at 8.24 m/s, 1.7 % at 8.98 m/s, and 2.0 % at 9.64 m/s. The experimental data show that the temperature increases almost linearly with increasing velocity and consequently increasing energy.

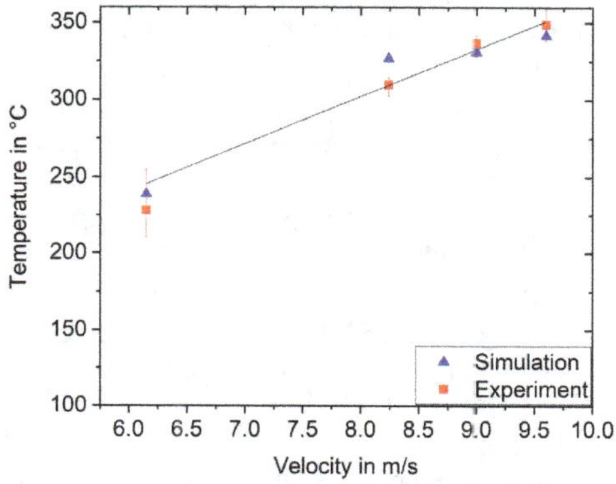

Fig. 7: Comparison of experimentally and numerically determined maximum temperatures in the shear zone at HSIC with four different punch velocities. The results are in very good agreement.

Summary and Conclusion

In the present work, a comprehensive numerical model was validated against a real HSIC process. Since the temperature in the shear zone plays a dominant role at high process speeds, validation was realized for the first time using a temperature in the shear zone measured by a high-speed pyrometer during HSIC. Numerically and experimentally, the good cold-formable steel DC06 was used. Extensive preliminary work [11]and the experimental velocity measurement from HSIC served as input for the material and GISSMO damage model used. It was necessary to model the cutting process completely, due to the non-rotationally symmetrical geometry of the part. Such geometries cause inhomogeneous temperature distributions in the shear zone. The extensive simulation made it possible to match the numerically determined temperature with the measuring field of the pyrometer. The results of numerical and experimental temperature determination showed almost identical temperatures (in the range of 225 to 345 °C) at four different test velocities (6.15 to 9.64 m/s). So the experimentally and numerically determined temperatures are in good agreement (standard deviation between 1.7 % and 5.4 %). It can be concluded that the extensive numerical model used is very well suited to represent the HSIC for the DC06 steel. The consideration of the local temperature in the shear zone enables a significantly higher accuracy. Therefore, the present results will significantly improve the numerical process design of the HSIC in future work.

References

[1] S.F. Golovashchenko, N. Wang, Q. Le, Trimming and sheared edge stretchability of automotive 6xxx aluminum alloys, J. Mater. Process. Technol. 264 (2019), pp. 64–75. https://doi.org/10.1016/j.jmatprotec.2018.09.001

[2] M.-L. Li, D.-S. Bi, J.-K. Hao, Automotive sheet metal SAPH440 and Q235 formability of comparative study, 2015 International Conference on Material Science and Applications (icmsa-15) : Atlantis Press, pp. 1027–1030

[3] K. Sugimoto, T. Hojo, A.K. Srivastava, Low and medium carbon advanced high-strength forging steels for automotive applications. Metals 9 (2019), Nr. 12, pp. 1263. https://doi.org/10.3390/met9121263

[4] A.K. Perka, M. John, U.B. Kuruveri, P.L. Menezes, Advanced High-Strength Steels for Automotive Applications: Arc and Laser Welding Process, Properties, and Challenges, In: Metals 12 (2022), Nr. 6, pp. 1051. https://doi.org/10.3390/met12061051

[5] S. Winter, M. Nestler, E. Galiev, F. Hartmann, V. Psyk, V. Kräusel, M. Dix, Adiabatic Blanking: Influence of Clearance, Impact Energy, and Velocity on the Blanked Surface, J. Manuf. Mater. Process. 5 (2021), Nr. 2, pp. 35. https://doi.org/10.3390/jmmp5020035

[6] F. Schmitz, S. Winter, T. Clausmeyer, M.F.-X. Wagner, A. E. Tekkaya, Adiabatic blanking of advanced high-strength steels, CIRP Ann. 69 (2020), Nr. 1, pp. 269–272. https://doi.org/10.1016/j.cirp.2020.03.007

[7] D.M. Neto, J.R. Barros, M.C. Oliveira, P.V. Antunes, A. Ramalho, R.L. Amaral, A.D. Santos, J.L. Alves, L.F. Menezes, Heat generation when forming AHSS: experimental and numerical analysis of tensile and draw-bead tests. IOP Conference Series: Materials Science and Engineering 967 (2020), Nr. 1, pp. 12086. https://doi.org/10.1088/1757-899X/967/1/012086

[8] P. Knysh, Y. P. Korkolis, Determination of the fraction of plastic work converted into heat in metals, Mech. Mater. 86 (2015), Nr. 3, pp. 71–80. https://doi.org/10.1016/j.mechmat.2015.03.006

[9] N. Abedrabbo, F. Pourboghrat, J. Carsley, John, Forming of AA5182-O and AA5754-O at elevated temperatures using coupled thermo-mechanical finite element models, Int. J. Plast. 23 (2007), Nr. 5, pp. 841–875. https://doi.org/10.1016/j.ijplas.2006.10.005

[10] F. Andrade, M. Feucht (Eds.): A comparison of damage and failure models for the failure prediction of dual-phase steels, 2017

[11] E. Galiev, S. Winter, F. Reuther, V. Psyk, M. Tulke, A. Brosius, V. Kräusel, Local Temperature Development in the Fracture Zone during Uniaxial Tensile Testing at High Strain Rate: Experimental and Numerical Investigations, Appl. Sci. 12 (2022), Nr. 5, pp. 2299. https://doi.org/10.3390/app12052299

[12] M. Feucht, F. Neukamm, A. Haufe, A phenomenological damage model to predict material failure in crashworthiness applications. Recent developments and innovative applications in computational mechanics: Springer, 2011, pp 143–153. https://doi.org/10.1007/978-3-642-17484-1_17

[13] V. Psyk, C. Scheffler, M. Tulke, S. Winter, C. Guilleaume, A. Brosius, Determination of Material and Failure Characteristics for High-Speed Forming via High-Speed Testing and Inverse Numerical Simulation, J. Manuf. Mater. Process. 4 (2020), Nr. 2, pp. 31. https://doi.org/10.3390/jmmp4020031

[14] G.I. Taylor, H. Quinney, The latent energy remaining in a metal after cold working. In: Proceedings of the Royal Society of London. Series A, Containing Papers of a Mathematical and Physical Character 143 (1934), Nr. 849, pp. 307–326. https://doi.org/10.1098/rspa.1934.0004

[15] C.K.C. Lieou, C.A. Bronkhorst, Thermomechanical conversion in metals: dislocation plasticity model evaluation of the Taylor-Quinney coefficient, Acta Mater. 202 (2021), Nr. 849, pp. 170–180. https://doi.org/10.1016/j.actamat.2020.10.037

Incremental forming

SheMet 2023

Sheet Metal 2023 | Materials Research Forum LLC
Materials Research Proceedings 25 (2023) 37-44 | https://doi.org/10.21741/9781644902417-5

Die-less forming of fiber-reinforced thermoplastic sheets and metal wire mesh

Jan-Erik Rath[1, a *] and Thorsten Schüppstuhl[1, b]

[1]Hamburg University of Technology, Institute of Aircraft Production Technology, Denickestr. 17, 21073 Hamburg, Germany

[a]jan-erik.rath@tuhh.de, [b]schueppstuhl@tuhh.de

* corresponding author

Keywords: Fiber Reinforced Plastic, Free Forming, Incremental Sheet Forming

Abstract. The growing market for fiber-reinforced thermoplastics (FRTP) requires new flexible production processes for prototype and small series production, as conventional forming techniques involving molds are not cost efficient in these cases. Inspired by incremental sheet metal forming (ISF), an alternative manufacturing processes for the forming of FRTP with just two robot guided standard tools is outlined. To maintain a locally formed shape in the heated, flexible fabric, auxiliary wire mesh metal is used as it has similar deformation mechanisms, especially shearability, while being sufficiently self-supporting. Feasibility of the approach is discussed and investigated in basic experiments.

Introduction & state of the art

For lightweight applications such as in aviation, the automotive industry or sports equipment manufacturing, fiber-reinforced thermoplastics are of increasing interest. While carbon, glass or aramid fibers provide high strength, the embedding thermoplastic polymer matrix has low weight and, in contrast to thermosets, is reshapeable, weldable and recyclable [1–3]. Furthermore, thermoplastic composites show better impact resistance and fracture toughness while manufacturing cycle times and costs are lower in comparison to thermoset composites [4,5]. Pre-impregnated and consolidated fiber-reinforced thermoplastic sheets, so called organo sheets, are commonly used as semi-finished product. Due to their high reinforcing effect and comparably good drapability, woven fiber fabrics are predominant [6]. After heating a sheet above melting temperature of the thermoplastic matrix, it is formable into the desired shape. This is usually achieved by placing the sheet between two heated metal molds in a process called thermoforming [7,8].

Due to the cost and effort of designing, producing and storing these molds, this standard process is not effectively suitable for prototype and small-batch production. Changes to the product geometry are difficult as existing molds need to be modified or new molds produced [9,10]. Rapid tooling approaches such as rapid machining [11], additive manufacturing [12] or incremental sheet forming [13] try to mitigate these effects, lowering the cost of mold production. However, time, effort and costs are either still too high, or mold quality too low for the processes to be successfully used for prototype mold production without restrictions, especially considering the comparably high pressure needed for FRTP forming and reconsolidation.

Therefore, directly forming an organo sheet without usage of a part-specific mold has been of interest to researchers since the late eighties, when Strong and Hauwiller [14] used a modular press and Miller et al. [10] an array of individually controllable rollers to locally form variable cross-sections into long endless fiber-reinforced organo sheets. Another approach to more flexibly form FRP is multi-point forming, where an array of individually adjustable pins constitutes a mold surface, which is covered by a smoothing diaphragm [15,16].

Content from this work may be used under the terms of the Creative Commons Attribution 3.0 license. Any further distribution of this work must maintain attribution to the author(s) and the title of the work, journal citation and DOI. Published under license by Materials Research Forum LLC.

Several authors investigated the application of single-point incremental forming (SPIF) on FRTP sheets. In SPIF, a simple forming tool such as a hemispherical punch is moved along a metal sheet surface by a CNC-machine or a robot. As the sheet is securely clamped on its edges, the tool introduces strains into the material and deforms it, thinning the sheet [17]. While incremental forming of thermoplastic sheets has been successfully demonstrated [18], direct SPIF of a short-fiber-reinforced thermoplastic is more challenging [19,20]. Therefore, Conte et al. [20] and Ambrogio et al. [21] used SPIF to deform a heated short-FRTP sheet sandwiched between two metal sheets in order to provide stability to the FRTP. Al-Obaidi et al. [22] as well as Emami et al. [23] used a very similar setup to deform unidirectional endless-FRTP sheets. However, endless fibers prevent the effect of stretching underlying conventional SPIF due to their very limited elongation [24], so that fiber breakage can occur, especially when drawing in of the fibers is limited by clamping or high fiber-volume-content [22,23,25]. Therefore, Al-Obaidi et al. [26] and Hou et al. [25] deformed the heated metal sheets by classic SPIF while a woven fiber-reinforced organo sheet was not clamped but “floating” between the metal layers. As a consequence, draping of the textiles was not determinate and defects such as wrinkles occurred [23,25,26].

To conclude, a real die-less process for forming woven FRTP into complex, doubly-curved shapes while considering the actual material behavior and draping requirements would be desirable. Preliminary investigations on manufacturing options [27] and the development of a die-less draping strategy [28] showed that feasible process concepts exist. In the following, we will briefly outline this forming strategy and present a new die-less forming concept for woven FRTP. After a practical investigation, overall feasibility and possible further advancements are discussed.

Forming strategy

Forming a flat woven textile into a three-dimensional shape is mainly achieved by bending and, if the shape is doubly-curved, shearing of the initially rectangular warp and weft yarns. This in-plane deformation is the main challenge for draping, as the amount of shear achievable in a fabric is limited and wrinkling can occur. Other draping mechanisms such as fiber displacement or textile stretching only play a minor role, and especially fiber elongation is negligible due to the high strength and stiffness of the fibers in axial direction [29,30]. Depending on the application and the geometry, different techniques are used for composite draping. Automated preforming processes allow pre-shearing of the fabric by globally applying tension in fiber or bias direction before placing it on the mold surface. Additionally, different tools such as rollers can be used [31,32]. In compression molding, single or segmented punches perform the final draping [8].

A die-less forming process necessarily involves local forming operations, in this case conducted by two standard tools such as hemispherical tool tips or small rollers, one on each side of the fabric and each individually guided by an industrial robot or similar kinematics. While the basic idea is comparable to Double-Sided Incremental Forming (DSIF), the organo sheet is clamped at a single point fixation only, to allow the fibers to be movable for the draping operations [27]. As any additional actuated gripper systems for the application of tension would add further complexity, we developed a draping strategy which mainly relies on the two tools deforming the fabric out-of-plane relative to a single point fixation [28]. The location of this fixation point on the sheet as well as the initial fiber orientations are selected by minimizing the global amount of shear in a kinematic draping simulation of the final geometry. During forming, the tools follow the resulting fiber orientations as calculated by the simulation, originating from an already formed starting point and running towards the edge of the fabric. Thereby, as the fabric is locally bent, compressive in-plane stresses can be introduced into the area between already formed paths. These compressive stresses lead to in-plane deformation in the form of shear [28]. Therefore, the desired shear distribution is the main sequencing criterion in determining the order of forming. As shear is added while the curvature increases and it is hardly reversible, forming paths requiring least shear are formed first [28]. The length of the forming paths can be set according to two different principles, depicted in

Fig. 1a). Either, full paths are formed from a rigid starting point until the edge of the fabric, or the mesh of formed warp and weft yarns is grown for a certain amount of cells at a time in each direction with respect to the fixation [28]. Whether the former path forming or the latter layering strategy is more appropriate, depends on the individual application as outlined in the following.

Manufacturing process concept

Before any forming operation, the thermoplastic matrix of an organo sheet needs to be heated above melting temperature for example by hot air, infrared or contact heating in order to regain drapability of the included fabric. However, already formed paths or areas need to maintain their new shape. As a heated FRTP is not self-supporting, the paths must either be instantly cooled during forming, and kept cool for the remainder of the process [27], or an auxiliary material must be used to support the flexible fabric. For best realization of the fabric draping, this auxiliary material should have similar deformation mechanisms but at the same time must be self-supporting at the required temperature. While metal sheets are therefore not ideal, metal wire mesh could be eligible. With the same mechanisms of bending and shearing of the woven warp and weft wires, simultaneously deforming a heated woven organo sheet and a metal wire mesh can in theory be realized by the same forming strategy outlined above.

Since the wires are significantly better at transmitting compressive stresses, the shearing of the reinforcing fabric can thus be promoted. In order for the two materials to deform together, or rather for the deformation of the wire mesh to be instantly transferred to the organo sheet, some kind of bond is required between the two. At the same time, imprints of the wire mesh into the molten matrix must be prevented.

When bending the wires, as conceptually shown in Fig. 1b), they must be deformed beyond the yield strength, as only the plastic component of the deformation is retained. The elasticity of the intersecting wires will further increase the amount of deflection necessary to achieve a permanent deformation of a forming path. These effects have to be considered when generating the tool paths and either the *path forming strategy* or the *layering strategy* might be more successful in simultaneously and permanently forming an organo sheet and a wire mesh into a desired geometry.

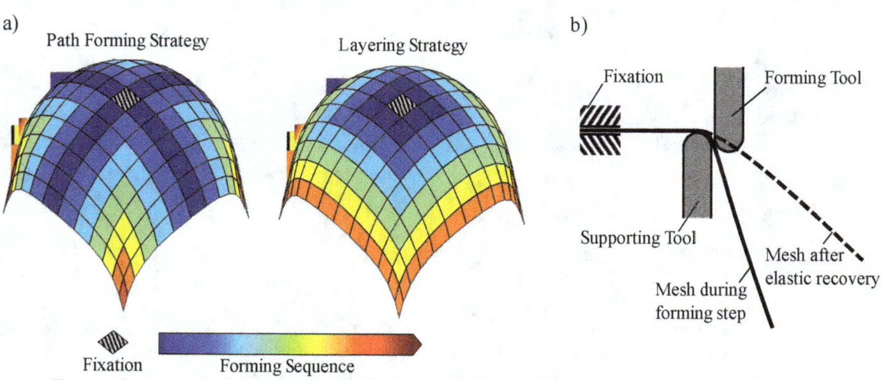

Fig. 1 a) Forming Sequences for a hemisphere according to path forming and layering strategies, b) schematic sketch of die-less forming of a wire mesh with two forming tools during and after a forming step

Practical investigation

After initial tests of the deformation behavior of wire meshes with different wire diameters and mesh sizes, a plain weave of stainless steel 1.4301 wires with a diameter of 0.56 mm and a mesh size of 3.15 mm was chosen for further experiments.

Setup 1. First, only the mesh with a size of 200 x 200 mm was centrally fixed to a pole and incrementally deformed into a hemispherical frustum of ca. 120 mm diameter and 40 mm depth with two handheld ISF-tools with hemispherical tool tips of 20 mm diameter, shown in Fig. 2. The wires were bent by the tools according to the *path forming strategy,* as described above. Afterwards, in order to find the most suitable forming strategy, a second mesh was deformed into the same shape, but following the *layering strategy*, growing the formed area in each direction by about 5 cells in each step.

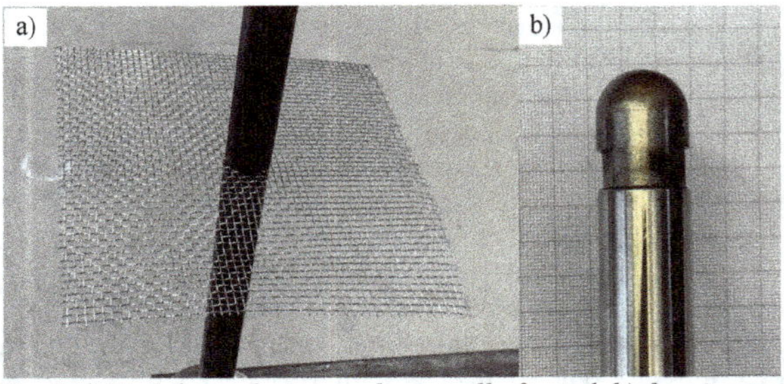

Fig. 2 a) Undeformed wire mesh centrally fixated, b) forming tool

Results. As depicted in Fig. 3, both strategies were able to introduce sufficient shear into the mesh in order to form a hemisphere. However, using the *layering strategy* not only produced a smoother and more precise hemispherical shape, but also proved simpler in use. As assumed above, especially towards the beginning of the process, forming a full path requires higher forming forces since the intersecting wires must be strongly deformed as well. Furthermore, the high degrees of forming in one pass in combination with a more asymmetric forming can lead to undesired deformation/distortion in the rest of the mesh.

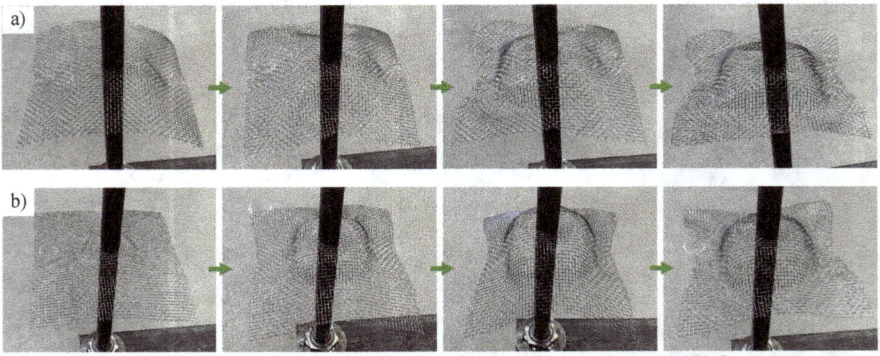

Fig. 3 a) Forming sequence and result following the path forming strategy, b) forming sequence and result following the layering strategy

Setup 2. To investigate basic feasibility of simultaneous organo sheet and wire mesh forming, a 250 x 250 x 0.6 mm two-layered twill weave carbon fiber organo sheet (INEOS Styrolution, Germany) with styrene-acrylonitrile matrix, an areal weight of 245 g/m^2 and 45 % fiber volume content was placed on a wire mesh of 270 x 270 mm. To prevent permanent adhesion of the organo sheet to the wire mesh, silicone release film was placed between them. The stack was centrally clamped between two poles and two infrared heaters heated the layup as shown in Fig. 4. After the

Materials Research Proceedings **25** (2023) 37-44 https://doi.org/10.21741/9781644902417-5

matrix reached a temperature of approximately 190 °C, the heating was disabled and the sheet remained formable for ca. 10 s. During this time, following the *layering strategy*, as many paths as possible were formed using the same handheld tools as above. Afterwards, the process was repeated, until the whole sheet was deformed into a hemispherical frustum of ca. 120 mm diameter and 40 mm depth. Finally, the organo sheet was separated from the wire mesh and the release film.

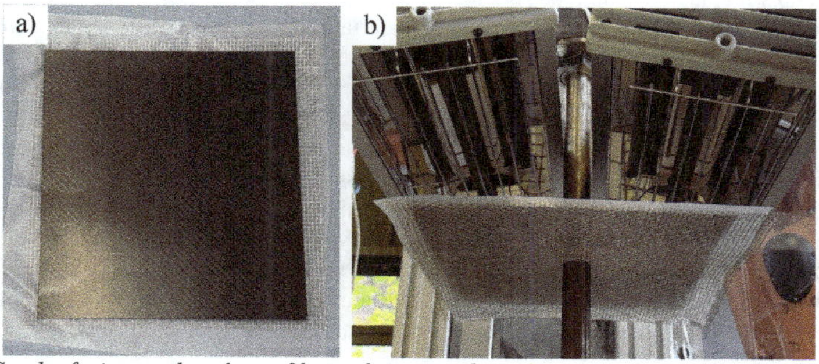

Fig. 4 a) Stack of wire mesh, release film and organo sheet, b) setup of centrally clamped stack and infrared heaters

Results. During performing an individual bending operation, the desired geometry was well producible through die-less forming and the organo sheet kept its shape as it rapidly cooled down. However, due to the missing bond between wire mesh and organo sheet, they detached from each other and their resulting geometries did not accurately match. Thus, as the sheet was heated again and another area was formed, a certain reverse deformation occurred so that some paths had to be repeatedly formed. Using only one of the infrared heaters was able to limit this problem by keeping half of the organo sheet cold and rigid. As shown in Fig. 5, the final geometries of the mesh and the organo sheet generally correspond to the targeted hemispherical frustum, while not exactly matching each other. Wrinkles occurred in the FRTP especially towards the edge of the frustum due to inhomogeneous heating and missing contact to the wire mesh. Due to the increased time at elevated temperatures, the silicone release film partially melted and permanently stuck to the inner organo sheet surface.

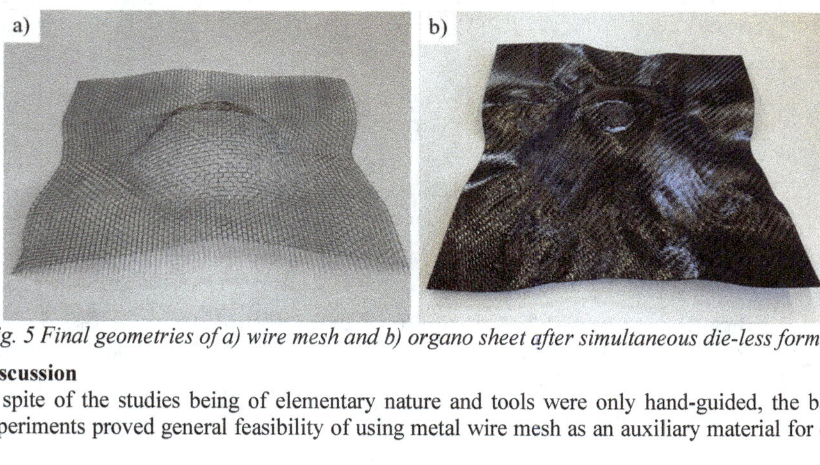

Fig. 5 Final geometries of a) wire mesh and b) organo sheet after simultaneous die-less forming

Discussion

In spite of the studies being of elementary nature and tools were only hand-guided, the basic experiments proved general feasibility of using metal wire mesh as an auxiliary material for die-less forming of organo sheets. As discussed in the theoretical considerations, the experiments showed that realizing a temporary bond between wire mesh and FRTP is one of the keys to successful forming. Instead of a simple release film, an elastic and durable sheet placed between the mesh and the FRTP could aid in maintaining a smooth organo sheet surface and realizing a bond between the two materials for example through later soluble adhesive. Alternatively, a flexible and stretchable vacuum bag enclosing wire mesh, smoothing sheet and organo sheet could ensure sufficient contact pressure between the materials and simultaneously aid in reconsolidation.

In order to reduce cycle times and thermal stresses through multiple localized reheating, keeping the whole organo sheet at forming temperature during the entire process would be desirable. Thus, the possibility of using the metal wires as resistance heaters for melting the FRTP matrix can be an advantage and should be investigated. For uniform heating and electrical isolation, an organo sheet protected by a smoothing sheet on either side, sandwiched between two wire meshes and finally enclosed in a vacuum bag might be the setup of choice.

The deformation characteristics of this entire layer structure must not only be investigated experimentally, but also mapped in a simulation. For only an accurate prediction of the elasto-plastic behavior will enable a universal and automated process chain for the generation of the appropriate forming paths to produce the desired shapes with sufficient accuracy.

Summary & outlook

As current FRTP forming processes involving molds are not cost efficient for prototype and small batch production, different options for die-less forming are currently under investigation. One of these options is forming a heated woven organo sheet together with a metal wire mesh as a supporting structure for the flexible fabric. Due to the similar deformation mechanisms, namely shear and bending, the stack of FRTP and wire mesh can be formed by two individually robot-guided standard tools following a specially developed draping strategy. Two variants of this strategy were tested in simple elementary experiments, proving general feasibility of this approach. In the future, the best option for creating a temporary bond between organo sheet and wire mesh needs to be identified and heating the thermoplastic matrix through wire resistance heating should be investigated. To gain valuable insight into the deformation behavior, material properties of the layup should be determined in standard tests. For further experiments concerning the forming strategy and rendering reproducible results, an automated manufacturing cell is to be set up.

Acknowledgements

Research was funded by the German Federal Ministry for Economic Affairs and Climate Action under the program LuFo VI-1 iFish.

References

[1] P. Davies, W.J. Cantwell, P.-Y. Jar, P.-E. Bourban, V. Zysman, H.H. Kausch, Joining and repair of a carbon fibre-reinforced thermoplastic, Compos. 22 (1991) 425–431. https://doi.org/10.1016/0010-4361(91)90199-Q

[2] M. Biron, Thermoplastics and thermoplastic composites: Technical information for plastics users, Elsevier BH, Amsterdam, 2006. https://doi.org/10.1016/B978-185617478-7.50005-2

[3] P. Mallick, Fiber-Reinforced Composites: Materials, Manufacturing and Design, CRC Press, 2008. https://doi.org/10.1201/9781420005981

[4] D.M. Bigg, D.F. Hiscock, J.R. Preston, E.J. Bradbury, High Performance Thermoplastic Matrix Composites, J. Thermoplast. Compos. Mater. 1 (1988) 146–160. https://doi.org/10.1177/089270578800100203

[5] I. Martin, D. Del Saenz Castillo, A. Fernandez, A. Güemes, Advanced Thermoplastic Composite Manufacturing by In-Situ Consolidation: A Review, J. Compos. Sci. 4 (2020) 149. https://doi.org/10.3390/jcs4040149

[6] M. Flemming, G. Ziegmann, S. Roth, Faserverbundbauweisen: Fertigungsverfahren mit duroplastischer Matrix, Springer Verlag, 1996. https://doi.org/10.1007/978-3-642-61432-3

[7] G. Ehrenstein, Faserverbund-Kunststoffe, Hanser Verlag, 2006. https://doi.org/10.3139/9783446457546.fm

[8] B.-A. Behrens, A. Raatz, S. Hübner, C. Bonk, F. Bohne, C. Bruns, M. Micke-Camuz, Automated Stamp Forming of Continuous Fiber Reinforced Thermoplastics for Complex Shell Geometries, Procedia CIRP 66 (2017) 113–118. https://doi.org/10.1016/j.procir.2017.03.294

[9] T. Hennige, Flexible Formgebung von Blechen durch Laserstrahlumformen, Meisenbach, Bamberg, 2001.

[10] A.K. Miller, M. Gur, A. Peled, A. Payne, E. Menzel, Die-Less Forming of Thermoplastic-Matrix, Continuous-Fiber Composites, J. Compos. Mater. 24 (1990) 346–381. https://doi.org/10.1177/002199839002400401

[11] A. Rosochowski, A. Matuszak, Rapid tooling: the state of the art, J. Mater. Process. Technol. 106 (2000) 191–198. https://doi.org/10.1016/S0924-0136(00)00613-0

[12] T.Z. Sudbury, R. Springfield, V. Kunc, C. Duty, An assessment of additive manufactured molds for hand-laid fiber reinforced composites, Int. J. Adv. Manuf. Technol. 90 (2017) 1659–1664. https://doi.org/10.1007/s00170-016-9464-9

[13] D. Afonso, R. Alves de Sousa, R. Torcato, L. Pires, Incremental Forming as a Rapid Tooling Process, Springer International Publishing, Cham, 2019. https://doi.org/10.1007/978-3-030-15360-1

[14] A. Brent Strong, P.B. Hauwiller, Incremental Forming of Large Thermoplastic Composites, J. Thermoplast. Compos. Mater. 2 (1989) 122–132. https://doi.org/10.1177/089270578900200204

[15] S.G. Kaufman, B.L. Spletzer, T.L. Guess, Freeform fabrication of polymer-matrix composite structures, in: Proceedings of International Conference on Robotics and Automation, 1997, pp. 317–322.

[16] D.F. Walczyk, J.F. Hosford, J.M. Papazian, Using Reconfigurable Tooling and Surface Heating for Incremental Forming of Composite Aircraft Parts, J. Manuf. Sci. Eng. 125 (2003) 333–343. https://doi.org/10.1115/1.1561456

[17] T. Trzepieciński, Recent Developments and Trends in Sheet Metal Forming, Metals 10 (2020). https://doi.org/10.3390/met10060779

[18] H. Zhu, H. Ou, A. Popov, Incremental sheet forming of thermoplastics: a review, Int. J. Adv. Manuf. Technol. 111 (2020) 565–587. https://doi.org/10.1007/s00170-020-06056-5

[19] S. Bagheri, A. Kami, M. Shakouri, Single point incremental forming of polyamide/30 wt% short glass fiber composite, J. Thermoplast. Compos. Mater. (2022) https://doi.org/10.1177/08927057221083497

[20] R. Conte, G. Ambrogio, D. Pulice, F. Gagliardi, L. Filice, Incremental Sheet Forming of a Composite Made of Thermoplastic Matrix and Glass-Fiber Reinforcement, Procedia Eng. 207 (2017) 819–824. https://doi.org/10.1016/j.proeng.2017.10.835

[21] G. Ambrogio, R. Conte, F. Gagliardi, L. de Napoli, L. Filice, P. Russo, A new approach for forming polymeric composite structures, Compos. Struct. 204 (2018) 445–453. https://doi.org/10.1016/j.compstruct.2018.07.106

[22] A. Al-Obaidi, A. Graf, V. Kräusel, M. Trautmann, Heat supported single point incremental forming of hybrid laminates for orthopedic applications, Procedia Manuf. 29 (2019) 21–27. https://doi.org/10.1016/j.promfg.2019.02.101

[23] R. Emami, M.J. Mirnia, M. Elyasi, A. Zolfaghari, An experimental investigation into single point incremental forming of glass fiber-reinforced polyamide sheet with different fiber orientations and volume fractions at elevated temperatures, J. Thermoplast. Compos. Mater. (2022). https://doi.org/10.1177/08927057221074266

[24] M. Fiorotto, M. Sorgente, G. Lucchetta, Preliminary studies on single point incremental forming for composite materials, Int. J. Mater. Form. 3 (2010) 951–954. https://doi.org/10.1007/s12289-010-0926-6

[25] C. Hou, X. Su, X. Peng, X. Wu, D. Yang, Thermal-Assisted Single Point Incremental Forming of Jute Fabric Reinforced Poly(lactic acid) Biocomposites, Fibers Polym. 21 (2020) 2373–2379. https://doi.org/10.1007/s12221-020-1016-0

[26] A. Al-Obaidi, A. Kunke, V. Kräusel, Hot single-point incremental forming of glass-fiber-reinforced polymer (PA6GF47) supported by hot air, J. Manuf. Process. 43 (2019) 17–25. https://doi.org/10.1016/j.jmapro.2019.04.036

[27] J.-E. Rath, R. Graupner, T. Schüppstuhl, Die-Less Forming of Fiber-Reinforced Plastic Composites, in: K.-Y. Kim, L. Monplaisir, J. Rickli (Eds.), Flexible Automation and Intelligent Manufacturing: The Human-Data-Technology Nexus, Springer International Publishing, Cham, 2023, pp. 3–14. https://doi.org/10.1007/978-3-031-18326-3_1

[28] J.-E. Rath, L.-S. Schwieger, T. Schüppstuhl, Robotic Die-Less Forming Strategy for Fiber-Reinforced Plastic Composites Production, Procedia CIRP 107 (2022) 1281–1286. https://doi.org/10.1016/j.procir.2022.05.145

[29] F. Heieck, Qualitätsbewertung von Faser-Kunststoff-Verbunden mittels optischer Texturanalyse auf 3D-Preformoberflächen. Dissertation, 2019.

[30] T. Gries, D. Veit, B. Wulfhorst, Textile Fertigungsverfahren: Eine Einführung, Carl Hanser, 2014. https://doi.org/10.3139/9783446440579.fm

[31] M.P. Elkington, A. Sarkytbayev, C. Ward, Automated composite draping: a review, in: SAMPE 2017, 22.5.-25.05.2017.

[32] M.P. Elkington, C. Ward, A. Chatzimichali, K. Potter, Studying effects of preshearing on hand layup, Advanced Manufacturing: Polymer & Composites Science 1 (2015) 80–93. https://doi.org/10.1179/2055035914Y.0000000007

Development of a collaborative online knowledge management system for incremental sheet forming

Thomas Bremen[1, a *], David Bailly[1, b], Marthe Vanhulst[2, c], Hans Vanhove[2, d], Joost R. Duflou[2, e] and Gerhard Hirt[1, f]

[1]Institute of Metal Forming, RWTH Aachen University, Intzestr. 10, 52072 Aachen, Germany

[2]Department of Mechanical Engineering, Katholieke Universiteit Leuven / Member of Flanders Make, Celestijnenlaan 300B, B-3001 Leuven, Belgium

[a]thomas.bremen@ibf.rwth-aachen.de, [b]david.bailly@ibf.rwth-aachen.de, [c]marthe.vanhulst@kuleuven.be, [d]hans.vanhove@kuleuven.be, [e]joost.duflou@kuleuven.be, [f]gerhard.hirt@ibf.rwth-aachen.de

Keywords: Incremental Sheet Forming, Knowledge Management, Accuracy

Abstract. Incremental sheet forming (ISF) has a high potential to produce individualized products and prototypes or small series from sheet metal. The universal forming tool, as well as the digital process control, in principle, allow a fast realization from design to a finished component. However, a high level of testing and planning is often required to achieve the necessary geometric accuracy. A large number of scientific studies on process parameters and process variants have led to significant progress in the specific studied cases. Nevertheless, this progress can often not be transferred to arbitrary component geometries. Despite the many valuable detailed scientific findings, there is a lack of a holistic overview. The translation of case specific findings to a broad range of real application components is missing. Due to the large number and complexity of process parameters as well as a wide variety of possible component geometries, a systematic collection of knowledge and experience is required. Therefore, this paper describes the development of a web-based database for process data of incremental sheet forming. This database should first of all allow component data and associated process data to be recorded, completely and reproducibly. By sharing the database with a large number of research teams and end users, an intensive exchange of experience can take place systematically and new knowledge can be generated collaboratively.

Introduction

A high dynamic market with a trend towards mass-customization and short time to market for new products, requires agile production technologies. In this regard, incremental sheet forming (ISF) has a high potential for sheet metal components. Due to the flexibility of a universal forming tool and the determination of a component geometry by numerical controlled movements of the tool, this process can in principle react very quickly to new or changed component geometries in small batch sizes [1]. In order to enable rapid realization from design to finished component for small quantities and prototypes, the planning and preparation processes in particular must also be carried out very quickly and with accurate parts as a result.

Allwood et al. [2] analysed specifications in terms of geometrical accuracy for many different sheet metal products. This shows that tolerances of less than ± 0.5 mm are required in most applications. Despite great research efforts, geometric tolerances required by industry have, so far, only been achieved with ISF in individual cases or specific applications [3]. The geometric deviations and the cumbersome planning and testing effort required to compensate for them, are one reason for the low prevalence of ISF in the industrial environment. Another reason are the stringent process limits, resulting in failure of the component outside a specific forming range.

Content from this work may be used under the terms of the Creative Commons Attribution 3.0 license. Any further distribution of this work must maintain attribution to the author(s) and the title of the work, journal citation and DOI. Published under license by Materials Research Forum LLC.

Many studies have been carried out over the last decades to increase geometric accuracy and process limits, such as investigating the influence of different process parameters, as well as optimizing tool paths to compensate for unwanted geometric deviations and applying a multi-stage approach [1,4]. This also led to the development of new process variants [3]. However, many of these investigations are limited to very simple component geometries or to very specific cases. Considering that the geometry has a significant influence on the process, as has been shown by Ambrogio et al. [5], a transfer of findings between different component geometries is only possible to a very limited extent.

The advantage of the flexibility of ISF is unfortunately challenged by its complexity. The process has a large number of degrees of freedom in process control and a very broad spectrum of theoretically possible component geometries. Measured against this, the process understanding and empirical knowledge that has been built up so far is insufficient to enable rapid process planning and thus broad industrial application. Therefore, the aim of the research work described here is to systematically gather knowledge and experience about the process, especially with regard to the large number of possible component geometries.

Platform for Knowledge Sharing

Documenting experience requires systematic and extensive gathering of data about components, corresponding processes and process results. Experiments and processes must be recorded in a structured way so that they are reproducible by other researchers or users. The shared experience documents which components can be manufactured and which process parameters are required. However, great value is also attributed to the failed process strategies, as they enable important knowledge to be gained. In this regard, it is also possible to document which components cannot be manufactured up to now and, ideally, why this is the case.

The documentation of process experience can be a benefit for each research team, but should not be limited to that. In order to generate a real impact, the data of as many research groups and end users as possible should be collected and shared. For this purpose, a web-based database has been implemented to allow users to document their own component and process data, to analyse the data from other users and compare different experiments in an easy and structured way. An additionally linked forum is intended to enable direct discussion on this data. This facilitates a direct exchange of knowledge and experience across different researchers and end users. As an addition to scientific publications, this form of dissemination enables an exchange on specific and detailed data that can also be directly referenced in discussions. It is a work and collaboration tool closing the gap between different research teams and PhD research generations.

The presented database can also be used to systematically document and moderate benchmark studies. Such benchmarks can trigger a collective learning process and further advance knowledge. Another motivation behind a platform for data collection is the prevailing evolution of data-driven methods. For example, a wide and detailed collection of process data and annotated metadata can enable machine learning to extract process boundaries for a broad range of geometry/material combinations. In the long term, the shared ISF data on the platform shall be used to generate new insights into the causes of geometric deviations and potential compensation measures.

Web-based Interface and Database for ISF Process Data

The collaborative database motivated in the previous paragraph was realized with a web-based interface. This enables the desired worldwide access of many different researchers and users. Moreover, it allows for the structured and visually responsive processing of the data to be documented. A major difficulty in the process optimisation of ISF is the complexity of the input and output variables of the process. The component geometry, tool path, and resulting accuracy are complex three-dimensional data that cannot be simplified to individual parameters without severe loss of information. This problem does not only apply to process optimisation, but also causes difficulties in the pure documentation of the processes. Other complex parameters are the (partial) supporting dies and the occurrence of intermediate geometry data in multi-stage processes and possible preliminary process steps. They all have to be documented as 3D data.

A well-designed user interface enables an efficient workflow for the documentation of the process data, is not prone to errors and still ensures a high information content. In order to reach this, the web-based platform was realized using Django, which is an open source framework written in Python that can be used to design web applications. MySQL is used as the database management system for the storage of the ISF process data. Within the framework, an ISF project/process is represented using a model. Fields are assigned to this model, which represent the variables to be stored.

In addition to a large number of quantifiable process parameters, CAD data of the component, tool paths and 3D scanned surfaces can also be uploaded as STL or STP files via the web interface. The uploaded data are visualized directly to the user and can thus be reviewed and analysed. An exemplary visualization of 3D data sets on the platform is shown in Fig. 1. In this figure, both the tool path (red), various scanned geometries after the forming process, and the actual target geometry can be seen.

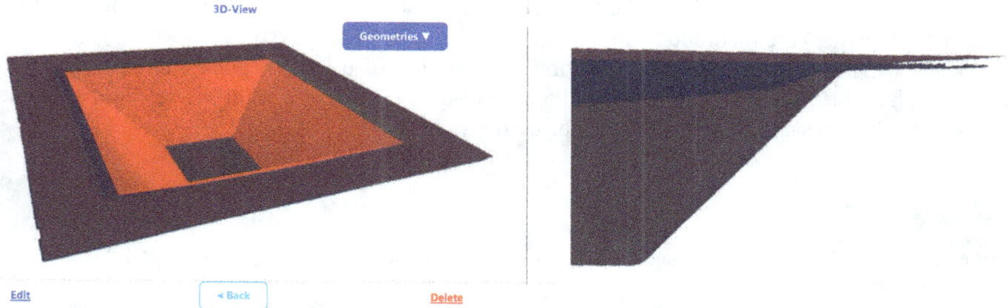

Figure 1. Screenshot of the visualization of 3D data in the user interface of the platform (left), side view to show different geometries (right)

In addition to the systematic documentation of ISF processes, the database also offers the possibility to search in the already stored component and process data. Search fields and filters can be used to find specific process plans that match a current problem. The database thus already offers the possibility of systematically documenting and exchanging process data. In order to be able to conduct this exchange in the sense of specific discussions, a forum has also been integrated. This way, the relevant process data can be referenced directly in technical discussions.

Structure of Process Data Acquisition

For the systematic acquisition of component and process data, it is necessary to have all parameters consolidated and clearly defined. The overall goal is to get assessable, searchable and comparable data that describe the components and processes accurately and completely, in such a way that the process can be reproduced easily. In order to achieve a clear description with a high depth of information, 3D data and process parameters are therefore recorded in their entirety. Furthermore, to enable searching and comparing the data, fuzzy criteria (e.g. geometry category) are also recorded. The processes considered in the database are limited to conventional SPIF, TPIF and multi-stage SPIF, with an additional specification of preform operations. Fig. 2 shows the structure of the data acquisition in the web interface, where the data are split up into six main parameter categories.

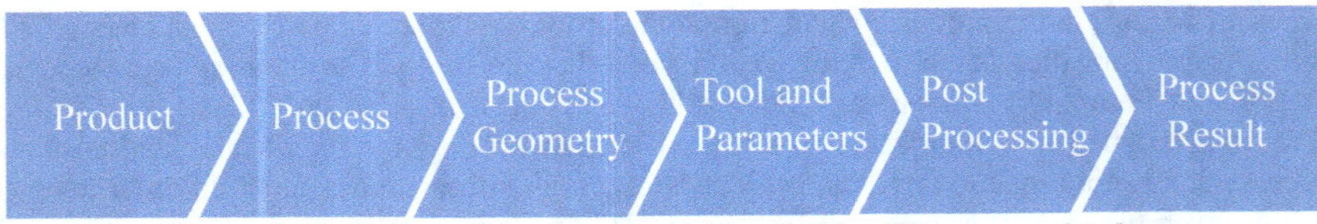

Figure 2. Work flow of the data acquisition for the ISF process database

In the first step (Product), the information about the sheet metal component is recorded, since the product specification is the basis for everything else and usually cannot be changed. In the second step (Process), the basic process configuration is queried, i.e. process variant, machine, etc. This information is used to adapt all further queries to the specific case. In the third step (Process Geometry), the geometries that define the process are recorded, i.e. component geometry, clamping, pre-geometry, etc. This was deliberately decoupled from the product description, since process-specific adaptation occurs here. In the fourth step (Tool and Parameters), the variable process parameters are described, which can be easily adapted in terms of process optimisation. In the fifth step (Post Processing), the possible processes after ISF operations are queried, such as heat treatment and trimming. In the last step (Process Result), result data (e.g. resulting geometry and surface quality) are recorded.

The next paragraph gives a more elaborate description of these categories and the data they contain. The type of bullet point reflects the type of query: boxes for multiple selections, circles for single selections and dots for other input fields.

Product

3D Part Geometry	**Geometry Size**	**Geometry Category**			**Geometry Features**	**Limiting Geometry Parameters**	**Material**	**Initial Sheet Thickness** (before ISF)
		Global Curvature	**Global Shape**	**Symmetry**				
• [.stl/.stp]	• Length [mm] • Width [mm] • Surface Area [mm²]	❑ Flat ❑ Uniaxially Curved ❑ Biaxially Curved ❑ Uniaxially Curved with Curvature Change ❑ Freeform	❑ Round ❑ Rectangular ❑ Squared ❑ Oval	❑ Asymmetry ❑ Rotational Symmetry	❑ Corner ❑ Cavity ❑ Fold ❑ Flange ❑ Undercut	• Maximum Wall Angle [°] • Tightest Radius [mm] • Deepest Cavity [mm]	• Alloy • Yield Point Rp0,2 [MPa] • Young's Modulus [GPa]	• [mm]

Figure 3. Product data query with related parameters

In a first step, the information about the desired sheet metal component is recorded (see Fig. 3). For each ISF process, basic information about the component to be manufactured must be queried, such as the 3D geometry and its bounding box dimensions. In addition, a fuzzy description of the geometry (curvature, shape, symmetry, features, etc.) helps to search and compare the different part geometries. Some parameters are also important to interpret process results. In particular, the data of limiting geometrical influences (e.g. wall angles), sheet material, as well as the input sheet thickness are relevant process information. Jeswiet et al. [6] demonstrated the influence of the wall angles on the final thickness distributions after forming and approximated this relation with the sine law. In addition, Ambrogio et al. [7] showed that sheet thinning is significantly influenced by the wall angle of the geometry to be formed and the material used. Li et al. [8] explained that the sheet thickness has a significant effect on the overall geometrical accuracy.

Materials Research Proceedings **25** (2023) 45-52
https://doi.org/10.21741/9781644902417-6

Process

Process Type	Type of Preform Fabrication	Machine Description
❑ Multi-Stage ○ SPIF ○ TPIF ○ Other	❑ Flat ❑ Stretch Forming ❑ Deep Drawing ❑ Bending ❑ Other	• Type • Axis-Configuration • Max. Force [kN] • Working Space (Length, Width, Depth) [mm]

Figure 4. Process data query with related parameters

In the second step, the basic process configuration is queried, i.e. process variant, machine, etc. (see Fig. 4). This information is used to adapt all further queries to the specific case. Information about the type of process and the equipment used is documented. For example, Taleb Araghi et al. [9] were able to show that not only classical ISF process parameters have a significant influence on the process, but also preceding process steps. The combination with a preceding stretch-forming process also reduced the geometric deviations strongly. Liu et al. [10] showed the dependence of the results after forming on the selected process strategy, where multi-stage strategies proved to increase the formability and thickness distributions compared to single stage forming. In the platform, multi-stage strategies can be uploaded as a chain of individual processes that reference each other. This allows each individual process step to be analysed separately and as process chain.

Process Geometry

Input Sheet Geometry	Input Sheet Thickness	Part Geometry for Process Planning	Clamping Geometry and Position
• [.stl/.stp]	• [mm/.txt]	• [.stp/.stl]	• [.stp/.stl] ○ Rectangular ○ Circle ○ Ellipse ○ Near-Shape ○ Other

Figure 5. Process Geometry data query with related parameters

In the third step, the geometries that define the process are recorded, i.e. component geometry, clamping, pre-geometry, etc. (see Fig. 5). The choice of process type and preform fabrication results in additional process parameters. If a preform is used, the CAD file of the preform as well as information about the sheet thickness distribution as a txt file can be uploaded. If a modified part geometry is used for process planning, such as for overbending strategies, this modified geometry can also be uploaded. Additionally, the geometry and position of clamping tools can be defined.

Materials Research Proceedings **25** (2023) 45-52 https://doi.org/10.21741/9781644902417-6

Tool and Parameters

Type of Forming Tool	Tool Tip Geometry	Tool Material	Tribology	Tool Path Design	Pitch Strategy
o Rigid o Rolling o Pulsing o Free Rotating o Driven Rotating o Other	o Hemispherical o Other	• Alloy	❑ Coating • Type ❑ Lubrication • Type • Application Method	• [.txt]	o Step Down o Scallop o Step Over • Pitch [mm]

Forming Speed	Process Temperature	Temperature Distribution	Forming Force
• [mm/min]	**Temperature Level:** o Room Temperature o Hot Forming **Type of heating:** o Local Heating o Global Heating • Forming Temperature [°C]	• [.txt]	• [.txt]

Figure 6. Tool and Parameters data query with related parameters

In the fourth step (Tool and Parameters), the variable process parameters are described, which can be easily adapted in terms of process optimisation (see Fig. 6). Detailed information about the forming tool as well as about the planned tool path is summarized. The increased focus of the current research on adaptations or optimisation of the used tool path confirms its significant influence on the process result, so that a detailed query of the tool path is reasonable. For example, Skjoedt et al. [11] showed the influence of the forming direction of the toolpath, where upwards and downwards forming have a significant influence on the geometric accuracy in a multi-stage approach. However, according to the literature, information about the forming temperature and speed also are essential. Al-Obaidi et al. [12] showed in induction heating assisted ISF that the temperature distribution in the sheet during forming has an important role in improving geometric accuracy.

Post Processing

Trimming	Heat Treatment
o Trimmed to Part Contour o Untrimmed **Type of trimming:** o Waterjet Cutting o Laser Cutting o Wire Cutting o Milling	• Time [s] • Temperature [°C]

Figure 7. Post Processing data query with related parameters

In the fifth step, extra processing operations after forming with ISF are queried, such as heat treatment and trimming (see Fig. 7). After the actual forming process, many components require post-treatment to finalize the component. Trimming processes are mainly used here, as well as heat treatment, which can be used, for example, to reduce residual elastic stresses. Due to their influence on the residual stresses remaining in the material, both processes also have an influence on the process result and are therefore queried under the table "Post Processing".

Process Result

General Producibility	Process Pictures	Result Description	Accuracy	Sheet Thickness	Surface Quality
o Producible o Not Producible	• [.image]	• free text	• Measurement Device • Measurement Accuracy [mm] • Scanned Geometries [.stl/.stl] • Deviation Plots [.image] • Max. Deviation [mm] • Average Deviation [mm]	• Minimum Thickness [mm] • Thickness Distribution [.txt]	• Surface Pictures [.image] • Optical Appearance [free text]

Process Time	Additional Information	Process Goal Description	Related Publication
• [HH:MM:SS]	• Process Goal ❑ Industrial • Scientific ❑ Geometrical Accuracy ❑ Process Limits ❑ Feasibility	• free text	• DOI

Figure 8. Process Result data query with related parameters

In the last step, result data such as measured geometry and surface quality are recorded (see Fig. 8). Up to this point, process parameters were queried from which, according to the literature, there is a possible influence on the process result. As already mentioned at the outset, quantitative and qualitative process results must be queried in a next step in order to be able to find relationships with the data uploaded so far. For this purpose, information such as the basic manufacturability is queried in the "Process Result" table, since ISF processes that have not been carried out successfully also represent an important element for further process understanding. In addition, process images, a qualitative description of the process performed and the experiment duration can be uploaded. For a detailed analysis of the results, scanned component data can be uploaded as well as descriptions and images of the resulting surface quality.
So far, the focus has been on uploading experimental data, an extension to numerical models and simulation results is intended in the near future.

Summary

ISF has a high potential for prototyping and small batch production. Nevertheless, the process is complex and compared to the large variety of possible component geometries there is a lack of process knowledge and experience. In order to be able to successfully and economically apply ISF in industry, it is necessary to ensure fast and successful process planning, since the planning and testing costs can only be allocated to a small number of components. In particular, avoiding undesired geometric deviations leads to considerable effort in process planning for complex application components and often also to a large number of trial and error tests.

The paper presented an approach and an implementation platform to systematically collect and analyse data of ISF processes. The realization as a database with a web-based user interface enables the exchange of ISF researchers and users around the world. This allows for an intensive exchange of knowledge and experience about components and ISF processes based on specific and precise data. The intensive exchange between the authors about the selection and definition of the process parameters has already led to a close exchange between two research teams. From the collected data and the intensive discussions of these data, important findings about the ISF process can be generated and verified in the future. This will enable collaborative learning and systematic progress for the ISF process and its planning procedures.

Acknowledgments
Funded by the Deutsche Forschungsgemeinschaft (DFG, German Research Foundation) under Germany's Excellence Strategy – EXC-2023 Internet of Production – 390621612. Marthe Vanhulst was supported by a Predoctoral Strategic Basic Research Fellowship of the Research Foundation—Flanders (FWO) with project 1S47622N.

References

[1] J. Jeswiet, F. Micari, G. Hirt, A. Bramley, J. Duflou, J. Allwood, Asymmetric Single Point Incremental Forming of Sheet Metal, CIRP Annals 54 (2005) 88–114. https://doi.org/10.1016/S0007-8506(07)60021-3

[2] J.M. Allwood, G.P.F. King, J. Duflou, A structured search for applications of the incremental sheet-forming process by product segmentation, Proceedings of the Institution of Mechanical Engineers, Part B: Journal of Engineering Manufacture 219 (2005) 239–244. https://doi.org/10.1243/095440505X8145

[3] H. Lu, H. Liu, C. Wang, Review on strategies for geometric accuracy improvement in incremental sheet forming, Int J Adv Manuf Technol 102 (2019) 3381–3417. https://doi.org/10.1007/s00170-019-03348-3.

[4] J. Duflou, J. Verbert, B. Belkassem, J. Gu, H. Sol, C. Henrard, A.M. Habraken, Process window enhancement for single point incremental forming through multi-step toolpaths, CIRP Annals 57 (2008) 253–256. https://doi.org/10.1016/j.cirp.2008.03.030

[5] G. Ambrogio, L. de Napoli, L. Filice, A novel approach based on multiple back-drawing incremental forming to reduce geometry deviation, Int J Mater Form 2 (2009) 9–12. https://doi.org/10.1007/s12289-009-0498-5

[6] J. Jeswiet, E. Hagan, A. Szekeres, Forming parameters for incremental forming of aluminium alloy sheet metal, Proceedings of the Institution of Mechanical Engineers, Part B: Journal of Engineering Manufacture 216 (2002) 1367–1371. https://doi.org/10.1243/095440502320405458

[7] G. Ambrogio, I. Costantino, D.N. Luigi, L. Filice, L. Fratini, M. Muzzupappa, On the influence of some relevant process parameters on the dimensional accuracy in Incremental Forming: A numerical and experimental investigation, Journal of Materials Processing Technology 153 (2004) 501–507. https://doi.org/10.1016/j.jmatprotec.2004.04.139

[8] Y. Li, H. Lu, W.J.T. Daniel, P.A. Meehan, Investigation and optimization of deformation energy and geometric accuracy in the incremental sheet forming process using response surface methodology, Int J Adv Manuf Technol 79 (2015) 2041–2055. https://doi.org/10.1007/s00170-015-6986-5

[9] B. Taleb Araghi, G.L. Manco, M. Bambach, G. Hirt, Investigation into a new hybrid forming process: Incremental sheet forming combined with stretch forming, CIRP Annals 58 (2009) 225–228. https://doi.org/10.1016/j.cirp.2009.03.101

[10] Z. Liu, Y. Li, P.A. Meehan, Tool path strategies and deformation analysis in multi-pass incremental sheet forming process, Int J Adv Manuf Technol 75 (2014) 395–409. https://doi.org/10.1007/s00170-014-6143-6

[11] M. Skjoedt, N. Bay, B. Endelt, G. Ingarao, Multi Stage Strategies for Single Point Incremental Forming of a Cup, Int J Mater Form 1 (2008) 1199–1202. https://doi.org/10.1007/s12289-008-0156-3

[12] A. Al-Obaidi, V. Kräusel, D. Landgrebe, Hot single-point incremental forming assisted by induction heating, Int J Adv Manuf Technol 82 (2016) 1163–1171. https://doi.org/10.1007/s00170-015-7439-x

Experimental investigation on thinning and forming force acting on multi-stage single point incremental forming

Nikhil Bari[1, a] and Shailendra Kumar[1, b*]

[1] Department of Mechanical Engineering, Sardar Vallabhbhai National Institute of Technology, Surat, India

[a]nikhilbari1665@gmail.com , [b*]skbudhwar@med.svnit.ac.in

Keywords: Incremental Sheet Forming, Predictive Models, Optimization

Abstract. This paper describes a study of thinning and forming force in multi-stage single point incremental forming. Four process parameters namely pitch, tool diameter, blank thickness and initial draw angle are considered to study their influence on maximum thinning and peak forming force. Experiments are designed using face-centered composite design (CCD). Experimental results are analyzed using analysis of variance (ANOVA). It is found that initial draw angle is the most influencing parameter for maximum thinning, while peak forming force is most influenced by blank thickness. Maximum thinning decreases with decrease in initial draw angle and increase in blank thickness. Peak forming force decreases with decrease in blank thickness. Moreover, predictive models are developed for maximum thinning and peak forming force. Also, optimization of parameters is carried out to minimize thinning and forming force. The confirmation tests are performed to validate predictive model and optimization results.

Introduction

Multi-stage single point incremental forming (MSPIF) is one of the efficient ways to resolve the limitations like steeper wall angle and un-uniform thickness distribution over SPIF. In MSPIF, deformation zone is expanded to larger areas therefore higher formability and uniform thickness distribution is observed. To achieve a more uniform thickness distribution and more formability, a MSPIF with accurate tool path is important [1]. Nowadays, MSPIF process is used in industries to overcome the limitations of SPIF process. Some researchers have investigated the applications like composite spherical pressure vessel mold, Carnio facial implants [2], bottom surface of the wing in ground (WIG) ship [3] etc. Consequently, MSPIF also has some limitations like high surface roughness, forming time, thinning, power consumption etc. [4].

Researchers have worked on thickness distribution and forming force in MSPIF process. Nirala and Agrawal [5] developed a thinning determination model and verified it with finite element analysis (FEA) and experimental results. Gheysarian et al. [6] investigated the influence of parameters on maximum thinning and found that diameter of tool has most significant effect on maximum thinning. An et al. [7] investigated the effect of process parameters on MSPIF process using 4 stage and concluded that reduction in pitch, tool diameter and feed rate results in better thickness distribution. Few researchers have studied the forming force acting in MSPIF. Ghafoor et al. [8] investigated the effect of different toolpath strategies on forming force and they reported that varying wall angle constant depth toolpath requires minimum forming force. Gandla et al. [9] studied the effect of number of passes on forming force acting in MSPIF and they reported that resultant forming force decreases with increase in number of passes. Aerens et al. [10] experimentally studied the influence of blank thickness, tool size, wall angle, and pitch on X, Y and Z components of forming forces and they also generated the predictive model for PFF.

Researchers [11, 12] have observed excessive thinning in parts made by MSPIF process. Few researchers studied forming forces in SPIF process to determine power consumption and tool life of the process, but limited literature [13, 14] is available on studying the forming force in MSPIF

Content from this work may be used under the terms of the Creative Commons Attribution 3.0 license. Any further distribution of this work must maintain attribution to the author(s) and the title of the work, journal citation and DOI. Published under license by Materials Research Forum LLC.

process. Further efforts should be made to investigate influence of process parameters on sheet thinning and forming force in MSPIF. Present study aims to study the influence of process parameters on maximum thinning and peak forming force acting in MSPIF process, to develop prediction models for the responses, and to optimize the sets of process parameters for minimum thinning and peak forming fore.

Experimental plan

Sheet of aluminum alloy 1050 is taken for the present study. High speed, high chromium steel of grade M2 material is used to fabricate the tool. EP-90 hydraulic oil is used as a lubricant to minimize friction and wear. The varying wall angle - constant depth multi-stage tool path technique is employed to produces good thickness distribution as compared to other strategies. In all test runs 110×110×35 mm square pyramid is formed using 4 stage tool path strategy. The starting draw angle is used as one of process parameters, and the wall angle is changed by 5^0 degrees at each stage. As illustrated in Fig. 1, with a middle level of initial draw angle, 40^0, 45^0, 50^0, and 55^0 degree wall angles are formed in first to fourth stages respectively. Toolpath for the operation are extracted using CATIA and MATLAB software [4].

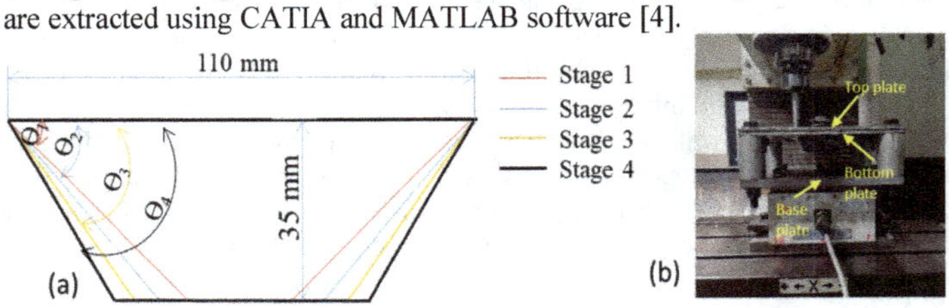

Fig. 1. (a) Truncated pyramid with varying wall angle (b) holding fixture

For the present study, four process parameters are considered namely pitch, tool diameter, blank thickness, and initial draw angle. Each parameter has three levels as given in Tab. 1. All process parameters and their levels are selected as per available setup, literature review, and trial experiments. The experiments are performed according to centered composite design i.e. CCD of response surface methodology (RSM). Overall 30 experiments (24 non-center + 6 center point) are designed.

Table 1. Parameters and their levels

Process parameters	Levels		
	-1	0	+1
Pitch [mm]	0.25	0.50	0.75
Tool diameter [mm]	8	10	12
Blank thickness [mm]	0.70	0.95	1.20
Initial draw angle [^{0}]	35	40	45

Siemens controlled 3 axis milling machine (M/s Batliboi, India) is used to perform all the experiments. The holding fixture is designed and assembled on the machine bed as shown in Fig. 1(b). Maximum thinning (%) is computed using Eq. 1,

$$\text{Maximum thinning (\%)} = 1 - \frac{t_{min}}{t_i} \times 100. \tag{1}$$

Where, t_i = Initial blank thickness, t_{min} = Minimum observed thickness. t_{min} is measured using digital micrometer (make MITUTOYO, Japan) whose accuracy is 0.001mm and repeatability of 95%. To measure thickness at corner of part, the parts are cut into two halves. Peak forming force (PFF) readings are taken from milling tool dynamometer (M/s Syscon Instruments, India). Dynamometer readings are converted to Newton (N) by multiplying 9.806 to the 'Kgf' reading.

Results and discussion

Total test runs and the respective values of responses are given in Tab. 2. The maximum thinning is ranging from 50.30% to 71.00%, whereas, PFF ranges from 567.76N to 2441.24N. Significant process parameters for the responses are determined by performing analysis of variance (ANOVA) on experimental data. Selected confidence interval is 95% therefore terms having p-value is less than 0.05 are considered as significant terms.

Table 2. Test run design and response values

S.N.	Pitch	Tool dia.	Blank thickness	Initial draw angle	Max. thinning	PPF	S.R.	Pitch	Tool dia.	Blank thickness	Initial draw angle	Max. thinning	PPF
1	0.75	8	0.70	45	71.00	724.70	16	0.75	8	0.70	35	55.80	636.42
2	0.25	8	1.20	45	65.70	1862.52	17	0.50	10	1.20	40	60.20	1999.85
3	0.75	8	1.20	35	52.70	1911.57	18	0.50	10	0.95	40	62.80	1323.04
4	0.50	10	0.95	40	68.20	1323.04	19	0.25	10	0.95	40	63.80	1234.76
5	0.25	8	0.70	35	55.20	567.76	20	0.50	10	0.95	40	62.80	1323.04
6	0.50	10	0.95	40	62.80	1323.14	21	0.75	12	0.70	35	54.70	812.98
7	0.50	10	0.95	40	62.80	1323.04	22	0.75	12	0.70	45	68.90	840.32
8	0.25	12	1.20	45	66.00	2078.32	23	0.25	8	1.20	35	50.30	1656.54
9	0.50	12	0.95	40	62.20	1479.98	24	0.75	8	1.20	45	65.40	2097.94
10	0.25	12	1.20	35	51.00	1941.12	25	0.75	10	0.95	40	61.40	1460.36
11	0.25	8	0.70	45	70.50	620.11	26	0.50	10	0.95	40	62.80	1323.04
12	0.50	10	0.95	35	57.90	1260.65	27	0.50	10	0.95	40	62.80	1323.04
13	0.25	12	0.70	45	68.50	790.61	28	0.50	10	0.70	40	62.90	744.32
14	0.75	12	1.20	45	64.60	2441.24	29	0.25	12	0.70	35	59.20	685.47
15	0.50	8	0.95	40	62.20	1264.19	30	0.75	12	1.20	35	52.70	2264.69

The ANOVA for maximum thinning is given in Tab. 3. The R^2 value for the model is 0.9551, adjusted R^2 and predicted R^2 values are 0.9131, 0.8250 respectively, these values are close to 1. The p-value of lack of fit is 0.8662 which is greater than 0.05 hence the lack of fit is not significant. The model is significant which is satisfying the basic requirements of the ANOVA model. The adequacy of the model is 16.842 which is greater than 4 hence the model is selected to examine the design space. From the ANOVA, blank thickness and initial draw angle are found significant because the p-value of these terms is less than 0.05. Moreover, no interaction term is found significant.

Main effect of significant process parameter is shown in Fig. 2. It is noted that maximum thinning of MSPIF parts decreases with increase in blank thickness (Fig. 2a). It is due to increase in stiffness of blank with increase in blank thickness. Similar results for SPIF was reported by Oleksik et al. [12]. Initial draw angle is also significant process parameter. As initial draw angle increases, maximum thinning also increases as shown in Fig. 2b. This is due to increase in initial draw angle, stretching in upcoming stages increases which increases the maximum thinning.

The ANOVA for PFF is given in Tab. 4. R^2, adjusted R^2, and predicted R^2 values of the models are 0.9989, 0.9980, and 0.9924. adjusted R^2, and predicted R^2 is in reasonable agreement with each other because the difference is less than 0.2. Adequate precision is 108.028 implies acceptable model. Overall p-value of the model is less than 0.05 means model is significant. Similarly, all process parameters and interaction effect of pitch-blank thickness (AC), tool diameter-blank thickness (BC), and blank thickness-initial draw angle (CD) are found significant since their p-value is less than 0.05.

Table 3. ANOVA for maximum thinning

Source	Sum of Squares	DoF	Mean Square	F-value	p-value	Remark
Model	887.9	14	63.42	22.78	< 0.0001	significant
A-Pitch	0.5	1	0.5	0.1796	0.6777	
B-Tool diameter	0.0556	1	0.0556	0.02	0.8895	
C-Blank thickness	80.64	1	80.64	28.97	< 0.0001	significant
D-Initial draw angle	756.09	1	756.09	271.57	< 0.0001	significant
AB	3.06	1	3.06	1.1	0.3109	
AC	1.82	1	1.82	0.6546	0.4311	
AD	0.0625	1	0.0625	0.0224	0.8829	
BC	0.1225	1	0.1225	0.044	0.8367	
BD	4.2	1	4.2	1.51	0.2382	
CD	0.0625	1	0.0625	0.0224	0.8829	
Residual	41.76	15	2.78			
Total	929.66	29				

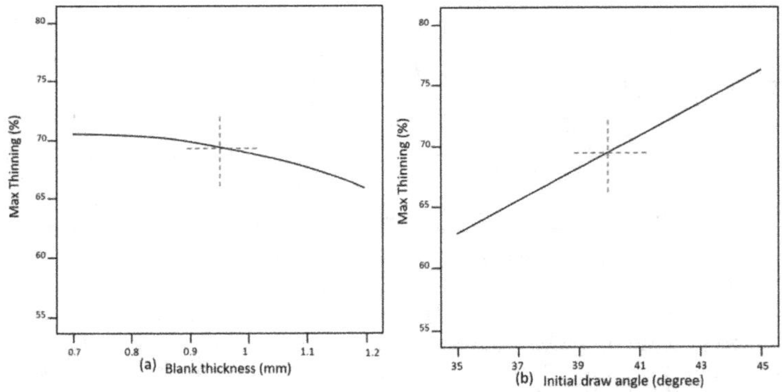

Fig. 2. Main effect plot for maximum thinning

The main effect graph for PFF is depicted in Fig. 3. From Fig. 3 (a) it is observed that PFF slightly increases with increase in pitch value. This is due to the fact that a large pitch deforms more material and consumes a larger amount of deforming energy. Similar results are reported for by Liu et al. [13] and Sisodia et al. [14] for SPIF process. As tool diameter increases, PFF increases marginally as shown in Fig 3 (b). The increasing trend is observed due to increase contact area between sheet and tool tip, hence more material is pushed and more force is required. Aerense et al. [10]; Kumar and Gulati [15] observed the same trend. It is observed that increase in blank thickness, PFF increases substantially as seen in Fig. 3 (c). The upward trend is observed because as blank thickness increases, stiffness of the blank increases and requires more energy to deform and hence PFF increases. It is also noted that PFF slightly increases with increase in initial draw angle (Fig. 3 (d)).

Table 4. ANOVA for peak forming force

Source	Sum of Squares	DoF	Mean Square	F-value	p-value	Remarks
Model	8.33E+06	14	5.95E+05	1013.35	< 0.0001	significant
A-Pitch	1.71E+05	1	1.71E+05	290.81	< 0.0001	significant
B-Tool diameter	2.21E+05	1	2.21E+05	375.87	< 0.0001	significant
C-Blank thickness	7.78E+06	1	7.78E+06	13246.01	< 0.0001	significant
D-Initial draw angle	58579.49	1	58579.49	99.78	< 0.0001	significant
AB	2500.2	1	2500.2	4.26	0.0568	
AC	42689.76	1	42689.76	72.72	< 0.0001	significant
AD	30.61	1	30.61	0.0521	0.8225	
BC	23748.35	1	23748.35	40.45	< 0.0001	significant
BD	470.33	1	470.33	0.8011	0.3849	
CD	11717.85	1	11717.85	19.96	0.0005	significant
Residual	8806.08	15	587.07			
Total	8.34E+06	29				

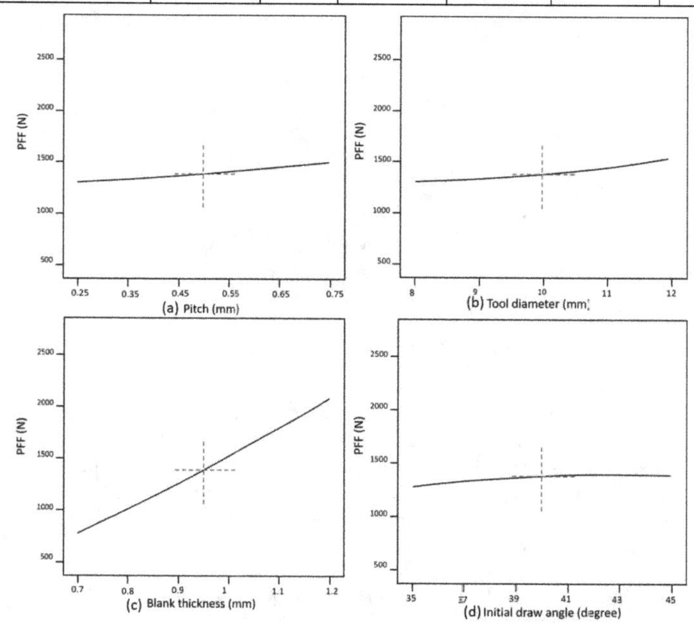

Fig. 3. Main effect plot for PFF

From ANOVA, three interactions are also found significant for PFF. The interactive effect of process parameters is studied through 3-D response surface graph while other process parameters are kept at middle level. The combined surface plot for pitch and blank thickness is depicted in Fig. 4 (a). The PFF rapidly increases with increase in both pitch and blank thickness. At 0.75mm pitch and 1.2mm blank thickness, maximum PFF value is observed. The contour lines of the response are parallel to each other and also parallel to the axis of pitch which reveals that blank thickness has major influence than pitch. The response plot for interaction of tool diameter and blank thickness is shown in Fig. 4 (b). It is observed that combined increase in blank thickness and tool diameter, increases the PFF. Blank thickness has major effect than tool diameter. At 0.95mm blank thickness, PFF increased by 16.22% (1257.21N to 1461.16N) with increase in tool diameter. With simultaneous increase in initial draw angle and blank thickness, the value of PFF increases rapidly (Fig. 4c). The response lines are nearly parallel to the axis of initial draw angle which infers that blank thickness has more influence than initial draw angle. At 0.95mm blank thickness, there is 9.27% (1241.81N to 1361.61N) increase in PFF. The present experimental study reveals that most influencing process parameter is blank thickness subsequently, tool diameter, pitch and initial draw angle.

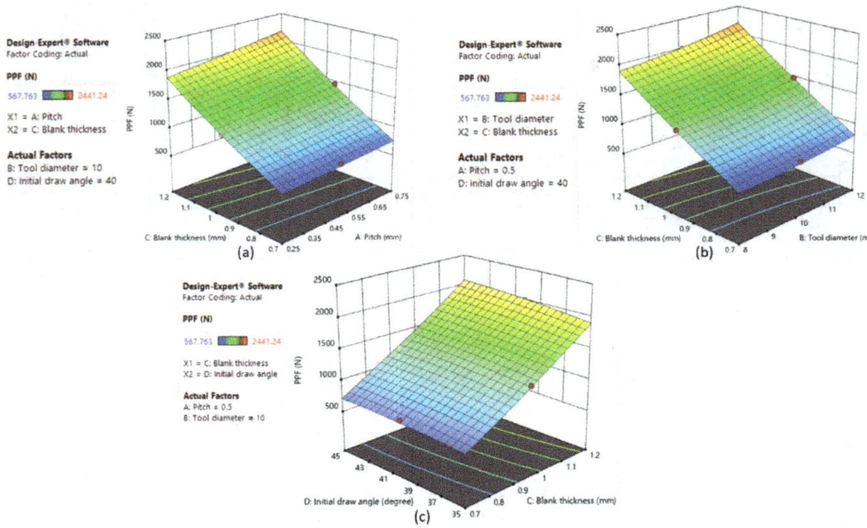

Fig. 4. Response surface graph for PFF

Predictive models

Quadratic model is best fitted for the present experimental data. The predictive models in terms of process parameters for maximum thinning and PFF are given in Eq. 2 and Eq. 3 respectively. Predictive models are generated to determine the value of responses at certain values of each process parameter. For better results, the values of process parameters must be bounded by original range. Predicted R^2 value for maximum thinning and PFF is 0.8250 and 0.9924 respectively which are closed to 1 hence the models are accepted.

$$\text{Maximum thinning [\%]} = 17.903+18.9222\times(A)+8.658\times(B)+43.544\times(C)-2.303\times(D)-0.875\times(A)(B)+5.400\times(A)(C)-0.050\times(A)(D)+0.175\times(B)(C)-0.051\times(B)(D)+0.050\times(C)(D)-13.968\times(A^2)-0.318\times(B^2)-30.768\times(C^2)+0.051\times(D^2). \quad (2)$$

$$\text{PFF [N]} = -1163.860-713.685\times(A)-166.391\times(B)-379.694\times(C)+91.129\times(D)+25.001\times(A)(B)+826.460\times(A)(C)-1.106\times(A)(D)+77.052\times(B)(C)-0.542\times(B)(D)+21.649\times(C)(D)+112.361\times(A^2)+7.886\times(B^2)+504.777\times(C^2)-1.171\times(D^2). \quad (3)$$

Optimization

Optimization is carried to get minimum thinning and PFF. The desirability approach is selected for the study. The desirability function is bounded by 1 and 0, where 1 is acceptable and 0 is unacceptable response. The optimum levels of pitch, tool diameter, blank thickness and initial draw angle is determined, which also satisfies the required criteria as given in Tab. 5.

Table 5. Criteria for optimization

S.N.	Parameters	Criteria	Lower bound	Upper bound	Optimum value
1	Pitch [mm]	Within range	0.25	0.75	0.25
2	Tool diameter [mm]	Within range	8	12	8
3	Blank thickness [mm]	Within range	0.700	1.200	0.854
4	Initial draw angle [^{0}]	Within range	35	45	35
5	Max. Thinning [%]	Within range	50.30	71.00	55.78
6	PFF [N]	Within range	567.76	2441.24	877.08

The desirability value of maximum thinning and PFF is 0.735 and 0.835 respectively. Also, the average desirability of responses is 0.783, which is close to 1, and hence acceptable.

Confirmation tests

The confirmation tests are performed to validate predictive models and optimization results. The results of confirmation tests are given in Tab. 6.

Table 6. Results of confirmation tests

Results of predictive model										
S.N.	Parameters				Responses					
	A	B	C	D	Max. Thinning			PFF		
					Predicted	Actual	Deviation	Predicted	Actual	Deviation
1	0.65	10	0.91	37	70.34	68.77	2.24%	1640.95	1580.04	3.71%
2	0.40	9	0.79	42	69.64	65.87	5.42%	1290.34	1327.25	2.86%
Results of optimized values										
1	0.25	8	0.854	35	55.78	54.83	2.80%	877.08	858.38	2.13%
2	0.25	8	0.854	35	55.78	54.98	1.43%	877.08	861.27	1.80%

The results of confirmation tests are within 10% of acceptable range therefore predictive models and optimization results are valid.

Conclusion

The present paper describes the experimental study on maximum thinning and peak forming force acting on multi-stage single point incremental forming process. It is observed that initial draw angle is the most influencing factor for maximum thinning followed by blank thickness. Maximum thinning decreases with decrease in initial draw angle and increase in blank thickness. While, blank thickness is the most influencing parameter for peak forming force followed by tool diameter, pitch, and initial draw angle. It reduces with decreasing all four process parameters. Also predictive models of second order (quadratic) are developed to predict maximum thinning and peak forming force. Optimization of parameters has been carried out to get minimum thinning and forming force and same are verified using confirmation tests.

References

[1] M. J. Mirnia, M. Vahdani, M. Shamsari, Ductile damage and deformation mechanics in multistage single point incremental forming, Int. J. Mech. Sci. 136 (2018) 396–412. https://doi.org/10.1016/j.ijmecsci.2017.12.051

[2] J. R. Duflou, A. K. Behera, H. Vanhove, L. S. Bertol, Manufacture of accurate titanium cranio-facial implants with high forming angle using single point incremental forming, Key Eng. Mater. 549 (2013) 223–230. https://doi.org/10.4028/www.scientific.net/KEM.549.223

[3] H. Choi, C. Lee, A mathematical model to predict thickness distribution and formability of incremental forming combined with stretch forming, Robot, Comput. Integr. Manuf. 55 (2017) 164–172. https://doi.org/10.1016/j.rcim.2018.07.014

[4] N. Bari, S. Kumar, Multi-stage Single-Point Incremental Forming: An Experimental Investigation of Surface Roughness and Forming Time, J. Mater. Eng. Perform. 24 (2022). https://doi.org/10.1007/s11665-022-07183-8

[5] H. K. Nirala, A. Agrawal, Sheet Thinning Prediction and Calculation in Incremental Sheet Forming. Springer Singapore, 2018. https://doi.org/10.1007/978-981-10-8767-7_15

[6] A. Gheysarian, M. Honarpisheh, Process Parameters Optimization of the Explosive-Welded Al/Cu Bimetal in the Incremental Sheet Metal Forming Process, Iran. J. Sci. Technol. - Trans. Mech. Eng., 43 (2019) 945–956. https://doi.org/10.1007/s40997-018-0205-6

[7] Z. G. An, D. Yan, J. J. Qie, Z. L. Lu, Z. Y. Gao, Effect of Process Parameters on Formability of a AZ31 Magnesium Alloy Thin-Walled Cylindrical Part Formed by Multistage Warm Single-Point Incremental Forming, Front. Mater. 7 (2020) 1–15. https://doi.org/10.3389/fmats.2020.00151

[8] S. Ghafoor, Y. Li, G. Zhao, J. Li, I. Ullah, F. Li, Deformation characteristics and formability enhancement during ultrasonic-assisted multi-stage incremental sheet forming, J. Mater. Res. Technol., 18 (2022) 1038–1054. https://doi.org/10.1016/j.jmrt.2022.03.036

[9] P. K. Gandla, V. Inturi, S. Kurra, S. Radhika, Evaluation of surface roughness in incremental forming using image processing based methods, Meas. J. Int. Meas. Confed., 164 (2020)108055. https://doi.org/10.1016/j.measurement.2020.108055

[10] R. Aerens, P. Eyckens, A. Van Bael, J. R. Duflou, Force prediction for single point incremental forming deduced from experimental and FEM observations, Int. J. Adv. Manuf. Technol. 46 (2010) 9–12. https://doi.org/10.1007/s00170-009-2160-2

[11] J. Li, P. Geng, J. Shen, Numerical simulation and experimental investigation of multistage incremental sheet forming, Int. J. Adv. Manuf. Technol. 68 (2013) 9–12. https://doi.org/10.1007/s00170-013-4870-8

[12] V. Oleksik, Influence of geometrical parameters, wall angle and part shape on thickness reduction of single point incremental forming, Procedia Eng., 81 (2014) 2280–2285. https://doi.org/10.1016/j.proeng.2014.10.321

[13] Z. Liu, Y. Li, P. A. Meehan, Experimental investigation of mechanical properties, formability and force measurement for AA7075-O aluminum alloy sheets formed by incremental forming, Int. J. Precis. Eng. Manuf.14 (2013) 1891–1899. https://doi.org/10.1007/s12541-013-0255-z

[14] V. Sisodia, S. Kumar, Experimental study of single point incremental forming with dummy sheet, Int. J. Mater. Eng. Innov. 10 (2019) 60–82. https://doi.org/10.1504/IJMATEI.2019.097915

[15] A. Kumar, V. Gulati, P. Kumar, H. Singh, Forming force in incremental sheet forming: a comparative analysis of the state of the art. Springer Berlin Heidelberg 2019. https://doi.org/10.1007/s40430-019-1755-2

Sheet Metal 2023 Materials Research Forum LLC
Materials Research Proceedings **25** (2023) 61-68 https://doi.org/10.21741/9781644902417-8

Investigation on the shear cutting of functional components manufactured in an orbital forming process

Andreas Hetzel[1,a *], Michael Lechner[1,b] and Marion Merklein[1,c]

[1]Institute of Manufacturing Technology, Friedrich-Alexander-Universität Erlangen-Nürnberg, Egerlandstraße 13, 91058 Erlangen, Germany

[a]andreas.hetzel@fau.de, [b]michael.lechner@fau.de, [c]marion.merklein@fau.de

Keywords: Metal Forming, Cutting, Orbital Forming

Abstract. Applying forming operations for the manufacturing of complex functional components instead of conventional shearing or milling is an effective way to increase the material efficiency of the process as well as the mechanical properties of the part. Nevertheless, in many cases subsequent cutting operations are mandatory to reach the final geometry or to add functional surfaces for the later assembly. During forming, hardening effects or different strain states can cause difficulties for the subsequent cutting operation. These characteristics are commonly determined for sheet metal forming processes like deep drawing or bending. However, the complex stress and strain states during sheet-bulk metal forming operations and their influence on the cutting parameters or the part properties have not been investigated comprehensively so far. Furthermore, different assigned processes, like for example orbital forming, allow a local material distribution, realizing a gradient in sheet thickness. Therefore, this contribution focuses on the establishment of a fundamental process understanding on the influence of a sheet-bulk metal forming process on the cutting parameters and the resulting part properties. Functional components are manufactured from C10 sheet metal by orbital forming and subsequently shear cut to generate the final contour. During cutting, specific force-stroke diagrams are evaluated, in order to analyze the influence on the process parameters. The resulting properties of the parts are investigated regarding the quality of the cutting surface and a potential geometrical distortion in consequence of elastic spring back. The influence of the forming process is outlined by a direct comparison with conventionally cut components.

Introduction

Global trends, like an increased functionality or a reduction in greenhouse gas emissions, are facing the manufacturing industry with emerging challenges [1]. Functional integration and lightweight design can be possible answers to overcome current limitations. Nevertheless, an increase in geometric complexity as well as a wide variety of possible materials are accompanying these approaches. Thereby, conventional manufacturing processes like shearing are reaching their limits due to a lack in material efficiency and geometrical limitations [2]. In consequence, the development of innovative products as well as efficient forming operations is mandatory. One possible approach is the application of bulk forming operations to sheet metal, commonly defined as the process class of Sheet-Bulk Metal Forming (SBMF) [3]. Within this process class, a complex three-dimensional stress and strain state can be used to locally increase the sheet thickness or form functional elements like teeth or carriers out of the sheet plane [4]. One assigned process to manufacture functional components with a gradient in sheet thickness is orbital forming [5]. This process is derived from conventional upsetting by tilting of one tool component around the angle Θ, which is typically between 0 - 1°. A significantly reduced contact area between tool and work-piece results in a reduced forming force down to 10%, compared to conventional upsetting [6]. The consequence is a predominantly radial material flow from the center of the component towards the outside, thereby filling cavities attached to the punch in form of rotational or cyclic symmetric

Content from this work may be used under the terms of the Creative Commons Attribution 3.0 license. Any further distribution of this work must maintain attribution to the author(s) and the title of the work, journal citation and DOI. Published under license by Materials Research Forum LLC.

thickenings [5]. Nevertheless, the change in sheet thickness due to a three-dimensional material flow as well as an alternating punch movement results in tension inside the component. These residual stresses are hard to determine due to the complex forming zones [3].

These circumstances become challenging when applying cutting operations to realize the final geometry or obtaining functional surfaces for a later assembly. The residual stresses may be released in form of spring back behavior or distortion of the component [7]. The effect of strain hardening during the forming is expected to influence the occurrence of the cutting edge, which is a significant factor for the quality of the joint [8]. Since the influence of SBMF processes on the geometrical appearance of the components as well as relevant parameters after a subsequent cutting has not been investigated comprehensively so far, this investigation focuses on the establishment of a fundamental process understanding. Therefore, orbital formed components are shear cut in two different stages and compared to conventional parts without forming history. Besides the influence on the force-stroke curve, the appearance of the cutting edge as significant factor for the part quality [9] as well as the geometrical properties in consequence of a possible spring back are evaluated.

Experimental setup

In the following chapter, the properties of the applied material as well as the used lubricant are described. Furthermore, the geometrical dimensions of the component, the process setup of orbital forming as well as the subsequent cutting operation are presented.

Investigated material and lubricant. Within this contribution, the unalloyed case-hardenable steel C10 (1.1121) with an initial sheet thickness of $t_0 = 2.0$ mm is investigated. This steel is typically applied for pressed or shear cut parts. The mechanical properties of the material were analyzed via tensile tests and are summarized in Tab. 1. For improved tribological conditions and in order to reduce wear, the lubricant Beruforge 120DL from Bechem is used.

Table 1: Material properties from tensile test of the used steel C10 (1.1121) in rolling direction

C10 (1.1121), samples n = 3	Specification
Yield strength *YS*	335.4 ± 4.93 MPa
Tensile strength *TS*	423.4 ± 6.31 MPa
Uniform elongation *UE*	0.14 ± 0.004
Lankford coefficient *r (0°)*	0.897 ± 0.027
Strain hardening exponent *n*	0.137 ± 0.003

Geometrical dimensions of the functional component. The investigated component is derived from a conventional clutch disc carrier plate from ZF Friedrichshafen AG and features an outer diameter of $d = 120.0$ mm. The dimensions of the different thickenings as well as the final cut geometry are depicted in Fig. 1. The desired thickening should reach a height of $h = 0.5$ mm in both segments. A slide bevel of 1.0 mm is defined, in order to facilitate the material flow into the cavities. To reach the final part geometry for the later assembly to a clutch disc, different cuts are necessary. The friction lining is mounted via rivets on the outer edge of the part, which requires drilling holes with a diameter of $d = 5.1$ mm. The carrier plate is mounted onto a flange in the center with a diameter of $d_i = 36.5$ mm and fixed with rivets featuring a diameter of $d = 8.1$ mm. The dimension of the lightweight windows can be found in the detail view Z.

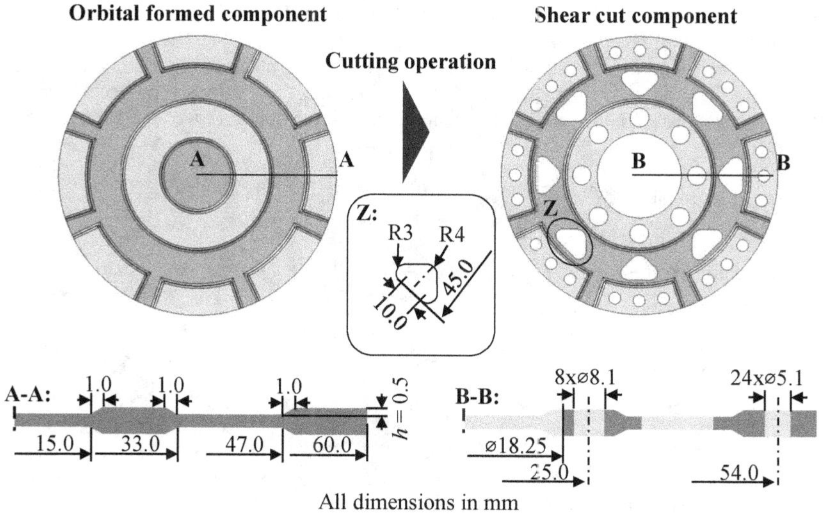

Fig. 1: Geometrical dimensions of the component after orbital forming and cutting

Orbital forming. The investigated orbital forming process was invented and characterized by Merklein et al. [5] and is schematically shown in Fig. 2. The blank with initial thickness of t_0 = 2.0 mm is placed onto the counterpunch. At the beginning of the process, the press ram of a hydraulic deep drawing press TZP400/3 from Lasco closes the tool and applies the forming force of F_{max} = 4000 kN. Subsequently, the desired tumbling kinematic is applied by the deflection of four hydraulic cylinders on the corner of a specific tumbling plate mounted onto a spherical calotte. A circular kinematic and a process sequence of $U_u = U_c = U_r = 5$, during which the angle is ramped up, held constant and is reduced to the initial position again, are chosen as uniform parameter settings to compare the results along with [5]. By tilting of the lower tool component, the specific tumbling angle of $\Theta = 1°$ is realized. The quasi-rolling movement of the upper punch over the part results in a predominantly radial material flow [6], which is responsible for the filling of the respective cavities in the punch as well as in the counterpunch. In consequence, a local adaption of the sheet thickness in form of a thickening and corresponding thinning can be realized.

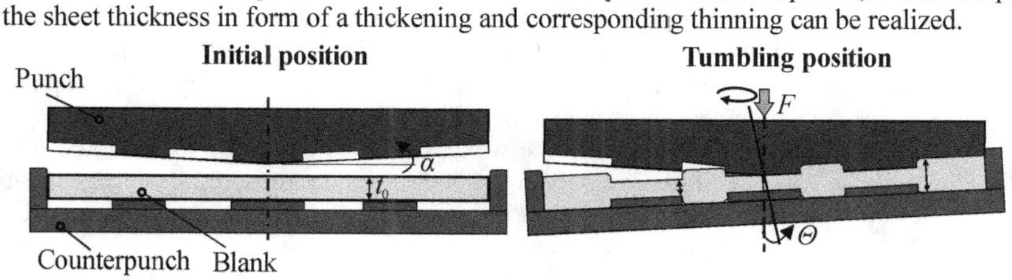

Fig. 2: Schematic process setup of orbital forming, adapted from [5]

Shear cutting setup. For the cutting process, a special tool setup had been developed and manufactured, which is schematically shown in Fig. 3. The component is placed onto the cutting plate and is positioned with circumferential pins and two adjustable levers. The press ram induces the cutting force via the upper tool plate. Gas pressure springs are responsible for the clamping of the orbital formed component trough the clamping plate. The process is divided into two single sequences, since the cutting stamps would be to close for a single stage process. Hence, the first step consists of cutting the lightweight windows and the flange hub. Afterwards, the cutting stamps are changed for both rivet drillings and the final contour is realized. The cutting clearance is defined as 8% from the maximum thickness of $t_{max} = 2.5$ mm, resulting in a clearance of $u = 0.2$ mm. The process is force controlled and a breaking of the stamps is prevented by the use of stopping plates. During the process, the characteristic force-stroke curve is recorded.

Shear cutting tool

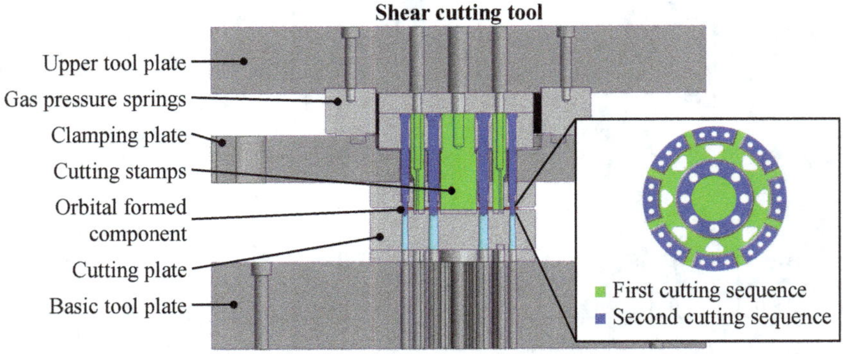

Fig. 3: Schematic process setup for shear cutting of the functional components

Influence on the cutting process

In order to evaluate the influence of varying part properties due to a different manufacturing process on the cutting parameters, the characteristic force-stroke diagram is compared between conventional (a) and orbital formed (b) components in Fig. 4. The characteristic sections can be divided into the tensioning of the gas pressure springs as clamping (1), the beginning of contact between punch and workpiece (2) with a subsequent increase of the required force. The maximum value is reached directly before the point of cut through in form of fracture of the residual sheet thickness (3). A significant drop in force is the consequence. The swinging increase on the end of the process (4) can be explained by the return stroke of the punch in combination with a partially stuck component.

The maximum cutting force for the conventional sheet with $t_0 = 2.0$ mm reaches values for $F_{c1} = 570.0 \pm 8.9$ kN and $F_{c2} = 681.8 \pm 7.6$ kN, thus tracing back to the different total equivalent cutting length of 455.2 mm and 588.1 mm for each sequence. The difference in stroke length of both peaks can be explained by an axial offset of the mounted cutting punches inside the tool.

When analyzing the diagram for the orbital formed component, a distinct appearance for both sequences can be detected. Due to the local gradient in sheet thickness, the engaged cutting line in the first sequence is varying with increasing stroke. First, the punch comes into contact with the inner segment, initiating the increase in force, which is lower compared to the force for the conventional blank. After the distance between the thickening is exceeded, the engaged cutting line increases promptly and the required force rises significantly (x). Since the inner segment is already partially cut, the maximum force reaches slightly lower values of $F_{o1} = 502.7 \pm 6.6$ kN. Due to the overall higher sheet thickness of $t_o = 2.5$ mm, the cut through is reached at a lower punch position and the punch force is increased. The influence of strain hardening during the forming operation influences the maximum cutting force as well, reaching values of $F_{o2} = 796.0 \pm 8.6$ kN for the second sequence. This value resembles an increase of 16.7%, compared to the conventional component.

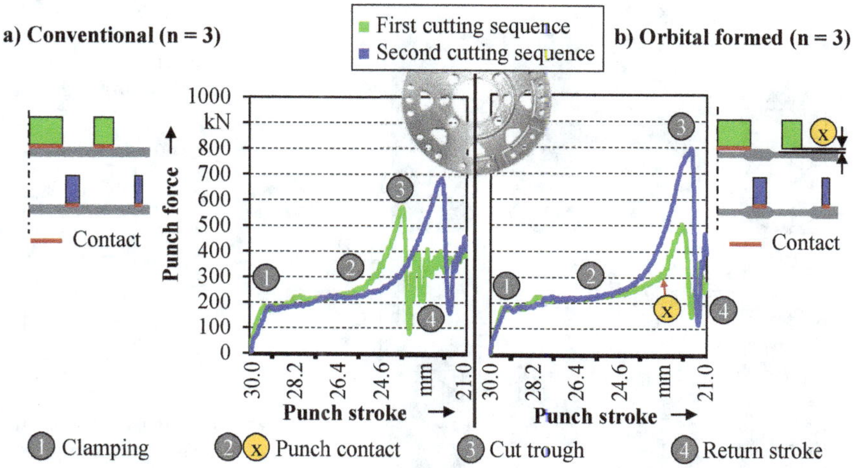

Fig. 4: Characteristic force-stroke diagram of a) conventional and b) orbital formed parts

Influence on the resulting part properties

Besides the process parameters, the resulting part properties are of importance to evaluate the influence of the manufacturing process on the shear cutting of functional components. Therefore, the appearance of the cutting edge as well as the geometrical properties of the manufactured parts are analyzed.

Shear cutting edge. The cutting edge resembles a relevant factor for the quality of the later joint. Since the material properties of the components vary during the orbital forming process, referring to the effect of strain hardening [5], the resulting cutting edge has to be analyzed. Therefore, the cutting edges are analyzed with a three-dimensional laser scanning microscope Keyence VK-X200 from Keyence. The resulting images with a perpendicular view on the edges are shown in Fig. 5. On the top side, the conventional component features burnish depths between 0.57 mm on edge 1, resembling 29% of the sheet thickness, up to 0.95 mm or 47% on edge 4, depicted in the diagram. This ratio is partially depending on the punch geometry but is in general in good accordance with the results in [10], which also show values of 37% - 47% for a comparable material strength class.

On the bottom side, the cutting edges for the orbital formed component are depicted. It can be seen, that the ratio of burnish zone to overall thickness in the diagram is significantly lower in all four segments. The values of this ratio reach from 15% in segment 1 up to 39% in segment 4, which is compared in the diagram as well. Due to the mentioned hardening during forming, the material strength increases as well. As presented by Levy and Tyne [10], the ratio of burnish zone to thickness is strongly depending on the tensile strength, reaching comparable values for the tensile strength of the initial and the orbital formed material C10 (1.1121).

Materials Research Proceedings **25** (2023) 61-68 https://doi.org/10.21741/9781644902417-8

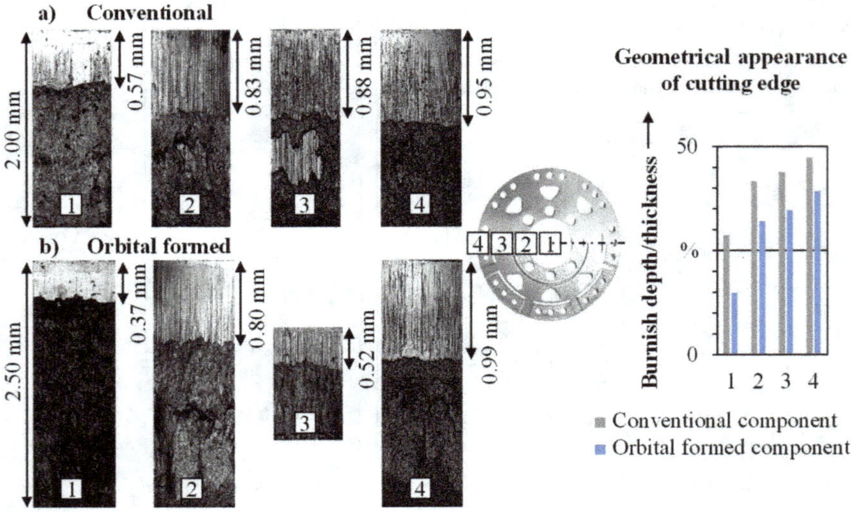

Fig. 5: Shear cutting edge and ratio burnish depth to thickness for both components

Besides the cutting edge itself, the influence of the cutting process on the mechanical properties of the component are evaluated, especially in the direct proximity of the cutting edge. Therefore, the micro hardness distribution of a radial cross section is evaluated, using the measuring system Fischerscope HM2000 from Fischer. The hardness plots for the different sections are depicted in Fig. 6. The effect of strain hardening close to the cutting edge due to the influence of the punch can be detected, since values of 288 HV0.05 could be reached. Achouri et al. [8] explain this influence by the plastic deformation of the material in the vicinity of the cutting edge. This plastic deformation is responsible for crack initiation and growth until fracture. This effect is visible for all segments, independent on the location of the cut or the respective thickness. To validate the influenced zone from the hardness plot, micro structural analysis of the material are prepared by different grinding and polishing operations and a subsequent edging. The microstructure is shown exemplarily for the outermost cross section in Fig. 6. The orientation of the grain structure indicates an influenced zone on the top side of the component of around 0.4 mm from the cutting edge. Furthermore, the rollover depth, the burnish zone as well as the fracture zone can be detected.

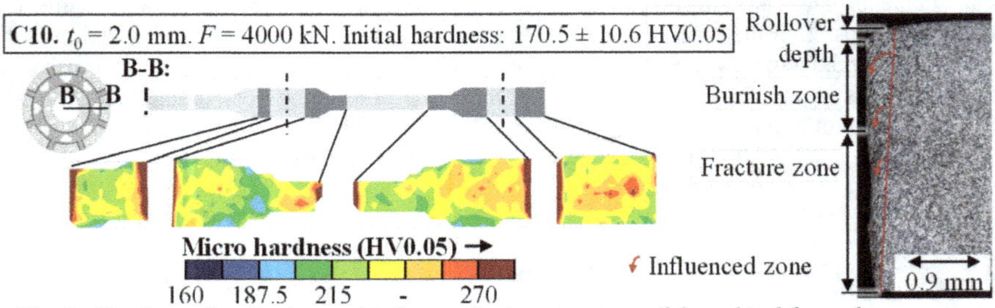

Fig. 6: Hardness distribution and influenced microstructure of the orbital formed component

Geometrical distortion. Another important factor for the later assembly is the resulting geometrical dimension in consequence of a possible distortion due to elastic spring back after the cutting process [11]. To evaluate this distortion, the part contour over the radius is measured using the three-dimensional measuring system ATOS from GOM, which is depicted in Fig. 7. When comparing both contours, the occurring distortion on the outer edge of the part with an offset of 0.28 mm and a resulting angle of 9° can be observed. This distortion can be explained by the elastic spring back due to a release of the residual stresses in combination with the missing support function of a die, as it exists for example during the orbital forming process. A distortion in upward direction is the consequence. This spring back is not a severe process failure but needs to be kept in mind for the later assembly. A possible solution could be a calibrating step with the orbital forming tool.

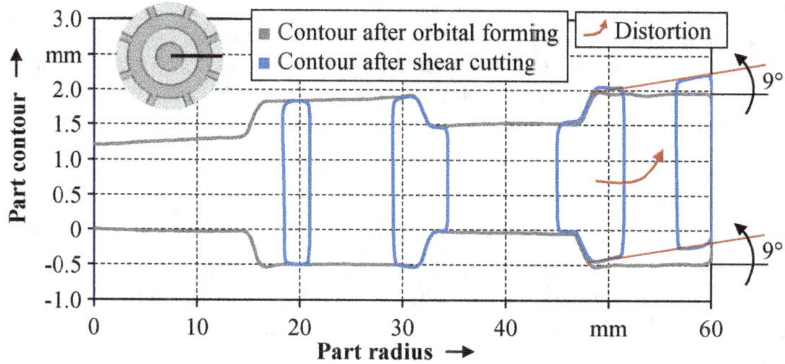

Fig. 7: Geometrical distortion of the shear cut orbital formed component

Conclusion and outlook

The aim of this contribution was to establish a fundamental process understanding on the influence of the three-dimensional stress and strain state during Sheet-Bulk metal forming processes on the resulting part properties after a subsequent shear cutting process. Therefore the characteristic force-stroke diagram was evaluated. Furthermore, the cutting edge and the strain hardening effect were analyzed. The geometrical distortion in consequence of elastic spring back outlines current limitations of the cutting process.

Summarizing the most important results within this contribution, the following conclusion could be derived:

- The effect of strain hardening during forming is responsible for an increase in required cutting force [7]. This increase is furthermore depending on the length of the engaged cutting line due to the locally varying sheet thickness and material properties.
- The increased hardness is responsible for a decreased ductility and consequently an increased fracture depth on the cutting edge. This results in an overall decreased ratio of burnish depth to total thickness, as already stated in [10].
- An occurring plastic deformation of the material near the cutting edge due to the penetrating punch can be obtained by microstructural analysis. Furthermore, this deformation is the reason for a significant strain hardening directly next to the cutting edge [9].
- The geometrical accuracy of the final part is limited with regard to an appearing distortion after the cutting process. This distortion offers values of 9° compared to the

initial formed component and can be explained by the elastic spring back [11] in combination with the missing support function of a die during the cutting process.

Since a fundamental process understanding on the influence of sheet-bulk metal forming processes on a subsequent cutting operation and the resulting part properties could be established, future research should focus on the expansion of this process knowledge. Therefore, the transferability of the presented results between different forming processes within SBMF as well as different material strength classes should be investigated. The implementation of a numerical model, representing the forming as well as the cutting process would be beneficial, in order to allow a deeper process analysis on the interdependencies between the forming and the cutting operation.

Acknowledgement
This work was supported by the German Research Foundation (DFG) within the scope of the Transregional Collaborative Research Center on sheet-bulk metal forming (SFB/TR73) in the subproject T10 - "Development of a manufacturing process for functional components with a varying sheet thickness profile" under grant number 457204991.

References

[1] H.A. Kishawy, H. Hegab, E. Saad, Design for Sustainable Manufacturing: Approach, Implementation, and Assessment, sustainability. 10,3604 (2018) 1-15. https://doi.org/10.3390/su10103604

[2] R. Neugebauer, W. Drossel, R. Wertheim, et al., Resource and Energy Efficiency in Machining Using High-Performance and Hybrid Processes, Procedia CIRP. 1 (2012) 3-16. https://doi.org/10.1016/j.procir.2012.04.002

[3] M. Merklein, J.M. Allwood, B.-A. Behrens, et al., Bulk forming of sheet metal, CIRP Annals - Manufacturing Technology. 61 (2012) 725-745. https://doi.org/10.1016/j.cirp.2012.05.007

[4] K. Mori, Y. Abe, K. Osakada, et al., Plate forging of tailored blanks having local thickening for deep drawing of square cups, J. Mater. Proc. Technol. 211 (2011) 1569-1574. https://doi.org/10.1016/j.jmatprotec.2011.04.010

[5] M. Merklein, R. Plettke, S. Opel, Orbital forming of tailored blanks from sheet metal, CIRP Annals - Manufacturing Technology. 61 (2012) 263-266. https://doi.org/10.1016/j.cirp.2012.03.130

[6] J. Nowak, L. Madej, S. Ziolkiewicz, et al., Recent development in orbital forging technology, Int. J. Mater. Forming. 1 (2008) 387-390. https://doi.org/10.1007/s12289-008-0076-2

[7] W. Volk, J. Stahl, Shear Cutting, in: L. Laperrière, G. Reinhart (Eds), CIRP Encyclopedia of Production Engineering, Springer, Berlin, Heidelberg, 2015, pp. 1-9. https://doi.org/10.1007/978-3-642-35950-7_16823-1

[8] M. Achouri, G. Germain, P. Dal Santo, et al., Experimental and numerical analysis of micromechanical damage in the punching process for High-Strength Low-Alloy steels, Materials and Design. 56 (2014) 657-670. https://doi.org/10.1016/j.matdes.2013.11.016

[9] A. Totre, R. Nishad, S. Bodke, An Overview Of Factors Affecting In Blanking Processes, Int. J. Emerging Technol. Advanced Engineering. 3-3 (2013) 390-395.

[10] B.S. Levy, C.J. Van Tyne, Review of the Shearing Process for Sheet Steels and Its Effect on Sheared-Edge Stretching, J. Mater. Engineering Performance. 21-7 (2012) 1205-1213. https://doi.org/10.1007/s11665-011-9997-x

[11] K. Mori, Review of Shearing Processes of High Strength Steel Sheets, J. Manuf. Mater. Process. 4-54 (2020) 1-14. https://doi.org/10.1016/j.jmapro.2020.02.041

Materials Research Proceedings **25** (2023) 69-76 https://doi.org/10.21741/9781644902417-9

Development of polygon forming processes for aerospace engineering

Philipp Müller[1,a *], Bernd-Arno Behrens[1,b], Sven Hübner[1,c], Jan Jepkens[1,d], Hendrik Wester[1,e] and Sven Lautenbach[2,f]

[1]Institute of Forming Technology and Machines, An der Universität 2, 30823 Garbsen, Germany

[2]Deharde GmbH, Am Hafen 14a, 26316 Varel, Germany

[a]mueller@ifum.uni-hannover.de, [b]behrens@ifum.uni-hannover.de, [c]huebner@ifum.uni-hannover.de, [d]jepkens@ifum.uni-hannover.de, [e]wester@ifum.uni-hannover.de, [f]s.lautenbach@deharde.de

Keywords: Sheet Metal, Aluminum, Polygon Forming

Abstract. The focus of this research lays on the further development of the Polygon Forming Technology, which is already successfully used for cold forming components in the aerospace industry. One example is the fuselage shell of the Airbus Beluga XL. According to the current state of the art it is possible to incrementally form large cylindrical or conical fuselage components by Polygon Forming. With the use of so-called infills, the Polygon Forming process can also be used to form components with pockets milled in the initial plane state. The limits of this technology exclude the creation of spherical geometries, such as those used in the front or rear fuselage sections of aircrafts. Presently, such components are produced by more complex stretch forming processes, which result in a considerable amount of scrap. In this work, a tool is developed to replicate the Polygon Forming process on experimental scale at the Institute of Forming Technology and Machines (IFUM) for materials commonly used in aerospace engineering. In addition, a downscaled pre-test tool is developed to investigate different tool geometries for incremental spherical forming inexpensive and easy according to the method of rapid prototyping.

Introduction

Since the mass of an aircraft has a decisive influence on fuel consumption, lightweight construction is of particular interest in aviation. Therefore, a reduction in weight has enormous economic and environmental benefits [1]. In addition to fiber-reinforced plastics and titanium alloys, mainly aluminum alloys are used due to their low density [2]. Various aluminum alloys come into operation in the aerospace industry. An application example for alloys of the 7xxx series is the use for highly stressed components in the fuselage structure [3]. Another alloy used for the production of fuselage parts is EN AW2024 - T351. Depending on the alloy, the mechanical properties such as strength or resistance to stress corrosion cracking can be adjusted with the aid of heat treatments. The respective conditions are defined according to DIN EN 515 [4].

To generate the desired geometries of the sheet metal parts, sheet metal forming processes are an essential technology in aircraft production [5]. The manufacturing of large-area fuselage structural parts in the aerospace industry is usually carried out by means of roll-forming or stretch-forming. Roll-forming is a process for the production of curved sheet metal parts. A possible design of roll-forming lines is a symmetrical three-roll arrangement [6]. This technology provides favorable loading conditions with comparatively low forces and is usually used for forming thick plates [7]. In stretch forming, a flat sheet blank is usually clamped on two opposite sides and is then formed into the desired shape by the action of a forming punch. Depending on the punch geometry, stretch forming can also be used to produce spherical sheet metal components. However, this technology is associated with a high material waste, limited dimensional accuracy

Content from this work may be used under the terms of the Creative Commons Attribution 3.0 license. Any further distribution of this work must maintain attribution to the author(s) and the title of the work, journal citation and DOI. Published under license by Materials Research Forum LLC.

and an undesired reduction in sheet thickness [7]. Another method to produce aircraft fuselage parts is the incremental bending process Polygon Forming, developed and patented by the company Deharde GmbH [8]. Here, aluminum sheets are bent by means of incremental forming. According to DIN 8586, bending serves to plastically form solid components (e.g. metal sheets), whereby the plastic state is essentially generated by bending stresses [9]. Usually special machines such as presses, bending or straightening machines are used for this purpose [10]. Polygon Forming represents a special form, which is suitable for the production of curved, turned, and conical sheet metal parts, where relevant geometric elements, such as pockets, drill holes or recesses are generated before the bending process. This eliminates the need for time-consuming post-processing, which usually require cost-intensive five-axis milling machines.
In the cooperative joint project "Aggregated Polygon Forming based Processes for large Fuselage Components" (AgaPolCo) involving the Deharde GmbH and the IFUM, the technology of Polygon Forming, which has so far only been applicable to cylindrical and conical components, is to be transferred to the production of spherical components. Spherical components are mainly used in the front or backside area, while cylindrical components are mainly used in the middle area of aircraft fuselages (see Fig. 1). In order to make Polygon Forming usable for the new spherical use case, a tool with which the Polygon Forming process of Deharde can be reproduced on a hydraulic press at the IFUM is developed. The tool will be designed, built and set up. In addition, another downscaled prototype tool is used to investigate forming strategies and tool geometries for incremental spherical forming. The experimental investigation provides a preselection of suitable tool geometries. Here, the experimental investigation is faster than a numerical one due to the simulative complexity. The identified optimal tool geometries will be further optimized iteratively in the further course of the project by means of numerical investigations.

Fig. 1: Fuselage sections of an aircraft on the example of an Airbus A320 (image source: [11])

Materials and Methods

The overall aim of the project "AgaPolCo" is the production of spherical components by incremental bending from the aluminum alloy EN AW2024 - T351. For this purpose, a polygon forming demonstrator tool is built with which blanks are initially bent cylindrically and in the evolving project it is aimed to form spherically on a hydraulic press. EN AW2024 - T351, in the following named AA2024, is one of the most popular high strength aluminum alloys. Due to its high strength and fatigue resistance, it is commonly used on components and structures in aircrafts. AA2024 has a yield strength of approx. $R_{p0,2} = 290$ MPa and a tensile strength of approx. $R_m = 440$ MPa [12].

Tool geometries for bending spherical components are to be investigated in a preliminary testing tool. For this purpose, several tool concepts are additively manufactured using the rapid prototyping method. Since the tools are not printed in massive form to save time and costs, they cannot stand high forming forces. Therefore, a more ductile material than the high-strength AA2024 must be used for the forming experiments. The AlMg alloy EN AW5754, short AA5754, has a very high formability, which allows the production of components with complex geometries. It has a yield strength of approx. $R_{p0,2}$ = 80 MPa and a tensile strength of approx. R_m = 240 MPa [12] and is therefore taken as pre-testing material.

Scaled Polygon Forming Tool

Based on the forming tool of the company Deharde, a tool is designed with which Polygon Forming can be investigated at the IFUM. The tool can be installed in a hydraulic double-column press type HDZ400 of the company Dunkes GmbH. The basic tool consists of the active component sword and bed as well as one adapter each for installation on the machine bed and on the press ram (see Fig. 2). The sword is screwed interchangeably to the upper die and is pressed incrementally into the sheet blanks during the process. The bed serves as a support for the blanks and is designed to be adjustable in width. The sword and bed are made of polyamide (PA6) to ensure a certain elasticity in the contact and thus prevent imprinting contours on the blanks. Three layers of spring steel 1.1274 in the thickness of t = 1.5 mm are placed on the tool bed to provide a smooth ground for the sheet blanks and to enable the bending of the outer contour of the sheet. The sheet blanks are rectangular plates with a dimension of 500 mm x 500 mm and a thickness of t = 3.2 mm made out of aluminum alloy AA2024. The rolling direction of the sheets is oriented perpendicular to the feed direction.

Initial forming experiments are conducted to investigate the accuracy and reproducibility of the process. Therefore, the aluminum plates are manually fed step by step with a feed width (FW) of FW = 20 mm after each stroke perpendicular to the press ram movement. After each feed the sword is pressed path-controlled into the sheet plates up to an indention depth (ID) of ID = 0.8 mm beneath the bed surface. The bed width (BW) is set constant to BW = 110 mm. The experiments are repeated three times with these parameters.

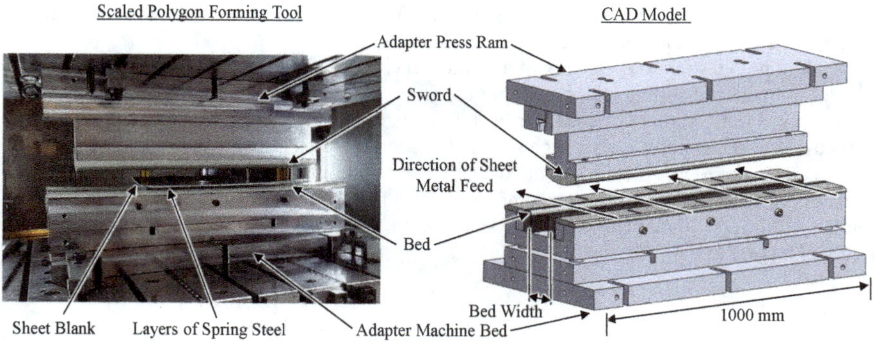

Fig. 2: Scaled Polygon Forming Tool at the IFUM

After incremental bending, aluminum plates are measured with an optical measuring device Atos 2 400 of the company Carl Zeiss GOM Metrology GmbH. The bending of the sheet plates is evaluated by determining the bending lines from the three-dimensional scan data. The bending lines are determined for each of the formed plates at five positions. Averaged bending lines are determined for the inside radii of the plates and compared with each other (see Fig. 3). A comparison of the three formed plates (Test 1-3 in Fig. 3) show, that the process provides a good reproducibility. The relative standard deviation of the three averaged inside radii is SD = 2.76 %.

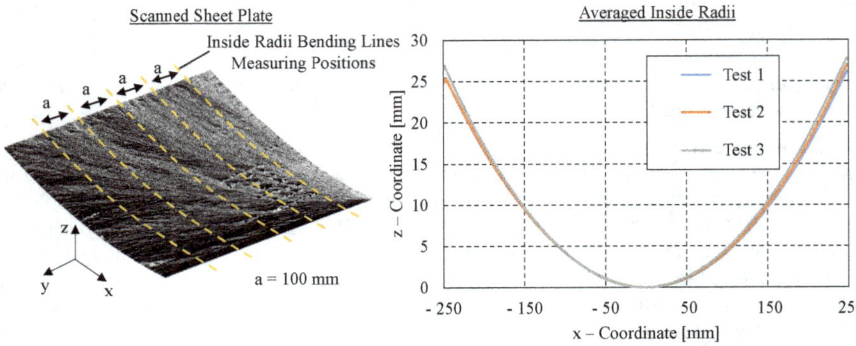

Fig. 3: Measurements of the profile bending of incremental formed cylindrical AA2024 plates

Therefore, the experiments show that the scaled Polygon Forming tool is suitable to reproduce and to analyse the Polygon Forming process executed on industrial scale by Deharde. In order to identify tool geometries for spherical forming, a pre-testing tool was developed.

Pre-Testing Tool

In order to identify suitable tool concepts for spherical forming, a preliminary experimental study is carried out in which various sword and bed geometries are investigated. Thus, the applicability of the developed concepts will be validated experimentally. For this purpose, different sword and bed variations are additively manufactured on a 3-D printer (type RF2000 of the company Conrad Electronic SE) and tested for spherical forming in a pre - testing tool on a manual press. The additively manufactured components are printed from the material polylactic acid (PLA) in segments, which are joined together to form total tool components (see Fig. 4). The segmental production is necessary because the 3D printer has a limited production space and the complete length of the components cannot be produced in one print.

The pre - testing tool consists of a sword which can be moved vertically along two guide profiles by means of sliding slides and return springs. The guide profiles are connected to a base frame from which the supports (bed) for the blanks extend. The sheets to be formed are placed on the bed and feed manually in after each stroke. The forming force is applied via a hydraulic punch, which can be moved to a fixed indentation depth (see Fig. 4 left). As plate material AA5754 in the thickness of t = 1 mm with a rectangular shape of 400 mm x 400 mm is used. The indentation depth is set to ID = 10 mm beneath the bed surface, the bed width is set to BW = 100 mm and the feed width is set to FW = 10 mm.

Materials Research Proceedings **25** (2023) 69-76
https://doi.org/10.21741/9781644902417-9

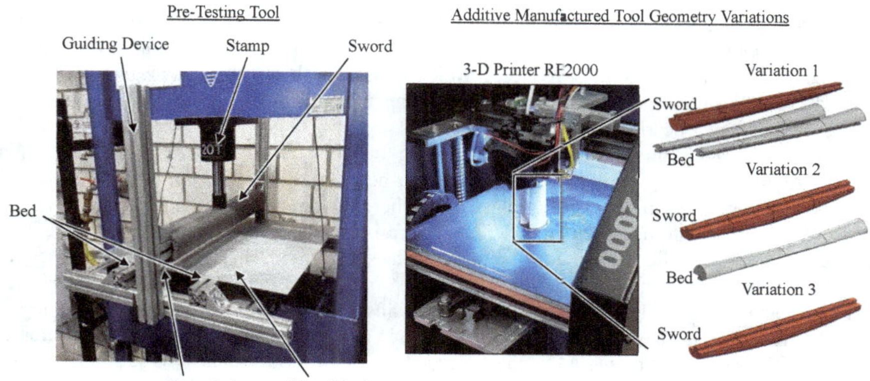

Fig. 4: Pre - testing tool (left), additive manufacturing of tools for spherical forming (right)

Three concept variations of sword and bed geometries are investigated (see Fig. 4 right). In variation 1, supports tapering transversely to the feed direction are provided as a bed and a matching laterally tapering centerboard is provided as a sword. The curved shape perpendicular to the feed direction is intended to generate a spherical bend in addition to the cylindrical bend in feed direction. In variation 2, a sword with a cambered shape and a one-sided support, which is the opposite shape to the cambered sword is used. The cambered form of the sword is intended to additionally press the sheet into a spherical shape. The shape of the bed is intended to provide a forming constraint on the underside of the sheet in the spherical direction. In variation 3, only a cambered sword is investigated for pressing a bend perpendicular to the feed direction. This variation thus provide the lowest forming constraint in the forming zone.

Variation 1. The forming behavior of sheets using the tool geometry of variation 1 is shown in Fig. 5. It can be seen that with the laterally curved sword and bed, bending occurs both in the feed direction and perpendicular to the feed direction. Thus, the desired spherical shape can be generated with variation 1.

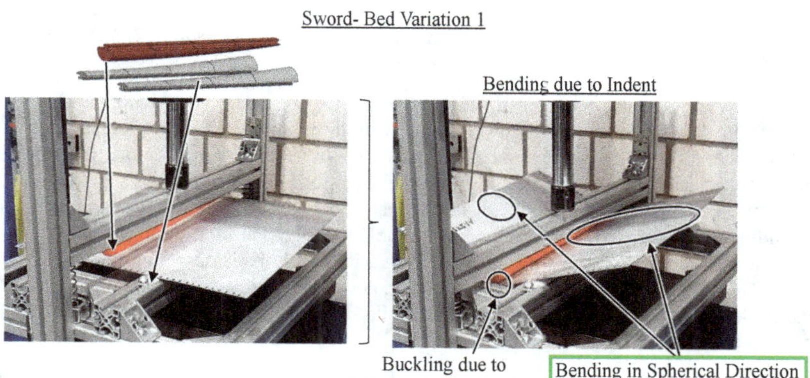

Fig. 5: Concept with curved bed and sword geometries (variation 1) for spherical forming

It is noticeable that the sheet under the sword buckles downwards on one side (see Fig. 5), since it does not recieve any counterpressure in this area. The form constraint coming from the bending geometry of the sword is thus reduced, which causes problems in dimensional accuracy. Nevertheless, this concept turns out to be very promising and can possibly be further optimized using elastic counter bearings under the sheet.

Variation 2. The forming behavior of sheets using the tool geometry of variation 2 is shown in Fig. 6. At the beginning of the forming process, a bending contrary to the targeted spherical direction appears in the middle of the plate. The cambered sword first comes into contact with the plate with its elevation in the center. When the sword is pressed further into the sheet, it stiffens over the length perpendicular to the feed direction. The sheet cannot change shape in the spherical direction. However, when the sheet is pushed further, it is observed that the initial stiffened area does not move along the length of the plate and remains at the edge of it (see Fig. 6 right). In the center of the plate, a spherical area with a curvature in and perpendicular to the feed direction is created. This is probably promoted by the concave bed. However, it can also be seen that an area is created in front of the feed area of the blank, in which the blank is strongly buckling. The concept of variation 2 is therefore not as suitable for spherical forming as variation 1.

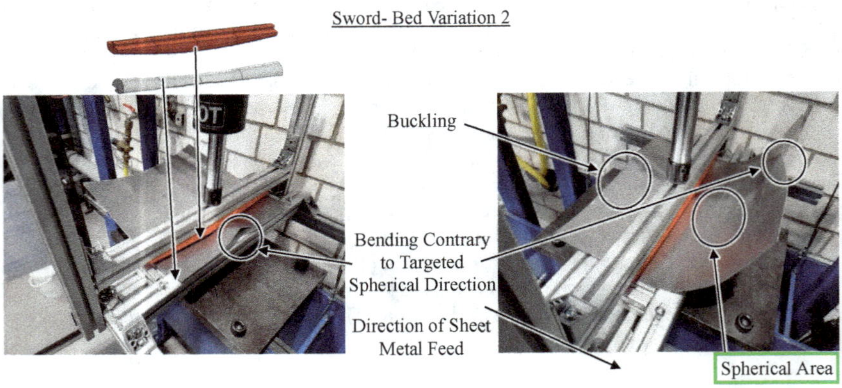

Fig. 6: Concept with cambered sword and concave bed geometry (variation 2) for spherical forming

Variation 3. The forming behavior of sheets using the tool geometry of variation 3 is shown in Fig. 7. It can be seen that comparable to variation 2 with the cambered sword, a bending occurs in the opposite direction to the targeted spherical geometry. In contrast to variation 2, the area bent against the spherical direction moves along with the sheet feed and does not remain stationary. A spherically shaped area can therefore not be generated only with a cambered sword without a corresponding bed. Variation 3 is therefore not suitable for spherical forming.

The different bending of the sheets caused by variant tool shapes can be explained by the occurrence of different stress states. During bending, compressive stresses prevail on the inside of the blade, where the sheets are compressed and tensile stresses on the other side where the sheets are widened [13].

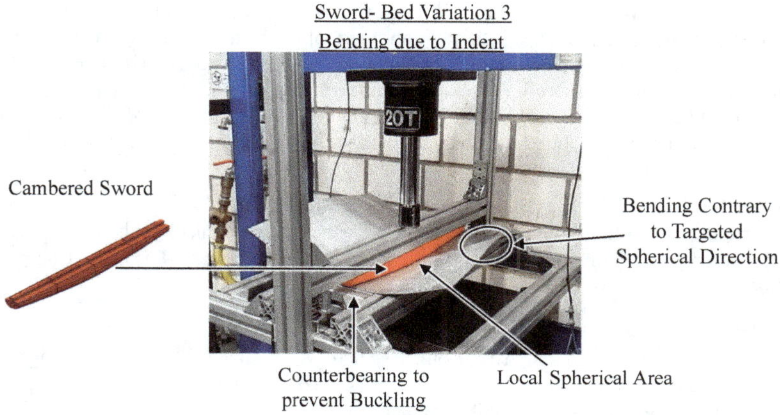

Fig. 7: Concept with cambered sword geometry (variation 3) for spherical forming

If the sheets are bent in the feed direction and perpendicular to it, compressive stresses occur with a positive superposition and maximum stress peaks in the area where bulging occurs (see Fig. 8).

In Variation 1, the opposing shape of the blade and bed probably results in lower maximum compressive stresses compared to Variation 2 where the maximum compressive stresses are positively amplified. The addition of the compressive stresses on the upper side of the sheet could initiate a bending in the opposite direction when the sheet is no longer under forming constraint. For Variation 3, the local stress state might differ compared to Variation 2 due to the symmetrical two-parted bed. The lower shape constraint leads to lower maximum compressive stresses and thus to lower stiffening bulging.

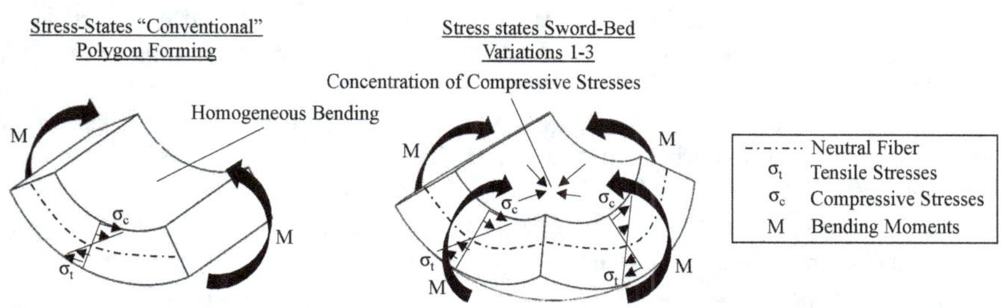

Fig. 8: Stress-states for cylindrical and spherical Polygon Forming

Summary and Outlook

In this research, an experimental tool for investigation of the polygon forming was developed. Cylindrical components were produced successfully from an aluminum alloy EN AW2024 - T351 commonly used in aircraft construction. In addition, different tool geometries for incremental spherical forming were investigated with a pre-testing tool. It could been shown that a curved sword with an associated bed on both sides leads to the most promising results regarding spherical forming. An elastic counter-bearing in the bed area could further improve the forming constraint and thus the dimensional accuracy of the spherical blanks.

In following works, a simulation will be carried out with which the polygon forming on the IFUM demonstrator can be mapped. In addition, the tool geometry according to variation 1 will be used in a numerical simulation for the forming of higher-strength AA2024, with increased sheet thickness of t = 3.2 mm. Therefore, iterative optimizations are made to compensate material springback and avoid buckling. After an iterative numerical optimization of the geometry, tool components will be manufactured with which spherical forming could be performed on the scaled polygon forming IFUM demonstrator.

Acknowledgments

Funded by the Investitions- und Förderbank Niedersachsen (NBank) – collaborative research project “Aggregated Polygon Forming based Processes for large Fuselage Components” (AgaPolCo)

References

[1] J-P. Immarigeon, R.T. Holt, A.K. Koul, L. Zhao, W. Wallace, J.C. Beddoes, Lightweight materials for aircraft applications, Materials Characterization, Volume 35, Issue 1, pp. 41-67, 1995. https://doi.org/10.1016/1044-5803(95)00066-6

[2] T. Dursun, C. Soutis, Recent developments in advanced aircraft aluminium alloys, Materials & Design (1980-2015) Volume 56, pp. 862-871, 2014. https://doi.org/10.1016/j.matdes.2013.12.002

[3] D. Uffelmann, Take-Off für hochfestes Aluminium im Automobilbau, ATZextra Karosserie Werkstoffe (2010) H.10. https://doi.org/10.1365/s35778-010-0451-8

[4] DIN EN 515:2017-5: Aluminium und Aluminiumlegierungen - Halbzeug –Bezeichnung der Werkstoffzustände. Beuth Verlag, 2017

[5] W. Koehler, B. Plege, K.F. Sahm, N. Padmapriya, Metal Forming: Specialized Procedures for the Aircraft Industry, Reference Module in Materials Science and Materials Engineering, 2017. https://doi.org/10.1016/B978-0-12-803581-8.01939-1

[6] F. Vollertsen, A. Sprenger, J. Kraus, H. Arnet, Extrusion, channel, and profil bending: a review, Journal of Materials Processing Technology Volume 87, Issues 1–3, pp. 1-27, 1999. https://doi.org/10.1016/S0924-0136(98)00339-2

[7] B. Heller, S. Chatti, M. Schikorra, A.E. Tekkaya, Blechbiegen. In: Siegert K. Blechumformung. VDI-Buch, Springer, Berlin, Heidelberg. S.141-221. 2015. https://doi.org/10.1007/978-3-540-68418-3_5

[8] E. Wilken, S. Lautenbach, H. Frerichs, et al.: Verfahren und Anordnung zur Formänderung eines Plattenartigen Werkstücks. WO2020/147935A1. 2020

[9] DIN 8586:2003-09: Fertigungsverfahren Biegeumformen – Einordnung, Unterteilung, Begriffe. Beuth Verlag, 2003

[10] Doege, E.; Behrens B.-A.: Handbuch Umformtechnik, 2. Auflage, Springer Verlag, Berlin, Heidelberg, 2010. https://doi.org/10.1007/978-3-642-04249-2

[11] Airbus, Commercial Aircraft, A320 | The most successful aircraft family ever, https://aircraft.airbus.com/en/aircraft/a320/a320ceo, access at 14.09.2022

[12] DIN EN 485-2: Aluminium and aluminium alloys–Sheet, strip and plate – Part 2: Mechanical properties; German version EN 485-2:2016+A1:2018

[13] Altan, T., Tekkaya, A. E.: Sheet Metal Forming – Processes and Applications, ASM International, 2012. https://doi.org/10.31399/asm.tb.smfpa.9781627083171

Materials Research Proceedings 25 (2023) 77-84 https://doi.org/10.21741/9781644902417-10

Investigation of the suitability of a tool element manufactured by fused filament fabrication for incremental sheet metal forming

Verena Kräusel[1], Tobias Klinger[1], Jakub Kořenek[1], Alexander Pierer[1], Dieter Weise[1,a,*], Peter Scholz[1], Jana Petrů[2], Jiří Koziorek[2], Michal Prauzek[2], François Rosoux[3] and Łukasz Madej[4]

[1] Fraunhofer Institute for Machine Tools and Forming Technology IWU, Reichenhainer Straße 88, 09126 Chemnitz, Germany

[2] VSB - Technical University of Ostrava, 17. listopadu 2172/15, 708 00 Ostrava-Poruba, Czech Republic

[3] Sirris, 80 Bld Reyers, 1030 Brussels, Belgium

[4] AGH University of Science and Technology, Al. Mickiewicza, 30 30-059 Kraków, Poland

[a,*]dieter.weise@iwu.fraunhofer.de

Keywords: Incremental Sheet Forming, Rapid Tooling, Additive Manufacturing

Abstract. Incremental sheet forming (ISF) constitutes a flexible production process for sheet materials for small to medium batch sizes, in which the geometry of the part is created via the movement of a stylus. ISF can be carried out with or without support from the opposite side. With the use of dies, the geometry deviation can be reduced. In order to be able to guarantee an overall flexible process, the ability to produce dies quickly and individually from the lot size one is necessary. In addition to milling, which has been the primary method used up to date, additive manufacturing (AM) also meets the requirements for flexible die production. To investigate the suitability of additive manufacturing to produce dies for ISF, a pyramid-shaped die was fabricated from polylactic acid (PLA) using the fused filament fabrication (FFF) process. This die was used for three incremental sheet metal forming operations using pure aluminum sheets and was compared with an identical milled tool. Based on the measurement results, the suitability of 3D printed dies for ISF is examined, and opportunities as well as application limits for such dies are discussed.

Introduction

Incremental sheet forming (ISF) is one of the sheet metal forming processes, which are not dependent on a part-specific die. This category includes mechanical processes such as hammer forming and forming by means of shot peening or water jets, as well as thermal processes such as thermal straightening and forming by means of laser beams. All these processes are characterized by high flexibility due to the easily controllable local forming zone. Furthermore, adjustments to the sequence are possible, and the investment costs for dies are significantly lower or completely omitted. However, while using these processes, only low productivities can be achieved, therefore they are preferably used for small to medium lot sizes.

ISF has evolved from the process of spinning and was originally used for prototyping [1]. In the meantime, a wide range of applications has emerged, and an almost unlimited variety of sheet geometries can be produced [2]. In contrast to conventional sheet metal forming processes, the forming is carried out via the continuous movement of a stylus. The simplest variant without a counter die is the so-called single-point ISF. In addition, other variants with an underlying die or support have been continuously developed to further improve the geometrical accuracy of the ISF [3]. The die supports the already formed geometry and reduces the subsequent deformation of these areas. ISF with a die can achieve the highest geometrical accuracy [4]. Due to the recorded

Content from this work may be used under the terms of the Creative Commons Attribution 3.0 license. Any further distribution of this work must maintain attribution to the author(s) and the title of the work, journal citation and DOI. Published under license by Materials Research Forum LLC.

lower global tool load also, easy-to-machine and low-cost materials (e.g. plastic, wood, foam or plywood) are suitable for the technology [2]. Nowadays, the dies are most commonly manufactured by milling from a full block or bonded material.

However, to conduct an overall flexible ISF process, a fast and adaptable strategy for manufacturing of the dies is required. Due to the small lot sizes, the production of the die for the ISF is an important factor in the time and cost incurred. This challenge could have been overcome by the development in the field of rapid tooling (RT); here, the tool is generated by adding material layer-by-layer [5]. Different RT approaches, including the generation of plastic tools, have been intensively studied for conventional sheet metal forming processes [6]. At first, RT with plastics focused on patterns for casting and indirect tooling, which do not have to bear the high loads that occurred in direct tooling applications [7]. Already 20 years ago [5] stereolithography was used to produce plastic AM dies, which have been used as masters for casting tools from aluminum reinforced epoxy resin. Direct RT among others was the domain of metal tools additively manufactured by selective laser sintering (SLS). This process produces stiff tools with a rough surface, which increases wear and friction [6]. One of the first applications of the SLS tool for sheet metal forming was realized by Cheah et al. [8].

Costs and time for generating FFF tools. FFF has been used to produce tools for conventional sheet metal forming, especially because the low cost of materials and printers offers high potential cost savings compared to metal tools [9]. The tools for conventional sheet forming "are generally expensive and the lead time is relatively long, drawbacks in cost and time are not problematic in comparison with direct manufacturing of parts" [10,]. Using FFF with the same material can reduce the cost for manufacturing a deep drawing tool consisting of a die, a counter die, and a blankholder by 78 % by milling polyamide 6 (PA6)/PLA tools and by 93 %. Bergweiler [11] showed that if the die fabrication is not possible to perform in-house, costs and fabrication time can be saved by using online loan additive manufacturing compared to external loan milling.

FFF for generating dies for incremental sheet metal forming. While in the last years' much work has been done in researching the usability of FFF tools for conventional deep drawing, the use for ISF received only limited attention. Rieger [4] used robots for FFF printing acrylonitrile butadiene styrene (ABS) dies and for the subsequent ISF. At first, the die was printed horizontally on a wall using one robot and afterwards a second robot formed a metal sheet onto this die using ISF without repositioning the die in between. The experiments were performed with 0.8 mm thick DC04 steel sheets.

It was shown that with the use of a 3D-printed die for the ISF, a significant improvement in geometry accuracy could be achieved compared to the ISF without a die. The used test geometry resulted in a maximum deviation of the sheet of 17 mm when the ISF was conducted without a die, as well as with the use of a second stylus from the opposite side. This deviation occurred due to the subsequent deformation of the sheet. In the experiments with the use of a 3D printed die either in the form of a full support (tested with 77 % and 38 % infill degree) or as a partial support (38 % infill degree) the subsequent deformation could be completely eliminated. The full dies with 77 % and 38 % infill degree respectively, generated sheets with very similar geometric accuracy. With the use of the partial dies the subsequent deformation only occurred in parts of the sheet, where there was no support in place. The authors concluded, that for the dies used, the base surfaces of the test geometry must be supported, but it is not necessary to support the geometry itself. In addition, the use of FFF printed dies with an infill degree as low as 38 % proved suitable for the ISF. Following the work of Rieger [4], this research is going in a similar direction, by manufacturing a die in a FFF process and using it subsequently for the ISF-process. Due to the good availability and the potential cost savings by using a commercial desktop FFF printer, the suitability of such a printer for ISF shall be further investigated in this work. The motivation of this work is based on this. The novelty of pyramid-shaped, FFF-printed dies being suitable for incremental sheet metal forming and the deviations of tool and part at small batch sizes are to be investigated.

Methodology

Manufacturing and measurements of the ISF dies. For the geometry of the die a truncated square pyramid with an edge length of 125 mm, a height of 50 mm and an angle of the side walls of 45° was selected as a case study. Due to the geometry, a constant high effective strain in the sheet can be realized at the pyramid side faces, and in addition, the influence of corners on the resulting geometry can be investigated. The printing was performed using an Ultimaker S5 printer with the inexpensive, comparatively strong and good processible filament Ultrafuse PLA white from the company BASF as the printed material.

In comparison to dies for deep drawing, where the major load on the die is in the direction of the press movement, the dies for the ISF are subjected to the load from the stylus, which is mainly dependent on the formed geometry. Since the truncated pyramid has a wall angle of 45°, an infill pattern was looked for, which feature a high compressive strength in this direction. The octet pattern was chosen for this purpose based on the perpendicular and parallel orientation of the lines to the object edges, which are expected to increase the strength perpendicular to the pyramid side faces. A unit cell of the octet pattern is shown in Fig. 1 a. The degree of filling was deliberately selected to be very low to save printing time and to test the application limits of the filling structures. Since Rieger [4] used an infill degree of 38 %, which worked fine with a 0.8 mm thick DC04 sheet, an even lower infill degree of 20 % was investigated in this work. The printed die wall thickness is 1 mm. To minimize the stair stepping effect a 0.4 mm nozzle in combination with a layer height of 0.1 mm was used. To enable the printing a polyvinylacetat (PVA) support structure was applied, which was subsequently dissolved in a water bath. The complete printing time of the die was 41 h.

The finished print of the die made from white PLA is displayed in Fig. 1 b. For comparison, another die with the same geometry was milled from a polyurethane block (Fig. 1 c).

a) **b)** **c)**

Figure 1: a) unit cell of the octet infill pattern b) 3D printed die c) milled die

The printed and the milled dies were 3D measured after the fabrication using the fringe projection method with a COMET 6 measurement system from the company Steinbichler. The geometry deviations are displayed in Fig. 2. The largest deviation of the printed dies occurred in the area of the lower pyramid corners, which bulged upwards by up to one millimeter. This bulging is caused due to the warpage of the PLA during the cooling down. Furthermore, the stair-stepping effect of the FFF process is visible on the surface. In addition, artifacts of the filling pattern are visible in Fig. 2 b in the form of z-parallel rings on the pyramid side faces. The milled die (Fig. 2 c) showed higher geometric accuracy, and the maximum deviation amounts less than 0.1 mm.

Materials Research Forum LLC
https://doi.org/10.21741/9781644902417-10

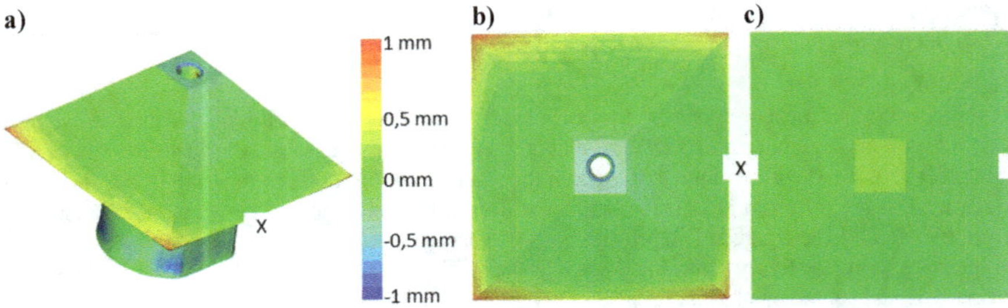

Figure 2: Geometry deviations of the a-b) printed die and c) milled die. The X points to the side with the flattened tool holder; the legend shows the deviation from the nominal geometry in mm.

Execution of the ISF. The ISF-process was executed on a fixture with a non-traversing clamping frame and a pure aluminum sheet (EN AW-1050A H11) with a thickness of 1.2 mm (Fig. 3 a). Due to the static clamping frame, the stylus always alternately moved on a path on the outside near the clamping frame and on a path on the inside at the die, so that from side-view a W-shaped part with wall angles of 45° is formed (Fig. 3 b). The desired geometry of the sheet contains only the inner pyramid.

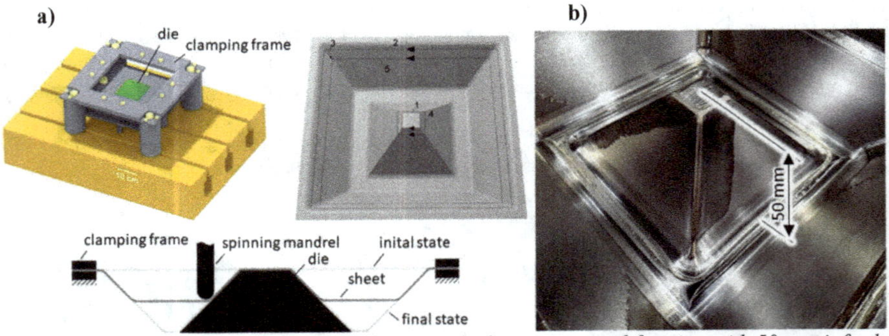

Figure 3: a) setup of the ISF trials b) Sheet metal after incremental forming with 50 mm infeed depth and light wrinkles at the bottom pyramid corners.

The tool used for the testing was a freely rotating, spherical hard metal stylus with a diameter of 30 mm. The sheet was lubricated on the upper side. The ISF was executed using a step increment of 0.5 mm and a movement speed of the stylus of 5 m/min, which resulted in a total forming time of 35 min. During the forming, the bottom of the sheet bulged upwards, which resulted in a small amount of wrinkling at the end of the base corners of the sheet. One sheet was formed using the milled die, and three sheets were formed using the printed die.

Results of the 3D measurements. The printed and the milled die as well as the formed sheets were 3D measured after each ISF. Only the inner pyramid of the sheets was measured because this part contains the desired geometry. The results are displayed in Fig. 4. The measurements of the sheets are fitted to the plane at the top of the truncated pyramid.

Materials Research Proceedings **25** (2023) 77-84 https://doi.org/10.21741/9781644902417-10

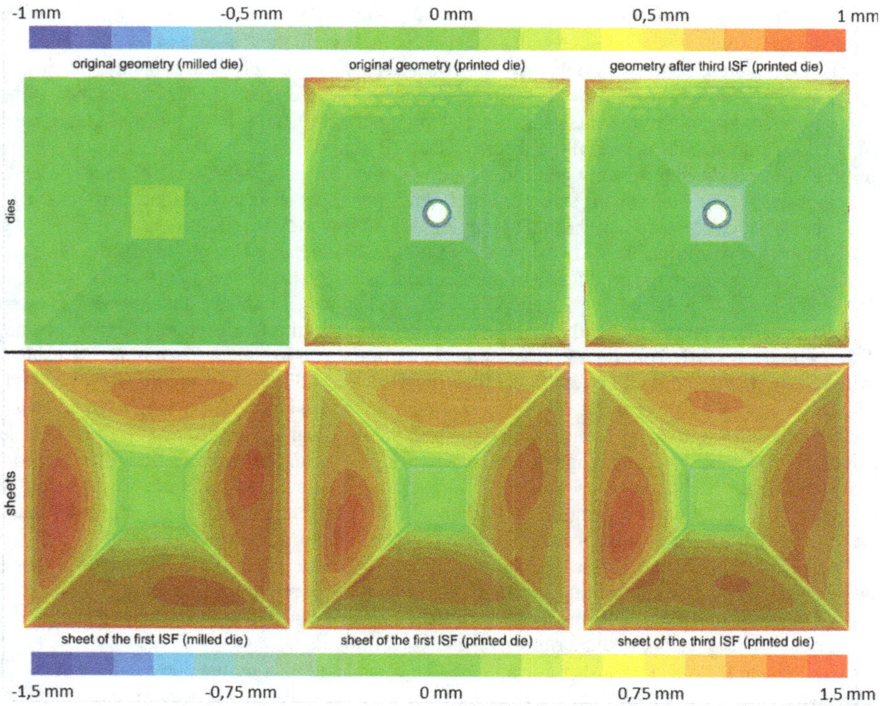

Figure 4: results of the 3D measurements of the dies and formed sheets in comparison to the nominal geometry. The deviations are in mm. The upper legend corresponds to the measured dies, and the lower legend to the measured sheets.

The results of the 3D measurements show that nearly no plastic deformation of the printed die occurred after three subsequent ISF. When comparing the printed die in the initial state and after the third ISF only the edges on the top plane of the truncated pyramid were minimally rounded and colored black based on the contact with the sheet. The top plane was slightly leveled after the third ISF, in comparison to the beginning, on the other hand, the edges of the pyramid side faces do not exhibit any deformation. Summing up, the die only faced insignificant plastic deformation. The fact that the die is deformed only elastically is an important requirement for the application of such dies in the ISF-process. The stability of the dies can also be seen in the resulting sheet geometries. The deviations between the three sheets, which were formed using the printed die, were negligible. This also shows the good reproducibility of the formed parts in this very small product series. Furthermore, it could be shown, that the resulting sheet geometries formed with the milled and the printed die are almost identical, which enables the printed die to be a viable substitute for the milled die.

However, the deviation from the nominal sheet geometry of up to 1.5 mm is clearly visible on the resulting sheet geometries using both dies. Thereby the deviations from the nominal geometry are independent of the type of die (printed or milled) but are an artifact of the ISF process, mainly caused by the sheet stringback, which was not considered in the used process control. The achieved geometric accuracy is comparable to the results of Rieger [4], who also used the completely 3D printed dies and likewise did not consider the sheet stringback in the ISF-process.

The layer height for 3D printing was deliberately chosen to be very low at 0.1 mm, as it was not possible to foresee how much the stair-stepping effect would affect the surface quality of the soft aluminum sheet. No imprints of the stair-stepping effect of the printed die could have been found on the downside surface of the formed sheets. By any further increase of the layer height, printing time can be significantly reduced. Thereby it must be ensured that the ISF does not leave any unwanted impressions on the sheet. Furthermore, when using sheets with higher hardness (e.g., steel sheets), it can be assumed that stair-stepping effects leave fewer impressions.

Comparison of milling and 3D printing for the dies production. The milling of the die took 73 min, in comparison, the production of the same die using 3D printing is significantly slower, it required 1 d and 17 h. Removing the support structures in a water bath lasted another 2 h.

Both programs for producing the dies were not optimized for the shortest possible process time. The milling was carried out on a CNC-milling center optimized for metal machining. By using a milling machine designed for plastics, the time required for milling can be significantly reduced. The same applies to 3D printing. Time can be saved by increasing the layer height. In addition, with a larger nozzle and a matching printer, the printing time can be reduced even further, but the print quality will be correspondingly lower. There is also potential for optimization of the support structure, which accounts for 15 h of the printing time. The support structure was generously designed, by adjusting the print, it will probably be possible to reduce this time by half. Nevertheless, the milling process can produce a functional die for the ISF considerably faster.

It should be noted that a 3D-printer can operate very well over a long time without supervision or tool changes like in the milling machine. For example, all printing times up to 16 h can be carried out very well overnight, and longer prints can be carried out over the weekend if necessary, so that the dies are ready in the morning of the next working day.

To calculate the material costs of dies it must be considered, that on the one hand the material mass needed for 3D printed dies is lower, because of the waste-free production and the lower material consumption through the use possible of filling patterns. On the other hand, the filament is more expensive than the block material. Calculated for to the used die geometry, 32 % of the material costs were saved compared to the milled die (Table 1). With a mass of only 181 g, the printed die is also 73 % lighter than the milled die with a mass of 675 g.

Table 1: Calculation of the material cost of the dies

Die type	Material	Mass [kg]	Price [€/kg]	Material costs [€]
	PLA	0,18	30	5,43
3D printed	PVA	0,08	80	6,48
				Σ 11,91
milled	Sika Block M945 (PUR)	2,39	7,4	**17,70**

Summary

It could be shown that FFF printed PLA dies with an infill density as low as 20 % are suitable for the ISF of fine pure aluminum sheets. With a use of a die fabricated by FFF the same sheet geometries could be reproducibly manufactured as with an identical milled die in a very small lot size. After three ISF operations, no plastic deformation of the dies could be detected. By using the 3D printer, it was possible to save almost one-third of the material costs for the dies used.

3D printed dies can thus be a viable alternative to milled dies. However, higher geometry deviations and printing times of up to several days must be accepted for the advantages of material cost savings and a high degree of design freedom.

The advantages of 3D printing regarding the production of dies for the ISF are the low investment cost for the 3D printer, the high freedom of design, the noncutting machining, the lower mass of the dies and the potential of saving material costs as well as the potential of printing new geometry on existing dies. On a contrary, there are the following disadvantagesdisadvantages of FFF 3D printing: the stair-stepping effect of varying magnitude depending on layer height, the slow production of the dies, the dies are less rigid and strong and the limitation that only thermoplastics can be processed. Furthermore, the 3D printing is only suitable for small dies and the warping increases with increasing die size also the larger geometry deviations occur (compared with milling) and support structures may be necessary and must be removed (depending on the model).By using FFF 3D printing concepts beyond a desktop printer, for example, by a robot with a mounted FFF print, the printing of the die and the forming of the sheet metal can be performed in the same machine in one setup. An example of such a system has already been successfully implemented by Rieger [4] with two robots. Furthermore, new possibilities evolve, like printing new geometries on existing dies without the need to fabricate a completely new die.

Acknowledgements
This paper was completed in association with the project Innovative and additive manufacturing technology — new technological solutions for 3D printing of metals and composite materials, reg. no. CZ.02.1.01/0.0/0.0/17_049/0008407 financed by Structural Funds of the European Union and project. The IGF project RapidSheet (209 EBG) of the research association "Deutsche Forschungsvereinigung für Mess-, Regelungs- und Systemtechnik e.V. – DFMRS", Linzer Str. 13, 28359. Bremen was funded by the Federal Ministry for Economic Affairs and Energy via the AiF within the framework of the program for the promotion of joint industrial research (IGF) based on a resolution of the German Bundestag. AGH was funded through RapidSheet by NCBiR. Reference is made to the availability of the final report of the IGF project 209 EBG for the interested public in the Federal Republic of Germany. This work has received funding from the European Union's Horizon 2020 research and innovation programme under grant agreement N° 856670.

References

[1] W.C. Emmens, G. Sebastiani, A.H. van den Boogaard, The technology of Incremental Sheet Forming – A brief review of the history. Journal of Materials Processing Technology, vol. 210 (2010) 981–997. https://doi.org/10.1016/j.jmatprotec.2010.02.014

[2] R.K. Ajay, Mittal, Incremental Sheet Forming Technologies, CRC Press, Boca Raton, FL. (2021). https://doi.org/10.1201/9780429298905

[3] A. Göttmann, Herstellung von Bauteilen aus Titan mittels Inkrementeller Blechumformung für Anwendungen aus der Medizin und Luft- und Raumfahrt, first ed. (2017)

[4] M. Rieger, D.D. Störkle, L. Thyssen, B. Kuhlenkötter, 2017. Robot-based additive manufacturing for flexible die-modelling in incremental sheet forming. In: Brabazon, D., Naher, S., Ahad, I.U. (Eds.), Proceedings of the 20th International ESAFORM Conference on Material Forming, ESAFORM: Dublin, Ireland, 26-28 April 2017. AIP Publishing, Melville, New York (2017) 40012. https://doi.org/10.1063/1.5008038

[5] Z.H. Du, C.K. Chua, Y.S. Chua, K.G. Loh-Lee, S.T. Lim, Rapid Sheet Metal Manufacturing. Part 1: Indirect Rapid Tooling. The International Journal of Advanced Manufacturing Technology, vol. 19 (2002) 411–417. https://doi.org/10.1007/s001700200042

[6] G. Schuh, G. Bergweiler, F. Fiedler, P. Bickendorf, C. Colag, A Review on Flexible Forming of Sheet Metal Parts, 2019 IEEE International Conference on Industrial Engineering and

Engineering Management (IEEM), Macao, Macao (2019) 1221–1225. https://doi.org/10.1109/IEEM44572.2019.8978879

[7] A. Rosochowski, A. Matuszak, Rapid tooling: the state of the art. Journal of Materials Processing Technology, vol. 106 (2000) 191–198. https://doi.org/10.1016/S0924-0136(00)00613-0

[8] C.M. Cheah, C.K. Chua, C.W. Lee, S.T. Lim, K.H. Eu, L.T. Lin, Rapid Sheet Metal Manufacturing. Part 2: Direct Rapid Tooling. The International Journal of Advanced Manufacturing Technology, vol. 19 (2002) 510–515. https://doi.org/10.1007/s001700200054

[9] L.B. Aksenov, I.Y. Kononov, Thin Sheet Forming with 3D Printed Plastic Tool. SSP, vol. 299 (2020) 705–710. https://doi.org/10.4028/www.scientific.net/SSP.299.705

[10] N. Nakamura, K. Mori, Y. Abe, 2020. Applicability of plastic tools additively manufactured by fused deposition modelling for sheet metal forming. The International Journal of Advanced Manufacturing Technology, vol. 108 (2020) 975–985. https://doi.org/10.1007/s00170-019-04590-5

[11] G. Bergweiler, F. Fiedler, A. Kampker, K. Lichtenthäler, Additiv gefertigte Umformwerkzeuge im Prototypenbau. Umformtechnik Blech Rohre Profile (2019) 18–19.

Sheet Metal 2023 | Materials Research Forum LLC
Materials Research Proceedings **25** (2023) 85-92 | https://doi.org/10.21741/9781644902417-11

Reverse rigid body motion in multi-stage single point incremental forming

Marthe Vanhulst[1, a *], Hans Vanhove[1,b], Yannick Carette[1,c] and Joost R. Duflou[1,d]

[1]Department of Mechanical Engineering, Katholieke Universiteit Leuven / Member of Flanders Make, Celestijnenlaan 300B, B-3001 Leuven, Belgium

[a]marthe.vanhulst@kuleuven.be, [b]hans.vanhove@kuleuven.be, [c]yannick.carette.duflou@kuleuven.be, [d]joost.duflou@kuleuven.be

Keywords: Incremental Sheet Forming, Analysis, Rigid Body Motion

Abstract. This paper provides an analysis of the Rigid Body Motion (RBM) in multi-stage Single Point Incremental Forming of a cylindrical cup. The RBM is studied using Digital Image Correlation to track the material in each contour of the tool, resulting in a better understanding of the RBM phenomenon than has been shown in literature. The study shows that the highest Z-translations of the midpoint of the cone take place in the last contours of each stage, where these RBM peaks also increase per stage. An interesting observation is a reverse rigid body motion, appearing in the contours close to the end of a stage, right before the sudden drop of the bottom in the last contours. This causes flattening of the bottom of the cone and results in a smaller RBM per stage for the later forming stages. Additionally, the stepover between two intermediate stages has been shown to have an important effect on the stepped features appearing in the bottom, but did not have a significant effect on the accuracy of the midpoint of the cone.

Introduction

As a flexible sheet metal forming technique, Single Point Incremental Forming (SPIF) shows great potential as an alternative for producing small batch sizes and customized parts due to the elimination of expensive dies. Additionally, the small plastically deformed zone located in the direct vicinity of the contact area between the tool and workpiece results in much higher strains and therefore an increased formability compared to conventional processes like deep drawing [1]. However, two important challenges remain: the low geometric accuracy due to unwanted deformations and the forming limits, that result in failure of the part at certain wall angles. As a method of increasing the forming limits of parts produced with SPIF, multi-stage forming can be considered [2], as well as dynamic local heating [3]. This paper will focus on the first strategy, which allows forming parts with much higher wall angles, but also introduces unwanted deformations such as Rigid Body Motion (RBM). In this study, the rigid body motion for forming a cylindrical cup in multiple stages will be analyzed. The comparison between different stages will be performed by analyzing the difference in depth of the bottom in between two contours, and will be addressed as the incremental RBM.

Rigid Body Motion (RBM) in Multi-stage SPIF

In the previous paragraph, multi-stage incremental forming has been proposed as one of the possibilities for overcoming the forming limits and preventing the part from failure. Despite its success in increasing the process limits, multi-stage forming also induces extra unwanted deformations, which already showed as stepped features in the bottom of the produced cone when multi-stage forming was described by Duflou et al. [2]. These features are induced by forming the part in multiple stages, when additional unwanted deformations occur due to the geometrical stiffness of the preform in each stage, and take place in the forming direction.

Content from this work may be used under the terms of the Creative Commons Attribution 3.0 license. Any further distribution of this work must maintain attribution to the author(s) and the title of the work, journal citation and DOI. Published under license by Materials Research Forum LLC.

The first attempts on multi-stage forming of a cylinder to increase the maximum achievable wall angle limits did not take the unwanted stepped features into account when designing the multi-stage strategy and intermediate shapes, resulting in large geometrical deviations [2]. Later, Malhotra et al. [5] firstly used the term Rigid Body Motion (RBM) to describe these unwanted stepped features induced by the multi-stage approach and described analytical formulas to calculate the rigid body translation for out-to-in (OI) and in-to-out (IO) toolpath strategies. OI toolpaths, where the tool starts close to the backing plate and increases in depth, are commonly used in multistage strategies. IO toolpaths on the other hand, start inside the shape and gradually move outwards, which results in a different displacement of the material and different deformations. Based on their predicted RBM Malhotra et al. proposed a new strategy that combines an OI and IO toolpath within one stage, in order to compensate for the predicted rigid body translations. With this strategy, they were able to form a 90 degree wall angle cylindrical cup with a near flat base. Xu et al. [6] updated these RBM predictions by relating the material constants to the blank properties. Lingam et al. [7] analyzed the existing prediction models of Malhotra et al. [5] and Xu et al. [6] and included force predictions in their model to take the elastic deflections of the tool and sheet during forming into account. Later, Lingam et al. [8] proposed a methodology that combines IO and OI stages based on the strains, where the best forming sequence of features can be selected and unwanted RBM features are avoided by predicting the rigid body translation. Ndip-Agbor et al. [9] on the other hand also studied RBM and came to the conclusion that previous studies showed little insight in why RBM occurs and tended to focus mostly on specific cases or empirical observations. To compensate for this literature gap, they related the RBM to the 3D contact region between tool and sheet and used this as the basis for their RBM predictions. Wu et al. [10] proposed a new methodology based on both OI and IO toolpaths together with design rules for constructing intermediate stages for a vase shape with the aim of improving thickness distributions and reducing RBM. Later, Wu et al. [11] extended their strategy for maximal thickness distributions and minimal RBM to a truncated pyramid. Gupta et al. [12] did manage to eliminate the RBM for a more complex aerospace component, but needed multiple trials before succeeding.

Despite significant reported research efforts on this topic, no in-depth empirical studies on the rigid body translations per individual tool contour (one pass of the tool at a constant depth) have been performed. This study aims to investigate the RBM in each stage of forming, by measuring the Z-translation between tool contours of the different stages using Digital Image Correlation (DIC). This will result in a broader understanding of the RBM phenomenon in multi-stage incremental forming.

Materials and methods

The studied workpiece shape is a cylindrical cup with a diameter of 142 mm and a depth of 35 mm, on which two strategies of multi-stage forming are applied. The first forming strategy (Fig. 1.a) was described by Duflou et al. [2], where intermediate conical shapes with increasing wall angles in steps of 10 degrees are formed (Fig. 1.a). The maximum stepover between two subsequent stages is not constant and is the largest between stage 1 and 2 with a distance of 9.16 mm (perpendicular to the tool). In a second strategy (Fig. 1.b), a smaller maximum stepover of 2.5 mm is considered and kept constant for all stages. This results in 13 shapes, starting from a cone with a 50 degrees wall angle up to the desired cylindrical cup of 90 degrees. The multi-stage experiments are conducted with a hemispherical tool of 10 mm diameter on a KUKA KR500MT robotic arm with a feedrate of 2000 mm/min on 1.5 mm sheets of Aluminum alloy 5754 (AlMg3). The sheet is clamped with a circumferential backing plate of 184 mm diameter and Nuto 46 lubrication oil is applied before forming. A contouring toolpath forms the part with a scallop width of 1 mm. This corresponds to a stepdown in the Z-direction of $sw * \sin(\alpha)$ with sw the scallop width of 1 mm and α the wall angle. For the first shape of both strategies, a cone of 50 degrees, the stepdown corresponds to 0.766 mm.

The same DIC setup (Fig. 1.c) is used as in [13] with two Mako G-507B Allied Vision cameras and Edmund Optics lenses with a focal length of 35 mm. The DC-LEDS together with a concave shield ensure diffuse lighting conditions. With the bottom half of the part as the field of view, the setup exploits the symmetrical properties of the part to make sure all areas in the region of interest (ROI) are visible by both cameras at all times during forming. The speckles are applied by screen printing. The DIC images are taken when the tool is outside the field of view, to eliminate capturing the high local elastic deformations around the tool tip. Additional images are taken before forming, in order to eliminate noise, and an extra image is taken after finishing a shape and retracting the tool. The software used for calculating the correlations is MatchID [14] with an update of the reference image in each time step, in order to cope with the high deformations induced by the process. After correlating the images in the time and space domain with the parameters indicated in Table 1, the coordinate system of each image is transformed to the local coordinate system of the workpiece-robot setup as presented in Fig. 1.c.

Table 1: Correlation parameters.

Correlation software	**MatchID**
Subset size [pixels]	39x39
Step size [pixels]	10
Correlation algorithm	Zero-Normalised Sum of Squared Differences (ZNSSD)
Interpolation	Bicubic Spline Interpolation
Shape function	Quadratic
Stereo transformation	Quadratic

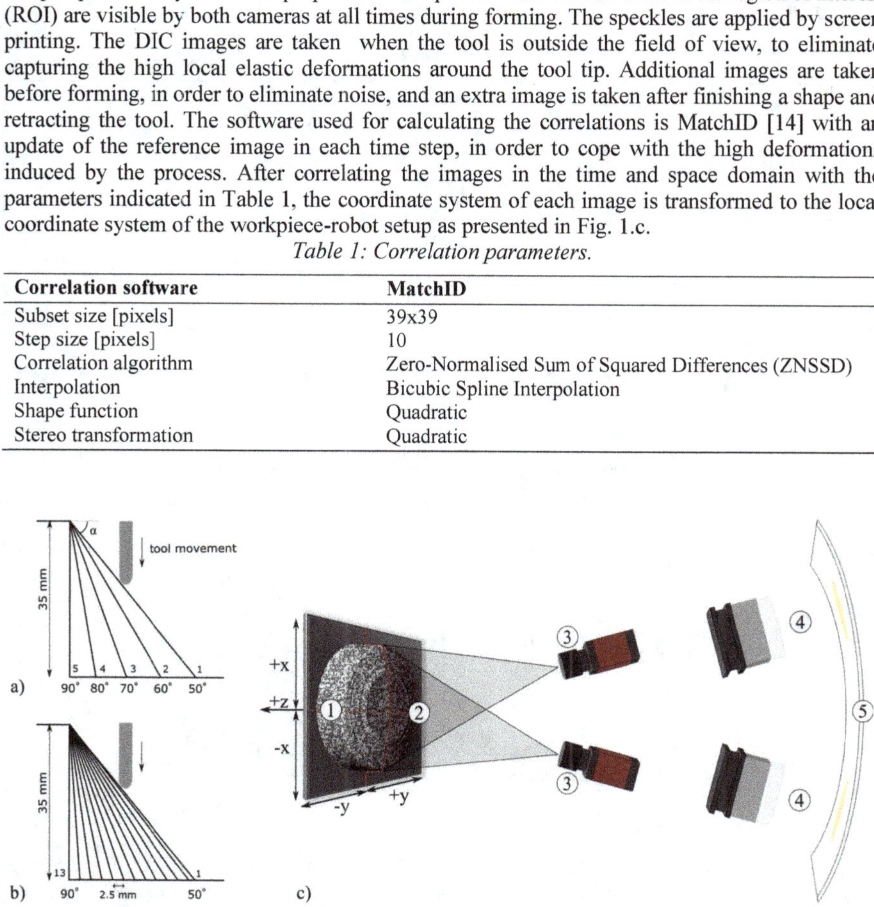

Fig. 1: a) Strategy 1, b) strategy 2 and c) DIC setup with (1) a clamped part, (2) the field of view, (3) the cameras, (4) the spotlights and (5) a concave shield to diffuse the light. The origin of the coordinate system is located in the middle of the (originally flat) sheet, with the X- and Y-axes as indicated in the figure and the Z-axis pointing opposite to the sheet forming direction.

Results and Discussion

The DIC results are compared by calculating the incremental rigid body deformation of the midpoint of the cone (located in the unformed bottom) between two subsequent contours of the tool. A contour is defined as one pass of the tool at a constant depth level. Each of these contouring tool passes can induce an unwanted rigid body deformation of the bottom of the cone due to the geometrical stiffness of a preformed part. The RBM per contour is defined as the difference in Z-coordinates between the current contour and the previous one and is visible in Fig. 2.a for the first strategy and in Fig. 2.c for the second. As can be seen, when forming the first stage starting from a flat sheet, the rigid body motion is first upwards (A) due to the bulging of the bottom in the first contours. Then, the RBM shows an overshoot downwards (B), which is larger than the stepdown in the Z-direction.

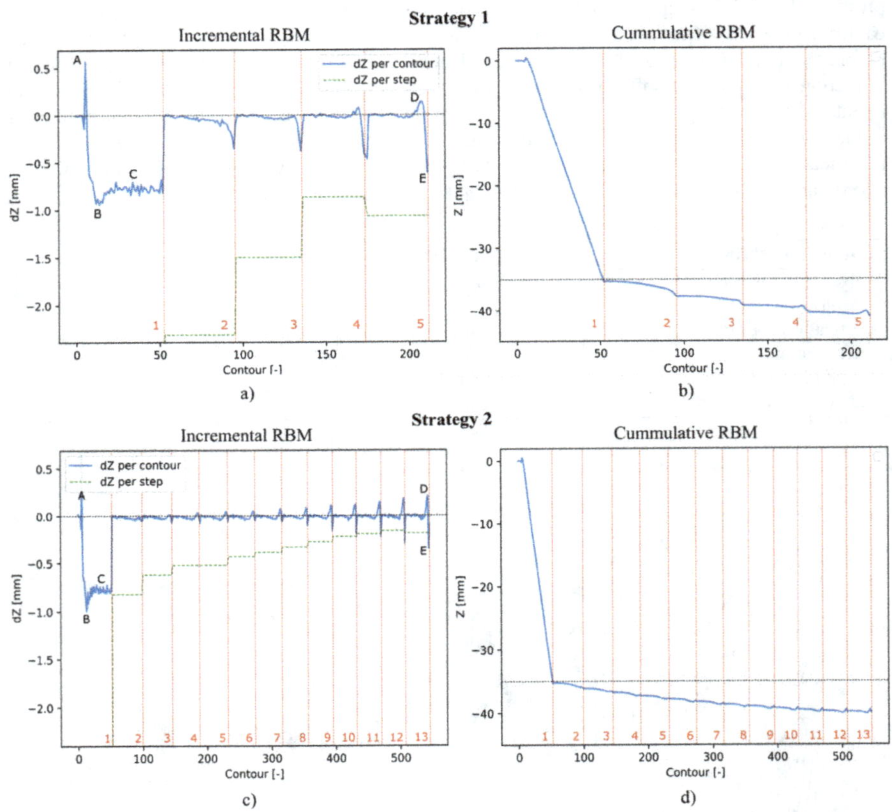

Fig. 2. Incremental and cummulative translations of the midpoint of the workpiece, respectively left and right. (a, c) Strategy 1 and 2, respectively: incremental Z-translation per contour (full line) for the midpoint of the cone and its overal Z-translation per stage (dashed line). (b, d) Strategy 1 and 2, respectively: Z coordinates of the midpoint per contour, with the horizontal line indicating the desired cone depth.

The incremental RBM in the first stage is targeted to be equal to the stepdown of 0.766 mm in each contour of the first stage and zero in the next stages, to keep the bottom flat. After the initial overshoot of the absolute incremental RBM (B), the RBM stays approximately constant for the remaining contours in the first stage and corresponds to the stepdown (C). In the second and subsequent stages, the RBM is desired to stay zero, in order to avoid further unwanted deformations. However, the stepped unwanted features are mostly formed in the last contours of the toolpath for each stage of forming, where the RBM suddenly increases in absolute value (E). Right before this RBM peak takes place, a reverse rigid body translation can be observed (D) in the later stages. Due to this reverse RBM, the bottom of the cone moves upwards and slightly decreases the unwanted deformation by flattening the bottom. Additionally, both the reverse RBM and RBM peak in the last contours get higher in absolute value in each multi-stage step. The fact that the RBM is reversed and slightly flattens the bottom again, has not been shown in literature yet and provides a necessary insight in the mechanics of the RBM.

The total RBM per stage is also visible in Fig. 2.a and 2.c (dashed line) and decreases in absolute value in each stage, except for the last stage, where a slight increase is visible. The highest rigid body translations therefore take place in the first stages of the multi-stage approach. As can be seen, both the incremental RBM (full line) and total RBM per stage (dashed line) of the second strategy are much lower in absolute value than the first strategy. This is due to the smaller stepover between two intermediate stages, which lowers the material movement and therefore the rigid body translation of the midpoint of the cone per contour.

The Z-translation of the midpoint of the cone is visible in Fig. 2.b and 2.d for the first and second strategy respectively. Here, the reverse RBM and the RBM peak from the previous images are also visible. When inspecting the Z-translation of the midpoint of the cone, the final Z-position, and therefore total RBM of the multi-stage cone after forming, is for both strategies almost equal (-41 mm for Strategy 1 and -40 mm for Strategy 2). This shows that with smaller stepovers between subsequent stages, the (incremental) RBM per stage is smaller, but this does not have a significant effect on reducing the total (cumulative) RBM after forming.

Although the total RBM of the midpoint of the cone after forming is almost equal for both strategies, Fig. 3 shows that the stepped features are less outspoken for the second strategy, thanks to the smaller stepover between two subsequent stages. The most outspoken stepped feature is located between the first and second stage of the first strategy, where the maximum stepover is the highest. This figure also shows the material trajectory for the two strategies. As can be seen, the material moves a little more upward in the second strategy. This strategy is characterized by a more horizontal material movement, which, in combination with the smaller stepover, leads to higher radial forces, resulting in an upwards lift of the bottom of the cone.

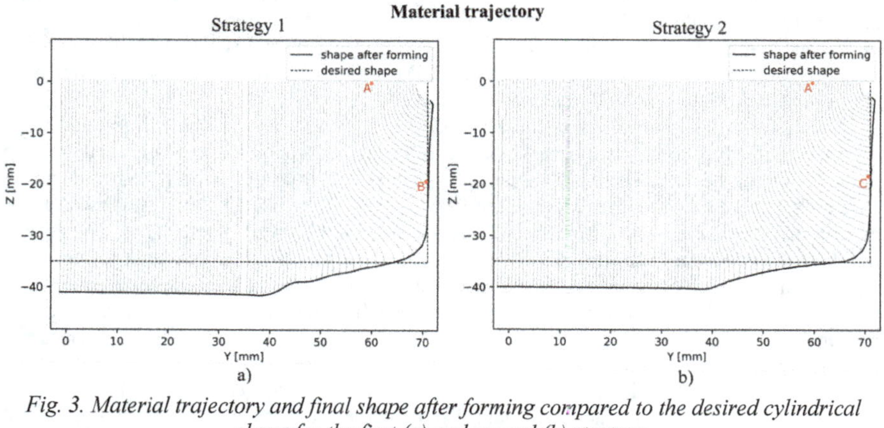

Fig. 3. Material trajectory and final shape after forming compared to the desired cylindrical shape for the first (a) and second (b) strategy.

Materials Research Proceedings **25** (2023) 85-92 https://doi.org/10.21741/9781644902417-11

Fig. 4 shows the difference between the reverse rigid body translation (D in Fig. 2.a) and the regular RBM peak in the last contours of a stage (E in Fig. 2.a). As can be seen, the material movement around the tool is directed axially (in the negative Z-direction) in Fig. 4.b, resulting in a downwards movement of the bottom of the cone. On the other hand, for the reverse RBM in Fig. 4.a, the material movement around the tool is directed more towards the walls (in the Y-direction). This results in a higher radial force and therefore a movement of the bottom in the positive Z-direction, due to the geometrical stiffness of the preformed features. An additional observation in this figure is the fact that the RBM and reverse RBM are not depending on the wall angles of the shape and are constant over the section, except close to the tool-workpiece contact zone.

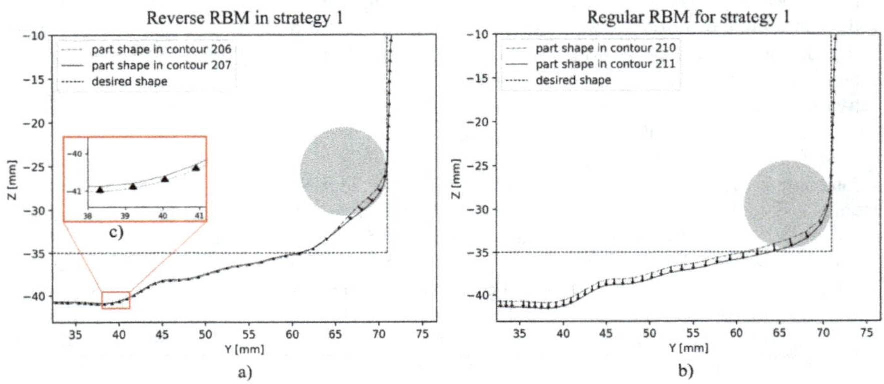

Fig. 4. (a) Material trajectory and reverse RBM between contours 206 and 207 of the tool (zone D) for the first strategy. (b) Material trajectory and peak RBM in the last contours 210-211 of the tool (zone E). (c) Close-up of the reverse RBM.

Conclusion and outlook

In this paper, an empirical study based on DIC measurements is carried out to analyze the rigid body translation per tool contour during multi-stage forming. A first observation is the bulging effect in the first contours of the first stage, which is a single-stage approach. Here, the midpoint of the cone first starts moving upwards due to the bulging effect in the first contours of the toolpath. After a few contours, a large overshoot of the RBM in the negative direction takes place right before an equilibrium equal to the stepdown is reached.

A second observation is visible in the next stages: after a few stages in the multi-stage approach, a reverse rigid body motion is visible in the later contours, followed by a sudden drop of the bottom of the cone. These effects can be explained by analyzing the tool-workpiece contact zone and material trajectory per contour: the RBM is induced by a downwards movement of the material in combination with a high geometrical stiffness of the already formed part. The reverse RBM on the other hand, appears when the material at the tool tip is mostly undergoing radial outward directed loading. Due to a high geometrical stiffness, this radial force induces an upwards movement of the already formed bottom part. Both the RBM and the reverse RBM peaks increase in each stage of the multi-stage forming, which is due to the shape of the intermediate stages and the tool-workpiece contact zone.

A third observation is the decrease in total absolute RBM per stage, except for the last stage, where the stage RBM slightly increases again. This decrease in total absolute RBM per stage might seem contradictory, since the absolute incremental RBM in the last contours of the tool increases in each stage. However, the increase in reverse RBM in each stage partly counteracts the effect of the RBM increase, which explains the decrease in total RBM per stage.

A last observation shows that the incremental RBM per contour and per stage is lower for strategy 2 with smaller stepovers between two intermediate stages. This can be explained by the material movement between two contours, which is smaller for smaller stepovers and therefore results in smaller rigid body translations. However, this does not have a significant effect on the total accuracy of the midpoint of the cone, since strategy 2 also consists of additional intermediate stages. Even though no significant difference in RBM of the midpoint of the cone can be observed, the smaller stepover did result in less outspoken stepped features in the bottom.

This study shows that the RBM indeed appears to be linked to the tool-workpiece contact zone, as was assumed in literature, and also reveals a reverse rigid body translation appearing in the later stages of a multi-stage approach. Possible future work entails a study on the relationship between the incremental RBM and the tool-workpiece contact zone, while also taking into account the geometrical stiffness of the part due to previous forming stages. Another aspect for future work includes a study of the geometric ratio to increase the robustness for knowledge generalization.

Acknowledgements

Marthe Vanhulst was supported by a Predoctoral Strategic Basic Research Fellowship of the Research Foundation – Flanders (FWO) with project 1S47622N.

References

[1] J. Jeswiet, F. Micari, G. Hirt, A. Bramley, J.R. Duflou, J. Allwood, Asymmetric Single Point Incremental Forming of Sheet Metal, CIRP Ann. – Manuf. Technol. 54 (2005) 88-114. https://doi.org/10.1016/S0007-8506(07)60021-3

[2] J.R. Duflou, J. Verbert, B. Belkassem, J. Gu, H. Sol, C. Henrard, A. M. Habraken, Process window enhancement for single point incremental forming through multi-step toolpaths, CIRP Ann. – Manuf. Technol. 57 (2008) 253–256. https://doi.org/10.1016/j.cirp.2008.03.030

[3] J.R. Duflou, B. Callebaut, J. Verbert, H. De Baerdemaeker, Improved SPIF performance through dynamic local heating, Int. J. Mach. Tools Manuf. 48 (2008) 543–549. https://doi.org/10.1016/j.ijmachtools.2007.08.010

[4] M. Skjoedt, N. Bay, B. Endelt, G. Ingarao, Multi Stage Strategies for Single Point Incremental Forming of a Cup, Int. J. Mater. Form. (2008) Suppl. 1:1199–1202. https://doi.org/10.1007/s12289-008-0156-3

[5] R. Malhotra, A.Bhattacharya, A. Kumar, N.V. Reddy, J. Cao, A new methodology for multi-pass single point incremental forming with mixed toolpaths, CIRP Ann. – Manuf. Technol. 60 (2011) 323–326. https://doi.org/10.1016/j.cirp.2011.03.145

[6] D. Xu, R. Malhotra, N.V. Reddy, J. Chen, J. Cao, Analytical prediction of stepped feature generation in multi-pass single point incremental forming, J. Manuf. Process. 14 (2012) 487–494. https://doi.org/10.1016/j.jmapro.2012.08.003

[7] R. Lingam, A. Bansal, N.V. Reddy, Analytical prediction of formed geometry in multi-stage single point incremental forming, Int. J. Mater. Form. 9 (2016) 395–404. https://doi.org/10.1007/s12289-015-1226-y

[8] R. Lingam, A. Bansal, O. Prakash, N.V. Reddy, Mechanics-Based Integrated Product and Process Design for Incremental Forming, J. Manuf. Sci. Eng. 140 (2018) 021016 (11 pages). https://doi.org/10.1115/1.4038600

[9] E. Ndip-Agbor, P. Cheng, N. Moser, K. Ehmann, J. Cao, Prediction of rigid body motion in multi-pass single point incremental forming, J. Mater. Process. Technol. 269 (2019) 117–127. https://doi.org/10.1016/j.jmatprotec.2019.02.007

[10] S. Wu, Y. Ma, L. Gao, Y. Zhao, S. Rashed, N. Ma, A novel multi-step strategy of single point incremental forming for high wall angle shape, J. Manuf. Process. 56 (2020) 697–706. https://doi.org/10.1016/j.jmapro.2020.05.009

[11] S. Wu, L. Gao, Y. Matsuoka, S. Rashed, Y. Zhao, N. Ma, Multi-step toolpath approach of improving the dimensional accuracy of a non-axisymmetric part in incremental sheet forming and its mechanism analysis, J. Mech. Sci. Technol. 36 (4) (2022) 1975-1985. https://doi.org/10.1007/s12206-022-0333-1

[12] P. Gupta, J. Jeswiet, Manufacture of an aerospace component by single point incremental forming, Procedia Manuf. 29 (2019) 112–119. https://doi.org/10.1016/j.promfg.2019.02.113

[13] M. Vanhulst, H. Vanhove, Y. Carette, S. Waumans, J.R. Duflou, Analysis of Thickness Distributions Calculated from Surface Strains Obtained through Digital Image Correlation for Incremental Sheet Forming, Key. Eng. Mater. 926 (2022) 875-882. https://doi.org/10.4028/p-14s93w

[14] Information on https://www.matchid.eu

Sheet Metal 2023 Materials Research Forum LLC
Materials Research Proceedings **25** (2023) 93-98 https://doi.org/10.21741/9781644902417-12

Numerical model design for accuracy prediction of parts made by hybrid incremental sheet forming

Giuseppina Ambrogio[1,a], Giuseppe Serratore[1,b], Luigino Filice[1,b] and Fabrizio Micari[2,d,*]

[1]Department of Mechanical, Energy and Management Engineering, University of Calabria – 87036 Rende (CS), Italy

[2]Department of Engineering, University of Palermo – 98128 Palermo, Italy

[a]giuseppina.ambrogio@unical.it, [b]giuseppe.serratore@unical.it, [c]luigino.filice@unical.it, [c]fabrizio.micari@unipa.it

Keywords: Incremental Sheet Forming, Finite Element Method (FEM), Hybrid Manufacturing

Abstract. Hybrid manufacturing is a promising approach which helps to efficiently improve the performance of conventional manufacturing processes, just contemporary considering different technologies. The suitable combination of Incremental Sheet Forming (ISF) with subtractive or additive processes has been already demonstrated in the past as alternative approach to compensate excessive thinning or low accuracy of the produced parts. Focusing on the last solution, based on the coupled use of ISF and Additive Manufacturing (AM), a new challenge related to the design of a robust numerical model able to predict the behavior of the hybrid-material is still open. In this paper, this issue has been considered and an experimental-numerical validation is proposed.

Introduction

Nowadays the mass customization (MC) is the right technological paradigm that companies have to implement as production strategy if they want to be competitive on the market, because it aims at providing customized products to consumers in large volumes and at costs reasonably low compared to conventional customization processes [1]. Naturally, only in the last twenty years the growth of MC became more evident and wider, thanks to the development of advanced manufacturing technologies and to the increased demand for diversity in the products' range. All these implied the increasing need of production methods focusing further on the individual [1]. During the past years scientific literature offers more technological solutions to implement rapid manufacturing strategy useful for mass customization: dieless forming, additive techniques but specially their hybridization represents a promising approach [2].

Dieless forming is a technique to form sheet metal into three-dimensional shape, firstly introduced in 1994 by professor Matsubara [3] and then engineered by AMINO [4].

As promoted by the founder, the dieless forming is based on the incremental forming principle, pursued by a numerical control machine which drive a hemispherical punch along an assigned and concentric 3D path. The result of this incremental forming methodology is the three-dimensional geometry which reproduce the desired shape without the requirement of additional dedicated and expensive equipment. The simplest configuration of this methodology, named Incremental Sheet Forming (ISF) requires just a CNC machine and clamping frame for the sheet [5]. The sheet is deformed by the punch action but it is also free to move during the repetitive passes, as a consequence rigid movement as well undesired springback are the main drawbacks. Nevertheless, the flexibility and the readiness of this technological solution is not negligible, thus it is a promising alternative for rapid manufacturing.

Content from this work may be used under the terms of the Creative Commons Attribution 3.0 license. Any further distribution of this work must maintain attribution to the author(s) and the title of the work, journal citation and DOI. Published under license by Materials Research Forum LLC.

Starting by a completely different way to interpret the part manufacturing, Additive Manufacturing (AM) techniques represent the modern solution for any type of industry. Within AM family are included all the techniques where objects can be fabricated directly from CAD design without tools or specially designed jigs/fixtures and involves minimal human intervention [6]. Implementing an AM process, the fabrication of unique parts or products with hollow and mould cavities are easily obtainable. Appreciable economic advantages, such as part lightening or reduced time to market, can be achieved by replacing conventional manufacturing with AM techniques in many applications [7]. Design cycle is shortened by a huge extent owing to which products can be quickly brought to market. The application of AM in aerospace sectors showed a mass saving exceeding 50% that combined with the reduced batch dimension typically required places this technique as the most promising. Naturally, AM has own drawbacks that still today penalise the application, such as the long execution time and the total part dimension [7].

More recently, an innovative approach has been introduced founded on the idea to emphasize the impact of the only positive aspects of the manufacturing processes by their integration. More specifically, hybrid manufacturing was formally introduced in 2009, when Schuh et al [8] proposed the following definition: "hybrid means the combination of processes and machines in order to produce parts in a more efficient and productive way" using the *"1 + 1 = 3"* effect and allowing to process materials or shapes which could not be manufactured before or at lower cost. Existing hybrid solutions were summarized in 2014 by Lauwers et al [9], but the development of hybrid process is continuously evolving driven by industrial needs of engineered products and by social aim of higher energy saving.

In 2019, Ambrogio et al. [10] introduced the additive-incremental forming by combining the high flexibility and high readiness of ISF with the versatility of AM. Specifically, the ISF hybridization was promoted pre-treating the blank by AM step with a double aim: for the product accuracy, in order to alter the thinning phenomena and increase the sheet rigidity, or for obtaining more complex 3D shape characterised by concave profile and undercut. Satisfactory results were found but the experimental investigation highlighted an evident weakness in the hybrid process calibration. According to this, the present study is based on the design of a numerical model for helping the additive-incremental forming set up of multi material blanks.

Material and method

The hybrid material model considered is an AISI 304 stainless steel blank thickened locally by an AISI 630 stainless steel crown performed by AM. Specifically, the thickening is a 0.4 mm layer performed by selective laser sintering (SLS). Thanks to the flexibility of the AM process is possible to design and produce tailored blanks without any shape limitation and then, thanks to the flexibility of the ISF process is possible to produce the desired parts with improved geometric accuracy and/or mechanical performances. The chemical and the mechanical properties are reported respectively in Table 1 and Table 2.

Table 1 - Chemical composition of the utilized stainless steels.

Material [%]	C	Mn	P	S	Si	Cr	Ni	Other
AISI 304	0.07	2.0	0.045	0.015	1.0	17-19	8-10	N (<0.11)
AISI 630	0.07	1.5	0.040	0.030	0.7	15-17	3-5	Cu (3-5)

Table 2 - Mechanical properties of the utilized stainless steels.

Materials	Hardness Vickers	Young's Modulus [GPa]	Tensile Strength Yield [MPa]	Tensile Strength Ult. [MPa]	Elong. at break %
AISI 304	205	195	295	660	40
AISI 630	310	197	869	993	20

The AISI 304 is characterized by a better formability while the AISI 630 is used as a reinforcement. The experimental methodology started with a preliminary step performed for thickening the blank by the material deposition, layer by layer, with single films of thickness 40 µm. A sintering laser system printer was adopted and a circular and homogeneous ring was overlapped on the blank. The total thickness of the sintered material is equal to 0.4 mm. Specifically, the sheet is characterized by a side of 180 mm and a thickness of 1.1 mm, while the circular crown is characterized by an internal and external diameter of 70 mm and 140 mm, respectively. Fig. 1 shows the back-side view of the manufactured part in the working area of 140 mm x 140 mm.

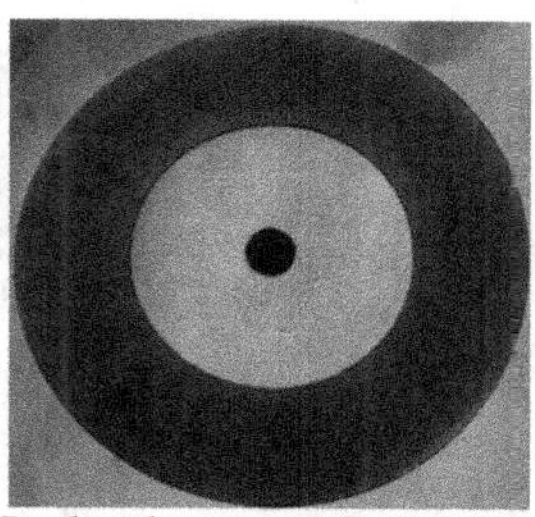

Fig. 1 - Back-side view of the reinforced sheet.

After this preliminary phase, a truncated cone with a base diameter of 70 mm, a slope angle of 50°, and a final height of 30 mm was manufactured by ISF. Three replications were performed by imposing the following parameters during ISF: tool diameter of 12 mm, depth step of 0.5 mm, tool rotation of 300 rpm, and feed rate of 2000 mm/min.

Finally, after the manufacturing phase, the obtained parts were postprocessed and scanned by a MINOLTA laser system in order to catch the geometrical accuracy. Specifically, to remark the suitability of the ISF hybridization the same geometry was also performed by the conventional approach and working on the simple material blank.

Numerical model

A numerical model (Fig. 2) with the commercial software Abaqus was developed to model the behavior of a multi-material sheet formed with the hybrid additive-ISF process. According to the experimental background, the numerical model on hybrid material considered an AISI 304 stainless steel blank thickened locally by an AISI 630 stainless steel. A linear hardening elastoplastic behavior was considered based on the mechanical properties reported in Table 2. The tool was modeled as rigid body with R3D4 and R3D3 shell rigid elements and the blank, composed of two different bodies, as a shell deformable body with S4R shell elements with 5 integration points along the thickness direction. A surface-to-surface finite-sliding contact model was used to the define the contact between the tool and the blank. The friction between the tool and the blanks was modelled according to the Coulomb's formulation considering a coefficient of 0.1. The periphery of the blank was pinned and the tool path was generated with a CAD/CAM software. The connection between the base blank and the sintered layer was modelled considering a fully constrained contact behavior. A tie constrain was employed to model the permanent bond of the circular crown on the base blank to obtain no relative motion between them. This constraint assume an ideal bond and it prevents the separating or relative sliding between the nodes of the tie constrained surfaces.

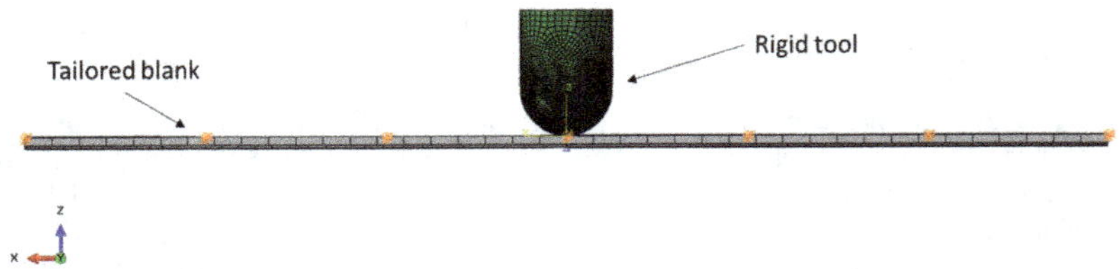

Fig. 2 - Lateral view of the simulation setup.

The numerical problem was solved with the explicit solver to evaluate the final profile and thickness. A semi-automatic mass scaling was employed with a target time increment equal to 2e-6 s. Fig. 3 depicts the modelled simulation assembly considering only the working area of 140 mm x 140 mm:

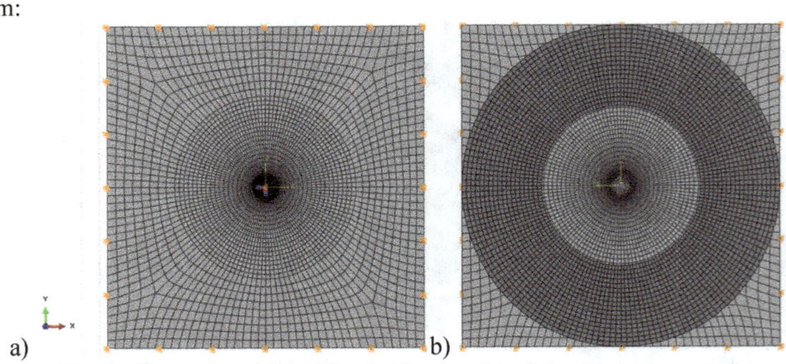

Fig. 3 - Front view (a) and Backside view (b) of the simulation setup.

Subsequently, a second step was simulated after the forming phase removing the tool and the boundary conditions to consider the spring back-effect. After to the model tuning, each simulation lasted 550 min running on an Intel Xeon W-2175 CPU.

Discussion of the results

This section deals with a comparison between the section profile of the truncated cone in terms of deviation from the ideal profile. In this way, the first experimental result was obtained comparing the sections along the part transversal section. As highlighted in Fig. 4 the hybrid and multi-material approach enhances the part accuracy.

Materials Research Proceedings **25** (2023) 93-98
https://doi.org/10.21741/9781644902417-12

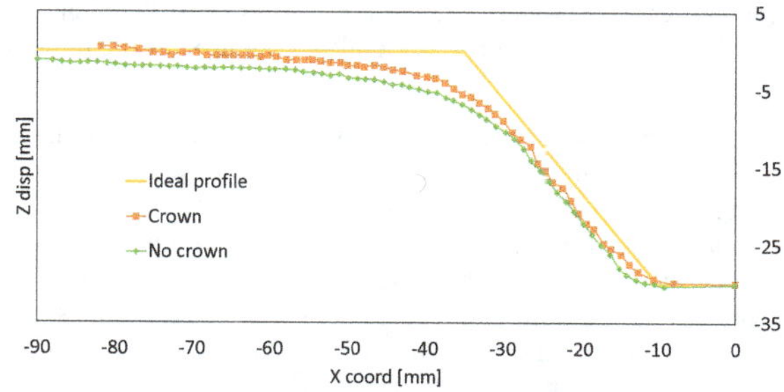

Fig. 4 Comparison along the transversal section between parts obtained by conventional ISF (i.e. no crown) and hybrid additive-incremental forming (i.e. crown).

Fig. 5 shows the results obtained with the circular crown while Fig. 6 without reinforcement. These results show that the developed model could be effectively used to predict the final shape of the manufactured parts even in the case of tailored blanks via AM technique. Although slight difference between numerical and experimental result, due for example at the deformation during the cutting of the sample, or due to the resolution of the 3D laser scanning system, an improvement is obtained with the addition of the circular crow, as expected from the previous experimental investigation [10].

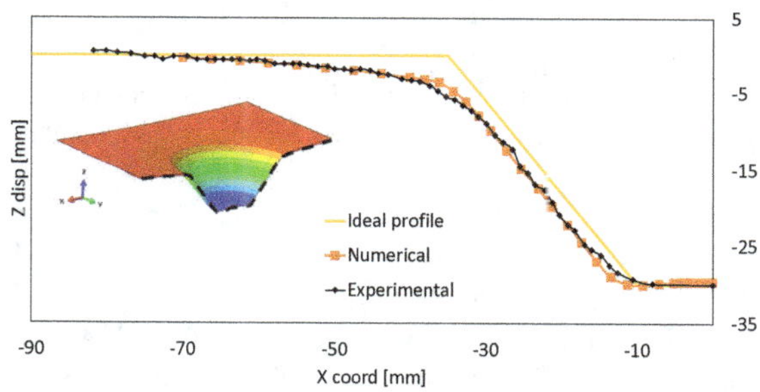

Fig. 5 - Section profile of the truncated cone with circular crown.

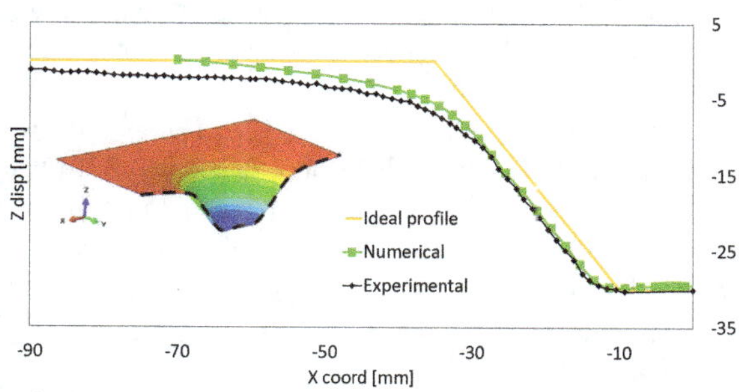

Fig. 6 - Section profile of the truncated cone without circular crown.

Other than what is shown Fig. 5, the deviation between numerical and experimental results in Fig. 6 is mainly due to the pinned boundary condition, since alone the working area is modelled. In the experimental results there is the stringback effect of the clamped zone which was not modelled. This emphasizes the beneficial effect of the reinforcement which improves the clamping condition of the blank, and which approaches the ideal condition of the numerical model.

Summary

In the present work, a simplified numerical model is proposed with the aim to develop a tool able to predict the final shape of a hybrid manufacturing process obtained combining ISF and AM. Incremental forming, in fact, is not yet a consolidated industrial process due to some drawbacks which reduce its suitability. Among them, accuracy is probably one of the most critical. The use of hybrid ISF-AM process may reduce substantially the final shape inaccuracy, but a proper design of the added volume has to be carried out, thus a robust numerical tool is strictly required.

In this paper the tailored blank has been simulated by Abaqus and a significant improvement of the final accuracy has been predicted.

As the future work, this approach will be further developed by means of cohesive elements to tune and experimental outcomes, in order to assess the process feasibility also avoiding any breaking on the interface between different materials.

References

[1] G. Da Silveira, D. Borenstein, F.S. Fogliatto, Mass customization: Literature review and research directions, Int. J. Prod. Econ., 72 (2001) 1–13. https://doi.org/10.1016/S0925-5273(00)00079-7

[2] D. Deradjat, T. Minshall, Decision trees for implementing rapid manufacturing for mass customisation, CIRP J. Manuf. Sci. Technol., 23 (2018) 156–171. https://doi.org/10.1016/j.cirpj.2017.12.003

[3] S. Matsubara, Incremental backward bulge forming of a sheet metal with a hemispherical head tool-a study of a numerical control forming system II, J. Japan Soc. Technol. Plast., 35 (1994) 1311–1316.

[4] M. Amino, M. Mizoguchi, Y. Terauchi, T. Maki, Current Status of “Dieless” Amino’s Incremental Forming, Procedia Eng., 81 (2014) 54–62. https://doi.org/10.1016/j.proeng.2014.09.128

[5] J. Jeswiet, F. Micari, G. Hirt, A. Bramley, J. Duflou, J. Allwood, Asymmetric Single Point Incremental Forming of Sheet Metal, CIRP Ann., 54 (2005) 88–114. https://doi.org/10.1016/S0007-8506(07)60021-3

[6] M. Srivastava, S. Rathee, V. Patel, A. Kumar, P. Koppad, A review of various materials for additive manufacturing: recent trends and processing issues, J. Mater. Res. Technol., (2022). https://doi.org/10.1016/j.jmrt.2022.10.015

[7] M. Srivastava, S. Rathee, S. Maheshwari, T.K. Kundra, Additive manufacturing: fundamentals and advancements, CRC Press., 2019. https://doi.org/10.1201/9781351049382

[8] G. Schuh, J. Kreysa, S. Orilski, Roadmap „Hybride Produktion“, Zeitschrift Für Wirtschaftlichen Fabrikbetr., 104 (2009) 385–391. https://doi.org/10.3139/104.110072

[9] B. Lauwers, F. Klocke, A. Klink, A.E. Tekkaya, R. Neugebauer, D. Mcintosh, Hybrid processes in manufacturing, CIRP Ann., 63 (2014) 561–583. https://doi.org/10.1016/j.cirp.2014.05.003

[10] G. Ambrogio, F. Gagliardi, M. Muzzupappa, L. Filice, Additive-incremental forming hybrid manufacturing technique to improve customised part performance, J. Manuf. Process., 37 (2019) 386–391. https://doi.org/10.1016/j.jmapro.2018.12.008

Joining by forming

Sheet Metal 2023
Materials Research Proceedings **25** (2023) 101-108

Materials Research Forum LLC
https://doi.org/10.21741/9781644902417-13

Innovative mechanical joining processes in versatile process chains - potentials, applications and selection procedures

Michael Lechner[1,a*], Fabian Kappe[2,b], David Römisch[1,c], Tim Rostek[3,d], Simon Wituschek[1,e]

[1]Institute of Manufacturing Technology (LFT), Friedrich-Alexander-Universität Erlangen-Nürnberg, Egerlandstr. 13, 91058 Erlangen, Germany

[2]Laboratory for Materials and Joining Technology (LWF), University of Paderborn, Pohlweg 47-49, 33098 Paderborn, Germany

[3]Forming and Manufacturing Technology (LUF), University of Paderborn, Pohlweg 53, 33098 Paderborn

[a]michael.lechner@fau.de, [b]fabian.kappe@lwf.upb.de, [c]david.roemisch@fau.de, [d]tr@luf.uni-paderborn.de, [e]simon.wituschek@fau.de

Keywords: Joining, Material, Adaptive Manufacturing

Abstract. Mechanical joining processes have gained extensive importance in recent years. In contrast to temperature-based processes, mechanical joining approaches allow the joining of dissimilar materials such as steel and aluminum. What most of the processes have in common is that they are often designed for a specific combination of materials and sheet thicknesses. Consequently, the versatility of these processes is limited. This article will therefore provide an overview of four innovative manufacturing processes, which allow a higher adaptability than conventional mechanical joining processes. In this context, the basic strategies, their applications as well as their advantages and limitations are presented. The contribution closes with a summarizing evaluation and an outlook.

Introduction

In many industrial sectors, there is a trend towards lightweight construction and the reduction of CO2 emissions [1]. One of the most important strategic approaches is the targeted combination of lightweight materials such as high-strength steel, aluminum and fiber-reinforced plastics. A major challenge, however, is the selection of a suitable joining technique. Due to the different melting points, it is often not possible to use thermally based processes [2]. Therefore, mechanical joining methods have gained extensive importance in recent years. In addition to approaches such as cold welding, self-pierce riveting, mechanical clinching and joining by forming in particular have a great potential for technological application [3]. Thereby, it has already been shown in the past that these methods are suitable for joining high-strength steels with aluminum [4]. Groche et al also show the great potential of the technologies, but point out that the influencing variables and process limits are not yet fully understood and it is therefore difficult to predict the final properties [5]. Nonetheless, these processes can be used to achieve highly productive joining processes, which are particularly important for automotive engineering [6]. Due to the different requirements, there are currently a large number of different process variants, which are compared, for example, in terms of their application possibilities [7]. To address these challenges, in the following four innovative processes are presented and evaluated, which allow an increase in the versatility.

Mechanical joining without auxiliary elements

Mechanical joining without auxiliary elements is a joining process consisting of a two-stage process in which pin structures, produced by forward extrusion from the sheet metal plane are used to join dissimilar materials. This process has proven to be suitable to join materials such as high-

Content from this work may be used under the terms of the Creative Commons Attribution 3.0 license. Any further distribution of this work must maintain attribution to the author(s) and the title of the work, journal citation and DOI. Published under license by Materials Research Forum LLC.

strength steel and aluminum [8], as well as steel and fibre-reinforced plastics. Pin structures are already used to increase the strength of metal/fibre reinforced plastic joints. However, often forming processes are not used to manufacture these pins. Instead, manufacturing processes such as powder bed fusion, direct energy deposition or moulding processes are used. Cold extrusion has several advantages compared to these processes. On the one hand, the pin structures have a high strength due to the strain hardening during the extrusion process and, on the other hand, the process can be integrated into existing manufacturing processes with additional weight savings as no auxiliary part is necessary. The two-stage process consists of the pin extrusion from the sheet metal plane and the subsequent joining process, whereby two different process routes are used, the direct pin pressing and the caulking. In the former, the extruded pin is pressed into an unperforated joining partner and forms an undercut due to the compression within the joining partner. In the latter, the pin structure is inserted through a pre-punched joining partner and thus the pin head is upset.

In the first step, the pin must be formed by extrusion from the sheet metal plane. This process requires a multi-acting tool in which the blank holder and punch can be controlled independently. The blank holder prevents the sheet from bulging during cold extrusion and reduces the amount of radial material flow. First, the blank holder pressure is applied, followed by an axial punch movement. The punch penetrates the sheet and displaces the material both axially downwards into the die cavity and laterally outwards into the sheet metal plane limited by a mechanical stop. The punch penetration depth thereby regulates the displaced material and consequently the pin height.

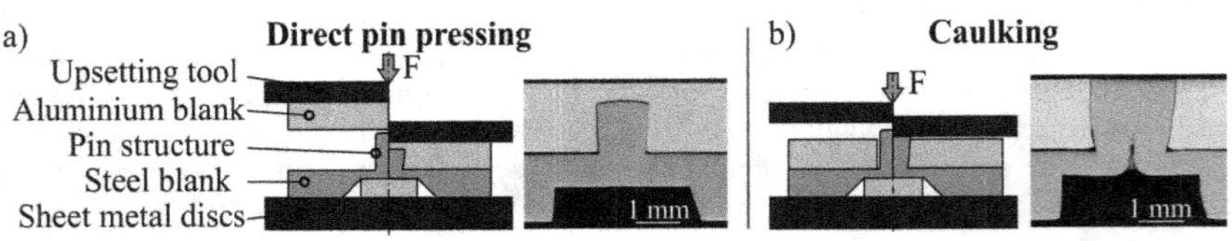

Figure 1: Schematic illustration of the cold extrusion process according to [9]

Furthermore, due to the higher plastic deformation of the material at a higher punch penetration depth, the strain hardening of the material increases, which has an influence on the mechanical properties of the pin and thus on the subsequent joint. For the same reason, the ratio of the punch cross-sectional area to the cross-sectional area of the pin structure also influences the strength of the pin structure. The pin structures are used in the subsequent step for joining using caulking or direct pin pressing (Fig. 1) [9]. Depending on the joining strategy, different pin heights are required, since caulking requires a pin height that is higher than the initial sheet thickness of the joining partner. With direct pin pressing, pin heights can be used that are both smaller and larger than the sheet thickness. However, the pin height plays a crucial role in the formation of the undercut, which is particularly important for the cross tension strength and prevents the joint from unbuttoning under shear loads. Fig. 1b) schematically shows the process sequence for caulking. The pin structure is placed in the pre-punched joining partner and placed between conventional upsetting tools. The punch indentation, which occurs during pin forming, needs to be supported to prevent damage or bending back of the pin. Subsequently, the upper upsetting track moves axially and upsets the pin, which results in a radial material flow and the pin initially expands in the head area. This creates an undercut and a therefore a form-fit connection. With direct pin pressing, shown in Fig. 1a), no pre-punched joining partner is required. Here, the punch indentation is first supported and the component is then placed between conventional upsetting tools. The joining partner is subsequently placed over the pin structure, followed by an axial movement of the upper upsetting tool downwards, causing the pin to penetrate the joining partner which displaces the material axially upwards. Due to the joining operation and the penetration of the pin into the joining partner, the pin is upset, whereby the pin creates an undercut. The greater the initial pin height, the more the pin is compressed, which has a positive effect on the cross tension strength. When considering connections in which continuous fibre-reinforced thermoplastics (CFRT) are used as joining partners, the general joining process changes depending on the joining strategy. In this case, particular attention must be paid to the fibre-friendly insertion of the pin. Therefore, when using fibre-reinforced plastics with a thermoplastic matrix for caulking, the joining partner is first heated locally to improve the fibre movability and is subsequently pierced with a mandrel. In contrast to drilling, this significantly reduces fibre damage, which has a positive effect on strength. Subsequently, the pin is inserted through the perforated joining partner and upset, analogous to the caulking of metal/metal joints. When metal pin structures are pressed directly into CFRTs, the matrix is also first heated thermally to increase the movability of the fibres. Different heating strategies such as infrared radiation, vibration or ultrasound can be used for heating. In this way, the fibres should rearrange around the pin structures and avoid pre-damage which could reduce the load capacity of the joint. In [10], fibre orientation mechanisms were investigated using different pin diameters and head geometries when joining metal pins with CFRTs using direct pin pressing. The pin structure is then pressed into the heated matrix and the joint is reconsolidated and cooled under pressure to fill the void zones created by the pressing process with matrix and fibers.

Joining with adaptive friction elements

Joining with adaptive friction elements is a process for the positive and non-positive joining of sheet metal semi-finished products. Individually manufactured friction elements as auxiliary joining parts allow the joint to be modified in terms of dimensions as well as geometric and mechanical properties. The joining process considered here is divided into two sub steps (Fig. 2a).

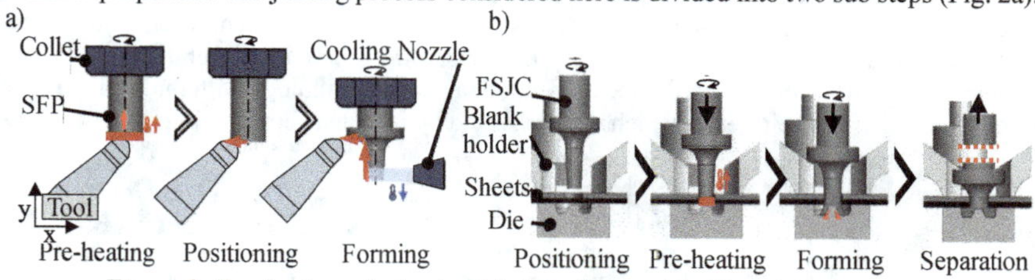

Figure 2: Production of adaptive friction elements/auxiliary joining part

First, the auxiliary joining part is produced. A wire section is set in rotation as the starting material. Contact with a tool generates friction and thus heat. This causes the material to be plasticized locally in the area of tool engagement. The contour of the shank and the head of the auxiliary joining part is formed by appropriate tool movements. Due to the kinematic shape generation, a high degree of changeability with regard to geometry and dimensions can be achieved. [11]

Since heat is introduced into the auxiliary joining part by means of friction, temperature profiles can be realized, if necessary in combination with cooling processes, which can be used in the sense of heat treatment. In this way, the microstructure and thus the mechanical properties of the material can be specifically adjusted, for example by martensitic hardening. The temperature control can be individually controlled, in particular by feed movements and the speed of the tool or semi-finished product as well as the use of coolants, which results in a high degree of variability. [12]

After the auxiliary joining part has been produced, it is fed to the joining point, as shown in Fig. 2b). By rotating the auxiliary joining part while the joining partners are stationary, an opening can be made in the joining partners as required to produce the joint, comparable to flow drilling. In this way, pre-punched or pre-hole-free sheet-metal semi-finished products can be joined. The rotating auxiliary joining part then comes into contact with a stationary die, so that heat is generated by local friction and plasticizes the lower area of the auxiliary joining part. Due to the now high forming capacity, an undercut can be formed by axial repositioning so that a form-fit joint is produced. The high variability of the joining technology described here results in particular from the individual design of the auxiliary joining parts. Wire-shaped semi-finished products made of various metallic materials with different diameters can be used. Furthermore, dimensions and geometry of the auxiliary joining part can be individually adapted by kinematic shape generation. The mechanical properties of the auxiliary joining part are determined by the range of materials that can be used, which can be further modified by individual thermal or thermomechanical treatments. Appropriately, customized auxiliary joining parts are suitable for covering a wide range of joining points. Thanks to an adjustable auxiliary joining part length, sheet metals of different thicknesses (t = 1-6 mm) made of numerous materials can be joined in any sequence, without any preferred direction, with positive locking and in some cases also with frictional locking. Pre-piercing of the semi-finished products can be largely dispensed, especially if hardened auxiliary joining parts are used. At present, however, accessibility of the joint on both sides is required. The load-bearing capacity of the joint can be specifically influenced in particular by the diameter, the head or undercut geometry and the strength of the auxiliary joining parts used.

Versatile self-piercing riveting (V-SPR)

Due to its high load-bearing capacities, a wide range of application and high process robustness, semi-tubular self-piercing riveting, which is also one of the mechanical joining processes, is being frequently used [13]. However, the increasing number of material-thickness combinations and the current rigidly designed tool systems, result in the need for a large number of rivet-die combinations to fulfil the different joining tasks [14]. In order to increase process efficiency and effectiveness, new versatile joining strategies are required, which can react to the changes mentioned above. One possibility to achieve this versatility is based on combining a new joining system technology with increased punch-sided tool actuator technology with multi-range capable semi-tubular self-pierce riveting which was presented by [15] and shown in Fig. 3a).

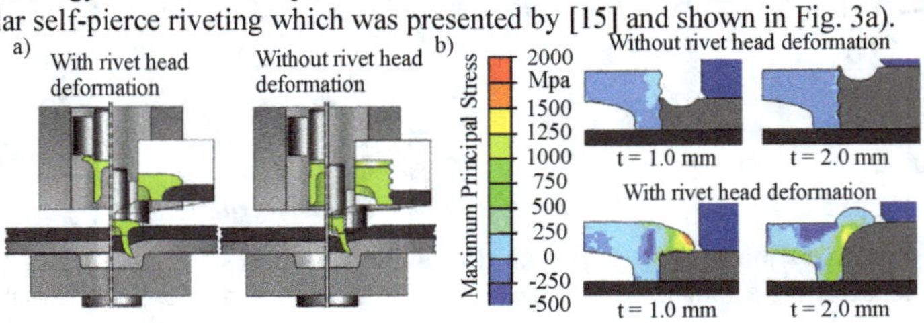

Figure 3: Process sequence for the versatile self-piercing riveting process

The joining system presented enables a joining process (inner punch) and a forming process (outer punch) to be carried out by means of independently controllable punch movements. The forming process is required in order to adapt the rivet to the respective sheet thickness combination. This can be done in two different ways: With and without rivet head forming. If a rivet with head forming is used, the protrusion of the rivet head is formed to the respective sheet thickness. The punch-sided material is formed into annular ring grooves which are located in the head area of the rivet, if a rivet without head forming is used. Both rivet concepts are presented in Fig. 3a). The feasibility of the new mechanical joining process was evaluated using a previously validated numerical axisymmetric 2D-substitute model, which focused the rivet head formation [16]. Therefore, only the upper part of the joint consisting of the punch-sided sheet and the rivet head is modeled. Subsequently, the two rivet concepts were tested with different punch-sided sheet thicknesses (1.0 mm and 2.0 mm) in order to verify their adaptability to changed boundary conditions (Fig. 3b)). Both rivet concepts enabled the formation of the joint for both thicknesses. However, the simulations also showed that the rivet with head deformation leads to an increased stress concentration in the head area.

Tumbling self-piercing riveting process

An industrially often used joining process is the semi-tubular self-piercing riveting, which is characterized by a very robust design with short process times. However, the joining process is also comparatively rigid and the tools used can only adapt to the process to a very limited extent in order to be able to react to disturbances and process variables [2]. A forming process with a large number of degrees of freedom and a great potential for increasing process control is tumbling. The bulk metal forming process is characterized by a tumbling punch with a conical shape that tilts out of the tool axis. This tilting causes a reduction of the contact area between the punch and the workpiece and can be used to intentionally control the material flow. One parameter of the tumbling strategy is the tumbling angle. It causes a reduction of the contact area with increasing angle and thus a lower required forming force [17]. Another component is the kinematic with which the tumbling punch is positioned on the rivet head [18]. This can be applied in the form of a circle, a spiral, by radial tilting and as multi-blade kinematics. The different kinematic models are characterized by a varying material flow in the forming operation and thus it can be used in a specific pattern. In addition, the number of revolutions, the ramp-up phases until the maximum tumbling angle is reached and the rotational or linear speed of the tumbling punch can be adapted [19]. The integrated tumbling process into the self-piercing riveting process allows an increased degree of reaction to process and disturbance variables and the production of tailored joints shown in Fig. 4.

Figure 4: Schematic tumbling self-piercing riveting and influence of varying tumbling angles

The combination makes it possible to combine the advantages of the two processes and to expand the process window. The increased number of process parameters in a tumbling riveting process can be used to react to influences such as varying mechanical and geometric properties with targeted material flow control. Furthermore, a freely configurable tumbling strategy enables an in-situ adaptation of the process route, which makes it possible to react during the joining [19].

Discussion and evaluation of versatility in innovative joining processes

The innovative joining processes shown previously are characterized by individual features in the process design. They provide varying potentials for extending the possibilities of influence before or during the joining process to increase the versatility. The investigations show that the versatility is achieved on different process levels and thus different approaches for applications are feasible. A distinction can first be made according to the process stage. In the pin joining, friction element joining and versatile self-piercing riveting processes, an adaptation to the joining task is performed in an earlier process stage. The versatility of pin joining for metal/metal connections arises in particular in the first process stage, the extrusion of the pin structures. Here, by varying the process parameters and boundary conditions, it is possible to directly influence, the subsequent mechanical joint properties and, in addition, the joinability of the pin joints. Thus, by adjusting the pin height, it is possible to adapt to variations in sheet thickness and changed joining partner properties, and also to react to the mechanical requirements of the joint by adjusting the number of pins.

Materials Research Proceedings **25** (2023) 101-108 https://doi.org/10.21741/9781644902417-13

Investigations of the influences of the process parameters have shown that the pin height can be directly influenced by varying different process variables [1]. Experimental investigations in [6] on the influence of different pin heights on the joint properties under shear load have shown that due to the increasing strain hardening during the extrusion process, the strength of the pin structures increases with the pin height and, analogously, the strength of the joint itself. The maximum load capacity of the joint depends directly on the diameter of the pin structure. However, it was shown based on investigations on different pin geometries with different pin cross-sections that under otherwise identical process conditions, a smaller pin cross-section in relation to the forming punch has a higher shear strength in relation to the cross-section compared to a pin with a larger cross-section. This shows that an increased flow resistance into the die or deformation resistance due to a reduced pin cross-section leads to an increase in strain hardening, especially in the pin base, where a metal/metal pin connection usually fails under shear load. These investigations show that an increase in the versatility of pin joining is achieved by the extrusion process. The individual adaptation of the extruded pin structures enables an extended range of possible joining applications.

Joining with adaptive friction elements also includes a preceding process stage. The high variability of the joining process is in particular due to the individual design of the auxiliary joining elements used. Wire-shaped semi-finished products made of various metallic materials (e.g. aluminum, steel) with different diameters (currently 5-14 mm) can be used for these. Furthermore, dimensions and geometry of the auxiliary joining part can be individually adapted by kinematic shape generation (diameter 3-12 mm, length 4-15 mm, head shape, etc.). The range of materials that can be applied, which can be further modified by individual thermal or thermomechanical treatments determines the mechanical properties of the auxiliary joining part. Appropriately, customized auxiliary joining parts are suitable for covering a wide range of joining points. Thanks to an adjustable auxiliary joining part length, sheet-metal semi-finished products of different thicknesses (t = 1-6 mm) made of numerous materials can be joined in any sequence, i.e. without a preferred direction, with positive locking and in some cases also with frictional locking. Thus, as in pin joining, the increase in versatility is achieved by adapting geometric or mechanical properties of the joining component. Pre-piercing of the semi-finished products can be largely dispensed with, especially if (in-process) hardened auxiliary joining parts are used. The load-bearing capacity of the joint can be specifically influenced in particular by the diameter, the head or undercut geometry and the strength of the auxiliary joining parts used. There are characteristic similarities to pin joining as well, with direct pin pressing and caulking with a pre-hole in the joining partner.

In semi-tubular self-piercing riveting, the versatility can also be increased by a geometric adaptation of the auxiliary joining elements. However, the auxiliary joining part is not adapted individually for the joint, but the spectrum of joining capabilities is enlarged by targeted adaptations of the geometry. In addition to the adaptation of the geometries of the joining components, an increase in the versatility is achieved by extending the influence from the process control side. The tool design is modified and the number of process parameters is increased by integrating additional sensors and actuators. With a targeted adjustment of these parameters, the material flow can be controlled and it is possible to react to disturbance variables. In the case of versatile semi-tubular self-piercing riveting, process-side control is provided not only by the adapted rivet geometry, but also by the internal and external punches. By adapting the process route, influence can be exerted on the resulting geometric joint formation and the joining range of multi-material systems can be extended [17]. The investigations show an influence of the tumbling strategy on individual characteristic features of the joint. The influence of the tumbling kinematics on the joining process is further investigated in [18], which shows a decrease of the joining forces with increasing tumbling angle. For the kinematic models applied, it can be stated that a more gradual adjustment of the tumbling angle causes the force-displacement curves to increase more uniformly. As a further process parameter, the tumbling velocity is investigated in [19] in the form of the rotation speed of the tumbling punch for a circular and spiral kinematics. The results show a uniform formation of the rivet head end positions with increasing rotation speed. Overall, the findings show that the process combination consisting of a self-piercing riveting process superimposed by tumbling kinematics can increase the possibilities for influencing the process. By adapting the tumbling strategy, the characteristic features relevant for the joint can be specifically controlled and adapted to the joint to be manufactured. The resulting increased material flow control can thus be applied to expand the range of feasible joints and significantly increase the adaptability of the process.

Two superordinate approaches to increase versatility can be identified in innovative joining processes. On the one hand, a wider range of joining tasks can be achieved through the targeted and individual adaptation of geometric and mechanical properties of the joining components. However, a further distinction must be made here whether the joining components are individually adapted for each joining task, as in pin joining and friction element joining, or whether the possible forming operations on the auxiliary joining element are increased, as in the case of versatile semi-tubular self-piercing riveting. On the other hand, the joining process itself can be influenced by the targeted control of adapted tool systems.

Summary and Outlook

In the context of the investigations, it was shown that it is possible to react to fluctuations in the process with the help of versatile joining processes. This is demonstrated by the three processes presented: mechanical joining without auxiliary elements, joining with adaptive friction elements, versatile and tumbling self-piercing riveting. The results of the investigations show the different characteristics of the versatility of the individual processes. It is demonstrated how the versatility is achieved and how it can be applied in the manufacture of the joints. For joining with pin structures, the versatility is particularly in the manufacturing process of the pins. By specifically controlling the geometry and mechanical properties of the pins, the joining spectrum can be expanded. Joining with adaptive friction elements is characterized by the individual design of the auxiliary joining parts. Thus, as with pin joining, the adaptability of the joining technology is also extended for this joining process in an earlier process stage. The same is applied to the use of adapted rivet geometries in versatile semi-tubular self-piercing riveting. Here, however, the influence on the process is additionally increased by a special tool design. In tumbling semi-tubular self-pierce riveting, the versatility is increased during the joining process by targeted control of the material flow. In future investigations, this adaptability is to be increased even further. In this context, for example, the development of real-time capable models appears to be of great interest in order to enable in-situ control of the joining processes. Furthermore, the analysis of joining technologies under industrial conditions appears to be of significant importance. On the one hand, it should be analyzed how the mass of the tools can be reduced in order to be able to use them in combination with an automated robotic system. On the other hand, it should be investigated how the production time can be reduced. Finally, the application properties with regard to corrosion and fatigue strength must be assessed.

Acknowledgment

Funded by the Deutsche Forschungsgemeinschaft (DFG)—TRR 285-Project-ID 418701707.

References

[1] K. Martinsen, S.J. Hu, B.E. Carlson, Joining of dissimilar materials, CIRP Annals 64 (2) (2015) 679–699. https://doi.org/10.1016/j.cirp.2015.05.006

[2] D. Li, A. Chrysanthou, I. Patel, G. Williams, Self-piercing riveting-a review. Int J Adv Manuf Technol 92 (5-8) (2017) 1777–1824. https://doi.org/10.1007/s00170-017-0156-x

[3] K.-I. Mori, N. Bay, L. Fratini, F. Micari, A.E. Tekkaya, Joining by plastic deformation. CIRP Annals 62 (2) (2013) 673–694. https://doi.org/10.1016/j.cirp.2013.05.004

[4] Y. Abe, T. Kato, K. Mori, Self-piercing riveting of high tensile strength steel and aluminium alloy sheets using conventional rivet and die. JMPT 209 (8) (2009) 3914–3922. https://doi.org/10.1016/j.jmatprotec.2008.09.007

[5] P. Groche, S. Wohletz, M. Brenneis, C. Pabst, F. Resch, Joining by forming—A review on joint mechanisms, applications and future trends. JMPT 214 (10) (2014) 1972–1994. https://doi.org/10.1016/j.jmatprotec.2013.12.022

[6] G. Meschut, V. Janzen, T. Olfermann, Innovative and Highly Productive Joining Technologies for Multi-Material Lightweight Car Body Structures. JMEP 23 (5) (2014) 1515–1523. https://doi.org/10.1007/s11665-014-0962-3

[7] J. Varis, Economics of clinched joint compared to riveted joint and example of applying calculations to a volume product. JMPT 172 (1) (2006) 130–138. https://doi.org/10.1016/j.jmatprotec.2005.09.009

[8] D. Römisch, M. Kraus, M. Merklein, Experimental Study on Joining by Forming of HCT590X + Z and EN-AW 6014 Sheets Using Extruded Pin Structures. JMMP 5 (1) 2021. https://doi.org/10.3390/jmmp5010025

[9] D. Römisch, M. Kraus, M. Merklein, Investigation of Different Joining by Forming Stratgies when Connecting Different Metals without Auxiliary Elements. KEM 883 (2021) 19–26. https://doi.org/10.4028/www.scientific.net/KEM.883.19

[10] J. Popp, T. Kleffel, D. Römisch, T. Papke, M. Merklein, D. Drummer, Fiber Orientation Mechanism of Continuous Fiber Reinforced Thermoplastics Hybrid Parts Joined with Metallic Pins. Appl Compos Mater 28 (4) (2021) 951–972. https://doi.org/10.1007/s10443-021-09892-0

[11] C. Wischer, E. Wiens, W. Homberg, Joining with versatile joining elements formed by friction spinning. Journal of Advanced Joining Processes 3 (2021) 100060. https://doi.org/10.1016/j.jajp.2021.100060

[12] C. Wischer, W.Homberg, A contribution on versatile process chains: joining with adaptive joining elements, formed by friction spinning. Prod. Eng. Res. Devel. 16 (2-3) (2022) 79–88. https://doi.org/10.1007/s11740-021-01094-8

[13] DVS/EFB, Self-pierce Riveting – Overview: Sonderverfahren (2019).

[14] X. He, I. Pearson, K. Young, Self-pierce riveting for sheet materials: State of the art. Journal of Materials Processing Technology 199 (1-3) (2008) 27–36. https://doi.org/10.1016/j.jmatprotec.2007.10.071

[15] F. Kappe, M. Bobbert, G. Meschut, New Approach for Versatile Self Piercing Riveting: Joining System and Auxiliary Part. KEM 883 (2021) 3–10. https://doi.org/10.4028/www.scientific.net/KEM.883.3

[16] F. Kappe, C.-R. Bielak, V. Sartisson, M. Bobbert, G. Meschut, Influence of rivet length on joint formation on self-piercing riveting process considering further process parameters. ESAFORM (2021). https://doi.org/10.25518/esaform21.4277

[17] S. Wituschek, M.Lechner, Investigation of the influence of the tumbling angle on a tumbling self-piercing riveting process. Proceedings of the Institution of Mechanical Engineers, Part L: Journal of Materials: Design and Applications 236 (6) (2022) 1302–1309. https://doi.org/10.1177/14644207221080068

[18] S. Wituschek, F. Kappe, M. Lechner, Investigation of the influence of varying tumbling strategies on a tumbling self-piercing riveting process. Prod. Eng. Res. Devel. 16 (2(2022) 353–362. https://doi.org/10.1007/s11740-021-01099-3

[19] S. Wituschek, M. Lechner, Versatile tool design for a tumbling self-piercing riveting process. The 12th Tooling Conference & Exhibition, Tooling 2022, (2022) 499–506.

Materials Research Proceedings 25 (2023) 109-116 https://doi.org/10.21741/9781644902417-14

Investigation on pin caulking as a versatile joining process

David Römisch[1,a *] and Marion Merklein[1,b]

[1]Institute of Manufacturing Technology, Friedrich-Alexander-Universität Erlangen-Nürnberg, Egerlandstraße 13, 91058 Erlangen, Germany

[a]david.roemisch@fau.de, [b]marion.merklein@fau.de

Keywords: Joining, Metal, Pin Caulking

Abstract. Lightweight design is increasingly used to combine the specific advantages of various dissimilar materials into multi-material systems. The aim is to make production more resource-efficient and reduce emissions. However, one challenge in the adaptation of multi-material systems is the lack of versatile joining technologies capable of joining these materials. Joining with cold formed pins is a two-step process with potential in joining metal to metal and metal to fibre-reinforced plastics (FRP). These pins can be joined using two joining strategies, direct pin pressing into an unperforated joining partner and caulking of pins, which are inserted through a pre-punched joining partner. For pin pressing, several studies have already been carried out regarding joinability of steel and aluminium, but caulking offers advantages such as a reduced joining force and the fact that the pin can transmit force over the entire sheet thickness of the joining partner, which can lead to increased strength under axial load. Therefore, the caulking of pins extruded from dual-phase steel with a 6000 aluminium alloy is investigated. The focus is on a fundamental investigation of the joining process and the joint formation when caulking pins of varying heights with sheets of different thicknesses. Subsequently, a process window is derived from these findings.

Introduction

Resource-efficient production and energy and emission savings in as many areas as possible are more relevant than ever in times of energy scarcity and climate change. This can also be seen in the transport sector, which accounts for around 26 % [1] of emissions in Europe. Here, lightweight design is increasingly being used, in which the specific advantages of different materials can be integrated into a single component. High-strength materials such as steel and aluminium as well as fibre-reinforced plastics (FRP) are often used, which are joined to form multi-material systems. However, this requires new versatile joining processes, since conventional joining processes tend to reach their process limits when joining these dissimilar materials. For this purpose, research is being conducted into various mechanical joining processes [2]. Joining with pin structures has potential in the area of joining dissimilar materials and is already being used in metal/FRP joints to increase the load-bearing capacity of bonded joints [3]. For the production of pins, processes such as additive manufacturing [4] or welding processes such as cold metal transfer [5] are used. However, there are also efforts to produce pins by forming processes, due to their good surface, strength and the possibility of integrating the process into existing process chains. Joining with cold-formed pin structures from the sheet metal plane is a new process that has already demonstrated its potential for joining steel/aluminium [6] and steel and fibre-reinforced plastics [7]. This method is a two-stage process, in which mostly cylindrical pins are first extruded from the sheet metal plane and used for joining in the next step. There are already studies on direct pin pressing, in which the pins are pressed into an unperforated joining partner after extrusion [6], causing the pin to be compressed inside the joining partner forming an undercut and thus leading to a form-fit and force-fit connection. Another joining strategy that can be used with extruded pins is caulking, in which the pin is inserted through a pre-punched joining partner and is upset at the pin head to form an undercut and a form-fit and force-fit joint. Compared with direct press-fitting,

Content from this work may be used under the terms of the Creative Commons Attribution 3.0 license. Any further distribution of this work must maintain attribution to the author(s) and the title of the work, journal citation and DOI. Published under license by Materials Research Forum LLC.

the force required for joining is reduced and, in addition, the pin can transmit force over the entire sheet thickness. At present, however, there are only a few papers dealing with joining by caulking. For this reason, caulking as a joining strategy for joining metallic pins with aluminium is investigated in this work. Different pin heights are used and caulked with the aluminium with three different sheet thicknesses. First, the joining process and then the joint formation will be investigated. Subsequently, a process window for the joining strategy is derived based on the results.

Methods and Materials

Materials. For pin extrusion, a dual phase steel HCT590X+Z (DP600), often used for structural parts in car body construction [8], with an initial sheet thickness t_0 = 1.5 mm is used. For the joint formation investigation specimens with a size of 45 x 45 mm^2 were used. The sheet material for the joining partner is a precipitation hardening aluminium silicon magnesium alloy of the 6000 series EN AW-6014-T4 with three different sheet thicknesses of 1.0 mm, 1.5 mm and 2.0 mm.

Cold forming of pin structures. The pin extrusion process is a sheet bulk metal forming process that uses a multi-acting tool design enabling a blank holder and a punch to be controlled independently. The process principle and important process parameters are shown in Fig. 1. The feasibility of the process to extrude cylindrical pins from the sheet metal plane was shown by Hirota [9] and Ghassemali [10]. The process used in this work begins with the application of the blank holder force, to prevent sheet bulging during pin extrusion and to reduce the radial flow of material into the sheet metal plane. Once the force is applied, the punch with a diameter d_P = 3 mm moves axially downwards with a constant speed of v = 5 mm/min and penetrates the sheet with a sheet thickness t_0 = 1.5 mm, whereas the forming work is carried out by a hydraulic cylinder. Due to the plastic deformation initiated by the punch, material is displaced axially downwards into the die as well as laterally outwards into the sheet plane and laterally inwards into the die with a diameter d_D = 1.5 mm. Dionol ST V 1725-2 was used for lubrication during pin extrusion. The punch penetration depth s and thus the pin height h is limited by mechanical stops with which the punch comes into contact as soon as the forming process is finished. Subsequently, the punch and blank holder move axially upwards and the specimen is ejected.

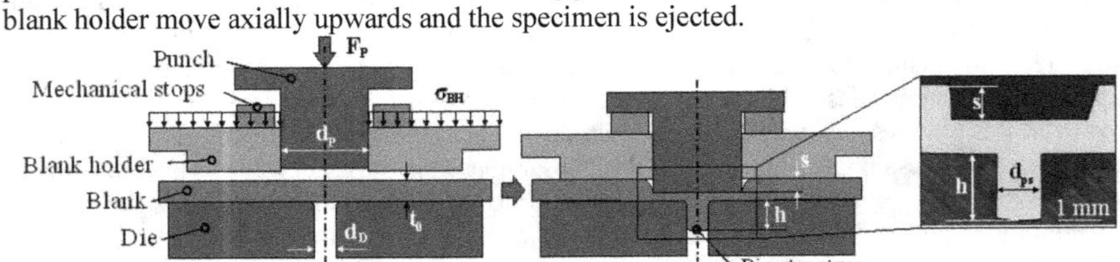

Pin extrusion - Process parameters	
Punch diameter d_P (mm)	3
Punch speed v_P (mm/min)	5
Blank holder pressure σ_{BH} (MPa)	280
Die diameter for forming d_D (mm)	1.5
Sheet thickness t_0 DP 600 (mm)	1.5

Fig. 1: Illustration of the pin extrusion process according to [6] and relevant process parameters

Joining by forming - pin caulking. For the joining of the extruded pins, caulking is investigated within the scope of this work, where a perforated joining partner is used. The hole is drilled using a 1.5 mm drill bit. The pin is subsequently positioned between two upsetting tools and the punch cavity is supported with sheet metal discs to prevent damage or a bending back of the pin during the joining process. The drilled joining partner is then positioned over the pin. In the next step, the

Materials Research Forum LLC
https://doi.org/10.21741/9781644902417-14

upper upsetting tool moves axially downwards until a pre-force of 20 N is reached. As soon as the initial force is achieved, the upsetting tool moves downwards at a constant speed of 5 mm/min and upsets the pin until a maximum force threshold of 8 kN is reached. The pin is upset until it is flush with the sheet metal surface of the joining partner and the upsetting tool has full-surface contact with the joining partner. This results in a steep linear increase in force, as the joining system is elastically deformed. This increase in force thus defines the end of the joining process. An illustration of the process is shown in Fig. 2. Within this work, pins with different heights, listed in Fig. 2, were examined with regard to their suitability for joining EN AW-6014-T4 with different sheet thicknesses (1.0 mm, 1.5 mm, 2.0 mm). Here, the process requires that the pin has a greater height than the sheet thickness being joined. The configurations investigated are listed in Fig. 2. For the joint characterisation micrographs of the different joints were used and three different geometric properties were measured. The pin head diameter, the pin base diameter and the undercut, which was calculated by averaging the left and the right undercut (c.f. Fig. 2)

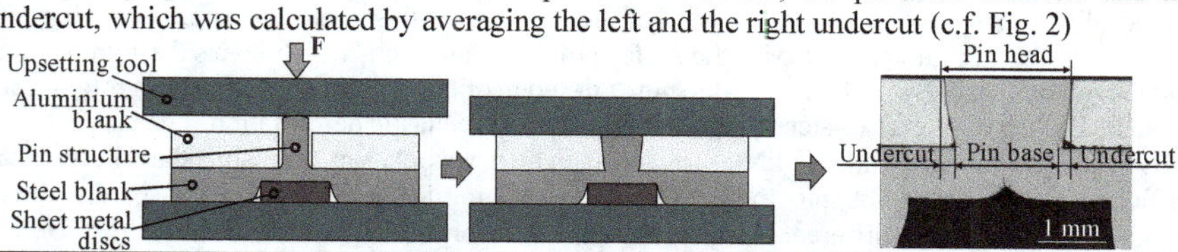

Pin caulking-Process parameters		Investigated configurations								
Joining speed (mm/min)	5	Pin height (mm)		1.50	1.75	2.00	2.25	2.50	2.75	3.00
Sheet thickness t_0 EN AW-6014-T4 (mm)	1.0, 1.5, 2.0	Sheet thickness Al (mm)	1.0	x	x	x	x	x	x	x
Max. joining force (kN)	8		1.5	-	x	x	x	x	x	x
Drill diameter (mm)	1.5		2.0	-	-	-	x	x	x	x

Fig. 2: Schematic illustration as well as parameters of the caulking process and investigated pin height/sheet thickness configurations

Results

Pin caulking. In the following, the pin caulking joining strategy with cold extruded pins will be discussed in more detail. Similar to the alternative joining strategy of direct pin pressing [6], caulking can also be divided into 3 phases. Fig. 3a shows an example of a force-displacement curve of a pin made of DP600 with a height of 1.75 mm, which was caulked with an aluminium sheet with a thickness of 1.0 mm. Here, the 3 phases of the process have been illustrated.

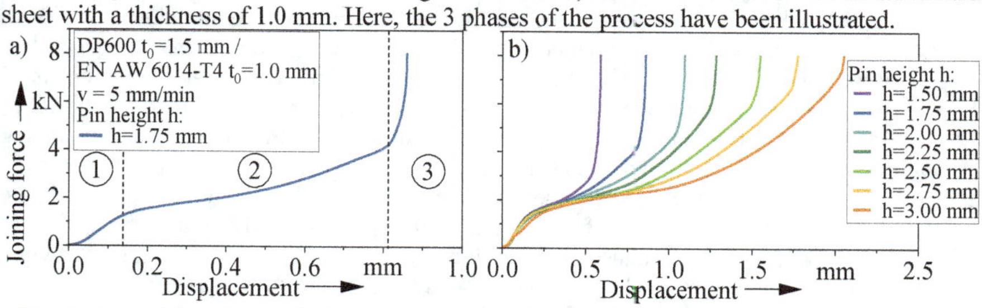

Fig. 3: Pin caulking force-displacement curves. (a) Exemplary curve with the different joining phases (b) Comparison of different pin heights joined with an EN AW 6014-T4 (t_0 = 1.0 mm)

The process begins with a setting process, which can be detected by the non-linear curve. This is caused by a curvature of the sheet due to residual stresses in the sheet caused by pin extrusion, which is flattened by the force applied by the upper upsetting tool, causing the sheet to rest fully

Materials Research Proceedings **25** (2023) 109-116
https://doi.org/10.21741/9781644902417-14

on the surface of the lower tool. Subsequently, a linear increase in force follows due to the elastic deformation of the joining system, which consequently transitions to the elastic-plastic deformation of the pin, which initiates phase 2 of the process. This is initially characterised by the compression of the pin, which can be seen from the continuous slope of the curve. However, it cannot be ruled out that the aluminium joining partner is already deformed in this area due to the expansion of the pin and thus a superimposed deformation of the pin and the aluminium is present. From approx. 0.45 mm, however, an increase in the slope of the curve can be seen. This occurs due to the increasing plastic deformation of both joining partners and the fact that the aluminium is displaced radially and axially upwards as a result of the upsetting and the expansion of the pin. The end of phase 2 can be recognised by the strong increase in force from approx. 0.8 mm. The displacement of the upsetting tool corresponds approximately to the difference between the height of the sheet and the height of the pin, but deviates slightly from the actual value, especially due to the non-linear component at the beginning of the process. At this point, the upper upsetting tool increasingly comes into contact with the surface of the aluminium, which levels the joining zone and the material, axially pushed up by the pin at the edge of the drill hole, until there is full-surface contact. The curve then rises steeply and linearly due to the elastic deformation. The process then ends when the predetermined 8 kN are reached, which was chosen to ensure that the force is sufficient for upsetting the pin to the surface of the joining partner. In Fig. 3b, the force-displacement curves for different pin heights are plotted for the 1.0 mm aluminium sheet. Here it can be seen that with increasing pin height, the displacement travelled by the upsetting tool in phase 2 of the joining process increases to approximately the respective difference between pin and sheet thickness and the start of phase 3 begins progressively at a higher force level. This can be explained by the increasing pin height and the resulting higher required upsetting ratio of the pin. Thus, the strain hardening of the pin and the surrounding material steadily increases and more material has to be displaced radially in the sheet plane due to the larger volume of the pin. As a result, the slope of the force-displacement curve increases continuously with increasing pin height. Examining the configuration with a pin height of 3 mm in Fig. 3b it can be seen that the elastic portion of phase 1 of the process shows a discontinuity, which can be attributed to a buckling of the pin due to the high ratio of the pin height to diameter of 2.

Geometric joint characterisation. Examining the corresponding micrographs in Fig. 4 for the joints with a sheet thickness of the aluminium of 1.0 mm, the increase in the ratio of upsetting of the pins with greater height can be seen in the expanding diameter of the pin head. Due to the axial compression of the pin, the joining process results in a radial material flow in the pin caused by the resulting biaxial tensile stress state in the pin. Due to the fixed connection of the pin base with the sheet, this occurs especially in the region of the pin head, resulting in the formation of a wedge-shaped pin geometry. The diameter of the pin head increases continuously for the pin height of 1.5 mm to 3.0 mm, as shown in the micrographs in Fig. 4. However, starting at the pin height of 2.50 mm a buckling and bending of the pin can be observed, which occurs due to joining instabilities resulting from the increasing height and the high upsetting ratio of 60 % for the 2.50 mm pin. Investigating the measured mean undercut values, with an increase in pin height the undercut can be increased by 88.2 % from a value of 0.17 mm for the 1.5 mm pin to a value of 0.32 mm for a 2.25 mm pin. The undercut further increases up to a value of 0.37 mm for the 3 mm pin, however the joint cannot be characterised as "okay", since the buckling of the pin occurred during joining.

Materials Research Forum LLC
https://doi.org/10.21741/9781644902417-14

Pin height						
1.50 mm	1.75 mm	2.00 mm	2.25 mm	2.50 mm	2.75 mm	3.00 mm

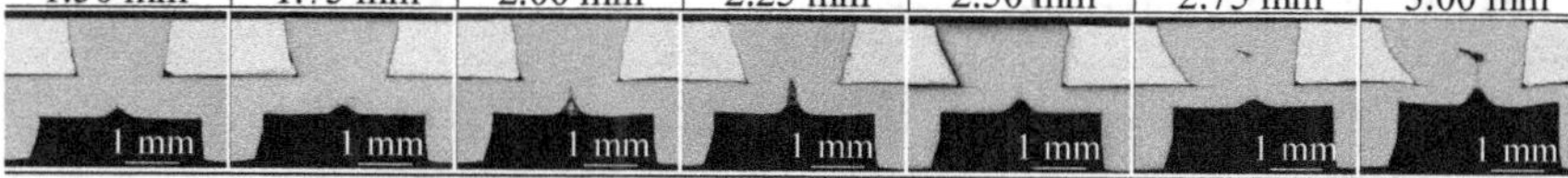

Geometric joint characteristics (t_0 = 1.0 mm)							
Pin height (mm)	1.50	1.75	2.00	2.25	2.50	2.75	3.00
Pin head diameter (mm)	1.90	2.09	2.16	2.35	2.46	2.64	2.73
Pin base diameter (mm)	1.56	1.64	1.71	1.73	1.85	1.93	1.99
Mean undercut (mm)	0.17	0.23	0.22	0.32	0.31	0.36	0.37

Fig. 4: Micrographs of the caulking joints for different pin heights with a 1.0 mm joining partner

Investigating the micrographs in Fig. 5 and the corresponding geometric characteristics of the joints caulked with the aluminium (t_0 = 1.5 mm), it first is obvious that the pin heights used compared to the 1.0 mm sheet only start at a height of 1.75 mm. This is attributed to the fact that with a pin height lower or at the sheet thickness of the joining partner, no load bearing joint can be produced. Furthermore, it can be seen that for the pin height of 1.75 mm an undercut for the pin of 0.1 mm could be measured. However, as it can be seen from the micrograph the undercut was not sufficient enough to join the two sheets adequately. One reason could be, that the drill hole got slightly larger than expected. Additionally, it is possible that the upset pin volume was too small to achieve sufficient undercutting. An increase in pin height to 2.00 mm increases the mean undercut of the joint by 50 % to a value of 0.15 mm. Further increasing the pin height leads to a continuously greater value up to a value of 0.29 mm for the 3.00 mm pin, which amounts to a 93.3 % increase compared to the 2.00 mm pin. Interestingly, for the joint with a 2.0 mm pin a lower undercut was measured compared to the 1.75 mm pin. It can be seen that the pin base diameter for the 2.0 mm pin increases by approx. 4.3 % from 1.64 mm to 1.71 mm compared to the 1.75 mm pin. A similar increase of 5.1 % can be observed for the 1.50 mm pin to the 1.75 mm pin. However, the pin head diameter increases by 10 % compared to only 3 % from the 1.75 mm pin to the 2.0 mm pin. Therefore, it can be said that the pin base increases more than the pin head, which results in a smaller wedge angle and thus also a smaller undercut.

Pin height					
1.75 mm	2.00 mm	2.25 mm	2.50 mm	2.75 mm	3.00 mm

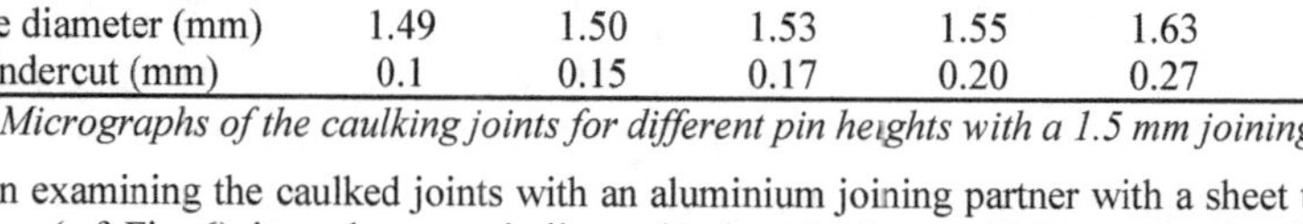

Geometric joint characteristics (t_0 = 1.5 mm)						
Pin height (mm)	1.75	2.00	2.25	2.50	2.75	3.00
Pin head diameter (mm)	1.68	1.77	1.86	1.91	2.16	2.29
Pin base diameter (mm)	1.49	1.50	1.53	1.55	1.63	1.70
Mean undercut (mm)	0.1	0.15	0.17	0.20	0.27	0.29

Fig. 5: Micrographs of the caulking joints for different pin heights with a 1.5 mm joining partner

When examining the caulked joints with an aluminium joining partner with a sheet thickness of 2.0 mm (c.f. Fig. 6), it can be seen, similar to the sheet thickness of 1.5 mm, that the pin heights used only start from 2.25 mm and thus 0.25 mm above the initial sheet thickness. Furthermore, it can be seen that the first joint also does not indicate a sufficient undercut. As the pin height increases, the pin head diameter and the measured undercut also increase, analogous to the other sheet thicknesses. However, the values achieved for the undercuts are lower than the undercuts for

Materials Research Proceedings **25** (2023) 109-116 https://doi.org/10.21741/9781644902417-14

the comparable pin height/sheet thickness combinations due to the higher sheet thickness and the resulting lower percentage upsetting of the pins.

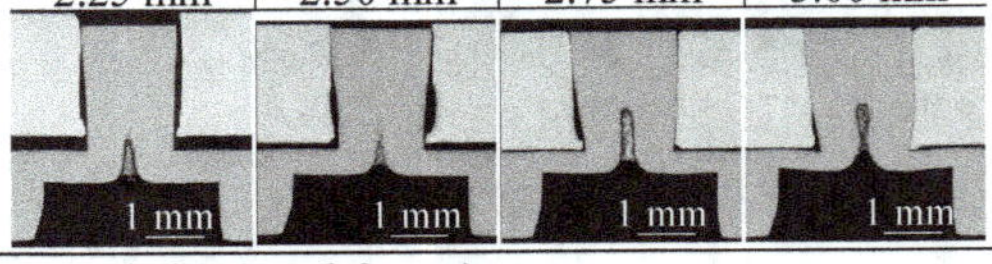

Geometric joint characteristics (t_0 = 2.0 mm)				
Pin height (mm)	2.25	2.50	2.75	3.00
Pin head diameter (mm)	1.62	1.75	1.84	1.86
Pin base diameter (mm)	1.49	1.49	1.51	1.52
Mean undercut (mm)	0.07	0.14	0.16	0.19

Fig. 6: Micrographs of the caulking joints for different pin heights with a 2.0 mm joining partner

For a comparison of the different pin height/sheet thickness configurations, the undercuts for the individual sheet thicknesses are plotted in Fig. 7. In addition, the joints are classified as "okay" and "not okay", which indicates whether the joint is load-bearing or whether there is an insufficient undercut or a joining instability such as a buckling of the pin. It can be seen that the joint with the 1.0 mm joining partner achieves a higher undercut with a comparable pin height. If we look at the respective joints with a pin height of 2.5 mm, which is present for all three sheet thicknesses, the 1.0 mm sheet thickness achieves an average undercut of 0.31 mm compared to an undercut of 0.2 mm (1.5 mm sheet) and 0.14 mm (2.0 mm sheet). This is not surprising, since the pins are significantly more compressed with a joining partner with a smaller sheet thickness than with a thicker sheet. Here the pin is compressed by 60 % for the 1.0 mm sheet compared to 40 % (1.5 mm sheet) and 20 % (2.0 mm sheet). Nevertheless, due to the fact that the compression increases with a higher pin, the measured undercuts generally tend to increase with the pin height regardless of the sheet thickness of the joining partner. It is also interesting to note that when comparing joints with different initial pin heights but the same percentage upsetting, similar undercut values can be measured. For example, an undercut of 0.17 mm was measured with a pin height of 1.5 mm (1.0 mm sheet metal), an undercut of 0.17 mm with a height of 2.25 mm (1.5 mm sheet metal) and an undercut of 0.19 mm with a pin height of 3.0 mm (2.0 mm sheet metal). Thereby, all three pin height/sheet thickness configurations showed a percentage upsetting of 33.3 %.

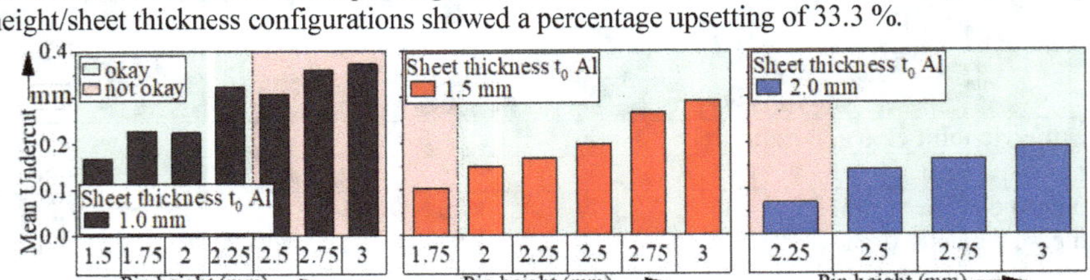

Fig. 7: Comparison of the mean undercut for the investigated joints

For this reason, the undercut was also normalised to the percentage upsetting of the pin and plotted as a dependent of the pin height in Fig. 8. It can be seen that, independent of the sheet thickness, similar dependencies of the undercut on the percentage upsetting of the pins with varying pin heights result and that these normalised values are at a similar level. Consequently, if the average value of the normalised undercuts is calculated, the average undercut per percent pin upsetting is 0.0057 ± 0.0007 mm/%, regardless of the pin height or the used sheet thickness of the joining partner. Therefore, it can be concluded that the pin height/sheet thickness ratio or the resulting ratio of pin upsetting can be used to influence the joint formation and to adjust the undercut.

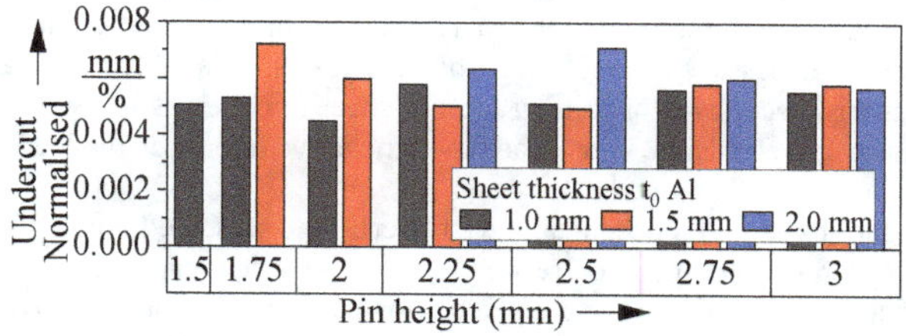

Fig. 8: Comparison of the undercut values for the different pin height/sheet thickness configurations normalised to the percentage upsetting of the respective pin

Based on the results presented, a process window can accordingly be derived that provides information about the joinability of the different sheet thicknesses investigated with different pin heights (c.f. Fig. 9). It can be seen that there are overlapping areas between the sheets in terms of joinability with the same pin heights, but that no pin height is able to join all three sheets. However, one advantage of pin joining is the posibility to react to varying boundary conditions by adjusting the pin height. This is also the case with pin caulking. Considering the areas where the joints turned out to be "not okay" in the investigations, it can be seen that in the case of caulking, the lower limit is initially determined by the sheet thickness of the joining partner, and an additional pin height of at least 0.25 mm is necessary to achieve a load-bearing joint. However, it can be assumed that with increasing pin height and correspondingly a larger undercut, a higher load-bearing capacity can be achieved. In addition, as explained above, with a joining partner sheet thickness of 1.0 mm, a pin compression of 60 % or more, which corresponds to a 2.5 mm pin, leads to joining instabilities that cause the pin to buckle laterally. This upsetting is not reached with the examined pin heights for the other sheet thicknesses, which is why no upper limit can be determined for the sheet thicknesses of 1.5 mm and 2.0 mm. However, it can be seen from the micrographs that with increasing pin height, the sheet thinning below the pin increases. This can lead to premature failure of the joints under load. For this reason, further investigations should be carried out to analyse whether and to what extent the sheet thinning has an effect on the joint strength.

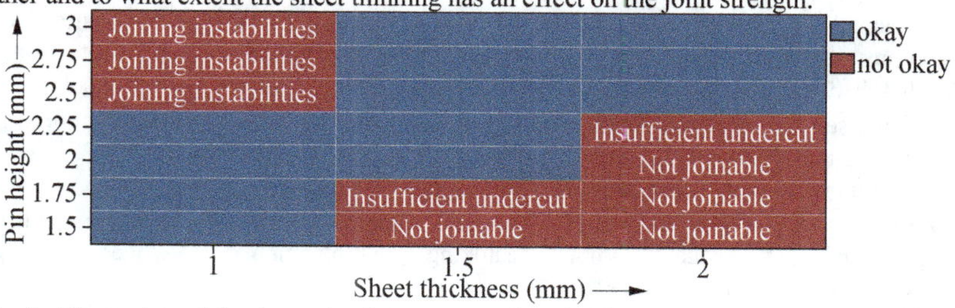

Fig. 9: Illustration of the determined process window for the three different sheet thicknesses

Summary and outlook

In this contribution, caulking, which has the potential for joining metal/metal and metal/FRP to reduce the component weight and therefore energy consumption, was fundamentally investigated as an alternative joining strategy to direct pin pressing. The influence of different pin heights/sheet thickness combinations on the joint formation and the resulting geometric joint properties was analysed. Thereby, an understanding of the joining process could be established. Additionally, it was shown that as the pin height increases, an improvement in the undercut occurs, which can be influenced by the percentage upsetting of the pin. Furthermore, with a joining partner sheet thickness of 1.0 mm, joining instabilities in the form of buckling pins occurred at a pin compression of 60 % or more, which led to joints that had to be classified as not okay. Moreover, with the 1.5 mm and 2.0 mm joining partner an additional minimum pin height of 0.25 mm above the sheet thickness is necessary to create load-bearing joints. In addition, the large pin heights, which are particularly necessary for joining greater sheet thicknesses, show that there is severe sheet thinning during pin extrusion, which can have a negative impact on the load-bearing capacity of the joint.

Future investigations should mechanically characterise the presented joints based on the results shown here, in order to determine corresponding load-bearing capacities with regard to the undercut values measured. Both the shear and axial load capacities should be investigated in order to analyse a possible influence of the sheet thinning.

Acknowledgment
This work was funded by the Deutsche Forschungsgemeinschaft (DFG, German ResearchFoundation)-TRR 285 C01 - Project-ID 418701707.

References

[1] Eurostat, Greenhouse gas emissions by source sector. https://ec.europa.eu/eurostat/databrowser/bookmark/1ffd5837-18ba-4e68-9634-1eaf2d6d4023?lang=de (accessed 2022.09.21).

[2] G. Meschut, M. Merklein, A. Brosius, D. Drummer, L. Fratini, U. Füssel, M. Gude, W. Homberg, P.A.F. Martins, M. Bobbert, M. Lechner, R. Kupfer, B. Gröger, D. Han, J. Kalich, F. Kappe, T. Kleffel, D. Köhler, C.-M. Kuball, J. Popp, D. Römisch, J. Troschitz, C. Wischer, S. Wituschek, M. Wolf, Review on mechanical joining by plastic deformation, Journal of Advanced Joining Processes, 177 (504-506), 2022 100113. https://doi.org/10.1016/j.jajp.2022.100113

[3] P.N. Parkes, R. Butler, J. Meyer, A. de Oliveira, Static strength of metal-composite joints with penetrative reinforcement, Composite Structures, 118, 2014 250–256. https://doi.org/10.1016/j.compstruct.2014.07.019

[4] E.E. Feistauer, J.F. Santos, S.T. Amancio-Filho, A review on direct assembly of through-the-thickness reinforced metal–polymer composite hybrid structures, Polym Eng Sci, 59 (4), 2019 661–674. https://doi.org/10.1002/pen.25022

[5] S. Ucsnik, M. Scheerer, S. Zaremba, D.H. Pahr, Experimental investigation of a novel hybrid metal–composite joining technology, Composites Part A: Applied Science and Manufacturing, 41 (3), 2010 369–374. https://doi.org/10.1016/j.compositesa.2009.11.003

[6] D. Römisch, M. Kraus, M. Merklein, Experimental Study on Joining by Forming of HCT590X + Z and EN-AW 6014 Sheets Using Cold Extruded Pin Structures, JMMP, 5 (1), 2021 25. https://doi.org/10.3390/jmmp5010025

[7] D. Römisch, J. Popp, D. Drummer, M. Merklein, Joining of CFRT-steel hybrid parts via hole-forming and subsequent pin caulking, Production Engineering, 2021. https://doi.org/10.1007/s11740-021-01093-9

[8] Salzgitter Flachstahl GmbH, Data sheet - HCT590X+Z (HCT600XD / HC340XD*/ CR330Y590T-DP**), 2017. https://www.salzgitter-flachstahl.de/fileadmin/footage/MEDIA/gesellschaften/szfg/informationsmaterial/produktinformationen/feuerverzinkte_produkte/deu/hct590x.pdf (accessed 2022.10.06).

[9] K. Hirota, Fabrication of micro-billet by sheet extrusion, Journal of Materials Processing Technology, 191 (1-3), 2007 283–287. https://doi.org/10.1016/j.jmatprotec.2007.03.024

[10] E. Ghassemali, M.J. Tan, A. Jarfors, S. Lim, Progressive microforming process: Towards the mass production of micro-parts using sheet metal, The International Journal of Advanced Manufacturing Technology, 66, 2012. https://doi.org/10.1007/s00170-012-4352-4

Sheet Metal 2023 Materials Research Forum LLC
Materials Research Proceedings **25** (2023) 117-124 https://doi.org/10.21741/9781644902417-15

Influence of geometrical parameters of conic pin structures in thermoplastic composite/steel hybrid joining

Julian Popp[1, a *], Christoph Zirngibl[2,b], Benjamin Schleich[3,c], Sandro Wartzack[2,d] and Dietmar Drummer [1,e]

[1]Chair of Polymer Technology, Friedrich-Alexander-Universität Erlangen-Nürnberg, Am Weichselgarten 10, 91058 Erlangen, Germany

[2]Engineering Design (KTmfk), Friedrich-Alexander-Universität Erlangen-Nürnberg, Martensstraße 9, 91058 Erlangen, Germany

[3]Product Life Cycle Management, Technical University of Darmstadt, Otto-Bernd Straße 2, 64287 Darmstadt, Germany

[a]julian.georg.popp@fau.de, [b]zirngibl@mfk.fau.de, [c]schleich@plcm.tu-darmstadt.de, [d]wartzack@mfk.fau.de, [e]dietmar.drummer@fau.de

Keywords: Composite, Joining, Hybrid Manufacturing

Abstract. Hybrid parts consisting of continuous fiber reinforced thermoplastic (CFRT) and steel components offer promising potential in lightweight construction. In this approach, the joining operation poses the major challenge due to different physical and chemical properties of both materials. This paper studies a joining method in which conical pin structures, protruding from the surface of the metal component, are inserted into the locally heated CFRT component to create a form fitting joint. The aim is to identify and quantify the influence of the parameters pin height and pin head diameter on the mechanical performance in single lab shear and cross head normal experiments. Based on the results it can be shown, that both an increased pin height and head diameter are beneficial under shear load in means of increased maximum load capacity while under normal load only an increased head diameter has clear benefits. Those effects are led back to the geometrical properties and the resulting stress distributions in the joining area.

Introduction

Lightweight construction is a widely applied approach in the transportation sector to achieve reductions in fuel consumption and consequent CO2 emissions. Thereby, a substitution of classic engineering materials, such as steel and aluminum alloys, with materials with superior weight related properties, such as fiber-reinforced polymers, is a common strategy to achieve substantial weight savings [1]. Thereby especially continuous fiber reinforced thermoplastics (CFRT) combine favorable weight related properties [2] with good processability and uncomplicated handling with unlimited shelf-live [3] and allow for functional integration through polymer welding and injection molding processes [4].

Despite these obvious advantages, CFRTs reach limitations under elevated temperatures in which the matrix begins to soften or melt and thus reduces the mechanical stability of the composite. Furthermore, under certain conditions like high abrasive wear [5] or under the influence of solvents and aggressive media CFRTs can be unsuitable. An approach to address these limitations is the use of intelligent hybrid structures where the lightweight CFRT component is combined with a metal component where the application exceeds the capabilities of the CFRT material. Thereby, the main challenge lies in a durable and cost-efficient joining technology [6]. Due to chemical incompatibilities, established joining methods like welding or brazing are not applicable [7] and while adhesive bonding is theoretically suitable and also leads to a fiber friendly force application into the composite, it oftentimes requires long curing cycles and specific surface

Content from this work may be used under the terms of the Creative Commons Attribution 3.0 license. Any further distribution of this work must maintain attribution to the author(s) and the title of the work, journal citation and DOI. Published under license by Materials Research Forum LLC.

pre-treatments [8]. Especially unpolar polymers with low surface energy, like polypropylene (PP) require elaborate surface modifications via e.g. plasma treatment to be adhesively joinable [9].

In the current industrial state of the art, typically mechanical joining via bolts or rivets is applied. Consequently, these joining methods are very well described in the literature and can be used to create durable joints [10]. However, these joining methods also come with disadvantages: first, an auxiliary element is needed which increases weight and costs of the structure. Second, a hole forming operation is required, which typically increases the process time and when realized via subtractive manufacturing, leads do destruction of load bearing fibers resulting in locally increased concentration of forces and weakens the composite. Consequently, a demand for a cost efficient, versatile and durable technology for joining CFRTs with metals is required.

A relatively new joining approach is the use of pin structures, that can be embedded into the locally heated CFRT composite as it is shown in [11] with cold formed pin structures while the fiber rearrange around the pin structures and thus avoids fiber breakage [12]. Other manufacturing technologies for pin structures are powder bed fusion [13], metal injection moulding [14], [15], and cold metal transfer welding (CMT) [16], [17]. Fig. 1 shows an example of a pin joint.

Fig. 1 Example of pin joint with pins embedded in composite component

The influence of the pin geometry on the joint characteristics has been investigated in several studies: In [12] the influence of different pin diameters and pin tip shapes on the fiber displacement and in [18] on the mechanical performance under shear load has been investigated, while the pins, due to a lack of an undercut failed through pin extraction. In [19] it could be shown that a joining with undercutting, conical pins is possible leading to good joint strengths under shear and especially normal load when compared to cylindrical reference pins. Thereby, the pins have been manufactured via a subtractive manufacturing process. That a manufacturing of undercutting, conical pin shapes is also possible via a cold forming processes could be shown in [20] which makes the conic pin shape a promising pin shape that combines good mechanical properties with easy manufacturing via cold forming.

Despite the above-described findings, a systematic and statistical investigation of different parameters of a conic pin shape and its influence on the joint characteristics has not yet been presented. Consequently, this study aims at creating an understanding of the influence of the parameters pin height and pin head diameter of conic shaped steep pins on the mechanical performance under shear and normal load of a CFRT-metal hybrid joint.

Material and Methods

Materials. The used CFRT material is a custom fabricated biaxial composite with an E-glass fiber reinforcement and a PP matrix. The matrix type is a BJ100HP from Borealis AG (Vienna, Austria) which is a special low-viscosity compound specifically for glass fiber reinforced applications. As glass fiber component a biaxial non crimp fabric from Saertex GmbH (Saerbeck, Germany) with an areal weight of 307 g/m^2 in 0° and 90° respectively and an aramide sewing thread with an aerial weight of 9 g/m^2 resulting in a total areal weight of 623 g/m^2. Thereby, the titer of both the 0° and 90° rovings is relatively low with 300 tex. The resulting thickness of the composite is 2 mm +/- 0.05 mm with a calculated fiber volume content of approximately 45 %. As a sheet metal a HCT590X steel type has been used. In order to be able to investigate a large variation of geometries the pins have been manufactured on a CNC lathe from a 42CrMo4 steel round. Subsequent of the turning process the pin structures have been joined with the steel component via welding. In future serial manufacturing, it is planned to manufacture the pins via cold forming as it is described in [20].

Definition of pin and sample geometry. In the scope of this study, two geometry parameters of the conical pin have been varied. First, the pin height *h* is varied between 1.7, 1.8 and 1.9 mm. Second the pin head diameter *D* is varied between 1.4, 1.5 and 1.6 mm. These parameters have been fully factorially combined for a total of nine combinations. Furthermore, pins with head diameter of 1.7 and 1.8 mm and a height of 1.8 have been manufactured, to determine, if a limitation of the head diameter is given above which adverse effects on the fiber reorientation occur that negatively impact the mechanical performance of the joint. The pin foot diameter *d* is constant at 1.25 mm as in previous studies, it's effect on the shear strength of the sample has already been investigated [21]. In total, eleven different pin geometries with a sample size of three have been investigated (compare Fig. 2 A and Table 1).

Table 1 Summary of investigated geometries

Name	*D* [mm]	*h* [mm]	*d* [mm]
D14_h17_d125	1.4	1.7	1.25
D14_h18_d125	1.4	1.8	1.25
D14_h19_d125	1.4	1.9	1.25
D15_h17_d125	1.5	1.7	1.25
D15_h18_d125	1.5	1.8	1.25
D15_h19_d125	1.5	1.9	1.25
D16_h17_d125	1.6	1.7	1.25
D16_h18_d125	1.6	1.8	1.25
D16_h19_d125	1.6	1.9	1.25
D17_h18_d125	1.7	1.8	1.25
D18_h18_d125	1.8	1.8	1.25

As samples single lap shear and cross head samples have been manufactured in order to investigate load capacity under both shear and normal loads The samples are according to the technical bulletin DVS EFB 3480-1 and the sample dimension can be seen in Fig. 2 section B&C. The lap shear specimens have a length of 105 mm and the joining point is located 8 mm from the short edge of the sample, resulting in an overlap of 16 mm (compare Fig. 2 B). The width of the samples is 45 mm and the resulting length of the joined sample is 194 mm. The cross head specimens have a length of 150 mm and a width of 50 mm with clamping holes with a diameter of 20 mm placed symmetrically with a spacing of 100 mm (compare Fig. 2 C).

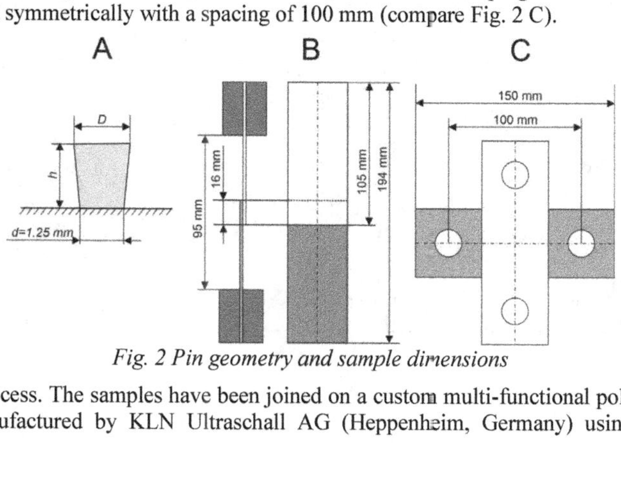

Fig. 2 Pin geometry and sample dimensions

Joining process. The samples have been joined on a custom multi-functional polymer welding machine manufactured by KLN Ultraschall AG (Heppenheim, Germany) using short wave infrared radiation emitted from an infrared spot from Optron GmbH (Garbsen, Germany). In order to define the irradiated area, a mask with a cylindrical opening with a diameter of 8 mm has been used. The heating parameters have been identified in pre trials to a power setting of 67.5 Watts and a heating time of 210 s. Thereby, special attention has been paid on avoiding overheating of the sample surface while ensuring a complete melting of the matrix through the thickness of the sample. After preheating the samples were joined with the pin being pressed into the CFRT concentrically with the radiated zone. The joining force is 3,000 N after which a consolidation and cooling phase of 30 s follows with applied pressure. Finally, the sample Is removed from the tool.

Mechnical characterization. The samples are tested on a universal testing machine type Z 1465 from Zwick&Roell AG (Ulm Germany) according to DVS EFB 3480-1 with a testing speed of 1 mm/min. Prior to testing, the samples are stored in normalized climate according to DIN EN ISO 139 of 23 °C and humidity 50 rel. %. For the lap shear test, the extension was measured via tactile extensometer with an initial measurement length of 50 mm. For the cross head test, the extension was measured via the machine traverse.

Results

When investigating the results from the single lap shear tests, it can be seen, that the maximum load capacity ranges between 325 ± 28.2 N and 448.7 ± 36.8 while the load capacity under normal load is lies significantly lower between 36.6 ± 3.9 N and 80.6 ± 8.0 N (compare Fig. 2 and Table 2). It is noticeable, that the standard deviation, especially for normal load results is comparably high and frequently exceeds 35 % of the base value. The highest individual measurements are 489 N (shear load) and 121 N (normal load) and can be interpreted as maximum achievable load capacities with single pin joints. Both measurements used a pin with a head diameter of 1.6 mm while the height is 1.8 mm (shear load) and 1.7 mm (normal load). Table 2 and Fig. 3 give an overview over the results. The failure mode for single lap shear specimens generally initially is a bearing failure of the CFRT component which is accompanied with a slight bending of the pin. One exception is a pin with D=1.7 mm and h=1.8 mm failed due to pin breakage at a force of 412N. A pin extraction like with non-undercutting structures [18] could not be observed. Under normal load, the samples failed due to elastic and plastic deformation of the CFRT component and subsequent pin extraction without pin deformation. It must be noted that of pins with a head diameter of 1.5 mm and height of 1.9 mm, only two lap shear and cross head samples could be tested because the pin structures were damage during the pin welding process.

*Table 2 Summary of single lap shear and normal tests; * indicates sample size of only 2*

Name	F max, shear [N]	F max, normal [N]
D14_h17_d125	346.7 ± 40.8	45.9 ± 12.4
D14_h18_d125	325.0 ± 28.2	36.6 ± 3.9
D14_h19_d125	394.3 ± 63.3	41.8 ± 8.2
D15_h17_d125	339.7 ± 58.7	61.2 ± 8.2
D15_h18_d125	416.7 ± 22.4	47.6 ± 18.5
D15_h19_d125	416.5 ± 38.5*	49.9 ± 7.9*
D16_h17_d125	350.7 ± 55.7	73.3 ± 33.9
D16_h18_d125	448.7 ± 36.8	72.0 ± 29.2
D16_h19_d125	422.0 ± 32.1	63.5 ± 22.6
D17_h18_d125	404.3 ± 61.2	53.9 ± 16.6
D18_h18_d125	410.3 ± 31.8	80.6 ± 8.0

Materials Research Proceedings **25** (2023) 117-124 https://doi.org/10.21741/9781644902417-15

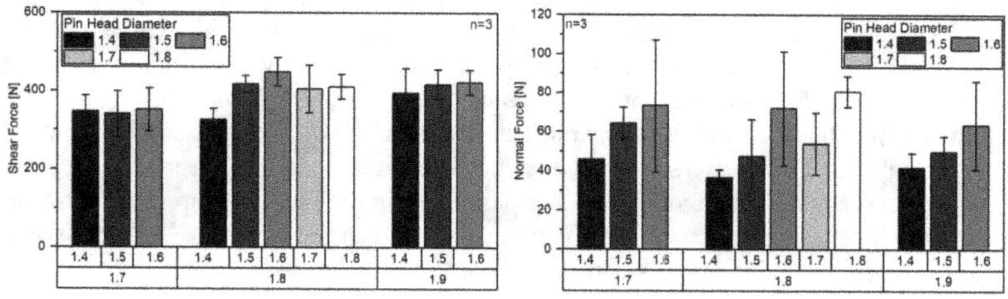

Fig. 3 Maximum shear (left) and normal (right) loads for different pin height and head diameter

Two different approaches can be used to explain the high standard deviation under normal load: first, the CFRT component on a macro scale is inhomogeneous with compact fiber rovings and matrix pockets (compare Fig. 4). Depending on the relative position, in which the pin is pressed into the composite, different joint morphologies can occur which are expected to influence the mechanical properties of the joint. Second, the manufacturing of the pin structures is complex and the repeatability is limited. A number of pins show residues of the welding process which partially fill the undercut of the pin structures presumably negatively influencing the joint properties. Fig. 3, right shows an example with very large welding residues. Typically, when residues could be observed, the phenomenon was less pronounced than it is shown in this image. In future Studies, it is planned to replace the combined turning and welding process with a two-step pin forming and caulking process, similar as it is shown in [22], which is expected to solve the repeatability issues.

Fig. 4 Microsection of undisturbed CFRT laminate (left) and photography of welded pin structure with large welding residues (right)

When comparing the results with a study, in which cold extruded pins are used [20], the standard deviation with cold formed pins is significantly lower which in part is led back to the high repeatability of the cold forming process. In [21] undercutting pin geometries where joined with unidirectionally reinforced CFRT components. When comparing the results of pins with a head diameter of 1.25 mm and a hight of 1.8 mm, the results in [21] are lower with 347 ± 57.4 N in comparison to 416.7 ± 22.4 N under shear and load and higher with 85.2 ± 5.6 N in comparison to 47.6 ± 18.5 N und normal load. In [21] the shear samples failed due brittle breakage of the pin which explains the comparable high standard deviation and the reduced load capacity while the normal load samples have a comparably low standard deviation. Since in this study, no welding residues could be detected, this is seen as the reason for higher repeatability. The increased load capacity in [21] however, can be caused by the different CFRT material which potentially is more suitable under normal loads.

Statistical analysis using Pearson's correlation coefficients. For the statistical analysis of the results, the calculation of Pearson's correlation coefficients (r) provide the opportunity to investigate the linear relationships between input and target variables and thus gain a deeper understanding of the joining process. Furthermore, the results offer an evaluation of which parameter variations indicate to have a significant effect on individual quality-relevant joint properties and which are less important. In this context, an r-value of -1 implies a strongly negative correlation at which the increase of one parameter results in the decrease of another parameter. In contrast, an r-value of +1 means a strong positive correlation and can be interpreted that the change in one parameter leads to a change of the other parameter in the identical direction. An r-value of 0 indicates that there is no relationship between the input and target variables. [30] As an overview, Fig. 5 shows the calculated Pearson's correlation coefficients.

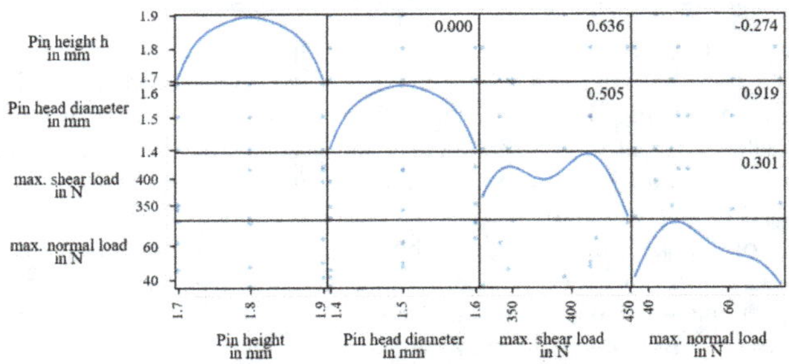

Fig. 5 Overview of the calculated Pearson's correlation coefficients

Focusing on the maximum shear load capacity, one can see that both input parameters show a positive correlation ($r = 0.505$ and $r = 0.636$) where the pin height tends to have a slightly stronger impact on the target variable. In this context, the increase of the pin height and to a lesser degree also of the pin head diameter result in an increase of the projection area of the pin structure. The increased projection area consequently leads to a decrease of the contact pressure between pin and CFRT laminate at a given load leading to a higher load capacity of the joint. In contrast, the Pearson's correlation coefficients of the maximum normal load capacity indicate a very strong positive linear relationship ($r = +0.919$) between the pin head diameter and the output variable. This bases on the increased necessary deformation in the CFRT component during the pin extraction with a larger pin head diameter in comparison to a smaller head diameter. This increased deformation leads to larger contact pressure between CFRT and pin flanks which due to the incline of the pin flank results in a counter force which is consequently measured as the load capacity. Compared to this effect, changes in the pin height result only in a weak modification of the resulting maximum normal load capacity showing a negative correlation value of -0.274. In summary, if the increase of both target variables is targeted, the selection of generally higher values of the pin head diameter is recommended.

Summary

In the present study, it could be shown that conical pins are suitable to create joints with biaxial CFRT sheets that can be loaded under both shear and normal load. Maximum load capacities for single pins were 489 N under shear and 121 N under normal load while maximum average values were 448.7 N (shear) and 80.6 N (normal) for series of 3 samples.

The high standard deviations in the measurments are explained with the morphology of the CFRT sample as well as the manufacturing process of the pin sample. It is expected, that with cold formed pins, the joining process will be more repeatable. The calculation of Pearson's correlation coefficients indicated a high impact of the selected pin head diameter on both the maximum achievable shear and normal load. An increase of this parameter leads to a larger contact zone with reduced contact pressure and thus to an increased load capacity. In contrast to this, the correlation indices of the pin height showed a positive effect on the shear load while simultaneously leading to slightly decreased maximum normal loads. Based on this, it is recommended to choose rather higher pin head diameters instead of a changed pin height. However, the parameter selections have to be in accordance with the permissible pin manufacturing and joining process limits.

In future studies, it is planned to investigate cold formed, undercutting pin structures where an increase in the load capacity under shear load is expected due to work hardening effects as well as an decrease of the standard deviation, due to the more repeatable forming process, is expected especially under normal loads. Furthermore, multi-pin arrays have to be investigated, both to increase the overall joint strength as well as to average effects of the inhomogeneous CFRT component which presumably leads to further increased repeatability. Finally, the findings of this study should also be verified for other material systems such as quasi-isotropic and unidirectional reinforcements as well as other reinforcement fibers, such as carbon or aramid fibers and for other mechanical tests, such as peel tests, to increase the overall understanding on this particular joining technique.

Fundings

This work was funded by the Deutsche Forschungsgemeinschaft (DFG, German ResearchFoundation)-TRR 285 C01-Project-ID 418701707.

References

[1] W. Siebenpfeiffer, Leichtbau-Technologien im Automobilbau, ATZ/MTZ-Fachbuch, SpringerVieweg, 2014. https://doi.org/10.1007/978-3-658-04025-3

[2] B.D. Agarwal, L.J. Broutman, K. Chandrashekhara, Analysis and Performance of Fiber Composites. 4th Edition, John Wiley & Sons, Inc., Hoboken, 2017.

[3] M. Sommer, K. Edelmann, A. Wöginger, M. Christmann, J. Mack, L. Medina, Thermoplastische Prepregs und Halbzeuge, in: M. Neitzel, P. Mitschang, U. Breuer, (Eds.), Handbuch Verbundwerkstoffe, Carl Hanser Verlag, Munich, 2014, pp. 147–199.

[4] A. Yousefpour, M. Hojjati, J.-P. Immarigeon, Fusion Bonding/Welding of Thermoplastic Composites. Journal of Thermoplastic Composite Materials 17 (2004) 303–341. https://doi.org/10.1177/0892705704045187

[5] N.-M. Barkoula, J. Karger-Kocsis, Effects of fibre content and relative fibre-orientation on the solid particle erosion of GF/PP composites, Wear 252 (2002) 80-87. https://doi.org/10.1016/S0043-1648(01)00855-9

[6] K. Martinsen, S.J. Hu, B.E. Carlson, Joining of dissimilar materials, CIRP annals – Manufacturing Technology 64 (2015) 696-699. https://doi.org/10.1016/j.cirp.2015.05.006

[7] Z. Dawai, Z. Qi, F. Xiaoguang, Z. Shengdung, Review on Joining Process of Carbon Fiber Reinforced Polymer and Metal: Methods and Joining Process, Rare Metal Materials and Engineering 47/12 (2018) 3686-3696. https://doi.org/10.1016/S1875-5372(19)30018-9

[8] H. Schürmann, Klebeverbindungen, in: H. Schürmann, Konstruieren mit Faser-Kunststoff-Verbunden, Springer Verlang, Berlin-Heidelberg, 2005, pp. 569-604. https://doi.org/10.1007/b137636

[9] W. Brockmann, Adhesive Bonding of Polypropylene, in: J. Karger-Kocsis, Polypropylene: An A-Z Reference, Kluwer, London, 1999, pp. 1-6. https://doi.org/10.1007/978-94-011-4421-6_1

[10] P. Mitschang, R. Velthuis, R. Rudolf, Fügeverfahren für FKV, in: Neitzel M., Mitschang P., Handbuch Verbundwerkstoffe, 2nd Edition, Carl Hanser, Munich-Vienna, 2014. pp. 469-482. https://doi.org/10.3139/9783446436978.015

[11] M. Kraus, P. Frey, T. Kleffe, D Drummer, M. Merklein, Mechnical joining without auxiliary element by cold formed pins for multi-material-systems, AIP Converence Proceedings 2113, 2019. https://doi.org/10.1063/1.5112570

[12] J. Popp, T. Kleffel, D. Römisch, T. Papke, M. Merklein, D. Drummer, Fiber orientation Mechanism of Continuous Fiber Reinforced Thermoplastic Hybrid Parts Joined with Metallic Pins, Applied Composite Materials 28 (2021) 951-972. https://doi.org/10.1007/s10443-021-09892-0

[13] R. Plettke, A. Schaub, C. Gröschel, C. Scheitler, M. Vetter, O. Hentschel, F. Ranft, M. Merklein, M. Schmidt, D. Drummer, A new process chain for joining sheet metal to fiber composite sheets, Key engineering material 611-612 (2014) 1468-1475. https://doi.org/10.4028/www.scientific.net/KEM.611-612.1468

[14] E.E. Feistauer, S.T. Amancio-Filho, Ultrasonic joining of Lightweight Alloy/Fiber-Reinforced Polymer Hybrid Structures, Joining of Polymer-Metal Hybrid Structures: Principles and Applications, (2018) 307-333. https://doi.org/10.1002/9781119429807.ch11

[15] E.E. Feistauer, E. Ebel, J.M. Santos, S.T. Amancio-Filho, Ultrasonic joining of through-the-thickness reinforced Ti-4 Al-6 V and polyetherimide hybrid joints, Society of Plastics Engineers: SPE ANTEC Anaheim (2017) 1718-1724.

[16] S. Ucsnik, M. Scheerer, S. Zaremba, D.H. Pahr, Experimental investigation of a novel hybrid metal-composite joining technology, Composites Part A 41 (2010) 369-374. https://doi.org/10.1016/j.compositesa.2009.11.003

[17] R. Thakkar, S. Ucsnik, Cost Efficient Metal to Fibre Composite Joining, 16th European Conference on Composite Materials (2014).

[18] J. Popp, T. Kleffel, D. Drummer, Influence of pin geometry on the joint strength of CFRT-metal hybrid parts with metallic pins, Joining plastics/Fügen von Kunststoffen 3 (2021) 177-183.

[19] J. Popp, D. Römisch, M. Merklein, D. Drummer, Joining of CFRT/Steel Hybrid Parts via Direct Pin Pressing of Cold Formed Non-Rotational Symmetric Pin Structures, Applied Sciences (2022) 12. https://doi.org/10.3390/app12104962

[20] D. Römisch, J. Popp, D. Drummer, M. Merklein, Joining of CFRT-steel hybrid parts via hole-forming and subsequent pin caulking, Production Engineering 16 (2022) 339-352. https://doi.org/10.1007/s11740-021-01093-9

[21] J. Popp, D. Drummer, Joining of continous fiber reinforced thermoplastic/steel hybrid parts via undercutting pin structures and infrared heating, Journal of advanced joining processes, 5, (2022). https://doi.org/10.1016/j.jajp.2021.100084

[22] R.J.S. Baptista, J.P.M. Pragana, I.M.F. Bragança, C.M.A. Silva, L.M. Alves, P.A.F. Martins: Joining aluminium profiles to composite sheets by additive manufacturing and forming, Journal of Materials Processing Technology 279 (2020). https://doi.org/10.1016/j.jmatprotec.2019.116587

Sheet Metal 2023 Materials Research Forum LLC
Materials Research Proceedings **25** (2023) 125-132 https://doi.org/10.21741/9781644902417-16

Combination of versatile self-piercing riveting processes

Simon Wituschek[1, a *], Fabian Kappe[2, b], Gerson Meschut[2, c] and Michael Lechner[1, d]

[1]Institute of Manufacturing Technology, Friedrich-Alexander-Universität Erlangen-Nürnberg, Egerlandstr. 13, 91058 Erlangen, Germany

[2]Laboratory for material and joining technology, University Paderborn, Pohlweg 47-49, 33098 Paderborn, Germany

[a]simon.wituschek@fau.de, [b]fabian.kappe@lwf.upb.de, [c]meschut@lwf.uni-paderborn.de, [d]michael.lechner@fau.de

Keywords: Joining, Material, Manufacturing

Abstract. Due to economic and ecological framework conditions, a resource-saving utilization of raw materials and energy is becoming increasingly important in particular in the mobility sector. For the reduction of moving masses and the resources consumed, lightweight construction technologies are part of modern production processes in vehicle manufacturing, for example in the form of multi-material systems. Challenging in the manufacture of multi-material systems especially in view of changing supply chains is the variety of materials and geometries that bring conventional joining processes to their limits. Therefore, new processes are required, which can react versatile to process and disturbance variables. A widely used industrial joining process is semi-tubular self-piercing riveting, which is however a rigid process. To increase the versatility, the two newly established processes multi-range self-piercing riveting and tumbling self-piercing riveting are combined and the capabilities for targeted material flow control are united. Therefore, an innovative two-stage process based on the combination is introduced in this paper. The rivet is set with the multi-range self-piercing riveting process with an overlap of the rivet head and then formed by a tumbling process. Further, a specific adaptation of the tumbling strategy is used to investigate the possibility of reducing cracks in the rivet head. Thereby, different tumbling strategies are used and similar geometric joint formations are achieved to compare the results.

Introduction

The ecological, economic and social developments of the last years require a major focus on resource-saving technologies and the increase of efficiencies in many areas [1]. In particular, the mobility sector is undergoing a major transformation, as this sector is responsible for approximately 26% of CO_2 emissions in the European Union, measured in 2019 [2]. One way to reduce energy consumption over the entire product life cycle, regarding both the production and the use of vehicles, is the deployment of lightweight construction technologies. This often involves the use of multi-material systems, which are characterized by the fact that they comprise components that have different mechanical and geometric properties [3]. The mechanical properties vary due to different materials used, and the geometric properties are dimensioned according to local requirements and consequently differ. This is reinforced by the vulnerability of global supply chains, which is also noticeable in lightweight materials due to the crisis. However, the manufacture of these systems requires the ability to join a variety of different components with varying materials and geometries and places all the greater demands on the flexibility of joining processes in more and more versatile process chains. [4]. These requirements are a major challenge for conventional joining processes currently in use, such as semi-tubular self-piercing riveting [5]. Due to its process design, this method is a rigid joining process that can only react to process and disturbance variables to a very limited extent [6]. In order to adapt the joining process to changed

Content from this work may be used under the terms of the Creative Commons Attribution 3.0 license. Any further distribution of this work must maintain attribution to the author(s) and the title of the work, journal citation and DOI. Published under license by Materials Research Forum LLC.

boundary conditions, a change of the tools or the auxiliary joining part used is necessary [7]. However, these changes also result in a decrease of the process efficiency. To avoid a change of tools, versatile processes that can react flexibly to changing conditions are required. In order to address the challenges of future production systems with a large number of variants and to increase the adaptability of the joining processes, the two processes multi-range semi-tubular self-piercing riveting [8] and tumbling self-piercing riveting [9] were investigated. The schematic process layouts of both processes are shown in Fig. 1.

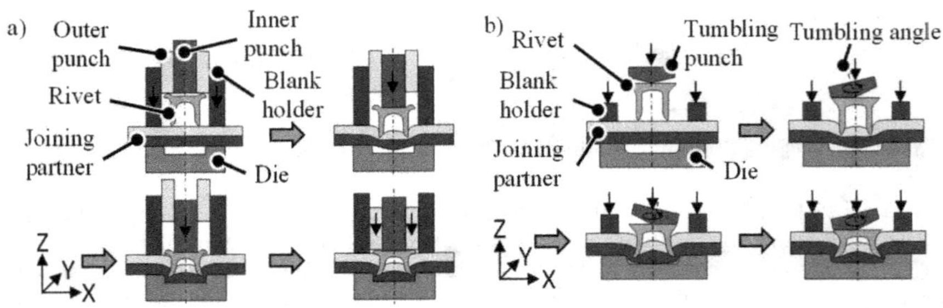

Fig. 1 Multi-range self-piercing riveting a) and tumbling self-piercing riveting b)

Conventional semi-tubular self-piercing riveting is a mechanical joining process for which double-sided access to the workpiece is required. The process design allows joining of similar and dissimilar materials such as steel, aluminum and fiber composites [10]. In addition, no pre-hole operation is necessary, the process can realize fast cycle times and achieves high static and fatigue joint strengths [11].

Using the process with multi-range tubular self-piercing rivet, shown in Fig. 1 a), the setting of the rivet is followed by a forming process, which forms the rivet head after setting onto the punch-sided sheet. Therefore, an inner and an outer punch is required. Due to the subsequent forming of the rivet head, the rivet can be adapted to the respective thickness of the joint. Just like the conventional self-piercing riveting process, a form-fit and force-fit connection is created, which allows high joint load-bearing capacities, which are comparable to conventionally produced joints, to be achieved for pure aluminum [4] and multi-material joints [12]. The studies mentioned, were able to show, that the induced linear material flow was able to join sheet thickness variations of more than 1 mm. However, it was found that, especially with low sheet thicknesses, air inclusions occurred between the rivet head and the punch-side joining partner. These might have a negative effect on following process steps such as a cathodic dip painting. Closing this gap was not possible due to cracking in the rivet head area that occurs at high degrees of deformation.

The tumbling semi-tubular self-piercing riveting process, shown in Fig. 1 b), is a combination of a tumbling process and conventional self-piercing riveting [9]. Tumbling is an incremental forming process that has its origins in cold forging. Characteristic of the process is the tumbling movement of the punch, which is described by an angular offset of the punch axis relative to the rest of the tool axis [13]. Together with a conical shape of the punch, the angular offset, which describes the tumbling angle, results in a reduction of the contact area between the punch and the workpiece. Thus, a higher surface pressure is achievable and large forming operations can be realized with a comparably low force level [14]. Furthermore, due to the reduced contact area, a kinematic model can be applied, which specifically controls the movement of the contact area on the workpiece in order to achieve increased material flow control and thus a process-side influence on the workpiece. The targeted movement enables both an influence on the geometry and on the mechanical properties by strain hardening. By combining both versatile joining approaches, more opportunities to influence the process are expected and a specific shaping of the rivet head can be achieved.

Materials and Methods

In the research, an aluminum EN AW-6014 is used, which is an alloy of aluminum, magnesium and silicon. The material is applied in many components of the body-in-white and critical outer skin applications in the automotive industry. Therefore, the material is a component of multi-material systems in many applications and thus adequately represents them. The material is characterized by good formability and hemming behavior, as well as good paint bake response. For the punch-side joining partner, a sheet thickness of $t_0 = 1.0$ mm and for the die-side joining partner, a sheet thickness of $t_0 = 2.0$ mm was selected.

The rivet element used is an adapted rivet geometry, which is modified for the multi-range joining process and has a material protrusion at the rivet head. This material protrusion is embossed by the outer punch in the conventional process and, thanks to the additional material, enables better joining suitability for greater total sheet packages. The selected die has a standard geometry from the industrial environment from type FM 095 2116 with a diameter of 9.5 mm and a die depth of 1.6 mm.

As a joining strategy based on the process combination, a two-stage process is selected. The first three process phases of conventional self-piercing riveting, consisting of clamping the joining partners, piercing the punch-side joining partner and flaring the rivet, are carried out with the multi-range tool, as shown in Fig. 2. The subsequent tumbling process is applied to completely form the joint and to shape the rivet head by increasing the radial material flow to the punch-side joining partner.

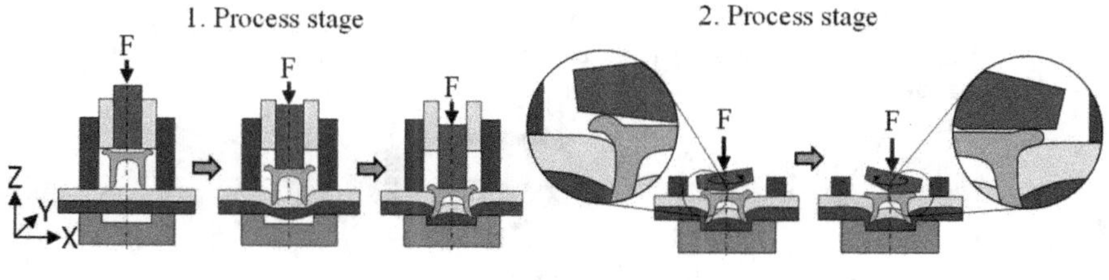

Fig. 2 Process combination of multi-range and tumbling self-piercing riveting

In order to investigate the joining process, which combines a linear and a tumbling kinematic, the specimens are joined successively using two joining systems. To set the joint, a joining system, which was presented in [15], with extended punch-sided tool actuator technology was used (see Fig. 3). It is based on a column guided frame. The lower tool contains the die, which provides the cavity needed to create the interlock. The upper tool includes an inner and an outer punch for the setting and forming process as well as a blankholder. Servoelectric drives are used to control the two punch movements with individual motions and independently parameterized force-displacement profiles. For the first process stage in this investigation, only the inner punch and the blank holder are used.

For the second process stage, a tool is used, which was presented in [9] and can be seen in Fig. 3 b). In this tool, which is mounted in a universal testing machine, the self-piercing riveting process can be superimposed by a tumbling kinematic of the punch. This is achieved by combining a rotational and a linear axis. The linear axis is installed on a rotating table, and thus the rotation and linear motion can be used to approach any point on a circle by any path. This makes the tumbling kinematics completely freely configurable and different tumbling strategies can be investigated.

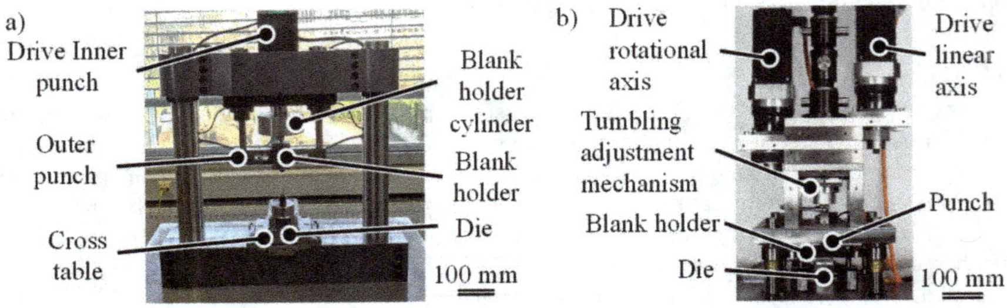

Fig. 3 Tool setup of multi-range self-piercing riveting a) and tumbling self-piercing riveting b)

The tumbling strategy consists of the parameters tumbling angle, tumbling kinematics, tumbling velocity and tumbling onset. The tumbling angle is defined by the inclination of the punch axis to the tool axis and a larger tumbling angle results in a reduced contact area. In Fig. 4 shown on the left, the tumbling angle is set by one rotation from the neutral position after the first contact between the punch and the rivet. Another possibility is to set the tumbling angle before the initial contact and thus reduce the contact pressure of the punch in the rivet head center, shown in Fig. 4 in the middle. The tumbling kinematic shows the movement of the contact surface on the rivet head and can be individually adapted. In a circular kinematic, the tumbling angle is kept constant after reaching the maximum, whereas in a spiral kinematic the tumbling angle increases over the entire joining process and reaches its maximum at the end of the process, depicted in Fig. 4. The tumbling velocity is not varied in the tests, since it only results in the number of rotations during the joining process. The tumbling onset at which the angle is set also plays a subordinate role because the onset should be after the first two process phases consisting of clamping and cutting.

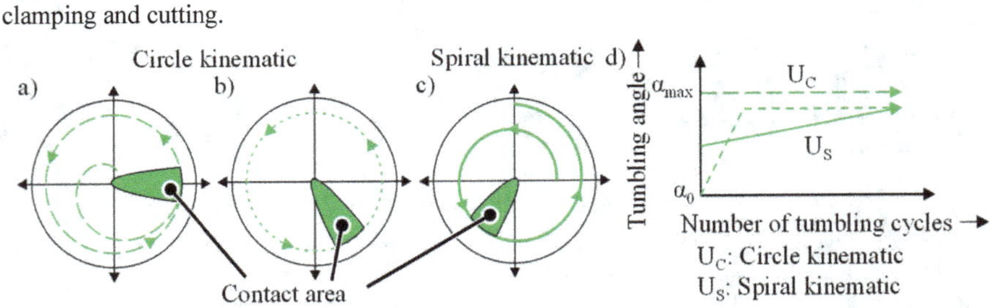

Fig. 4 Tumbling kinematic of the tumbling strategy used (a: Circle kinematic set by one rotation, b: circle kinematic set before joining, c: spiral kinematic, d: Representation of the tumbling angle over time for the kinematics used)

In order to investigate the influence of the tumbling strategy on the formation of cracks in the rivet head, macrographs of joints that are realized using different strategies are analyzed. Moreover, the achieved characteristic joint parameters, including the undercut, the rivet head end position and the residual sheet thickness are assessed to ensure the comparability of the evaluation of the crack formation depending on the tumbling strategy.

Results

In a first process stage, the rivet elements are set into the joint with the inner punch of the multi-range tool. The stroke of the punch is selected to ensure that the rivet head has a protrusion, as in previous investigations with the tool. In experiments with the multi-range tool, the outer punch would then form the protrusion of the rivet head on the sheet metal on the punch side, but this causes the challenges already described. Here, however, the effective mechanism of the tumbling punch is used to form the protrusion to the joining partner through the use of targeted material flow control and to completely form the joint. In this investigation, different configurations of the tumbling strategy are applied and different levels of sophistication of the adaptation of the tumbling strategy are presented to demonstrate the possibilities with the versatile joining process. First, a conventional tumbling strategy with a circle kinematic and a tumbling angle of 6° is applied and then the joint is evaluated using a macrograph. Fig. 5 shows the section through the center plane of the riveted joint. Significant cracks are in the center of the rivet head and at the protrusions. The crack in the rivet head is located centrally and is directed toward the tubular volume. In addition, a cavity can be identified on the upper side of the rivet head, as it is characteristic of tumbled joints. This deformation is caused by the conical punch, that due to its geometry, has a very small contact area between the punch and the rivet as the process begins. The contact is reduced only by adjusting the tumbling angle afterwards. The crack in the center is caused by the punch geometry, as this causes significant bending stresses on the rivet head.

At the start of the process, the rivet head first comes into contact with the punch tip due to its conical shape and applies a force. When the punch comes into contact, the adjustment mechanism begins to approach the tumbling angle, however, this requires a full revolution during which the punch is moved in the z-direction. From the tumbling velocity $\omega = 240$ °/s and the traverse speed, which describes the displacement of the punch in z-direction with $v = 10$ mm/min, a punch travel in z-direction of $h = 0.25$ mm results until the tumbling angle is completely adjusted. Thus, at the beginning of the process, the total force of the punch passes over the rivet head center into the joint. Furthermore, cracks can be identified on both sides at the rivet head protrusion. This damage to the rivet element at the rivet head is also due to the tumbling kinematics and the tumbling angle. The selected tumbling angle $\alpha = 6°$ causes a significant reduction of the contact area and a high surface pressure at the rivet head protrusion. Furthermore, a combination of radial and tangential material flow due to the tumbling motion and the large tumbling angle is causal for significant material deformation.

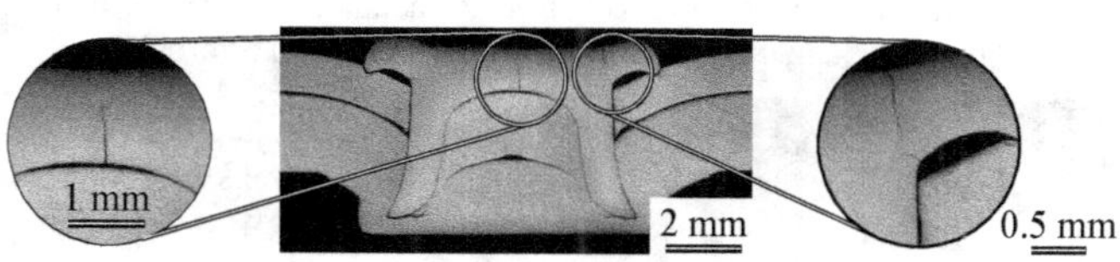

Tumbling strategy

Circle kinematic
$\alpha_0 = 6°$
$\alpha_{max} = 6°$
$\omega = 240°/s$

Fig. 5 Macrograph with detail views of joint set with standard tumbling strategy

To reduce the formation of cracks in the rivet head, the tumbling strategy is adapted based on the causes of the cracks that have occurred. With its mechanical design and customized control system, the tool offers all opportunities to adjust the tumbling strategy to the changing peripheral conditions. The cause of the cracks in the rivet center is the initial contact of the punch with the rivet head and the subsequent force applied. To prevent this effect of the conically shaped punch, the tumbling angle is set to the maximum angular position of the process before the initial contact between punch and rivet head. For a reduction of cracking at the rivet head projection, the maximum tumbling angle in the joining process is lowered to $\alpha = 5°$ to reduce the surface pressures. In addition, the punch is also accelerated to its maximum tumbling velocity $\omega = 240$ °/s before the onset of the process to prevent inhomogeneous deformations at the rivet protrusion. In Fig. 6, a macrograph is shown in the center plane of the riveted joint connected with the adapted tumbling strategy. The results show no cracks appearing at the rivet head center and a significant reduction in crack size at the rivet head protrusion. The lack of initial contact of the punch on the rivet head reduces the bending stress in the rivet head and no indentation occurs on the rivet head.

The 1° lower maximum tumbling angle has the effect of reducing the stress peaks at the rivet head protrusion and preventing a complete cracking of the auxiliary joining elements as seen before.

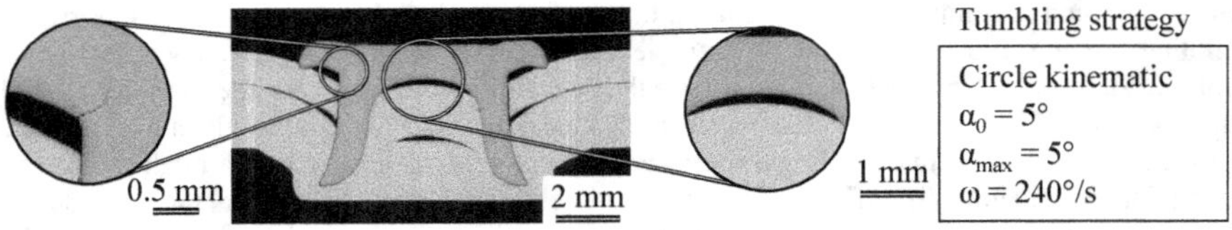

Fig. 6 Macrograph with detail views of joint set with adapted tumbling strategy

However, the tumbling strategy contains further parameters with which the joining process and thus the material flow and stress distribution can specifically be controlled. Alongside the tumbling angle, the tumbling kinematic has a significant influence on the resulting joint. Circular kinematics have a constant tumbling angle throughout the entire process, except for the setting phase of the angle. In the case of spiral kinematics, which are also predominantly rotating kinematics, the tumbling angle increases during the entire process and reaches its maximum deflection at the end of the process. This results in a constant increase in the tumbling angle and a lower deformation of the rivet head per revolution. Decreasing the deformation of the rivet head protrusion per increment is intended to reduce the probability of cracking and to form a defect-free joint.

Since initial contact at $\alpha = 0°$ in the rivet head center of the punch is also to be prevented in these experiments, the tumbling angle at the start of the process is selected at $\alpha = 4°$. This causes a spiral contour of the contact surface on the rivet head without causing initial contact in the rivet head center. Fig. 7 shows a macrograph of the joint of a connection set with spiral kinematics. The joint shows no cracks in the rivet head, either in the rivet head center nor at the rivet head protrusions. As with the adapted circular kinematics, there is no punch contact in the center, hence no cracking in this area occures. The results also show that moving the contact area in a radial direction in the form of a tumbling angle increase reduces the sensitivity to cracking. The maximum tumbling angle in the spiral kinematics is the same as in the adapted circular kinematics, but the comparatively slow buildup of the angle is less critical for the rivet head protrusion.

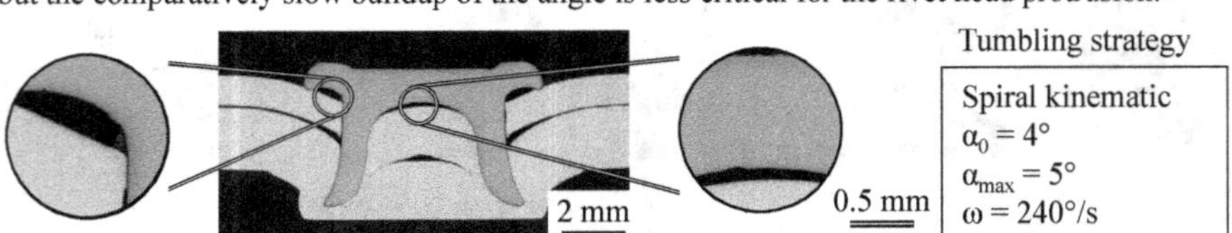

Fig. 7 Macrograph with detail views of joint set with versatile tumbling strategy

For a comparison of the results of the individual kinematic models, the geometric joint formation is considered because the individual parameters have specific influences on the joint and similar characteristics must be provided to allow a comparison of the formation of cracks. The values in Fig. 8 are averaged results over the right and left measured values because the macrographs do not show any angular offsets of the rivets. For each combination, only one test is shown, since the deviations of the results observed are negligible. The undercut is shown to be high for both kinematic models with a slightly reduced level for the spiral kinematics. This reflects previous findings [9] when comparing circular to spiral kinematics, but both kinematic models are at a similar level. The rivet head end position also shows results with a low maximum difference of 0.04 mm for the investigated variants. The differences can be explained by the variation of the tumbling onsets. In the 6° tumbling, the punch is in the 0° position at the beginning, whereas in tests with 5°, the punch is already at the maximum tumbling angle at the first contact between punch and rivet, thus causing a greater deformation of the rivet head protrusion. The difference in residual sheet thickness can also be explained by the adaptation of the tumbling strategy. With a continuously increasing angle, a larger radial material flow and a smaller axial material flow component are obtained. As a result, the rivet element is driven less into the joint and an increased residual sheet thickness results with a maximum difference to the other kinematic models of 0.04 mm. The results show similar geometric joint formations for all three tumbling strategies and are therefore suitable for comparison of the tumbling strategies.

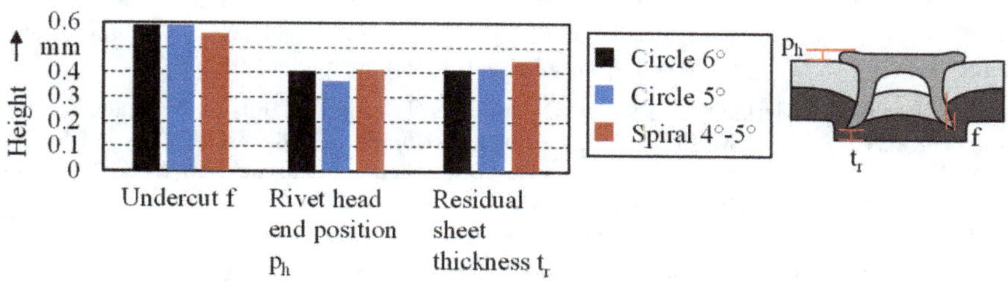

Fig. 8 Geometric joint formation of joints with varying tumbling strategies

The investigations show the increased versatility of the process combination by uniting the two processes into a two-stage joining process. The first process stage enables the rivet to be set precisely without deforming the rivet head in the radial outward area. Further, it is possible to increase the application options and the material flow control by adapting the process parameters of the tumbling strategy. The process combination allows a significant influence to be exerted on the shape of the joint without having to change the die or the rivet geometry. Additionally, the risk of the occurrence of cracks in the rivet head can be reduced and thus defect-free joints can be realized by adapting the tumbling strategy. By combining the two operating principles, the characteristic properties of the two processes can be combined. These are characterized in the case of multiple linear actuation by axial and in the case of tumbling semi-tubular self-piercing riveting by radial control of the material flow, which enables the tubular volume to be adjusted.

Summary and Outlook

In the investigations on a combination of versatile semi-tubular self-piercing riveting processes, a two-stage process consisting of a multi-range self-piercing riveting process and a tumbling self-piercing riveting process has been set up and evaluated. After setting the rivet to a constant rivet head height above the punch-side joining partner with the inner punch of the multi-range tool, the rivet head protrusions are formed onto the punch-side joining partner with a tumbling punch. For the second process stage, different tumbling strategies were investigated with a variation of several parameters. The results show that by adapting the tumbling strategy, the sensitivity of the rivet heads to cracking can be reduced while achieving nearly consistent geometric joint properties. Consequently, by combining the two processes, the versatility can be increased and new opportunities for joining technology can be created. However, closing the cavities under the rivet head protrusion is shown to be a challenge. Therefore, further investigations and adaptations of the tumbling strategy are necessary to realize the required crack-free joining. Further, the load bearing capacity is a very important feature of joints and should be investigated. This can be tested using the shear tensile and cross tensile test.

Acknowledgment

Funded by the Deutsche Forschungsgemeinschaft (DFG, German Research Foundation)—TRR 285-Project-ID 418701707. The authors thank the German Research Foundation for their organizational and financial support.

References

[1] J. Allwood, J.M. Cullen, Sustainable materials: With both eyes open. UIT Cambridge, Cambridge (2012) 373 pp.

[2] European Commission, European Green Deal (2021) Brussels.

[3] J. Dau, C. Lauter, U. Damerow, W. Homberg, T. Tröster, Multi-material systems for tailored automotive structural components. 18th Int. Conf. on Composites Mat., ICCM 2011 (2011).

[4] K. Mori, Y. Abe, T. Kato, Self-pierce riveting of multiple steel and aluminium alloy sheets. JMPT 214 (10) (2014) 2002–2008. https://doi.org/10.1016/j.jmatprotec.2013.09.007

[5] G. Meschut, M. Merklein, A. Brosius, D. Drummer, L. Fratini, U. Füssel, M. Gude, W. Homberg, P.A.F. Martins, M. Bobbert, M. Lechner, R. Kupfer, B. Gröger, F. Kappe, T. Kleffel, C.-M. Kuball, J. Popp, D. Römisch, J. Troschitz, C. Wischer, S. Wituschek, M. Wolf, Review on mechanical joining by plastic deformation. JAJP 5 (2022) 100–113. https://doi.org/10.1016/j.jajp.2022.100113

[6] D. Li, A. Chrysanthou, I. Patel, G. Williams, Self-piercing riveting-a review. Int J Adv Manuf Technol 92 (5-8) (2017) 1777–1824. https://doi.org/10.1007/s00170-017-0156-x

[7] R. Porcaro, A.G. Hanssen, M. Langseth, A. Aalberg, Self-piercing riveting process: An experimental and numerical investigation. JMPT 171 (1) (2006) 10–20. https://doi.org/10.1016/j.jmatprotec.2005.05.048

[8] F. Kappe, S. Wituschek, M. Bobbert, G. Meschut, Determining the properties of multi-range semi-tubular self-piercing riveted joints. Prod. Eng. Res. Devel. 16 (2-3) (2022) 363–378. https://doi.org/10.1007/s11740-022-01105-2

[9] S. Wituschek, F. Kappe, M. Lechner, Investigation of the influence of varying tumbling strategies on a tumbling self-piercing riveting process. Prod. Eng. Res. Devel. 16 (2-3) (2022) 353–362. https://doi.org/10.1007/s11740-021-01099-3

[10] X. He, I. Pearson, K. Young, Self-pierce riveting for sheet materials: State of the art. JMPT 199 (1-3) (2008) 27–36. https://doi.org/10.1016/j.jmatprotec.2007.10.071

[11] J. Varis, Economics of clinched joint compared to riveted joint and example of applying calculations to a volume product. JMPT 172 (1) (2006) 130–138. https://doi.org/10.1016/j.jmatprotec.2005.09.009

[12] F. Kappe, S. Wituschek, M. Bobbert, M. Lechner, G. Meschut, Joining of multi-material structures using a versatile self-piercing riveting process. Prod. Eng. Res. Devel. (2022). https://doi.org/10.1007/s11740-022-01151-w

[13] X. Han, L. Hua, Comparison between cold rotary forging and conventional forging. J Mech Sci Technol 23 (10) (2009) 2668–2678. https://doi.org/10.1007/s12206-009-0624-9

[14] P. Groche, D. Fritsche, E.A. Tekkaya, J.M. Allwood, G. Hirt, R. Neugebauer, Incremental Bulk Metal Forming. CIRP Annals 56 (2) (2007) 635–656. https://doi.org/10.1016/j.cirp.2007.10.006

[15] F. Kappe, M. Bobbert, G. Meschut, New Approach for Versatile Self Piercing Riveting: Joining System and Auxiliary Part. KEM 883 (2021) 3–10. https://doi.org/10.4028/www.scientific.net/KEM.883.3

Sheet Metal 2023 Materials Research Forum LLC
Materials Research Proceedings **25** (2023) 133-140 https://doi.org/10.21741/9781644902417-17

Influence of plastic orthotropy on clinching of sheet metal

Johannes Friedlein[1,a*], Christian Bielak[2,b], Max Böhnke[2,c], Mathias Bobbert[2,d], Gerson Meschut[2,e], Julia Mergheim[1,f] and Paul Steinmann[1,g]

[1]Institute of Applied Mechanics, Friedrich-Alexander-Universität Erlangen-Nürnberg, Egerlandstrasse 5, 91058 Erlangen, Germany

[2]Laboratory for Material and Joining Technology (LWF), Paderborn University, Pohlweg 47-49, 33098 Paderborn, Germany

[a]johannes.friedlein@fau.de, [b]christian.bielak@lwf.upb.de, [c]max.boehnke@lwf.upb.de, [d]mathias.bobbert@lwf.upb.de, [e]meschut@lwf.upb.de, [f]julia.mergheim@fau.de, [g]paul.steinmann@fau.de

Keywords: Anisotropy, Modelling, Clinching

Abstract. Clinching is a versatile mechanical joining method for assembling different sheet metal materials without auxiliary elements in short process times. The joint strength, however, solely relies on the material condition and its targeted interlock formation. Therefore, accurate material models are necessary incorporating all relevant phenomena to reliably predict the material behaviour. We extend a finite elastoplastic material model by incorporating the plastic orthotropy of the sheet metal in the joining process simulation. The anisotropy is captured by different variants of the 3D Hill 1948 yield function with associative and non-associative plastic flow. The constitutive models and the 3D clinching simulation are outlined and utilised to study the influence of plastic orthotropy focusing on the aluminium alloy EN AW-6014.

Introduction

Clinching is a mechanical joining technique for assembling different sheet metal materials by solely cold forming the material. Thus, it does not require auxiliary parts and hole pre-alignment. However, consequently the joint strength solely relies on the sheets and the sufficient formation of an interlock during the clinch joining process. Numerical investigations are essential for the development of the process route, to gain a deeper insight and for future optimisation. This requires highly accurate and robust material models to capture the large plastic deformations [1,2] and the complex non-proportional material flow. Especially the latter could change systematically due to anisotropic plastic flow, which can be accurately captured, for instance, by the r-values. Thus, this study focuses on the inherent plastic orthotropy of the metal sheets to be joined and its influence on the joint forming.

In [3] clinched joints of differently coated thin steel sheets H180Y exhibiting high Lankford (r-) values (1.4 ... 2.0) were studied. By using an anisotropic Hill yield locus for plasticity in ABAQUS for the quasi-static loading of the clinched joint, they showed a notable influence of anisotropy on the predicted shear strength. On the other hand, in [4] only a marginal effect of the plastic orthotropy for mild deep drawing steel DC05 on the strength of the joint was observed. Therein, the anisotropic plasticity model by Hill (1948, "Hill48") [5] was used under plane stress condition and equipped with parameters also determined from Lankford values. In addition to a von Mises and plane stress Hill48 yield function, [6] utilised the more advanced anisotropy model Yld2000-2d. They investigated equivalent models for clinched joints of deep-drawing-quality steel sheet proving a significant effect of the anisotropy on the calibration. In contrast, to the previous works studying the clinched joint, for instance [7] considers the preceding clinch joining process. Therein, only a minor influence of plastic orthotropy on the interlock was shown for DC05. For a

Content from this work may be used under the terms of the Creative Commons Attribution 3.0 license. Any further distribution of this work must maintain attribution to the author(s) and the title of the work, journal citation and DOI. Published under license by Materials Research Forum LLC.

dual-phase steel DP600 only a negligible effect of the planar anisotropy on the sheet contours after joining was found [8].

We focus on an aluminium alloy EN AW-6014 T4 with sheet thickness 2 mm. Its plastic orthotropy has already been characterised in [9] – by uniaxial tension tests and layer compression tests – being more pronounced than for a dual-phase steel HCT590X. Moreover, in contrast to the steel, the biaxial r-value of the aluminium alloy was less than 1, possibly rendering a different behaviour. The anisotropy is described by different variants of the Hill48 model including associative and non-associative flow rules. For the numerical investigations, a 3D clinching process model [10] is used. Moreover, in contrast to the state of the art, we focus on the importance of using 3D orthotropic plasticity to also capture the inherent difference between in-plane and out-of-plane plastic yielding and flow. This forms the basis to study the influence of plastic orthotropy on geometric and process-related parameters of clinch joining based on a flexible but still comparable group of anisotropy models.

Material Modelling

Clinching involves large plastic strains modelled herein by finite plasticity based on the logarithmic strain space [11]. After a short introduction into the framework and the elasto-plasticity model, the anisotropy formulations are introduced to describe the sheet metal.

Kinematics. The deformation of the body is captured by the deformation gradient $\boldsymbol{F} = \boldsymbol{I} + \nabla_X \boldsymbol{u}$ based on the unit tensor $\boldsymbol{I}$ and the gradient of the displacements $\boldsymbol{u}$. From the deformation gradient $\boldsymbol{F}$, the right Cauchy-Green strain tensor is defined as $\boldsymbol{C} = \boldsymbol{F}^T \cdot \boldsymbol{F}$.

Logarithmic strain space. As part of Seth-Hill's family of generalised strain measure, the logarithmic (Hencky) strain $\boldsymbol{H}$ is used within the logarithmic strain space and computed from the right Cauchy-Green strain tensor [11]. Due to the properties of the Hencky strain, the additive split $\boldsymbol{H} = \boldsymbol{H}^{\mathrm{e}} + \boldsymbol{H}^{\mathrm{p}}$ of the total strain into elastic and plastic part is applicable. Plasticity is herein based on a free energy density quadratic in the elastic strains $\boldsymbol{H}^{\mathrm{e}}$. Nonlinear isotropic hardening is described by a plastic potential, which is here defined by a tabular flow stress $\sigma_{\mathrm{f}}(H^{\mathrm{p,acc}})$, with the accumulated plastic strain $H^{\mathrm{p,acc}}$. The logarithmic stress tensor $\boldsymbol{T}$ is computed from the total Hencky strain $\boldsymbol{H}$ and total plastic Hencky strain $\boldsymbol{H}^{\mathrm{p}}$.

Isotropic and anisotropic yield functions. Plastic anisotropy is introduced by the classical 3D Hill48 anisotropic yield function [5] written in tensor notation [11] as

$$\Phi(\boldsymbol{T}, H^{\mathrm{p,acc}}) = \|\boldsymbol{T}\|_{\mathcal{H}_s} - \sqrt{2/3}\;\sigma_{\mathrm{f}}(H^{\mathrm{p,acc}}) \leq 0 \quad \text{with } \|(\cdot)\|_{\mathcal{H}_s} \equiv \sqrt{(\cdot)\colon \boldsymbol{\mathcal{H}}_s\colon (\cdot)} \tag{1}$$

This yield criteria is similar to isotropic von Mises (J2-) plasticity besides the generalisation of the stress norm using the deviatoric fourth order Hill tensor $\mathcal{H}_s$. Thus, for $\mathcal{H}_s = \mathfrak{I}^{\mathrm{dev}}$ (deviatoric fourth order unit tensor), we recover isotropic von Mises plasticity. The tensor $\mathcal{H}_s$ is set up from the Hill coefficients $h_{\alpha\beta}$ as described in [9, 11].

Evolution equations and algorithmic implementation. Starting from the yield condition in Eq. (1), the evolution equations are derived. This includes the symmetric plastic strain tensor $\boldsymbol{H}^{\mathrm{p}}$ and the accumulated plastic strain $H^{\mathrm{p,acc}}$ as

$$\dot{\boldsymbol{H}}^{\mathrm{p}} = \dot{\gamma}\boldsymbol{N} \quad \text{with } \boldsymbol{N}(\boldsymbol{T}) \equiv \frac{\boldsymbol{\mathcal{H}}_r\colon \boldsymbol{T}}{\|\boldsymbol{T}\|_{\mathcal{H}_r}}\;; \quad \dot{H}^{\mathrm{p,acc}} = \sqrt{2/3}\,\dot{\gamma} \tag{2}$$

based on the Lagrange multiplier $\dot{\gamma}$. The latter relation for the accumulated plastic strain is – for the non-associative model – an assumption proposed in [12], made for simplicity. Here, we have already generalised the relation for the plastic strain to incorporate a non-associative flow rule by introducing the Hill48-type tensor $\boldsymbol{\mathcal{H}}_r$. The special case of associative plasticity is recovered for $\boldsymbol{\mathcal{H}}_s = \boldsymbol{\mathcal{H}}_r$. To satisfy the yield condition also for a non-associative flow rule, we use a closest-point projection algorithm [13]. The implementation is carried out as an implicit user-defined material model in the commercial FEM-software LS-DYNA.

In this framework we identify four groups of material models that will be compared in the results section, namely isotropic plasticity "von Mises" ($\mathcal{H}_s = \mathcal{H}_r = \mathfrak{I}^{\text{dev}}$), associative anisotropic Hill48 plasticity ($\mathcal{H}_s = \mathcal{H}_r$) either based on r-values "Hill48-r/Hill48-rn" *or* yield stress ratios "Hill48-s", and non-associative anisotropic Hill48 plasticity identified based on r-values *and* yield stress ratios "Hill48-rs" ($\mathcal{H}_s \neq \mathcal{H}_r$). All these models share a common basis, which is beneficial for the comparison as differences can thus be clearly attributed generating insightful cause-effect relations. Here, non-associative flow is crucial to simultaneously capture the r-values and yield stress ratios, while still being comparable to the associative basis.

Parameter Identification

In the following, we first briefly outline the experimental setups and results. Secondly, the computation of the anisotropy coefficients for the different models is described in detail.

Layer compression test. This test is an adaption of the conventional compression test based on DIN 50106 (see e. g. [9] for details and further literature). With this setup the plastic flow curve up to true plastic strains of about 0.7. This data is taken from [2], where a good correlation for the application in clinching process simulations was achieved with this setup. However, it needs to be kept in mind that the sheet is loaded in thickness direction, therefore we obtain the flow stress in sheet normal direction, which is thus herein chosen as reference direction. This flow stress might differ from the rolling direction due to plastic anisotropy.

Tensile tests with different sheet orientations. Tensile tests were carried out in [9] according to DIN 6892-1. To investigate the anisotropy of the aluminium sheet metal, specimens were extracted under 0°, 45° and 90° to the rolling direction (RD) by laser cutting.

Firstly, from the tensile test the r-values r_θ can be determined from the strains measured by an optical measuring system. These were evaluated at an elongation of 10 mm (20% strain), compare [9], and are $r_{0°} = 0.8060$, $r_{45°} = 0.4637$ and $r_{90°} = 0.6105$. Moreover, it is possible to compute the average or normal anisotropy $r_\text{n} = [r_{0°} + 2\,r_{45°} + r_{90°}]/4 \approx 0.5860$ from these r-values [14]. This r-value is the basis for the Hill48-rn model, which corresponds to an isotropic in-plane behaviour. Here, anisotropy only affects the thickness direction, so with only one preferred direction (transverse isotropy). Thus, Hill48-rn could be solved using a 2D axisymmetric model saving computation time but still capturing the anisotropy in thickness direction. Nevertheless, for comparability the 3D simulation model was utilised.

Secondly, the yield stress ratios $\sigma_{\text{yield},\theta}/\sigma_{\text{yield},0°}$ (0°: 1; 45°: 0.9820; 90°: 0.9655) are determined from the true stress-strain curves based on the force-elongation curves in the different sheet orientations. Here, the reference yield stress $\sigma_{\text{yield},0°}$ corresponds to the rolling direction. However, our reference flow stress stems from the layer compression test and thus is determined in sheet thickness direction with $\sigma_{\text{yield,thickness}}$. Therefore, the yield stress ratio $s_\theta = \sigma_{\text{yield},\theta}/\sigma_{\text{yield,thickness}}$ is defined. To remain consistent with the Hill48 model

$$\frac{\sigma_{\text{yield,thickness}}}{\sigma_{\text{yield},0°}} = \left[-1 + \frac{2}{1 + r_{0°}} + \left[\frac{\sigma_{\text{yield},90°}}{\sigma_{\text{yield},0°}}\right]^{-2}\right]^{-0.5} \approx 0.9205 \tag{3}$$

is computed based on [15] to transform the ratios $\sigma_{\text{yield},\theta}/\sigma_{\text{yield},0°}$ (with $h_{11} = 1$) to s_θ ($h_{33} = 1$; 0°: 1.0864; 45°: 1.0668; 90°: 1.0489). In doing so, we further ensure that at least the r-value in RD $r_{0°}$ is exactly obtained by the Hill48-s model [16], see Fig. 1. Even though this alters the actual yield stress ratios, it is nevertheless regarded as sufficient for a fundamental study on the influence of plastic orthotropy on clinching and further improves the comparability to the isotropic von Mises model.

Anisotropy coefficients. We choose the sheet thickness direction as reference direction to be able to directly utilise the flow curve from the layer compression test as flow stress $\sigma_\text{f}(H^{\text{p,acc}})$. For the Hill48 anisotropy model this corresponds to the definition $(h_{33} = 1) \simeq (F_{yz} + G_{zx} = 1)$.

Moreover, we assume the out-of-plane anisotropy coefficients h_{23} and h_{31} to be identical to the isotropic case. From the r-values and yield stress ratios, the remaining anisotropy coefficients $h_{\alpha\beta}$ can be determined. The classical associative Hill48 anisotropic plasticity model can either solely be based on the r-values as Hill48-r/-rn models or be based on the yield stress ratios as Hill48-s model.

Hill48-r. Transforming the definitions given in [9,16], the Hill-coefficients are computed from the r-values for $h_{33}^r = 1$ as

$$h_{11}^r = \sqrt{\frac{r_{0^\circ} + r_{90^\circ}}{r_{90^\circ}[r_{0^\circ} + 1]}} \qquad h_{22}^r = \sqrt{\frac{r_{0^\circ} + r_{90^\circ}}{r_{0^\circ}[r_{90^\circ} + 1]}} \qquad h_{33}^r \equiv 1$$
$$h_{12}^r = [2r_{45^\circ} + 1]^{-0.5} \qquad h_{23}^r = h_{31}^r = 1/\sqrt{3} \tag{4}$$

Hill48-rn. Eq. (4) is also applicable for the Hill48-rn model replacing r_{0°, r_{45° and r_{90° by r_n.

Hill48-s. From the relation between the yield stress ratios and the anisotropy coefficients for Hill48 [15] with the assumption that $h_{33}^s = 1$, the following equations result

$$h_{11}^s = s_{0^\circ} \qquad h_{22}^s = s_{90^\circ} \qquad h_{33}^s \equiv 1$$
$$h_{12}^s = s_{45^\circ}[4 - [s_{45^\circ}]^2]^{-0.5} \qquad h_{23}^s = h_{31}^s = 1/\sqrt{3} \tag{5}$$

The resulting Hill anisotropy coefficients ($\boldsymbol{h_{11}}$; $\boldsymbol{h_{22}}$; $\boldsymbol{h_{12}}$) with $h_{33} = 1$ and $\boldsymbol{h_{23}} = \boldsymbol{h_{31}} = 0.5774$ are for Hill48-r (1.1335; 1.0446; 0.7203), Hill48-rn (1.1230; 1.1230; 0.6785) and Hill48-s (1.0864; 1.0489; 0.6306). Hill48-rs combines the parameters from Hill48-r and Hill48-s, thus utilises two sets of Hill-coefficients. As well-known, Hill48 anisotropy can typically not simultaneously capture the experimental r-values and yield stress ratios (marked by x) [14], as shown in Fig. 1. In contrast, the non-associative Hill48-rs model combines both sets and thus describes both well. Hill48-rn (dash-dotted) captures none of the in-plane ratios exactly, but only the (not shown) thickness direction.

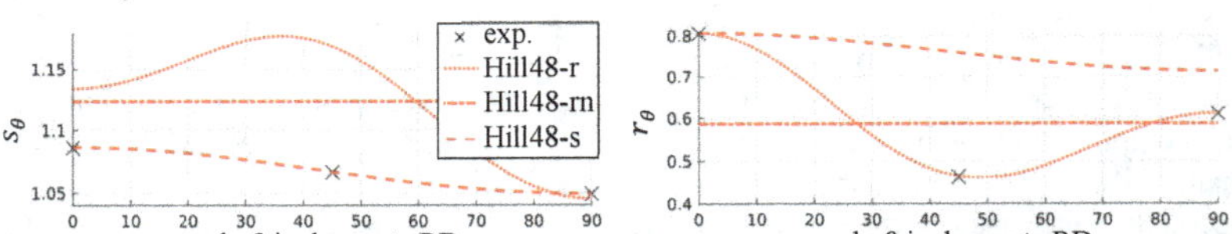

Fig. 1. Yield stress ratios s_θ (left) and r-values r_θ (right) as function of the angle θ to the rolling direction (RD). Hill48-s (dashed) captures the experimental yield stress ratios (marked by x) precisely, but not the r-values. In contrast, Hill48-r (dotted) only captures the r-values exactly. The ratios for Hill48-rn (dash-dotted) do not depend on the in-plane orientation, but nevertheless reproduce r-values and yield stress ratios different from 1.

3D Clinching with Isotropic and Anisotropic Plasticity

Subsequently, the 3D clinching process model is introduced followed by the numerical results.

3D clinching model. The 3D clinching process model and the essential settings are shown in Fig. 2 based on the validated model from [10]. To keep the meshes symmetric and be able to account all asymmetries to the influence of plastic orthotropy, eight-node hexahedral elements (ELFORM=1) without remeshing were chosen. Nevertheless, especially due to the complicated contact states, some asymmetry can result from the numerical model. Thus, all results are averaged by using the x-direction as RD and separately solving the model for RD along the y-direction (90° rotated) to account for the mentioned numerical asymmetries. The differences between the results are indicated by error bars in Fig. 5. Even though the values might scatter, the shown ratios and trends between rolling (RD) and transverse direction (TD) were consistent for all models.

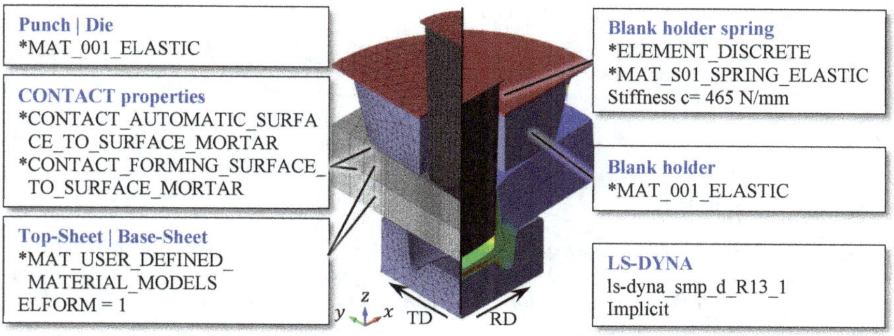

Fig. 2. 3D clinching process model and essential simulation settings. The rolling direction (RD) is aligned along the x-direction. Both sheets are symmetrically meshed by hexahedral elements without the need for remeshing (based on [10]).

Results. We study the influence of plastic orthotropy by analysing its effect on the overall sheet deformation, the characteristic parameters (interlock i, neck thickness n, shown in Fig. 3) and the process force. The sheet contours of the yz-plane (normal to x-axis, red, 90°, TD) and xz-plane (normal to y-axis, green, 0°, RD) for selected models are shown in the following.

For isotropic von Mises plasticity, the sheet contours from both planes match well and are shown as black reference contours in Fig. 3 and 4. Thus, the 3D model with isotropic plasticity is indeed symmetric. The geometric parameters, depicted in Fig. 5, confirm this symmetry.

Hill48-r obtains a slightly asymmetric deformation mainly visible by a higher interlock in rolling direction (green) in Fig. 3. This might be similar to the flow behaviour observable in the layer compression test, where the material flow is less pronounced in rolling direction. The higher interlock is compensated by a thinner neck in rolling direction, as summarised in Fig. 5. Moreover, the asymmetry and difference to isotropic plasticity (black) can also be seen in the upper corner A in Fig. 3, where Hill48-r predicts less vertical flow into the joint.

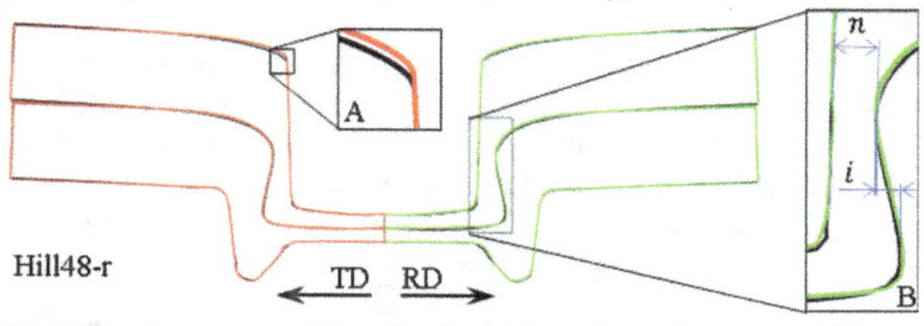

Fig. 3. Hill48-r. Sheet contours of the clinched joint. A small asymmetry is visible for the interlock in TD (red) and RD (green). Moreover, also the vertical flow into the joint (zoom A) is affected by anisotropy compared to the isotropic model (black). The zoom window B shows the definition of the interlock i and neck thickness n.

Hill48-rn well represents the symmetry in in-plane direction, thus is not shown in detail, but only summarised in Fig. 5. Hill48-s renders only a small asymmetry in the deformation, thus is also not shown in detail and is on average similar to Hill48-r, see Fig. 5.

Materials Research Proceedings **25** (2023) 133-140 https://doi.org/10.21741/9781644902417-17

Hill48-rs in Fig. 4 predicts a larger asymmetry of the interlock and neck thickness than the previous models, however with ratios $i_{\mathrm{RD}}/i_{\mathrm{TD}} = 1.07$ and $n_{\mathrm{RD}}/n_{\mathrm{TD}} = 0.96$ still relatively low and within the experimental scatter indicated in Fig. 5 by the shaded areas.

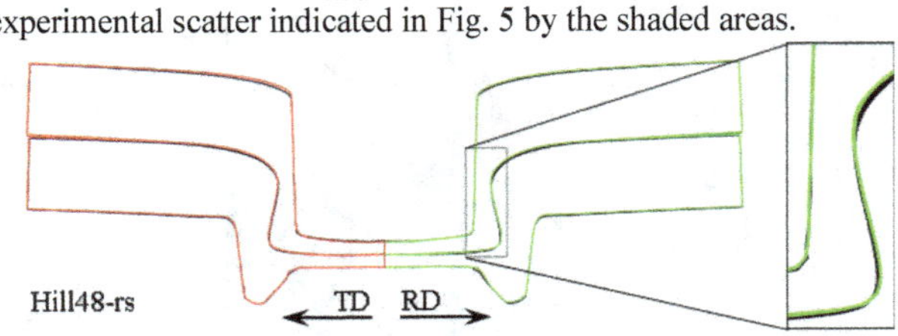

Fig. 4. Hill48-rs. Sheet contours of the clinched joint. The asymmetry between rolling (RD, green) and transverse direction (TD, red) is more pronounced visible by a larger interlock and thinner neck in rolling direction (green, zoom window).

The interlock i and neck thickness n are extracted from the numerical results as shown in Fig. 3 (right) and summarised in Fig. 5. All anisotropy models show in rolling direction a higher interlock and a thinner or almost equal neck thickness. The non-associative model Hill48-rs exhibits the highest geometric asymmetry, where it appears to combine the results for Hill48-r and -s. Thus, it appears to be important to capture both anisotropy effects. Nevertheless, in view of the experimental scatter, the asymmetry is negligible, so the in-plane anisotropy could be omitted. Moreover, the anisotropy does not only relocate the material between rolling and transverse direction, but also increases the interlock overall. This might be accounted to the changing flow behaviour that is introduced by the r-values. The Hill48-rn renders a nearly symmetric deformation with a higher interlock than the von Mises model and a slightly lower neck thickness. This confirms that the anisotropy in sheet thickness direction already notably influences the characteristic parameters and further supports the use of a 3D anisotropy model.

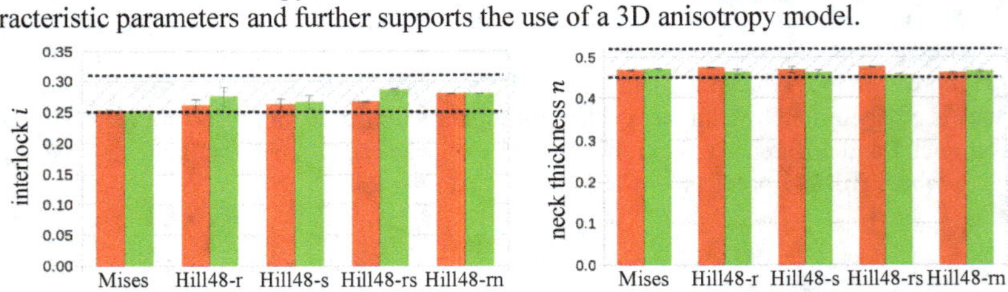

Fig. 5. Comparison of interlock (left) and neck thickness (right) for all models. Red: transverse direction (TD). Green: rolling direction (RD). The shaded areas with dashed upper and lower bounds indicate the experimental range. Anisotropy relocates the material between RD and TD and moreover increases the interlock.

Further analysis shows a continuous transition of these results between rolling and transverse direction. Thus, the extrema are well captured by considering only the yz- and xz-planes as depicted in the figures and diagrams above. Also changing the relative orientation of the die-sided and punch-sided sheet revealed no different trends.

The comparison of the process force for all models shows no clear tendency in Fig. 6 (left) as all match the experimental results (black) well. All forces evolve similarly with a certain scatter mainly due to unsmooth changes in the contact states. The process force seems to be dominated by the compression of the two sheets between the punch and the die, which is governed by the flow stress in sheet thickness direction being equal for all models. Fig. 6 (right) confirms that the joint contours for von Mises (blue) as well as Hill48-rs (red) match well to experiments (black).

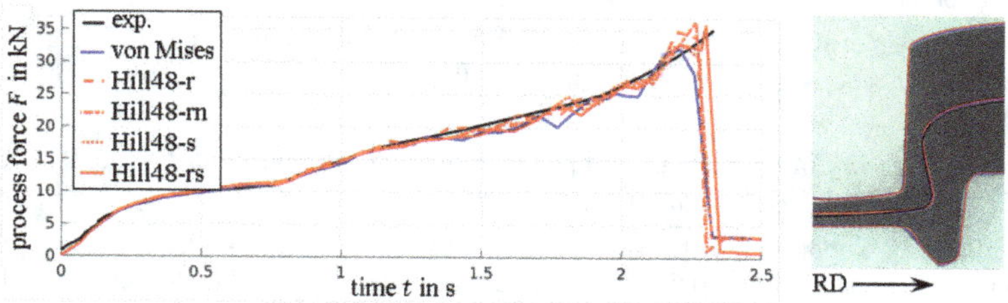

Fig. 6. Process force vs process time for different material models (left). All models match the experimental result (black) well. The different anisotropy models show no clear trend, such that the process force is rather independent and mainly affected by the flow stress in sheet thickness direction. The simulated joint contours for von Mises (blue) and Hill48-rs (red) plasticity match well to the experimental results (black).

Summary and Outlook

We implemented different 3D anisotropic plasticity models in LS-Dyna for the simulation of 3D clinching. The goal was to set up robust material and simulation models to study the influence of plastic orthotropy on the clinching process. Thereby, we compared different variants of the Hill48 yield function including associative and non-associative flow rules being equipped with different experimental input. The shown results are confined to an aluminium alloy EN AW-6014 T4, which revealed only a minor influence and asymmetry of geometric and process parameters due to plastic orthotropy. However, it was shown that the anisotropy between in-plane and sheet thickness direction can alone notably affect the clinched joint. Moreover, the use of a non-associative plasticity model that accurately captures more features of the sheet orthotropy led to a more pronounced asymmetry, however still well within experimental scatter. Even though the non-associative model is therefore recommended for higher predictability, it is seen as more important to accurately capture the ratio between in-plane and out-of-plane plastic flow. Especially for aluminium alloys a non-quadratic yield function is necessary, for instance Yld91, which however possesses the same anisotropy properties as the herein applied Hill48 model. This work lays the basis for a parametric study on the influence of advanced plasticity characteristics to understand when certain phenomena are relevant for reliable and versatile clinching of sheet metal. To quantify the effect of the geometric changes on the joint strength, for instance pull-out tests can be utilised in future work as part of a continuous simulation chain.

Acknowledgements

The funding by the Deutsche Forschungsgemeinschaft (DFG, German Research Foundation) – Project-ID 418701707 – TRR 285, subproject A01 and A05 is gratefully acknowledged. Moreover, we want to thank our colleagues in the TRR285 Lars Ewenz (B02) and Simon Wituschek (C02) for their support.

References

[1] M. Jäckel et al., Process-oriented flow curve determination at mechanical joining, Procedia Manufacturing, 47 (2020) 368-374. https://doi.org/10.1016/j.promfg.2020.04.289

[2] M. Böhnke, F. Kappe, M. Bobbert, G. Meschut, Influence of various procedures for the determination of flow curves on the predictive accuracy of numerical simulations for mechanical joining processes, Materials Testing, 63(6) (2021) 493-500. https://doi.org/10.1515/mt-2020-0082

[3] S. Saberi et al., Influence of plastic anisotropy on the mechanical behavior of clinched joint of different coated thin steel sheets. Int. J. Mater. Form. 1(1) (2008) 273-276. https://doi.org/10.1007/s12289-008-0349-9

[4] S. Coppieters et al., Numerical and experimental study of the multi-axial quasi-static strength of clinched connections. Int. J. Mater. Form. 6(4) (2013) 437-451. https://doi.org/10.1007/s12289-012-1097-4

[5] R. Hill, A theory of the yielding and plastic flow of anisotropic metals, Proc. R. Soc. Lond. 193(1033) (1948) 281-297. https://doi.org/10.1098/rspa.1948.0045

[6] A. Breda, S. Coppieters, T. Kuwabara, D. Debruyne, The effect of plastic anisotropy on the calibration of an equivalent model for clinched connections. Thin-Walled Structures, 145 (2019) 106360. https://doi.org/10.1016/j.tws.2019.106360

[7] S. Coppieters et al., Reproducing the experimental pull-out and shear strength of clinched sheet metal connections using FEA. Int. J. Mater. Form. 4(4) (2011) 429-440. https://doi.org/10.1007/s12289-010-1023-6

[8] S. Jónás, M. Tisza, Finite element modelling of clinched joints, Adv. Technol. Mater. 43(1) (2018) 1-6. https://doi.org/10.24867/ATM-2018-1-001

[9] J. Friedlein et al., Inverse parameter identification of an anisotropic plasticity model for sheet metal. In IOP Conference Series: Mater. Sci. Eng. 1157(1) (2021) 012004. https://doi.org/10.1088/1757-899X/1157/1/012004

[10] C.R. Bielak, M. Böhnke, M. Bobbert, G. Meschut, Numerical investigation of a friction test to determine the friction coefficients for the clinching process. Proc. Inst. Mech. Eng. L: Journal of Materials: Design and Applications, 14644207221093468 (2022).

[11] C. Miehe, N. Apel, M. Lambrecht, Anisotropic additive plasticity in the logarithmic strain space: Modular kinematic formulation and implementation based on incremental minimization principles for standard materials, Comput. Methods Appl. Mech. Engrg. 191(47-48) (2002) 5383–5425. https://doi.org/10.1177/14644207221093468

[12] T.B. Stoughton, A non-associated flow rule for sheet metal forming. Int. J. Plast. 18(5-6) (2002) 687-714. https://doi.org/10.1016/S0749-6419(01)00053-5

[13] J.C. Simo, T.J.R. Hughes, Computational inelasticity, 7, Springer Science & Business Media, 2006.

[14] A.M. Habraken, Modelling the plastic anisotropy of metals. Arch. Comput. Methods Eng. 11(1) (2004) 3-96. https://doi.org/10.1007/BF02736210

[15] O. Ghorbel et al., Coupled anisotropic plasticity-ductile damage: Modeling, experimental verification, and application to sheet metal forming simulation. Int. J. Mech. Sci. 150 (2019) 548-560. https://doi.org/10.1016/j.ijmecsci.2018.10.044

[16] P. Dasappa, K. Inal, R. Mishra, The effects of anisotropic yield functions and their material parameters on prediction of forming limit diagrams. Int. J. Solids. Struct. 49(25) (2012) 3528-3550. https://doi.org/10.1016/j.ijsolstr.2012.04.021

Sheet Metal 2023
Materials Research Forum LLC
Materials Research Proceedings **25** (2023) 141-146
https://doi.org/10.21741/9781644902417-18

Approach to transferring force-based fatigue curves into stress-related fatigue curves for clinch joints

Lars Ewenz[1,a]*, Martin Kuczyk[1,b] and Martina Zimmermann[1,2,c]

[1]Institute of Materials Science, TU Dresden, 01062 Dresden, Germany

[2]Fraunhofer Institute for Material and Beam Technology IWS, 01277 Dresden, Germany

[a]lars.ewenz@tu-dresden.de, [b]martin.kuczyk1@tu-dresden.de, [c]martina.Zimmermann@iws.fraunhofer.de

*corresponding author

Keywords: Fatigue, Joining, Clinching

Abstract. Cyclic strength is essential in many lightweight design concepts where the design is to be pushed to the limits of strength. While joining dissimilar metals such as aluminum and steel is a challenge of its own, fatigue life prediction for this joining type is all the more challenging. Here, clinching as a mechanical joining process offers many advantages. However, a generalized evaluation of the fatigue properties is complex since many influencing factors, such as the joint's geometry, the high plastic deformation, the proportion of bonding mechanisms, have to be considered. Force versus number of cycles (F-N) curves are the established basis to describe the fatigue behavior of clinch joints. However, a generalized evaluation of the service life requires stress versus number of cycles (S-N) curves. This research gives a first approach to transferring F-N curves to S-N curves on the basis of nominal stress determination in the damage relevant clinch cross section. The material combination used, EN AW-6014 and HCT590, offers excellent practical relevance since both materials are widely used in the automotive industry.

Introduction

Clinching is a mechanical joining process that has the potential to save immense amounts of energy. As a form-closure and force-closure process, clinching offers the advantage that it can be used for hybrid material combinations [1]. Especially in the mobility sector where clinching is used, i.e., in skin sheet applications. Here joining aluminum and steel can contribute to a significant weight reduction and thus help saving fuel consumption. Clinching also offers the advantage that materials can be joined without the need for auxiliary parts or high temperatures during the joining process.

Clinching is a versatile process allowing a vast variety of joining constellations. Clinching allows the joining of different sheet thicknesses, materials and surface conditions. Kalich and Füssel [2] have analyzed the influence of the surface condition on the properties and the binding mechanisms of the joints. In addition, joining more than two sheets is possible, and the corresponding layer architecture can be chosen individually. In another work, Kalich and Füssel [3] found that the joining direction however influences the mechanical behavior. In addition, the toolset geometry can change the mechanical properties of the joints, too, as was shown by Ewenz et al. [4]. The versatility of clinching makes a classical fatigue evaluation difficult since the stresses in the most critical area are unknown. Furthermore, it was shown in [5] that the failure mode is load-dependent. As a result, clinch joints are typically evaluated on a joint-to-joint basis by looking at the force versus number of cycles (F-N) curves.

However, a stress versus number-of-cycles curve (S-N curve) is required to enable a quantitative comparison between different clinch joints. In this work, an aluminum alloy EN AW-6014 and a HCT590X+Z dual-phase steel were used in different sheet thicknesses to

Content from this work may be used under the terms of the Creative Commons Attribution 3.0 license. Any further distribution of this work must maintain attribution to the author(s) and the title of the work, journal citation and DOI. Published under license by Materials Research Forum LLC.

produce clinch joints and evaluate the fatigue behavior and a first approach is presented, which allows a transfer from a F-N curve to a S-N curve.

Material and Method

The aluminum alloy EN AW-6014 and the dual-phase steel HCT-590X+Z were used for the experimental tests. A 2 mm thick sheet was used for the Al-Al clinch joint and a 1.5 mm thick plate for the St-St joint. The Al-St mixed joint consisted of a 1 mm aluminum sheet and a 1.5 mm steel sheet.

EN AW-6014 is a precipitation hardening alloy that has good formability in the T4 condition (solution annealed and quenched) and high strength after heat treatment to adjust the artificially annealing condition (T6). The clinch joints were designed to correspond to the established quality standards, i.e., no initial cracks, a symmetrical formation of the joint and no apparent gaps. Furthermore, all connections were processed with a conical punch, and an 8 mm fixed round point die to ensure good comparability. However, the punch diameters varied slightly between 4.8 mm and 5.6 mm to achieve a good quality of the joints. Fig. 1 schematically shows the clinching process. First, the sheets were fixed between the punch and the die. Subsequently, the punch moved towards the die until a predefined position was reached. This procedure achieves characteristic clinch point geometries according to the tool geometry (punch and die). The bottom thickness was used as a non-destructive quality parameter because this parameter can be used for a known geometry to conclude other geometric features such as undercut and neck thickness [6]. After the joining process, the joints containing aluminum are heat treated for 20 min at 185°C (T6) to increase the strength of the material.

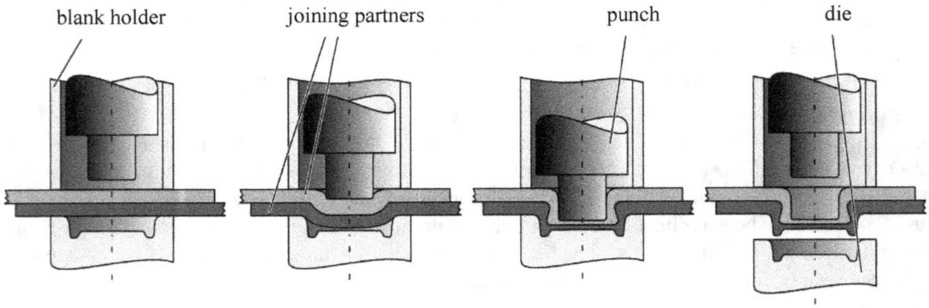

Fig. 1 Tool design and general process for clinching with closed die [7]

Metallographic cross-sections were prepared for the three different clinch joint geometries. For this, the clinch joints were cut, grinded and polished exactly to the middle of the joint. These cross-sections served to determine the geometrical features. Fig. 2 (a) shows a schematic cross-section with the geometrical features, neck thickness n, undercut u and the bottom thickness b.

https://doi.org/10.21741/9781644902417-18

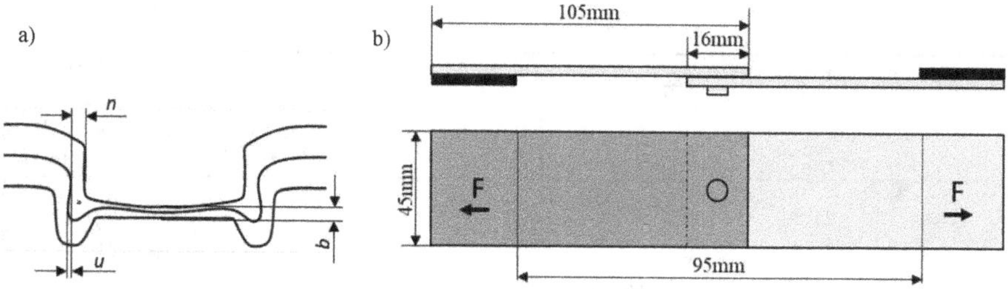

Fig. 2 (a) Schematic cross-section of a clinch joint with the geometrical features neck thickness n, undercut u and bottom thickness b, (b) geometry of single-lap shear specimens with alignment plates (black) as used for the mechanical testing

Flow curves were taken from the base materials using a layer compression test and are characterized by a biaxial stress condition. A Hocket-Sherby extrapolation was applied to calculate plastic strains higher than 0.5. Fig. 4 shows the flow curves of the base materials. The HCT590X+Z reaches higher stress levels than the AW-6014 in the T4 condition.

Fatigue tests were carried out using single lap-shear specimens made by two 105 mm x 45 mm sheets with a 16 mm overlap area (Fig. 2 (b)) using the resonance pulsation test system, Testronic by Rumul. About 20 specimens for each series were tested in a range of 10^5 to 10^7 load cycles. The load ratio applied was 0.1 in order to avoid buckling or bending of the specimens. The test frequency was about 70 Hz. When the specimens were broken or a frequency change of 5 Hz occurred, the test stopped. Specimens with more than 10^7 load cycles are marked as run outs. Visual inspection of the fracture surfaces in a scanning electron microscope revealed the corresponding failure modes.

Results

Cross-sections of the three different clinch joint geometries are shown in Fig. 3 and the corresponding values for the geometrical features can be found in Tab. 1. It can be seen, that each clinch series differs with regard to their geometrical features, which is a result of the different diameters of the conical punch applied. Nevertheless, they are all validated as joints with sufficient quality related to conventional standards. This means, no cracks could be observed after the joining process, the neck thickness-undercut ratio was acceptable and all connections were symmetrical.

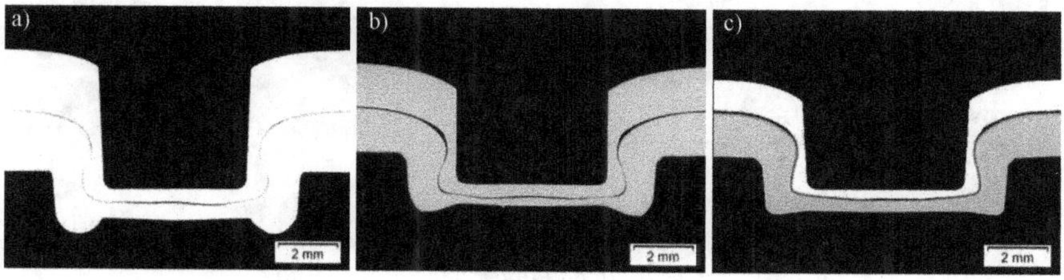

Fig. 3 Cross-section of the different clinch joints (a) Al-Al clinch joint, sheet thickness 2 mm; (b) St-St clinch joint, sheet thickness 1.5 mm; (c) Al-St clinch joint, sheet thickness Al-sheet 1 mm and St-sheet 1.5 mm

Table 1. Geometrical features of the different clinch joints

Parameter	Al-Al	St-St	Al-St
Neck thickness [mm]	0.42±0.03	0.36±0.01	0.23±0.04
Undercut [mm]	0.26±0.05	0.21±0.02	0.20±0.00
Bottom thickness [mm]	1.07±0.01	0.75±0.01	0.70±0.06

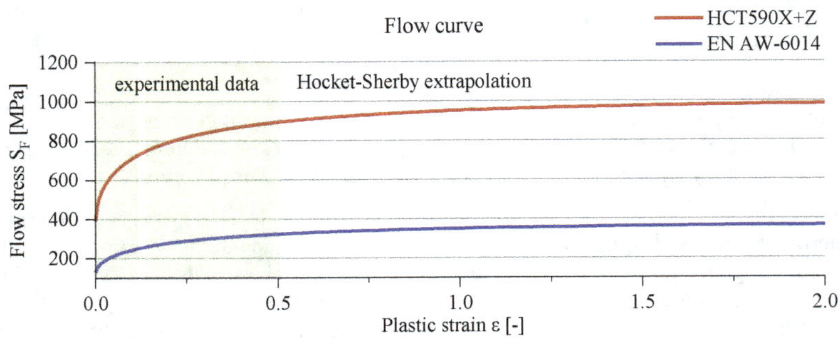

Fig. 4 Flow curve of the base materials

The results of the fatigue test show pronounced differences between the cyclic strength of the three clinch joint configurations. The St-St joints have the highest fatigue strength, while the Al-St joints show the lowest fatigue strength (Fig. 5 (a)). All clinch joint series display the same main failure mode, i.e. neck fracture, as depicted in Fig. 5 (b). Furthermore, it should be pointed that in the case of the Al-St joints, failure always occurred in the aluminum sheet.

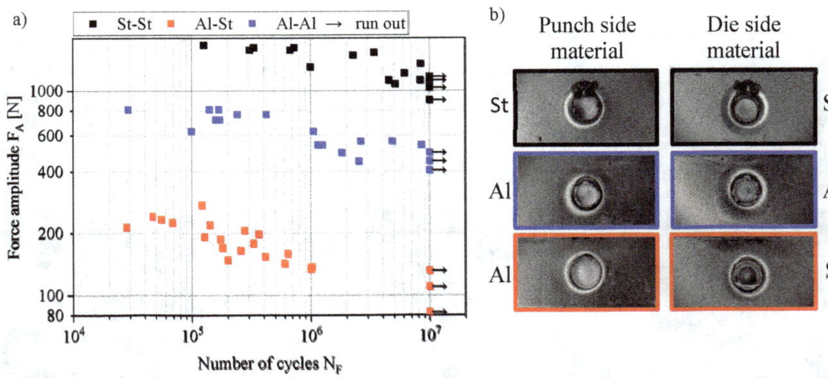

Fig. 5 a) F-N Curve of three different clinch joints and b) exemplary images of the main failure behavior of these clinch joints after fatigue testing

Development of an S-N-Curve

Two main factors significantly contribute to the fatigue behavior of the individual clinch joint configurations. On the one hand, an initial deformation is induced through the forming process during clinching, which can be interpreted as a kind of residual stress stored in the severely deformed microstructure. On the other hand, the cyclic load amplitude introduced during fatigue testing causes a local stress peak due to the complex geometry of the clinch joint.

The post-mortem fractographic analyses of the fatigue test specimens showed that irrespective of the different cyclic strengths, all clinch joints show failure starting from the neck area. Based on the metallographic and the fractographic analyses the nominal area A assumed as ring area in the neck can be determined. As a consequence, the nominal stress introduced in this area through the fatigue loading can be calculated, too. To obtain the afore defined residual stress introduced by the plastic straining during the clinch process, the plastic strain was calculated at most critical area. These calculations resulted in plastic strain values of about 1.5 for all three series. The corresponding flow stress SFS is obtained from the Hocket-Sherby extrapolation. These values result in a normalization factor for the fatigue results according to the superposition principle (see Eq. 1).

$$S = \frac{F_A}{A} + S_{FS} \quad (1)$$

In Fig. 6 the fatigue results from Fig. 5 are transferred to the stress versus number of cycles format, applying the aforementioned superposition principle to define the correlating stress amplitudes. While the St-St joints still show a much higher fatigue strength, the Al-Al joints and the Al-St joints no longer show such a pronounced difference than is the case of the F-N curves. Hence, the transfer of the F-N curves to S-N curves clearly corresponds to the expected cyclic strength of the failure relevant base material, the aluminum sheet. However, a remaining discrepancy in fatigue behavior can neither be explained by the clinch joint geometry nor the predeformed and hence strengthened material.

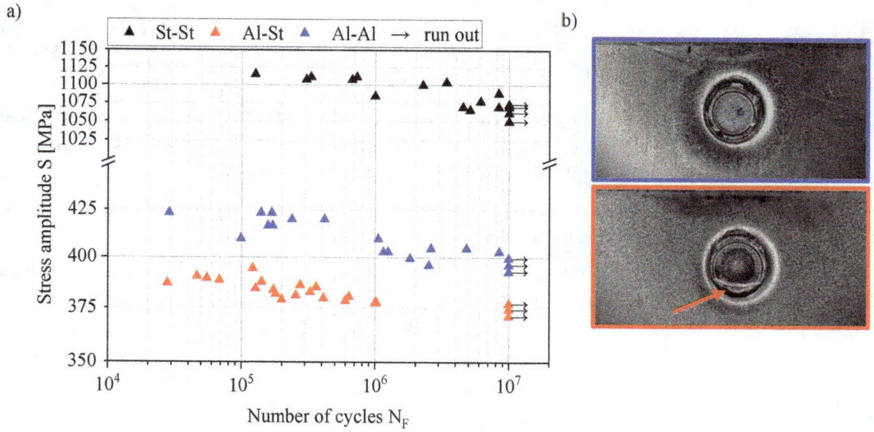

Fig. 6 a) S-N curve of all clinch joint constellations, b) detailed image of the main failure behavior of Al-Al (blue) and Al-St joint (red) the arrow marks the plastic deformation on the remaining cup.

Fig. 6 (b) shows that the remaining aluminum cup in the lower steel sheet of the Al-St joint has a pronounced plastically deformed zone (marked by an arrow) compared to the Al-Al joints. This difference in failure mode proves that the local loading situation must be influenced by additional factors. As a matter of fact, the current approach did not consider the likely influences of the binding mechanisms which will be the focus of future research. However, the investigations presented, provide a first approach to obtain S-N curves for clinch joints and thus pave the way for a more generalized lifetime prediction of clinch joints considering the particular needs of versatility.

Summary

St-St, Al-Al and Al-St clinch joints were manufactured. Fatigue tests of the joints were carried out, and the fracture surfaces were analyzed, subsequently. An approach to transfer the force versus number of cycles (F-N) curve into a damage-relevant stress versus number of cycles (S-N) curve was established for a neck fracture failure mode. The approach considered the damage-relevant nominal stresses calculated for the neck area based on the fatigue tests' forces and the clinch process's initial stresses.

Acknowledgement

„Funded by the Deutsche Forschungsgemeinschaft (DFG, German Research Foundation) – TRR 285 – Project-ID 418701707"

References

[1] Y. Abe, K. Mori, T. Kato, Joining of high strength steel and aluminium alloy sheets by mechanical clinching with dies for control of metal flow. Journal of Materials Processing Technology 212 (2012) 884–889. https://doi.org/10.1016/j.jmatprotec.2011.11.015

[2] J. Kalich and U. Füssel, Design of clinched joints on the basis of binding mechanisms. Prod. Eng. Res. Devel. 16 (2022) 213–222. https://doi.org/10.1007/s11740-022-01108-z

[3] J. Kalich, U. Füssel, Influence of the Production Process on the Binding Mechanism of Clinched Aluminum Steel Mixed Compounds, JMMP 5 (2021) 1–11. https://doi.org/10.3390/jmmp5040105

[4] L. Ewenz, J. Kalich, M. Zimmermann, U. Füssel, Effect of Different Tool Geometries on the Mechanical Properties of Al-Al Clinch Joints, KEM 883 (2021) 65–72. https://doi.org/10.4028/www.scientific.net/KEM.883.65

[5] L. Ewenz, C. R. Bielak, M. Otroshi, M. Bobbert, G. Meschut, M. Zimmermann, Numerical and experimental identification of fatigue crack initiation sites in clinched joints, Production Engineering (2022) 305–313. https://doi.org/10.1007/s11740-022-01124-z

[6] G. Meschut et al., Review on mechanical joining by plastic deformation, Journal of Advanced Joining Processes 5 (2022) 1–27. https://doi.org/10.1016/j.jajp.2022.100113

[7] R. Kupfer et al., Clinching of Aluminum Materials – Methods for the Continuous Characterization of Process, Microstructure and Properties. Journal of Advanced Joining Processes 5 (2022) 1–12. https://doi.org/10.1016/j.jajp.2022.100108

Sheet Metal 2023 | Materials Research Forum LLC
Materials Research Proceedings **25** (2023) 147-154 | https://doi.org/10.21741/9781644902417-19

Experimental investigation of the cause and effect relationships between the joint and the component during clinching

Christian Steinfelder[1, a*] and Alexander Brosius[1, b]

[1]Technische Universität Dresden, Chair of Forming and Machining Processes, 01062 Dresden, Germany

[a]christian.steinfelder@tu-dresden.de, [b]alexander.brosius@tu-dresden.de

Keywords: Sheet Metal, Joining, Stiffness

Abstract. The combination of the mechanical properties of a clinched joint and of the material surrounding the joint determine the resulting properties of the component and joint. The cause and effect relationships between the joint and the joint environment offers the possibility of a specific modification through an adaptation in the design process. In order to identify these cause and effect relationships and resulting interactions experimentally, numerous of experiments are required. In this publication, a concept for the automated manufacturing of head tensile test and shear tensile test specimens – from cutting to clinching – by using a punch laser machine is presented. Based on a full-factorial experimental design, the parameters change of the properties of the joint environment by beading and change of the punch displacement are addressed. The influence on the properties of the clinched specimen is evaluated based on the variables stiffness, force at the beginning of yielding and maximum force at head tensile loading and shear tensile loading. In addition, the geometric quality parameters of neck thickness, interlock and bottom thickness are evaluated. The relationships can be used to apply uniform loads to joints in joined structures to counteract oversizing.

Introduction

In joined structures, the joints are loaded unevenly. The aim in the design process is to achieve an even loading of the joints in order to avoid oversizing of the joints or an oversized number of joints. The loading of the joints can be influenced by adjusting the joints, the position of the joints or the components. For the targeted use of cause and effect relationships between components and joints in their design phase, it is absolutely necessary to understand them. For example, the joint environment can be influenced by the removal of material, the accumulation of material, or the forming of beads [1]. The loading of the joints can be influenced by such a change of the joint surroundings. In addition, there are influences at the joint level due to the joining process. For example, the punch displacement or the punch diameter can be changed there in order to modify the properties of the clinched joint. In order to identify these cause and effect relationships and resulting interactions experimentally, numerous tests are required. For this purpose, a concept for the automated production of head tensile test and shear tensile test specimens – from laser cutting up to clinching – by using a punch laser machine is presented. For the evaluation of the joint properties, the stiffness, the force at the beginning of yielding and the maximum force at head tensile and shear tensile loading are evaluated. In addition, the geometric quality parameters of the clinched joint are evaluated. The obtained interrelationships can be directly used in the load-oriented adaptation of the joining design of structures in order to adapt the loading in individual joints during the design process.

For lightweight construction purposes, thin sheets of higher strength can be used to transmit tensile and shear forces. However, they have low dimensional stiffness. As a result, they are susceptible to failure under other types of loading and require stiffeners for large-area parts in order to maintain the load-bearing capacity and safety against failure due to instability such as

Content from this work may be used under the terms of the Creative Commons Attribution 3.0 license. Any further distribution of this work must maintain attribution to the author(s) and the title of the work, journal citation and DOI. Published under license by Materials Research Forum LLC.

buckling. By forming beads into components, it is possible to change the stiffness locally. The influence of beads depends on both the geometry of the bead and the bead orientation and arrangement of multiple beads. Trzesniowski gives guidance for the design and arrangement of beads based on various examples [2]. Emmrich developed a finite-element-method (FEM) based method for the design of beads for support structures subjected to bending loads in the design process [3]. There are manby studies that are concerned numerically and experimentally with the design of clinching tools and their influence on the clinched joint properties. Ewenz et al. analyzed the influence of different punch geometries on the quasi-static and fatigue behavior of the clinched joints [4]. Mucha analyzed the lock-forming mechanism in the clinched joint in numerical and experimental studies [5]. The interlock can be influenced via the tool geometry and the punch displacement, whereby the punch displacement had almost no influence on the neck thickness in the investigated ranges. Studies that combine the influence of tool parameters, process parameters and the design of the joint environment are not known.

Experimental procedure

A TruMatic 1000 fiber punch laser machine from Trumpf is used for the automated production of head tensile and shear tensile specimens. This can be used to process sheet metal by laser cutting, punching and forming operations. The work on automated production and subsequent clinching of head tensile test and shear tensile test specimens requires adaptation of the existing tool systems for punching and forming operations. This is necessary because these are only designed for the processing of one sheet and not, as is usual for mechanical joining, for at least two sheets arranged one above the other. For this purpose, typical clinching tools were combined with the tools of the machine manufacturer. Fig. 1 b) shows an adapted clinching tool in a tool holder. Like a conventional clinching tool, it consists of a die and a punch. The die has a stripper which presses the sheet out of the die after the joining process, because the die is fixed and cannot be moved down on the system side. The punch has a polyurethane stripper. Fig. 1 c) shows the installed clinching tool with a stripper on the punch side. The punch can be moved in the y and z directions of the machine. The x-movement is realized by moving the sheet. Reverse arrangement of punch and die is also possible and offers advantages for other specimen configurations. However, for the results presented in this publication, only the tools shown in Fig. 1 are used. The processing sequences are determined when the NC program is created using the TruTops Boost software.

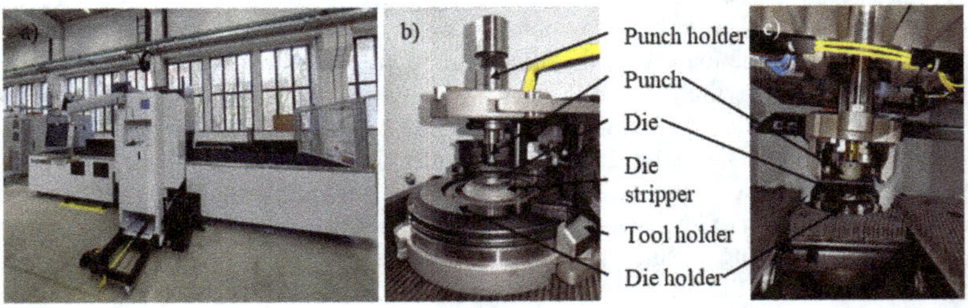

Fig. 1: a) Punch laser machine, b) Adapted clinching tool set (for visibility without punch stripper) fixed in the tool holder, c) Tool with punch stripper installed in the machine

The process is divided into two steps. Firstly, the preparation of the two sheets to be joined and secondly, the clinching itself. The sheets to be joined are first processed separately. Fig. 2 shows the design of the sheets to be joined using the example of shear tensile test specimens. The individual parts of the specimens are positioned in the sheets in such a way that when the sheets are stacked one on top of the other, the individual parts are in alignment with each other and can later be released from the sheet composite as a finished specimen after the clinching process. For this purpose, the individual parts of the specimens are fixed in the sheet with micro joints. Micro joints are a small connection between the blank and the workpiece, with the task of fixing the workpiece in the blank. In order to be able to extract the finished specimens from the sheet metal composite after joining, it is necessary to provide cutouts at the required locations. In addition to cutting operations performed by laser, beads can also be formed at this stage. The beads influence the properties of the specimens in the joint surroundings. The beads change the stiffness of the structure and changes the deformation behavior of the specimens. For the manufacturing of the beads, a standard tool for the production of endless beads with a semicircular profile was used, which was selected based on the sheet thickness. The beading tool has a height of 2 mm with a width of 4 mm. A change in bead geometry and its influence is not part of this investigation. The two sheets are then stacked and joined by clinching. The sheets are aligned by using the machine-side fixed point. Due to the automation, joints with different process parameters or different tools can be produced in a short time without any manual procedure. Finally, the specimens can be removed from the sheet and mechanically tested in tensile test or prepared for micrographs. The dimensions of the shear tensile test specimens and head tensile test specimens are according to EFB/DVS 3480-1 [6]. The dimensions of the shear tensile test specimens are 105 mm x 45 mm with an overlap length of 16 mm. The head tensile test specimens have the dimensions 150 mm x 50 mm.

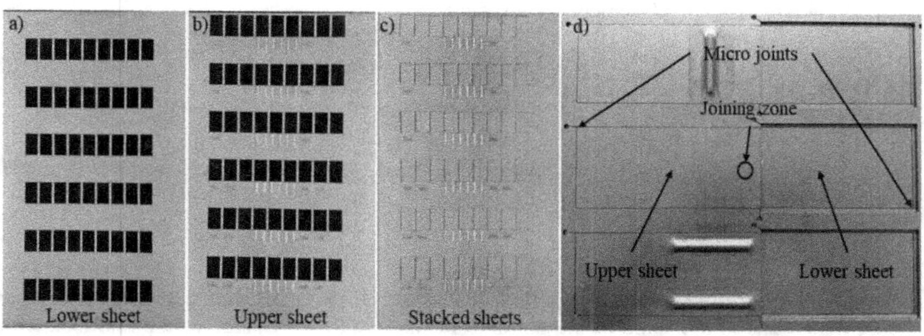

Fig. 2: Sheet metal layer before clinching, a) Lower sheet, b) Upper sheet, c) Stacked sheets, d) Detailed view of stacked sheet metal for clinching of shear tensile test specimen

Fig. 3 shows the specimen types used for investigations. Fig. 3 a) shows the shear tensile test specimens in the variants transverse bead, no bead and longitudinal bead. Fig. 3 b) shows the head tensile test specimens and the respective arrangement of the beads. The beads are arranged in the upper sheet in both variants. In addition, specimens for the preparation of micrographs are manufactured. Two connections were analyzed. 2 x s_0 = 1.5 mm HCT590X using a die with a diameter of 8 mm and depth of 1.6 mm (TOX BD8016) combined with a punch of 5 mm diameter (TOX A50100) and 2 x s_0 = 2.0 mm EN AW-6014 using a die with a diameter of 8 mm and a depth of 1.2 mm (TOX BE8012). The used punch has a diameter of 5 mm (TOX A50100).

 Materials Research Forum LLC
https://doi.org/10.21741/9781644902417-19

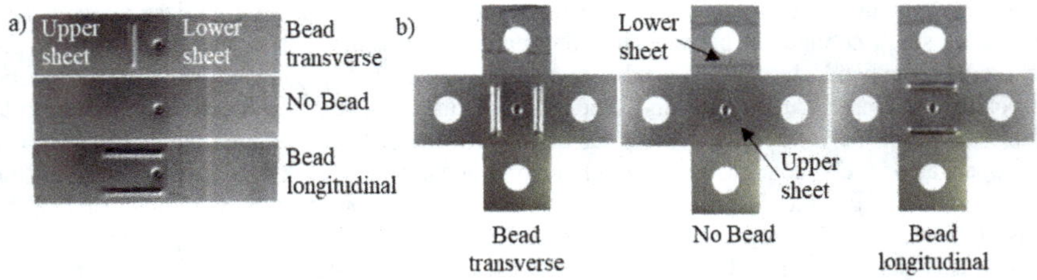

Fig. 3: Specimen for a) Shear tensile test, b) Head tensile test in the variants no bead, bead longitudinal and bead transverse

An experimental analysis is carried out based on a full-factorial design of experiments. The maximum forces under head tensile loading and shear tensile loading, the beginning of yielding, the stiffness and the characteristic parameters of neck thickness, interlock and bottom thickness are evaluated. The test design is based on the parameters given in Table 1 and is identical for both material combinations. For each variant, 7 specimens are manufactured and tested. For each of the two joint types 189 specimens were manufactured and tested. The head tensile test and shear tensile test are performed on a universal testing machine (Inspekt 250, Hegewald & Peschke) and the force and the crosshead displacement are measured.

Table 1: Varied parameters

Parameter level	Beads	Bottom dead center / (punch displacement change)
-1	Transverse	+0.1 mm (less punch displacement)
0	No	0 mm
+1	Longitudinal	-0.1 mm (more punch displacement)

Results

The results for the 2 x s_0 = 1.5 mm HCT590X connection are described below. Changing the punch displacement leads to a change of the bottom thickness and to a change of the interlock. If the bottom thickness is reduced, the formation of the interlock increases. The neck thickness, on the other hand, is not significantly affected by changing the punch displacement, see Fig. 4.

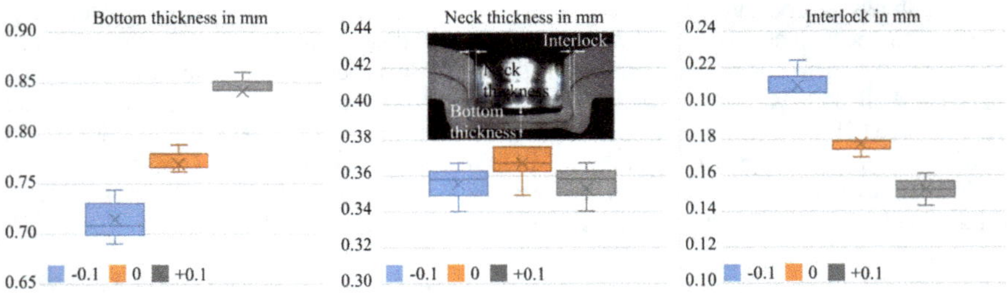

Fig. 4: Geometric quality parameters bottom thickness, neck thickness and interlock of the connection 2 x s_0 = 1.5 mm HCT590X, 7 specimens per series

Fig. 5 shows an example of the mean value curves from shear tensile tests and the scatter range of the maximum forces of the seven repetitions in each case. A change in the joint environment due to beads or a change in the punch displacement does not lead to any significant change in the maximum forces in the shear tensile test when considering the scatter width, see Fig. 5 a), b) and

c) - left. On average, an increase in the maximum force due to beads in the longitudinal direction can be observed, see Fig. 5 a), b) and c) - right. In all cases, the clinched joints failed due to neck cracking.

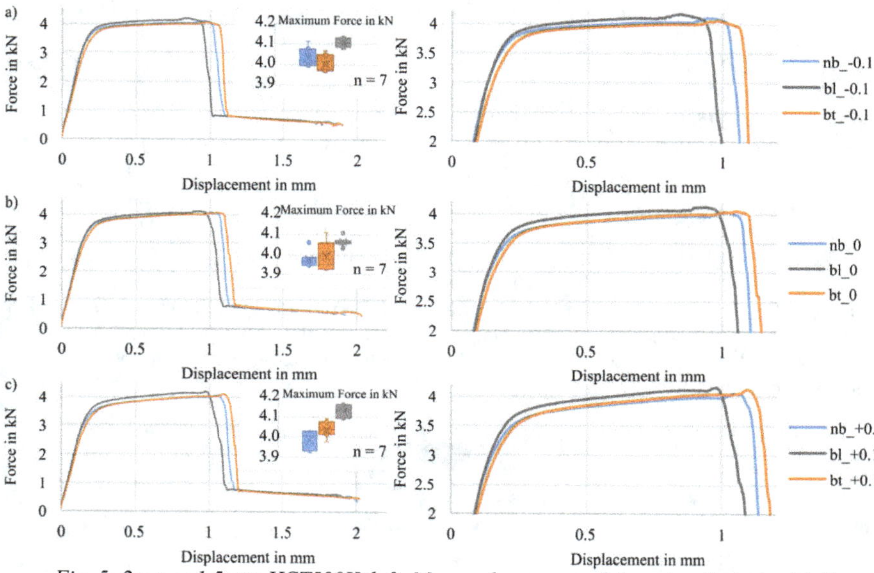

Fig. 5: 2 x s_0 = 1.5 mm HCT590X, left: Mean value curves of shear tensile tests with 7 repetitions and scatter of maximum forces, right: Detailed view: a) -0.1 mm (more punch displacement), b) Reference level, c) +0.1 mm (less punch displacement) in the variants no bead (nb), bead longitudinal (bl) and bead transverse (bt)

The stiffness of the specimen can be influenced by beads and changes the deformation behavior of the specimens. The change in deformation behavior has an influence on the loading of the joint. Thus, beads can be used to influence the loading of the joints. Compared to a specimen without beads, beads in the longitudinal direction increase the force at the beginning of yielding for all punch displacements in the shear tensile test and the head tensile test, see Table 2. By adding beads, the beginning of yielding can be shifted and the energy absorption of the clinched joint in the elastic range can be influenced. At the start of yielding, plastic deformation begins and thus the failure of the joint. The yield point and the stiffness can also be influenced by varying the punch displacement. In the shear tensile test beads in the transverse direction leads to a reduction in stiffness, while stiffness is not affected by beads in the longitudinal direction. In the head tensile test beads in the longitudinal direction leads to an increase in stiffness. Beads in the head tensile test have a greater influence on the properties, especially in the case of plastic deformation.

Materials Research Proceedings **25** (2023) 147-154
https://doi.org/10.21741/9781644902417-19

Table 2: Mean values of the Stiffness and yield force for the shear tensile test and head tensile test for the connection of 2 x s_0 = 1.5 mm HCT590X

Parameter	**Shear tensile test**		**Head tensile test**	
	Stiffness [N/mm]	**Yield force [N]**	**Stiffness [N/mm]**	**Yield force [N]**
nb_-0.1	23380	2586	1420	851
bl_-0.1	22930	2848	1680	1015
bt_-0.1	20620	2601	1750	870
nb_0	22500	2504	1670	659
bl_0	22330	2782	1730	913
bt_0	20300	2478	1430	849
nb_+0.1	21410	2404	1510	765
bl_+0.1	21670	2662	1700	925
bt_+0.1	19280	2447	1460	882

In the following the results for the 2 x s_0 = 2.0 mm EN AW-6014 connection are described. For the parameters bottom thickness, neck thickness and interlock, the same effects were found as for the investigated steel connections, see Fig. 6. But there the influences on the interlock formation were more clearly visible.

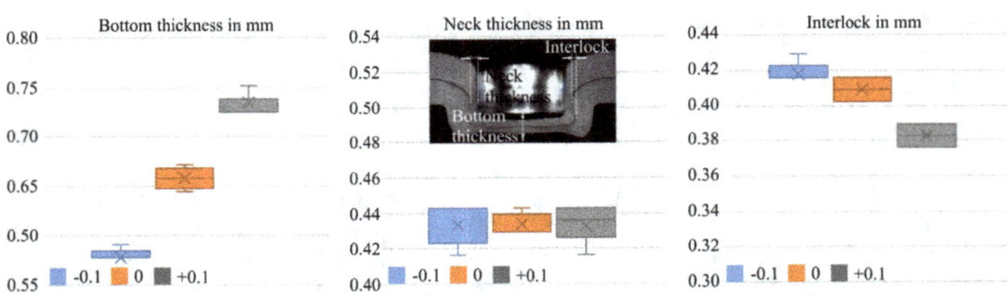

Fig. 6: Geometric quality parameters bottom thickness, neck thickness and interlock of the connection 2 x s_0 = 2.0 mm EN AW-6014, 7 specimens per series

For the connection 2 x s_0 = 2.0 mm EN AW-6014 Fig. 7 shows the mean value curves from shear tensile tests and the scatter range of the maximum forces of the seven repetitions in each case. Based on the measurement data, only a change of maximum forces for parameter less punch displacement and beads in a longitudinal direction can be detected. For the other parameter combinations, no significant changes in the maximum forces can be detected. A change in the joint environment due to beads or a change in the punch displacement does not lead to any significant change in the maximum forces in the shear tensile test when considering the scatter width, see Fig. 5 a), b) and c) - left. On average, an increase in the maximum force due to beads in the longitudinal direction can be observed, see Fig. 5 a), b) and c) - right. In all cases, the clinched joints failed due to neck cracking.

Materials Research Proceedings **25** (2023) 147-154 https://doi.org/10.21741/9781644902417-19

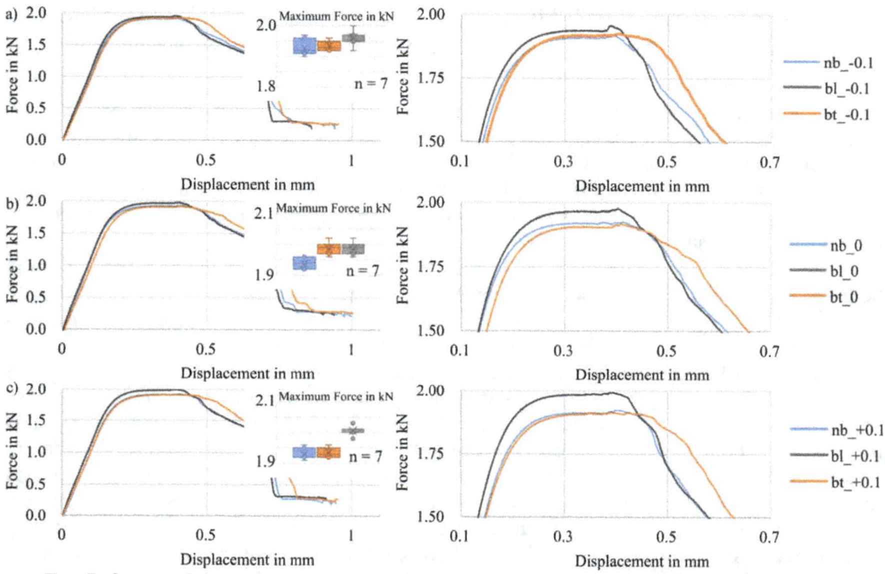

Fig. 7: 2 x s_0 = 2.0 mm EN AW-6014, left: Mean value curves of shear tensile tests with 7 repetitions and scatter of maximum forces, right: Detailed view: a) -0.1 mm (more punch displacement), b) Reference level, c) +0.1 mm (less punch displacement) in the variants no bead (nb), bead longitudinal (bl) and bead transverse (bt)

In the shear tensile test, only a minimal to no influence of the beads could be determined for the connection 2 x s_0 = 2.0 mm EN AW-6014 in total, see Table 3. In the head tensile test, a bead in the transverse direction reduces the stiffness and the beginning of yielding. No significant differences were found between the variants with no bead and with a bead in the longitudinal direction, see Tab. 3.

Table 3: Mean values of the Stiffness and yield force for the shear tensile test and head tensile test for the connection of 2 x s_0 = 2.0 mm EN AW-6014

Parameter	Shear tensile test		Head tensile test	
	Stiffness [N/mm]	Yield force [N]	Stiffness [N/mm]	Yield force [N]
nb_-0.1	10970	1486	1300	623
bl_-0.1	11360	1530	1340	650
bt_-0.1	10550	1526	1050	501
nb_0	11690	1447	1060	631
bl_0	11450	1468	1340	617
bt_0	11040	1398	1030	577
nb_+0.1	10810	1502	1260	565
bl_+0.1	11290	1476	1310	647
bt_+0.1	10410	1490	1070	526

Due to the large sheet thickness of 2.0 mm, high basic stiffness of the sheet is assumed, so that the beads have only a minor influence on the properties. Since stiffness is a geometrically based parameter and depends on the second moment of area, a greater influence of beads is to be expected with thinner sheets. This was also shown in the first tests and will be investigated in future work.

Summary and outlook

A concept and its implementation for the automated production of head tensile test and shear tensile test specimens using a punch laser machine were presented. By adapting the tooling systems and smart programming of the NC code, individual specimens with varying process parameters or tool geometries can be produced in a short time. This allows numerous test specimens with different process and tool parameters to be manufactured and experimentally investigated at a relatively low time and effort.

Based on the first evaluated data sets, it was possible to show that the resulting properties can be influenced by changing the joint environment and the process-side parameters. Beads can be used to influence the stiffness of the specimens. Changing the stiffness of the specimens modifies the deformation behavior and thus influences the loading of the joint. In particular, the properties in the linear-elastic range could be influenced in the case of the joints investigated. It can be assumed that the influence of the beads is strongly dependent on the sheet thickness and that the effects themselves are not dependent on the material. The ongoing expansion of the test plan is taking place in order to expand the database and thus analyze the cause and effect relationships in greater depth. This knowledge can be used in the design of components with multiple clinched joints in order to set a uniform loading at the joints. As the experimental investigations proceed, interactions between the individual parameters will also be evaluated so that they can be taken into account in the design or used specifically to influence the properties of the joint. This knowledge about the cause and effect relationships is used in further work in the development of a design method for the load-adapted design of joints and components and thus provides a direct contribution to increasing the versatility of the clinching process chain.

Acknowledgment

Funded by the Deutsche Forschungsgemeinschaft (DFG, German Research Foundation) – TRR 285 – Project-ID 418701707, subproject B01.

References

[1] S. Martin, K. Kurtusic, T. Tröster, Influence of the Surrounding Sheet Geometry on a Clinched Joint, Key Eng. Mater. 926 (2022) 1505-1515. https://doi.org/10.4028/p-09md1c

[2] M. Trzesniowski, Rahmen Chassis, in: Gesamtfahrzeug, Springer Fachmedien, Wiesbaden, 2019, pp. 267-353. https://doi.org/10.1007/978-3-658-26696-7_6

[3] D. Emmrich, Entwicklung einer FEM-basierten Methode zur Gestaltung von Sicken für biegebeanspruchte Leitstützstrukturen im Konstruktionsprozess, Dissertation (2004).

[4] L. Ewenz, J. Kalich, M. Zimmermann, U. Füssel, Effect of different tool geometries on the mechanical properties of al-al clinch joints, Key Eng. Mater. 883 (2021) 65-72. https://doi.org/10.4028/www.scientific.net/KEM.883.65

[5] J. Mucha, The analysis of lock forming mechanism in the clinching joint, Mater. Des. 32 (2011) 4943-4954. https://doi.org/10.1016/j.matdes.2011.05.045

[6] DVS/EFB-Merkblatt 3480-1: Testing of Properties of Joints: Testing of Properties of mechanical and hybrid (mechanical/bonded) joints.

Sheet Metal 2023 Materials Research Forum LLC
Materials Research Proceedings **25** (2023) 155-162 https://doi.org/10.21741/9781644902417-20

Comparison of ex- and in-situ investigations of clinched single-lap shear specimens

Daniel Köhler[1, a *], Aiting Yu[1, b], Robert Kupfer[1, c], Juliane Troschitz[1,d] and Maik Gude[1, e]

[1]Technische Universität Dresden, Institute of Lightweight Engineering and Polymer Technology, Holbeinstraße 3, 01307 Dresden, Germany

[a]daniel.koehler3@tu-dresden.de, [b]aiting.yu@mailbox.tu-dresden.de, [c]robert.kupfer@tu-dresden.de, [d]juliane.troschitz@tu-dresden.de, [e]maik.gude@tu-dresden.de

Keywords: X-Ray, In-Process Measurement, Clinching

Abstract. Force-displacement measurements and macrosections are commonly used methods to validate numerical models of clinching processes. However, these ex-situ methods often lead to springback of elastic deformations and crack-closing after unloading. In contrast, the in-situ computed tomography (CT) can provide three-dimensional images of the clinching point under loading conditions. So far, the quantity of elastic springback that causes measuring deviations between in- and ex-situ measurements is not determined. In this paper, a method is described to quantitatively compare the results of in-situ CT, ex-situ CT and CT scans of cut specimens, which are prepared for macrosectioning, among each other. The method is applied to a single-lap shear test of two clinched aluminum sheets. Here, the test is conducted to specific process steps, then the specimen is CT scanned in-situ (during loading) and ex-situ (after unloading). Subsequently, the specimens are cut for macrosectioning and CT scanned. Finally, the outer contours and the interfaces of cross section images are determined by digital image analysis and the deviations over the clinching point between ex- and in-situ methods are calculated.

Introduction

Clinching is a joining technology that has various advantages. Clinching points exhibit good mechanical properties, they are media-tight, various materials can be joined and they are cost efficient [1]. Since the joint is primarily characterized by a form- and force-fit, joining partners can be separated easily during recycling [2]. The validation of numerical models for example of shear tests [3] with clinching points is usually done by comparing the load-displacement behavior and the deformation state at discrete displacement levels. Here, the deformation state is usually extracted in a single cross section using destructive macrosectioning or – more recently – ex-situ computed tomography (CT) [4]. The basics for the CT measurement method are described in [5].

However, using these ex-situ methods, elastic deformations reset and cracks might close before the specimen is investigated. Additionally, during preparation for macrosectioning, the joint deforms due to residual stresses. These influences are usually not considered in the numerical model or can cause errors. Consequently, these methods only allow an imprecise validation. Moreover, only a single deformation state of the same specimen can be used for validation, leaving the development of the deformations over the process unrevealed. Thus, conclusions about the cause for possible deviations between numerical model and experiment are difficult to be drawn.

These disadvantages can be eliminated using the in-situ CT. Here, the investigated process is conducted inside a CT system. In order to conduct CT scans, the process needs to be interrupted and halted at displacement points of interest. Subsequently, the CT scans are used to reconstruct and create digital three-dimensional images of the specimen at the respective displacement level. This method is increasingly applied for joining and testing processes. For example, Füßel et al. benefitted from investigating the testing process under loading when analyzing the damage in

Content from this work may be used under the terms of the Creative Commons Attribution 3.0 license. Any further distribution of this work must maintain attribution to the author(s) and the title of the work, journal citation and DOI. Published under license by Materials Research Forum LLC.

Materials Research Proceedings **25** (2023) 155-162 https://doi.org/10.21741/9781644902417-20

single overlapping adhesively bonded riveted lap joints of fiber reinforced plastics and aluminum in lap shear tests [6]. Kunz et al. showed that CT visible markers inside an adhesive can be used for measuring the internal displacement field of the adhesive during loading [7]. Furthermore, in-situ CT was used by Pottmeyer et al. to investigate the damage of carbon fiber reinforced plastics due to the deformation of inserts under tensile loading [8]. Regarding clinching points, the in-situ CT method for shear tests with clinching points was introduced in [9]. Here, the force-displacement behavior and the cross section was compared qualitatively between numerical model and experiment. Furthermore, it was shown that due to the limited resolution and the partial cold welding, the interface between the joining partners of the same material was not visible in the CT scan. Therefore, radiopaque substances were placed in between the joining partners before joining. In [10] the feasibility of different substances to increase the visibility of the interface in CT scan was investigated.

Usually, in- and ex-situ CT scans are used to qualitatively assess the formation and geometry of clinching points. However, for validation of numerical models, a quantification is preferable. Thus, the objective of this paper is to present a method that enables the quantitative comparison of outer contours and interfaces from clinching points. The method is applied on a shear test with clinching points. Here, the specimens are investigated in in- and ex-situ CT scans (cf. Fig. 1 a) using a CT set-up described in Fig. 1 b and c. Additionally, after unloading, the specimens are cut as for macrosectioning. The introduced method is used to quantify the deviations between in-situ CT, ex-situ CT and macrosectioning. The deviations are discussed regarding the suitability of the different measurement methods for validating numerical models.

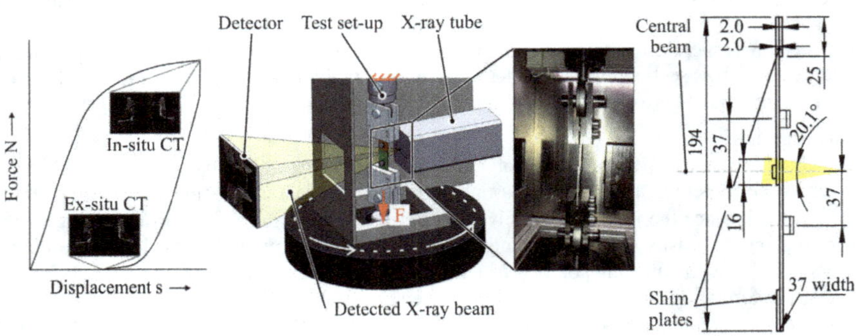

(a) (b) (c) (d)

Fig. 1: Schematic process of ex-and in-situ CT scans during the shear test (a), CT set-up (b), close-up view of the actual specimen (c) and dimensions of the specimen (d)

Materials and Methods

Specimen preparation and geometry. Two sheets made of EN AW-6014-T4 (2.0 mm thickness)[1] are overlapped and clinched with the punch A50100 and the related die BE8012[2] (cf. Fig. 1 d). As described in [10] a 0.01 mm thick tin foil is positioned in between the sheets prior clinching in order to enhance the visibility of the interface in the CT scan. After clinching, the specimens are heat treated at 458.15 K for 20 min. It represents a paint bake treatment commonly applied in automotive industry, whereby the aluminum material reaches T6 state. The testing device used for the in-situ CT shear tests requires slim clamps. Thus, the used specimens exhibit a width of 37 mm deviating from the standard ISO 12996 [3] (cf. Fig. 1 d.).

[1] Advanz™ 6F-e170, Novelis Inc., Atlanta, USA
[2] TOX PRESSOTECHNIK GmbH & Co.KG, Weingarten, Deutschland

Materials Research Forum LLC
https://doi.org/10.21741/9781644902417-20

Shear test and CT measurement. The shear specimens are tested with an universal testing machine ZwickRoell Z250[3]. The in-situ testing set-up CT-System FCTS 160 - IS[4] is situated in between the columns of the testing machine. It consists of a microfocus transmission tube FORE 160.01C TT (160 kV, 1 mA, 80 W, min. focal spot size < 3 µm) as X-ray source and a digital scintillation detector PaxScan 4030E DRZ-Plus[5] (pixel size = 127 µm, 3200 x 2300 pixels, 14 bit). Here, the X-ray-system rotates around the test set-up (cf. Fig. 1 b.). In a CT scan, X-ray projections are taken from the test set-up at defined angular steps. In order to investigate the influence of the testing machine's crossbeam displacement on the deviations, three different displacement levels are investigated using three specimens for each level. At the displacement level, the cross beam is held at constant position and an in-situ CT scan is conducted (ca. 15 min). Then, the test specimen is unloaded for an ex-situ CT scan. One specimen of each displacement level is prepared for macrosectioning. In order to facilitate the evaluation, the cut specimens are investigated by CT instead of microscopy. The CT and testing parameters are shown in detail in Table 1.

Table 1: Parameter of the CT measurement and the tensile testing

Domain	Parameter	Entity	Value
CT measurement	Acceleration voltage	[kV]	150
	Tube current	[µA]	30
	X-ray projections	[-]	1440 (4 per 1°)
	Exposure time	[ms]	625
	Voxel size	[µm]	7.58
	Magnification	[-]	16.8
	Focal spot size	[µm]	ca. 3
	Physical filter	[mm]	0.35 (carbon diamond)
Tensile testing	Displacement levels for CT scans	[mm]	0; 0.3; 0.65; 1.00
	Pre-load	[N]	60
	Testing speed	[mm/min]	2

The volumes are reconstructed with the Feldkamp algorithm in Cera eXplore[6]. Then, the volumes are edited in VG Studio Max[7]. Here, a cylinder is fitted to the outer surface of the die-sided sheet. Finally, for each specimen and displacement level, a cross section image is created with the section defined by the cylinder's symmetry axis and the axis along the testing direction.

Digital image editing. Using MATLAB[8] algorithms, the outer contour and the interface visible in the cross section images are extracted respectively. The contour is defined as a curve of one pixel thickness marking the boundary between material pixels of the punch-sided sheet (grey value >1) and air (Fig. 2 c).The extraction of the outer contours is described in the following and is visualized in Fig. 2.

[3] ZwickRoell GmbH & Co. KG, Ulm, Deutschland
[4] Finetec FineFocus Technologies GmbH, Garbsen, Deutschland
[5] Varian Medical Systems Inc, Palo Alto, USA
[6] Version 2.3.1 Siemens Healthcare GmbH, Erlangen, Deutschland
[7] Version 2.2, Volume Graphics GmbH, Heidelberg, Deutschland
[8] R2022a, The MathWorks, Inc., Natick, USA

1. Gaussian filtering (imgaussfilt(original_image,0.5)) to reduce image noise.
2. Binarization (original_image<=1) to assign a grey value to all material pixels (cf. Fig. 2 b).
3. Median filtering (medfilt2(original_image,[30 30])) to reduce the impact of artefacts.
4. Contour extraction (bwmorph(original_image,'remove')) to remove interior pixels leaving the outer contours of the sheets (cf. Fig. 2 c).
5. Manual removal of falsely assigned pixels, which are caused by artefacts.

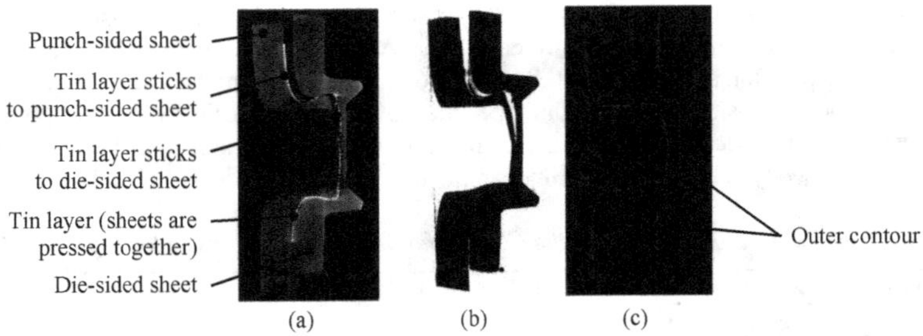

Fig. 2: The original cross section image from the CT scan (a) and the extraction steps binarization (b) and contour extraction and manual removal of false pixels (c)

The interface is defined as a one pixel thick curve which marks the center of the tin layer where the sheets are pressed together and where the tin layer sticks to the punch-sided sheet (cf. Fig. 2 a). When the tin layer sticks to the die-sided sheet, the contour of the punch-sided sheet is taken from the outer contour extraction and added to the interface curve manually. The extraction of the interface using the tin layer is described in the following and is visualized in Fig. 3.

1. Binarization (Original_image>175) to extract the tin layer pixels.
2. Removal of falsely assigned pixels, which are caused by artefacts.
3. Thickening (imdilate(original_image,strel('disk',20)) of the pixels to fill gaps and smooth of the curve of the tin layer.
4. Thinning bwmorph(original_image,'thin',Inf) of the thickened layer to a curve with a thickness of one pixel.

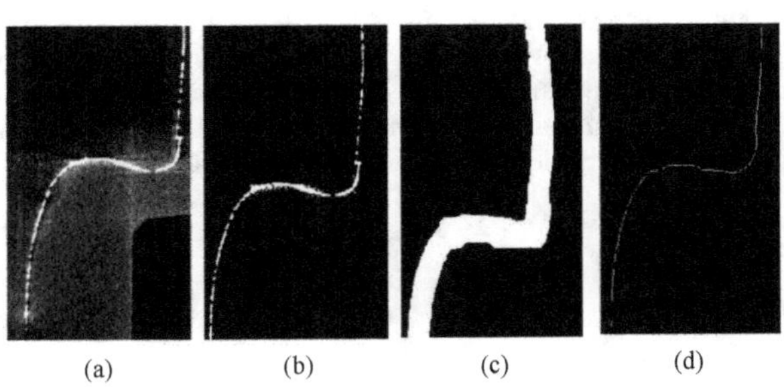

Fig. 3: The original cross section image with focus on a half of the interface (a) and the extraction steps binarization (b), thickening (c) and thinning (d).

Comparison procedure. The extracted contours and interface curves are compared quantitatively. For this purpose, both images have to be aligned. Therefore, the coordinate origin is defined by the symmetry curve and the outer contour of the die-sided sheet as this sheet barely deforms (cf. Fig. 4 a). The extracted contours and interfaces of an in- and ex-situ CT scan and the qualitative comparison of these curves can be exemplarily seen in Fig. 5.

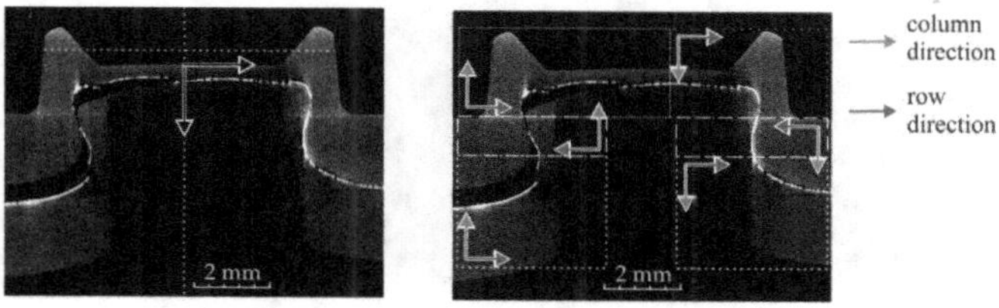

(a) (b)

Fig. 4: Definition of the coordinate system (a) and of the regions for the extraction of the measurement points (b)

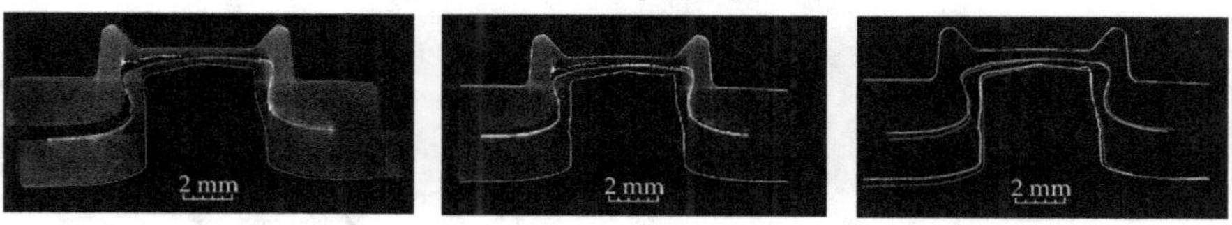

(a) (b) (c)

Fig. 5: Extracted contours (in-situ: red, ex-situ: yellow) and interfaces (in-situ: blue, ex-situ: green) from the specimen C04_A_CV_261 (displacement 0.65 mm): in-situ CT (a), ex-situ CT (b) and comparison (c).

For the comparison of the contours and interfaces, the respective pixels (measurement points) of the reference image need to be read in along the interface direction. Here, the algorithm detects the coordinates of the first measurement point in each row. In order to correctly read in the measurement points along the contour or the interface, the image is divided into six regions with specific row and column directions (cf. Fig. 4 b). Then, the minimum distance between one measurement point of the reference image and each measurement point of the comparing image is calculated. The minimum distance accounts for the deviation at this measurement point. This is repeated for all measurement points of the reference image.

Results

Visualization of deviations. The deviations can be visualized in two ways, which can be exemplarily seen for the interface in Fig. 6 a and b. In Fig. 6 a, the pixel coordinates of the measurement points of the contour are plotted. Additionally, the measurement points of the ex-situ CT are colored indicating the absolute deviation from the ex-situ CT measurement. Thus, the direction and size of the deviations can be concluded. In Fig. 6 b, the measurement points of the ex-situ CT are shown in sequence on the axis of abscissae and the absolute deviation at each measurement point is shown on the axis of ordinates. Thus, with this plot, the size of the deviations over the clinching point in relation to the average deviation can be compared more easily. In order to facilitate compatibility of both plots, marking points (green) indicating particular measurement points can be seen in both plots.

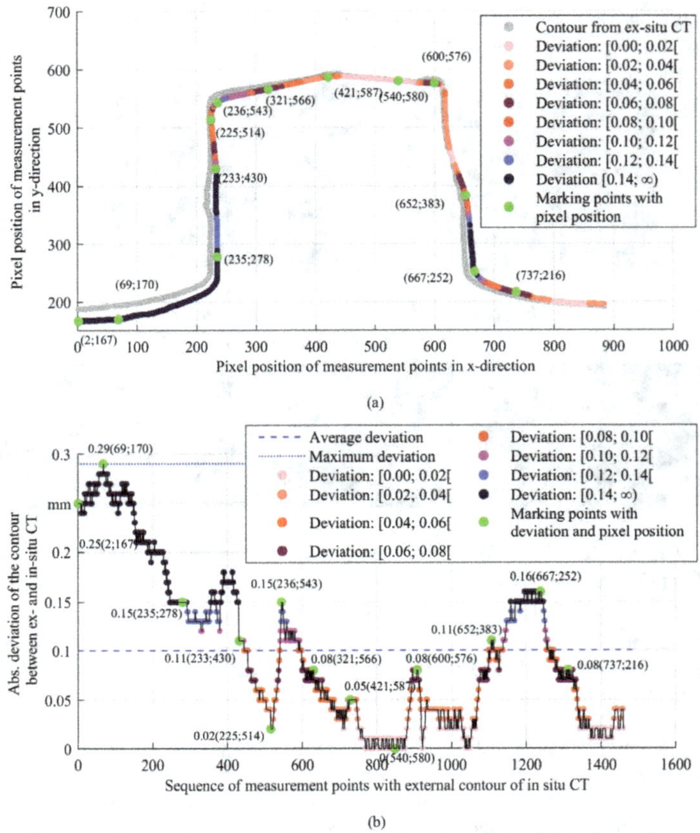

Fig. 6: Plots of the interface in the ex-situ CT (grey) and in the in-situ CT (coloured according to the absolute deviation from ex-situ CT) (a) and absolute deviation between ex-and in-situ CT against measurement points (b) for C04_A_CV_261 (0.65 mm displacement).

Evaluation of deviations. The method of comparing the contours and the interfaces is applied to all specimens and their in and ex-situ measurements. The average deviation is shown in Fig. 7. It can be stated, that the absolute deviations of contours and interfaces are within a range of 0.03 mm and 0.16 mm. Additionally, evaluating the regression curve, there is a tendency towards higher absolute deviations between ex- and in-situ CT measurements at higher displacements. For the macrosection comparison, the CT scan of the cut specimen is compared with the ex-situ CT measurement of the respective specimen for each displacement (Fig. 8). Here, the deviations are in a range 0.05 mm and 0.25 mm. In this case, too, according to the incline of the regression curve, the deviations tend to increase with higher displacements. Furthermore, the deviations between CT measurement of a cut specimen and ex-situ CT scan are almost always higher than the ex- and in-situ CT deviations. Considering the coefficients of determination, the linear model used for the regression curve only partially predicts the average deviation.

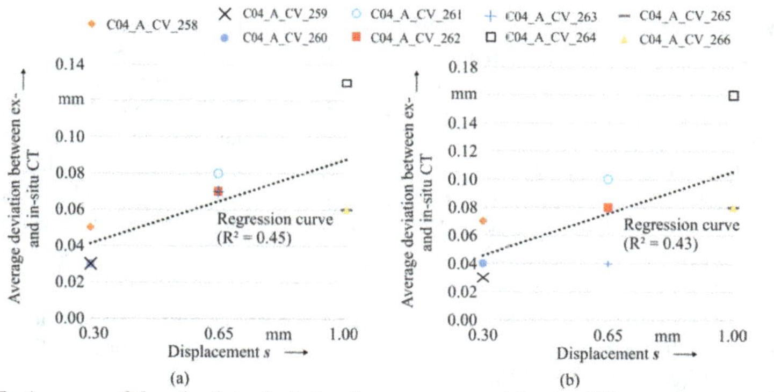

Fig. 7: Average of the absolute deviation between ex- and in-situ CT measurement of the interface (a) and the contours (b). Specimen labels according to TRR285 nomenclature.

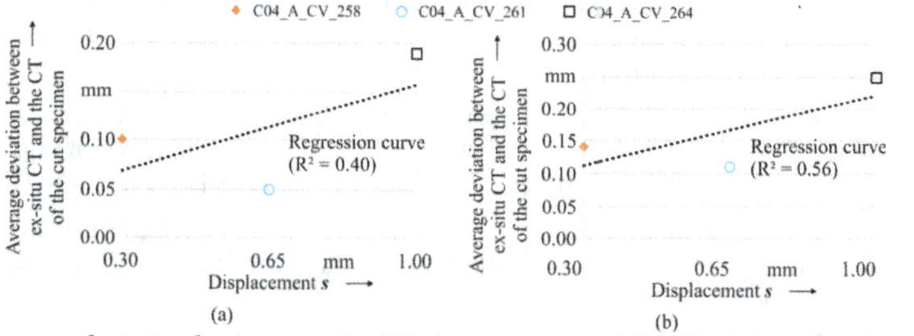

Fig. 8: Average deviation between ex-situ CT measurement and the CT of the cut specimen of the interface (a) and the contours (b). Specimen labels according to TRR285 nomenclature.

Summary and Conclusions

Due to elastic springback, ex- and in-situ measurements of clinching points deviate from each other, which is often not considered in the validation of numerical results. In this paper, a method is presented to extract the contour and the interface from the cross section image of a digital volume of a clinching point. Additionally, the deviations between the contours and interfaces of different volumes can be quantified. To show the application of the method, specimens are shear tested to different displacement levels. Then, the digital volumes of these specimens are created by ex- and in-situ CT measurements as well as by CT measurements of the cut specimen. Eventually, the method is used to calculate the deviations between the measurement methods for the different displacement levels.

The described method allows the quantification of deviations between contours and interfaces between different measuring methods. Thus, it can be used to measure the elastic springback effects and the deformation effects due to residual stresses, which occur during unloading and preparation for macrosectioning. Additionally, the method could be applied for measuring the deviations between individual specimens in order to quantify the variation in the manufacturing process. Furthermore, specimens without tin layer could be investigated. In this case, only the contour could be measured at low displacement levels. At higher displacement levels, the interface can be additionally measured where a gap occurs between the sheets.

It can be concluded that the springback effects can alter the deformation state notably. Moreover, average springback seems to increase with rising deformation level and is considerably higher at macrosectioning preparation. The measured springback is 5 to 25 times higher than the voxel size. However, the influence of the extraction of the interface procedure on the accuracy still needs to be investigated. Thus, this method showed that in-situ CT measurements can gain more precise information about the deformation state of clinching points during loading as there are clear influences of ex-situ measurement methods with increasing displacement level.

However, it is obvious that introducing a tin layer influences the tribological system during clinching and testing. Therefore, the results are not yet comparable to standard clinch points. In future investigations, the influence of the tin layer on joint formation and failure has to be analyzed and methods to minimize the influence have to be developed. Moreover, it should be investigated, if creeping effects occur during the CT scan and how they affect the results of the measurements.

Acknowledgements
This research was funded by the German Research Foundation (DFG) within the project Transregional Collaborative Research Centre 285 (TRR 285) (project number 418701707), sub-project C04 (project number 426959879).

References

[1] J. Varis, Economics of clinched joint compared to riveted joint and example of applying calculations to a volume product, J. Mater. Process. Technol. 172 (2006) 130–138. https://doi.org/10.1016/j.jmatprotec.2005.09.009

[2] M. Heibeck, M. Rudolph, N. Modler, M. Reuter, A. Filippatos, Characterizing material liberation of multi-material lightweight structures from shredding experiments and finite element simulations, Miner. Eng. 172 (2021) 107142. https://doi.org/10.1016/j.mineng.2021.107142

[3] International Organization for Standardization, ISO 12996:2013(en), Mechanical joining — Destructive testing of joints — Specimen dimensions and test procedure for tensile shear testing of single joints, Geneva 25.160.10, 2013.

[4] R. Kupfer, D. Köhler, D. Römisch, S. Wituschek, L. Ewenz, J. Kalich, D. Weiß, B. Sadeghian, M. Busch, J. Krüger, M. Neuser, O. Grydin, M. Böhnke, C.-R. Bielak, J. Troschitz, Clinching of Aluminum Materials – Methods for the Continuous Characterization of Process, Microstructure and Properties, Journal of Advanced Joining Processes 5 (2022) 100108. https://doi.org/10.1016/j.jajp.2022.100108

[5] S. Carmignato, W. Dewulf, R. Leach (Eds.), Industrial X-Ray Computed Tomography, Springer International Publishing, Cham, 2018. https://doi.org/10.1007/978-3-319-59573-3

[6] R. Füßel, M. Gude, A. Mertel, In-situ X-ray computed tomography analysis of adhesively bonded riveted lap joints, in: 17th European Conference on Composite Materials, Munich, 2016.

[7] H. Kunz, E. Stammen, K. Dilger, Local displacement measurements within adhesives using particle tracking and In Situ computed tomography, J. Adhes. 93 (2017) 531–549. https://doi.org/10.1080/00218464.2015.1113526

[8] F. Pottmeyer, J. Bittner, P. Pinter, K.A. Weidenmann, In-Situ CT Damage Analysis of Metal Inserts Embedded in Carbon Fiber-Reinforced Plastics, Exp. Mech. 57 (2017) 1411–1422. https://doi.org/10.1007/s11340-017-0312-0

[9] D. Köhler, R. Kupfer, J. Troschitz, M. Gude, In Situ Computed Tomography-Analysis of a Single-Lap Shear Test with Clinch Points, Materials (Basel, Switzerland) 14 (2021). https://doi.org/10.3390/ma14081859

[10] D. Köhler, R. Kupfer, J. Troschitz, M. Gude, Clinching in In-situ CT – Experimental Study on Suitable Tool Materials, ESAFORM 2021 (2021). https://doi.org/10.25518/esaform21.2781

Method for considering the process chain in the design process of clinched components

Sven Martin[1,a*] and Thomas Tröster[1,b]

[1]Paderborn University, Faculty of Mechanical Engineering, Chair of Automotive Lightweight Design, Warburger Straße 100, 33098 Paderborn, Germany

[a]sven.martin@uni-paderborn.de, [b]thomas.troester@uni-paderborn.de

Keywords: Joining, Modelling, Robustness

Abstract. To ensure a high quality design of joined components, it is essential to consider the joining process chain since deviations in the process chain can lead to failure of the component. Since eliminating the cause of process deviations can be very cost-intensive, it is desirable to ensure component and joining safety by making small changes. A method to regain all safety factors after a change in the process chain by making minor adjustments in the joining process and the design is presented here. The method is demonstrated on a clinched component consisting of a U-profile and a strike sheet.

Introduction

If any deviations and disturbances in the joining process chain occur, the real operational loads on the joints of joined components do not correspond to the calculated loads from the design and dimensioning phase [1]. For this reason, increasing efforts are made to optimize manufacturing tolerances [2]. It is further known that different joint stiffnesses lead to load redistribution between the joints [3]. Different joint stiffnesses are obtained for example, when clinching sheets with the same process parameters, but different deformation states in the joining area before joining [3]. In addition, robustness analyses are often carried out for safety reasons, in which it is assumed that a certain proportion of the joints are broken. Such analyses aim to achieve the desired component properties despite the statistically distributed failed joints. [4] Therefore, despite the use of simulation, the number of joints and the value of the safety factors tends to be higher than required. A more lightweight design with lower safety factors would be possible if the cause-effect relationships between the process parameters on the loads on the joints are describable and predictable. Additionally, the process and quality variables have to be continuously checked during manufacturing to identify deviations. When designing and dimensioning a high quality joining design with low safety factors for a versatile joining process chain, three challenges arise.

Firstly, the cause-effect relationships must be identified. Secondly, these relationships must be exploited to increase the degree of utilization as much as possible at every joint, and thirdly continuous monitoring over the whole process chain must be implemented. This paper demonstrates how deviations can be considered in virtual product development and countermeasures can be taken. Thus, a list of causes and countermeasures can be shared with the manufacturing personnel to ensure high product quality.

Method

The method shown in Fig. 2 c) is used to design a clinched component with consideration of a versatile joining process chain. The method is based on cause-effect relationships between influencing parameters on the loads on the joints. The geometry of the joint is an influencing factor and is affected by many parameters. For example, the existing deformation state and sheet thickness in the joining area before joining [3], the used tools [5] and the joining process parameters [6] influences the final joint geometry. The deviating joint geometry leads to a

Content from this work may be used under the terms of the Creative Commons Attribution 3.0 license. Any further distribution of this work must maintain attribution to the author(s) and the title of the work, journal citation and DOI. Published under license by Materials Research Forum LLC.

difference in stiffness, maximum elasticity and maximum force in destructive characterization such as shear lap or head tension tests [3]. Some of the influencing parameters on the clinched joint properties are mentioned in Fig. 2 a1). In addition, the loads on a joint also depend on external geometry-related component-dependent influences. In [7] it is shown that the load on a clinched joint depends on the geometry in the proximity of a clinched joint. In the numerical analysis of [3], it is illustrated that the variance of the load on a clinched joint in components of pre-deformed sheets depends on both, the global sheet thinning and the change in the stiffnesses of the clinched joints due to local sheet thinning. Further factors influencing the load on a joint are the number and distribution of the joints [8], as well as the components' geometry and the applied load and boundary conditions. Some influence values on the mechanical behavior of a clinched joint are listed in Fig. 2 a1). In Fig. 2 a2) some influences values on the joints' loads are named.

For this purpose, in numerical evaluations the variables of Fig. 2 b) were used to describe the loads on the joints and their scatter. In finite element simulations of joined components, joints are represented with equivalent elements in which the applied load is described with a three-dimensional force and three-dimensional moment vector. The equivalent force consists of the components f_r, f_s and f_t. Since the torsional moment of clinched joints is assumed to be zero, the bending moment leads to a normal force component which, when added to the force f_r, gives F_N. This equivalent force is shown in Fig. 2 b3). The sum of the shear force components f_s, f_t gives F_S. The load angle α is enclosed by F_S and F_N. The angle between the s-axis and F_S is the orientation angle β. With the absolute values F_S and F_N, a load on a joint can be drawn in a shear force normal force diagram, as shown in Fig. 2 b4). The maximum elastic force (black curve in Fig. 2 b4)) and the maximum force (grey curve in Fig. 2 b4)) that can be applied to the joint depending on the load angle α are defined with limit curves. These limit curves are either created by connecting measured points from experimental tests under shear, normal and mixed loading or are approximated by mathematical formulas.

The scatter in a joint can be shown by plotting the different loads on a joint in a shear force-normal force diagram, as in Fig. 2 b5). A hull curve can be derived for the resulting point cloud, which can be approximated by an ellipse. By expressing the ellipse shape and orientation by the center, the minor and major semi-axis and the ellipse orientation angle Θ, global statistical evaluations can be performed as shown in [1].

The cause-effect relationships from Fig. 2 a) and the evaluation measures from Fig. 2 b), are transferred to the design method shown in Fig. 2 c). The input of the method is an initial design of a clinched component, dimensioned with its nominal values. In the first step a change in the joining process chain is assumed. Either the properties of the sheet metal parts, the properties of some or all joints or the properties of joints and sheet metal parts are changed. Subsequently, virtual component testing is performed using finite element simulations with adjusted sheet and joint properties. After the calculation, it is checked whether the component meets its requirements during the entire test sequence. If the joined component does not meet its specifications, an adjustment is made. Either one or more joints (left cycle in Fig. 2 c)), the number and distribution of joints or the component geometry (right cycle in Fig. 2 c)) are changed.

Demonstration of the method on a clinched component

The method is demonstrated for elastic dimensioning. In this case, two sheet metal parts made of HCT590X with a sheet thickness of s_0=1.5 mm are joined. The geometry of the component and the load case are shown in Fig. 1. The corresponding values can be taken from Table 1.

For dimensioning, the clinched joint is characterized by experimental shear lap and head tension tests. The maximum force, the maximum elastic force, and the stiffness are determined from the obtained mean force-displacement curves for both set-ups. The coefficient of determination R^2 for the stiffness fit was set to 0.995. The obtained values can be taken from Table 2. In the simulation

Materials Research Proceedings **25** (2023) 163-170 https://doi.org/10.21741/9781644902417-21

model, depicted in Fig. 1 b), the component is meshed with shell elements with a mean edge length of 1 mm. The material behavior of the profile and the strike sheet is modeled with a Hocket-Sherby constitutive model, determined with tensile and layer compression tests. The clinched joints are represented by a beam and a rigid spider with a diameter of 10 mm. The equivalent stiffnesses are obtained by curve fitting and are also presented in Table 2. All simulations are calculated using the implicit solver of LS-DYNA R13.

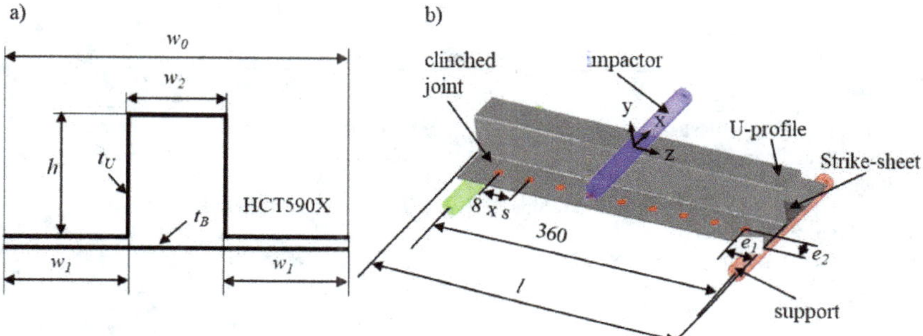

Fig. 1: a) Geometry of the joined component in section view b) FE-model and load case

Table 1: Information of the initial configuration

<table>
<tr><td colspan="2">Initial Geometry of U-Profile</td><td colspan="3">Joining design</td><td colspan="4">Clinching Tool</td></tr>
<tr><td></td><td>distance [mm]</td><td colspan="2">number of clinched joints</td><td>18</td><td colspan="2">punch diameter [mm]</td><td colspan="2">8</td></tr>
<tr><td>h</td><td>50</td><td colspan="2">e_1 [mm]</td><td>40</td><td colspan="2">die diameter [mm]</td><td colspan="2">10</td></tr>
<tr><td>w_1</td><td>50</td><td colspan="2">e_2 [mm]</td><td>25</td><td colspan="2">die depth [mm]</td><td colspan="2">1.6</td></tr>
<tr><td>w_2</td><td>40</td><td colspan="2">s [mm]</td><td>40</td><td colspan="4">TOX label: TOX A50100 / TOX BD8016</td></tr>
<tr><td>t_U</td><td>1.5</td><td colspan="7">Load Case</td></tr>
<tr><td>l</td><td>400</td><td colspan="2">support distance [mm]</td><td>360</td><td colspan="2">diameter of impactor and support [mm]</td><td colspan="2">10</td></tr>
<tr><td colspan="3">Geometry of Strike Sheet</td><td>w_0 [mm]</td><td>140</td><td>l [mm]</td><td>40</td><td>t_B [mm]</td><td>1.5 or 1.3</td></tr>
<tr><td colspan="2">Specification</td><td colspan="2">minimum impactor force [kN]</td><td>14</td><td colspan="2">minimum impactor depth [mm]</td><td colspan="2">3</td></tr>
</table>

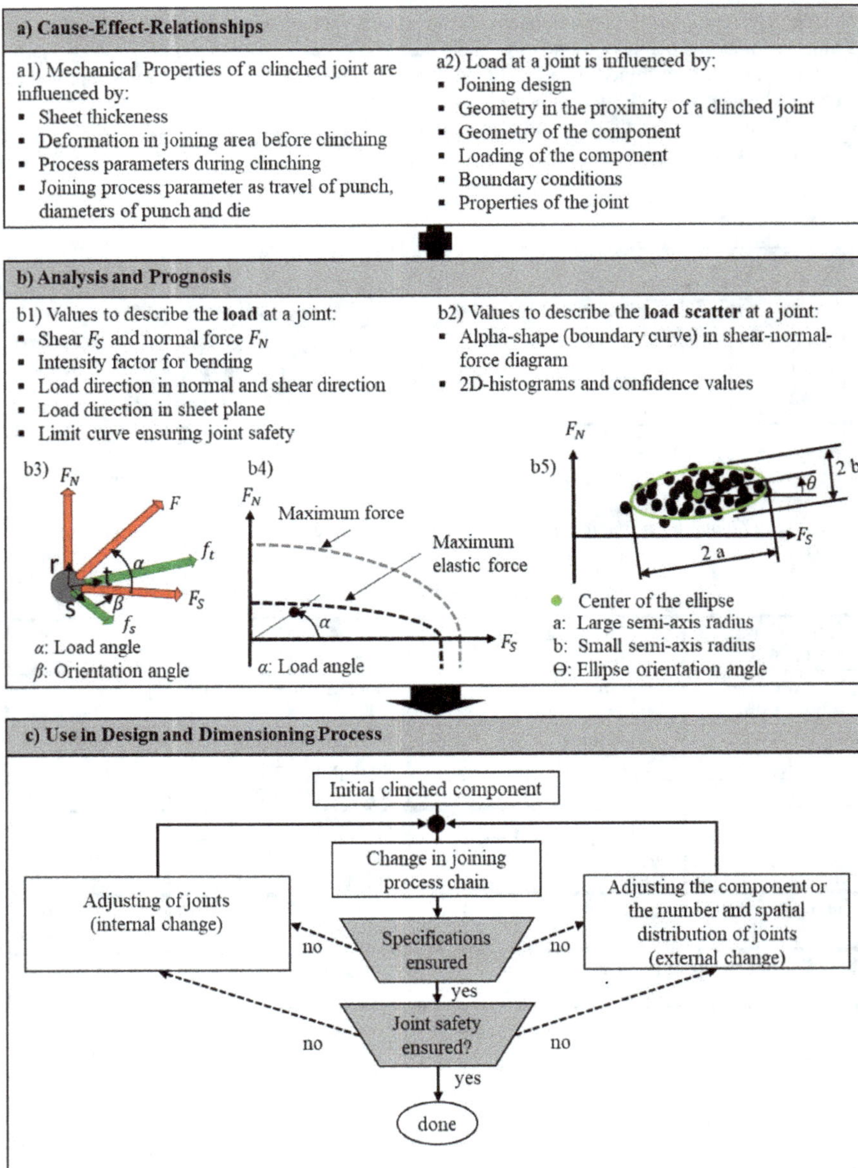

Fig. 2: Design Method for considering the process chain for a clinched component a) Cause-effect relationships b) Analysis and Prognosis c) Method

The property of the clinched joint for the combination of two 1.5 mm thick sheets are determined experimentally, as written before. In Table 3 the used equivalent stiffnesses and the maximum elastic force for combinations of two 1.3 mm thick sheets are listed. From the numerical investigation of [3] it can be seen that for a sheet thickness combination of 1.27 / 1.27 mm reduced by pre-stretching, the equivalent normal stiffness is reduced by approx. factor 0.4 and the equivalent shear stiffness by approx. factor 0.6. The prerequisite in this research was that the joining process parameters remained constant. It is also assumed that the maximum elastic forces are reduced by the same factor as the stiffnesses. Moreover, the same values are assumed for unequal pairings of 1.5 mm and 1.3 mm thick sheets as for 1.3 mm / 1.3 mm combinations. For a precise analysis these values must be determined experimentally for undeformed sheets and for the combinations of different sheet thicknesses for both joining directions separately.

Table 2: Properties of the clinched joint when joining two sheets with a thickness of 1.5 mm

Normal stiffness	~ 1670 N / mm
Shear stiffness	~ 22000 N / mm
Maximum elastic normal force	~ 700 N
Maximum elastic shear force	~ 3000 N
Maximum normal force	~ 2600 N
Maximum shear force	~ 4200 N
Equivalent normal stiffness	7700 N / mm
Equivalent shear stiffness	30250 N / mm

Table 3: Equivalent properties for the clinched joint in simulation for combinations with at least one sheet thickness of 1.3 mm

Equivalent normal stiffness*	3080 N / mm
Equivalent shear stiffness*	18150 N / mm
Maximum elastic normal force**	350
Maximum elastic shear force**	1800

* plausible estimation
** assumption

In Fig. 3, the limit curve for the maximum elastic force is approximated by the maximum elastic shear and the maximum elastic normal force. The maximum elastic shear is plotted on the abscissa and the maximum elastic normal force on the ordinate. Then, these points are connected in the first quadrant with an elliptic equation according to Eq. 1 and Eq. 2. In the fourth quadrant the limit curve is parallel to the ordinate and is described by Eq. 3 An internal change in a joint shifts the limit curve in the diagram and an external change leads to a different position of a point in the diagram.

$$F_S = F_{S,\ max,\ i} \cos(\alpha) \qquad \text{for} \qquad 0° \leq \alpha \leq 90° \tag{1}$$

$$F_N = F_{N,\ max,\ i} \sin(\alpha) \qquad \text{for} \qquad 0° \leq \alpha \leq 90° \tag{2}$$

$$F_N = F_{N,\ max,\ i} \qquad \text{for} \qquad -90° \leq \alpha \leq 0° \tag{3}$$

The method is demonstrated by considering sheet thickness thinning in the process chain. The requirement criteria for the component is, that it has to withstand an impactor displacement and an impactor force. Neither at an impactor force of 14 kN, nor at an impactor displacement of 3 mm the joints should be loaded plastically, but a local plastic flow of the material in the U-profile directly under the impactor is permissible. The requirements are stated in the last row of Table 1.

Materials Research Proceedings **25** (2023) 163-170 https://doi.org/10.21741/9781644902417-21

First, the load case is simulated with the initial design (Config. 1). Then, the sheet thicknesses of the strike sheet (Config. 2) and of the profile (Config. 3) are altered.

Configuration 1 (Initial Design):

U-profile and strike sheet thicknesses of 1.5 mm each; initial stiffness and limit curve of the clinched joint when joining 1.5 mm thick sheets (Table 2)

Configuration 2 (Thickness of strike sheet altered):

U-profile with a sheet thickness of 1.5mm; strike sheet with a sheet thickness of 1.3 mm; reduced stiffness and limit curve (Table 3)

Configuration 3: (Thickness of profile altered):

U-profile with a sheet thickness of 1.3 mm; strike sheet with a sheet thickness of 1.5 mm; reduced stiffness and limit curve (Table 3)

Configuration 1 is the reference design for which the requirements and the safety in the joint design should be verified. Configuration 2 represents a sheet thickness deviation of the strike sheet and Configuration 3 of the U-profile. The reduction of both sheet thicknesses is very similar to Configuration 3, since the worst clinched joint properties are assumed anyway and the strike sheet stiffness has hardly any influence, as will be shown. Therefore, this configuration is not discussed. To ensure that not only a quarter of the joints are visible in the shear force-normal force diagrams, the joints are not positioned exactly symmetrical.

Results

The forces are evaluated up to an impactor travel of 3.0 mm. In Fig. 3 the equivalent forces at a displacement of the impactor of 3.0 mm are shown. The calculation of Config. 1 show that the requirements have been met. In Fig. 3a) all points lie within the limit curve. The analysis of Config. 2 shows that the four outer joints are plastically loaded at an impactor depth of 3 mm which is depicted in Fig. 3 b). It can be also seen that the equivalent forces in the clinched joints are significantly lower in the design with the reduced the strike sheet thickness (red crosses) than in the initial design (grey crosses). Nevertheless, four points lie outside the reduced limit curve, after the sheet thickness reduction. According to the method, the outer joints must be positioned closer to the impactor. After shifting the four laterally joints 15 mm further towards the center in the y-direction, the joining safety is regained, which can be seen in Fig. 3 b) (green crosses). To show that the joint loads can also be reduced by changing the geometry within the proximity of the joint, the reduction of the loads with circular beads is shown in orange in Fig. 3 c). The bead has a height of 1.2 mm, a width of approx. 3 mm and is inserted in a diameter of 15 mm around the joint. In the given case, however, the insertion of those beads is not sufficient to regain joining safety. When reducing the sheet thickness of the U-profile (Config. 3), the specification (Table 1) is not met since the force of 14 kN is not reached. Therefore, the height of the profile is increased by 5 mm. As a result, the targeted force level is reached again. The joint design can remain unchanged despite the reduced joint properties because all points lie within the limit curve shown in Fig. 3 d). Since there is no implementation of cause-effect relations between global bending and the joints' load, the increase of the profile height was calculated manually.

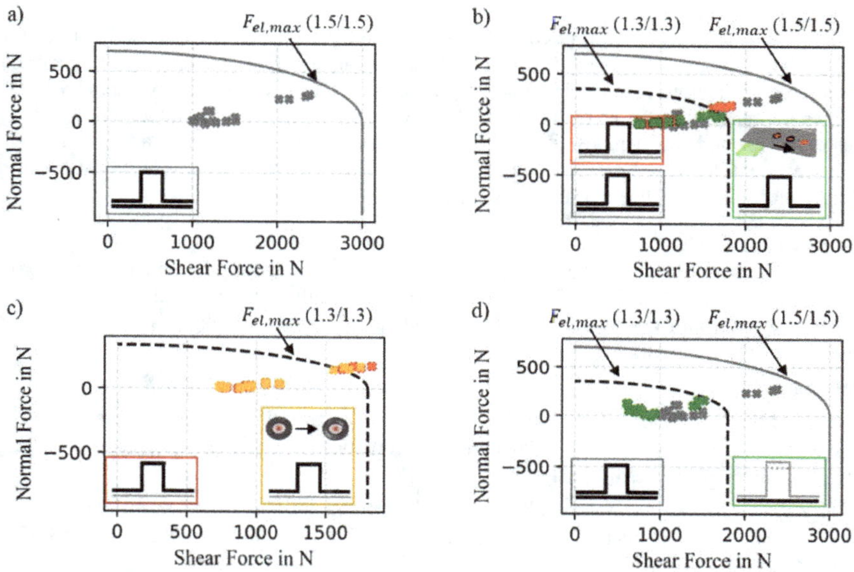

Fig. 3: Loads in the joints at an impactor depth of 3.0 mm a) Config. 1 b) Config. 2 - Optimized Design for a strike sheet with a thickness of 1.3 mm c) Config 2 - Inserting circular beads around the outer joints for a strike sheet with a thickness of 1.3 mm d) Config. 3 - Optimized Design for a U-profile with a thickness of 1.3 mm

Discussion

It was shown that with the presented method the versatile joining process chain can already be considered in the design phase. For the applicability of the method, it is essential that the cause-effect relationships are known and that the required data have been determined. In particular, the different mechanical properties of the resulting joints at different conditions in the joining area before clinching must be known. The method demands a huge amount of data, which increases when the method is extended for the non-linear dimensioning of joined components. As shown here, estimated assumptions can also be used, if no precise experimental data is available. The results are then, of course, more prone to error.

The geometry of the component can be changed either globally or locally. If utilizing the local component geometry in the proximity of the joint, linear beads can also be used in addition to circular beads. If linear beads are inserted, it must be considered that their effect depends not only on the load situation but also on their alignment to the joint. In addition to the option shown here of creating a list of countermeasures for manufacturing, when a deviation in the process chain is detected, the method can also be used to derive more robust designs, or to achieve more similar utilization rates in the joints.

Summary

With the presented method, it is possible to create a catalogue of countermeasures for production personnel on how to intervene in the component and joining design in the event of certain changes in the process chain to regain joining safety and to still met the component requirements. This ensures the manufacturing of high quality joined components despite deviations in the process chain. The method was demonstrated on a clinched component consisting of a U-profile and a strike sheet. At the same time, the method can also be used to obtain more robust designs, or to achieve more similar utilization rates in the joints.

Acknowledgement

The funding by the Deutsche Forschungsgemeinschaft (DFG, German Research Foundation) – TRR 285 – Project-ID 418701707, subproject B01 is gratefully acknowledged. Thanks, are also due to the Paderborn Center for Parallel Computing (PC2) for funding this project with the computing time provided.

References

[1] S. Martin, T. Tröster, Joint point loadings in car bodies – the influence of manufacturing tolerances and scatter in material properties, ESAFORM 2021 (2021). https://doi.org/10.25518/esaform21.3801

[2] H. Zheng, F. Litwa, M. Bohn, K. Paetzold, Tolerance optimization for sheet metal parts based on joining simulation, Procedia CIRP 100 (2021) 583-588. https://doi.org/10.1016/j.procir.2021.05.127

[3] S. Martin, C.R. Bielak, M. Bobbert, T. Tröster, G. Meschut, Numerical investigation of the clinched joint loadings considering the initial pre-strain in the joining area, J. Prod. Eng. 16(2) (2022) 261-273. https://doi.org/10.1007/s11740-021-01103-w

[4] S. Donders, M. Brughams, L. Hermans, N. Tzannetakis, The effect of spot weld failure on dynamic vehicle performance. Sound and Vibration 39(4) (2005) 16-25.

[5] J. Mucha, W. Witkowski, The clinching joints strength analysis in the aspects of changes in the forming technology and load conditions, Thin-Wall. Struct. 82 (2014) 55-66. https://doi.org/10.1016/j.tws.2014.04.001

[6] M.B. Tenorio, S.F. Lajarin, M.L. Gipiela, P.V.P. Marcondes, The influence of tool geometry and process parameters on joined sheets by clinching, J. Braz. Soc. Mech. Sci 41(2) (2019) 1-11. https://doi.org/10.1007/s40430-018-1539-0

[7] S. Martin, K. Kurtusic, T. Tröster, Influence of the Surrounding Sheet Geometry on a Clinched Joint, Key Eng. Mat. 926 (2022) 1505-1515. https://doi.org/10.4028/p-09md1c

[8] A.B. Ryberg, L. Nilsson, Spot weld reduction methods for automotive structures, Struct. Multidisc. Optim. 53 (2016) 923-934. https://doi.org/10.1007/s00158-015-1355-4

Numerical and experimental investigations of piercing fibre-reinforced thermoplastics

Benjamin Gröger[a*], Jingjing Wang[b], René Füßel[c], Juliane Troschitz[d], Daniel Köhler[e], Christian Vogel[f], Andreas Hornig[g], Robert Kupfer[h] and Maik Gude[i]

Institute of Lightweight Engineering and Polymer Technology, Technische Universität Dresden, Holbeinstraße 3, 01307 Dresden, Germany

[a]benjamin.groeger@tu-dresden.de, [b]jingjing.wang@tu-dresden.de, [c]rene.fuessel@tu-dresden.de, [d]juliane.troschitz@tu-dresden.de, [e]daniel.koehler@tu-dresden.de, [f]christian.vogel@tu-dresden.de, [g]andreas.hornig@tu-dresden.de, [h]robert.kupfer@tu-dresden.de, [i]maik.gude@tu-dresden.de

Keywords: Joining, Simulation, Fluid-Structure Interaction

Abstract. The joining of continuous fibre-reinforced thermoplastic composites (TPC) by means of plastic deformation often results in a complex material structure in the forming zone. Especially process- and material-related parameters have high influence on the local deformation behaviour and therefore on the properties of the joint. In this paper, the focus is on mechanical joining processes based on the principle of moulding holes by a tapered pin. For the investigations, a simplified test is used in which the pin is pushed through a heated TPC plate in the thickness direction. By the pin movement the fibres and molten matrix are displaced radially and along the tool motion direction. Detailed investigations of the resultant material structure by computed tomography and numerical simulations are performed with varying pin tool geometries with bidirectional TPC material. For numerical analysis, the Arbitrary-Lagrangian-Eulerian method combined with a multi-filament approach is used. The result show that the tool geometry has a strong influence on the piercing force, the resultant material structure, and the occurring phenomena. It could be shown, that the simulation is capable to predict the resultant material structure.

Introduction

With their high specific stiffness and strength thermoplastic composites (TPC) offer a good opportunity to achieve a high degree of lightweight and increase resource efficiency. Due to the inhomogeneous inner material structure of orientated brittle fibres and ductile isotropic thermoplastic matrix an anisotropic material behaviour results. Hence, a variety of novel joining technologies have already been developed to join TPC components to adjacent assemblies. Bonding, the state of the art joining method, requires high standards for preparation and long curing times. Therefore, new mechanical joining technologies are in the scope of research [1], especially in the field of clinching technology [2]. The local forming processes of this joining method lead to a change of the material structure [3]. The changed material structure directly influences the load bearing behaviour and the capacity of the joints [4].

In thermal-assisted mechanical joining processes for metal-TPC-joints a tool [5] or auxiliary element [6] is moved in thickness direction through the material generating a hole and a form closure. Depending on process conditions (e.g. tool concept and velocity, process temperature), material properties of the matrix (e.g. melting temperature) and fibres (e.g. Young's modulus, fracture strain) different material structure phenomena occur during hole forming [7]. Due to the softening of the matrix at higher temperatures, the viscous character of the matrix increases. Consequently, the fibre-matrix interactions (FMI) can be described as a fluid-structure-interaction

Content from this work may be used under the terms of the Creative Commons Attribution 3.0 license. Any further distribution of this work must maintain attribution to the author(s) and the title of the work, journal citation and DOI. Published under license by Materials Research Forum LLC.

(FSI). Besides fibre-fibre interactions (FFI) can be observed [7]. Joining by piercing based on moulded holes is characterized by a displacement of fibres in in- and out-of-plane direction caused by pin penetration [8] and independent matrix flows around the pin [9]. Such joints are often assessed by the resultant load bearing behaviour in structural tests [10]. The required joining force to displace fibres and matrix for a single pin often cannot be measured due to the test setup [11] or the application of multi-pin arrays [12]. In [8] a single pin geometry is used for joining and the influence of heating area and temperature on the joining force and resultant material structure is investigated. Due to the temperature gradients in most processes, the joining zone cannot be assumed to be isothermal. This leads to a different material behaviour within TPC because of the temperature dependency of the matrix.

By simulating such joining processes, the resultant material structure and load bearing capacity can be predicted. However, this requires the modelling of the textile architecture [13] for continuous fibres. For moulded holes with in-plane reorientation of the fibres, also an analytical approach is used in [14] to predict the fibre path.

In the present paper, the interaction of the piercing process and material structure is shown. Therefore, a simple test setup with different pin geometries is used. Due to a lack of information when using transient local heating, the whole test setup is heated up to process temperature and kept constant instead. The resultant material structure is investigated by computed tomography (CT). Based on the findings, numerical models are developed and evaluated. A multi-scale approach and the Arbitrary-Lagrangian-Eulerian (ALE)-method is used [15].

Experimental setup

The specimens are made of multi-layered uni-directional glass fibre (GF) reinforced polypropylene (PP) Borealis BJ-100HP with a melting temperature of 165 °C. The specimens are made of four unidirectional TPC-sheets with layup of $[0°/90°]_s$, and have a thickness of 2 mm with a dimension of (80 x 80) mm^2.

The test setup shown in Fig. 1a is integrated in a standard testing machine Zwick Z1465 (ZwickRoell GmbH & Co KG) equipped with a heating chamber and a 100 kN load cell. The isothermal conditions of 180 °C, controlled by three thermocouples at the specimen, the pin and within the chamber, are realized within a dwell time of 30 min. To prevent sticking between the heated specimen and the die/blank holder, thin perforated films (0.2 mm) made of polytetraflour-ethylene are fixed to the metal parts. The pin is pushed path-controlled with a velocity of 1000 mm/min into the TPC. After the piercing process, the pin rests within the TPC until the specimen temperature is cooled down to 150 °C by an air flow system near the pin. To investigate the process-structure-interaction (PSI) in terms of the tool-structure-interaction, two different pin geometries shown in Fig. 1b are used.

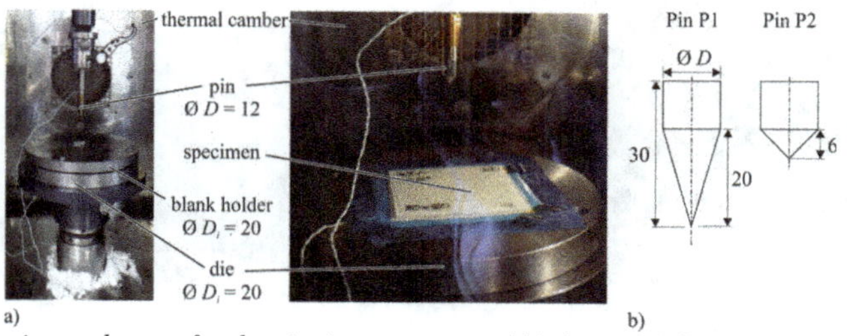

Fig. 1: a) Experimental setup for the pinning process and b) the two different pin geometries; all dimensions in mm.

Numerical setup

For the numerical simulations, the multi-filaments (MF) approach from [16] in combination with the ALE-method is used. The MF approach is based on a digital element chain [17], where the properties of the beam element summarize the properties of a certain amount of single fibres [18]. The setup is presented in Fig. 2 and the mesh information with the characteristic element length l_e are given in Table 1.

Due to the orthogonal four layers of MF, a half model is set up. The boundary conditions (BC) at the ends of all MF are defined to only allow a translation in fibre direction. Whereas, the nodes of the MF in the 90°-layers, which lie in the symmetry plane, can only move in-plane without rotational degrees of freedom. For both MF directions side walls with a penalty contact definition are implemented to avoid fibre movement out of the fluid domain. In accordance to the specimen dimension and the joining zone area, in 0°-direction the full length of 80 mm is modelled. For the 90°-layers, only 40 mm width of the layers is considered (Fig. 2 right). The tools and side walls are assumed to be rigid. Furthermore, except for the pin, they have fixed BC. Consequently, the clamping pressure between blank holder and die is not taken into account. Penalty contact definitions between MF, die and blank holder are defined. The pin has a displacement controlled translation definition in thickness direction through the layers. It is defined as a solid to avoid leakage in the ALE calculation. However, a special forming contact based on shell elements proved to be suited to model the sliding between MF and the pin. In order to apply this contact, a shell mesh is defined at the surface nodes of the solid. A constrained-based coupling algorithm between fluid and MF is defined. The beam element chain with a diameter of 0.5 mm representing the MF exhibits an elastic material behaviour with a Young's modulus of 57 MPa. For the matrix properties, a simplified viscous material model with a constant dynamic viscosity η of 530 Pas (based on experimental results) is used. The simulations are carried out with a mass scaling factor of 10,000.

Table 1: Mesh information of the numerical setup for pin P1.

Domain	l_e [mm]	Number of elements
Fluid	0.5	361,000
Multi-filament	1.5	4,988
Tools	-	54,904
Punch shell	-	8,204
Walls	-	21,120
Σ		450,816 elements; 480,678 nodes

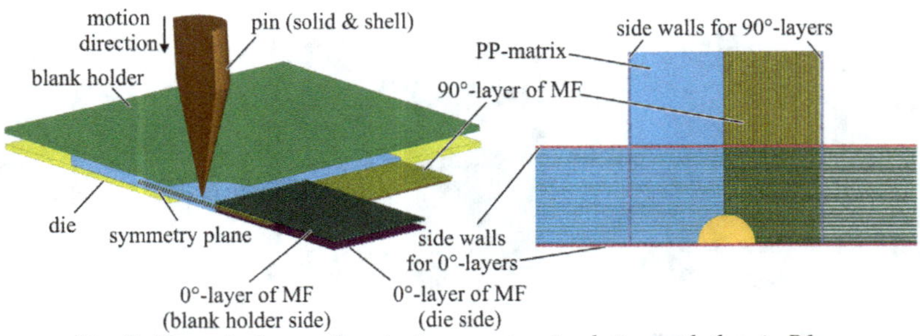

Fig. 2: Numerical setup for pinning process simulation with the pin P1.

Results and Validation

In Fig 3, the force-displacement curves of experiments where both pin geometries are investigated at two specimens (specimen number in brackets) are given. It can be seen that both the force peak and the descent of the curves of the blunt pin P2 are much higher than for the sharp pin P1. It is also evident that the force peak of P1 is achieved at higher displacements due to the long taper shape.

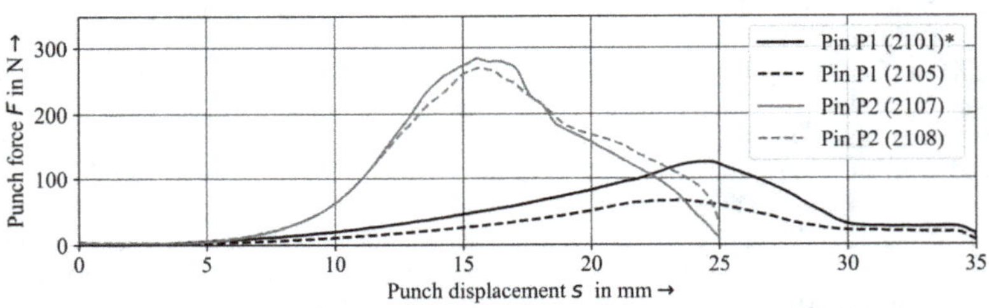

*Fig. 3: Force-displacement curves for the different pin geometries. *modified curve.*

The resultant material structure of the specimen '2105' is presented in Fig. 4. On the blank holder side, a bulk is formed in 0°-direction consisting of matrix and in 90°-direction of yarns and matrix (1) (Fig. 4a). The 0°-yarns, pointing directly to the pin, are bent in piercing direction (2). The nearby yarns are displaced in 90°-direction (3). This radial reorientation leads to a torsion of fibre bundles in the yarn and the bulk forming on the free surface on the blank holder side (3) (Fig. 3a-b). The 90°-yarns are also bent in thickness direction (4). The increasing pin movement leads to a compaction, shifting and twisting (torsion) of all yarns in each layer (5) (Fig. 4b). Especially, the yarns of the die-sided layers are shifted out-of-plane. Here, on the one hand, the yarns distant to the pin are more shifted than twisted (6). On the other hand, the yarns close to the pin are splayed which is caused by the tool-yarn-friction and the motion of the pin (7). The friction also causes yarn shifting in fibre direction of 0°-yarns. The interaction between the different layers (e.g. FFI) lead to a shifting and bending (ondulation) of the 90°-yarns in radial direction to the pin (8) (Fig. 4c). The tool-yarn-friction, FFI and the yarn tension, result in a compaction of the yarns around the pin (5). Hence, matrix flow in the outer area of the piercing zone occurs, leading to matrix rich zones (9). Because of the shifting, bending and splaying, voids occur around the forming zone (10).

Likewise, the resultant material structure of specimen '2107' in combination with pin P2 is investigated (Fig. 5). In contrast to pin P1, the bulk consists of matrix in both fibre directions (1) (Fig. 5a). As with pin P1, the yarns in all layers pointing directly to the pin are bent in thickness direction (2, 4). Due to the blunt shape of pin P2, no reorientation in 90°-direction or bulk forming consisting of matrix and yarns on the blank holder side occur. Instead, all fibres are compacted, distorted and shifted in out-of-plane direction (5) (Fig. 5b). Therefore, the height *s* with 12.35 mm of the displaced splayed fibres (7) is higher in comparison the configuration of P1 with 3.86 mm. Due to the high displacement of the 0°-yarn and the FFI, the ondulation of the 90°-yarns increases (8) (Fig. 5c). In contrast to the P1 configuration, all yarns of each layer are twisted and distorted similarly in the piercing direction (5). The compaction in the resultant material structure is less than in P1, which reduces the formation of matrix rich zones (9). Because of the more significant reorientation phenomena of shifting, bending and torsion, more voids in the forming zone can be observed (10).

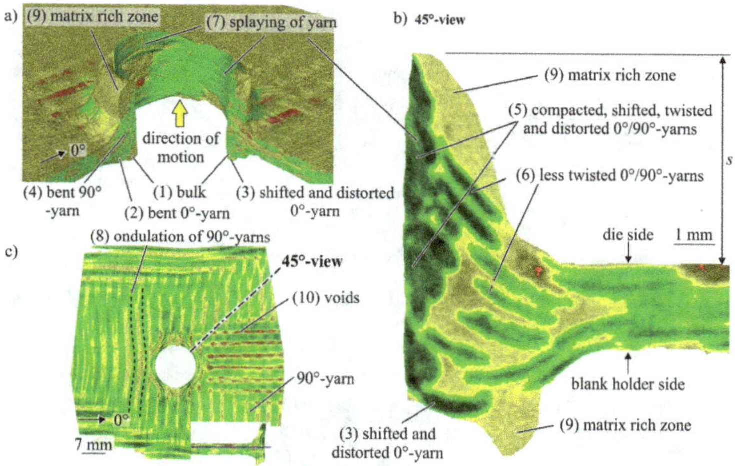

Fig. 4: CT-analysis of the resultant material structure for pin P1 (yarns highlighted in green, matrix in yellow and air volumes inside the TPC in red).

The results of the simulations for pin P1 with the resultant material structure are given in Fig. 6 and for P2 in Fig. 7. The modelling strategy is sensitive to the used pin geometry. Thus, the results show the influence of the pin sharpness on the resultant material structure. For pin P1, it can be seen that the MF of the blank holder sided layer are splayed (Fig. 6b). Also the bottom and mid layers are displaced continuously (Fig. 6b). Six MF next to the pin tip are excessively displaced and deformed, which can be caused by an unrealistic contact behaviour that results from the MF sticking to the pin and neglecting the clamping pressure (Fig. 6c). Excluding the "stuck" fibres, a displacement of approx. 7.4 mm is measured. Furthermore, the pin is not pushed through all layers despite a sufficient displacement (cf. Fig. 3).

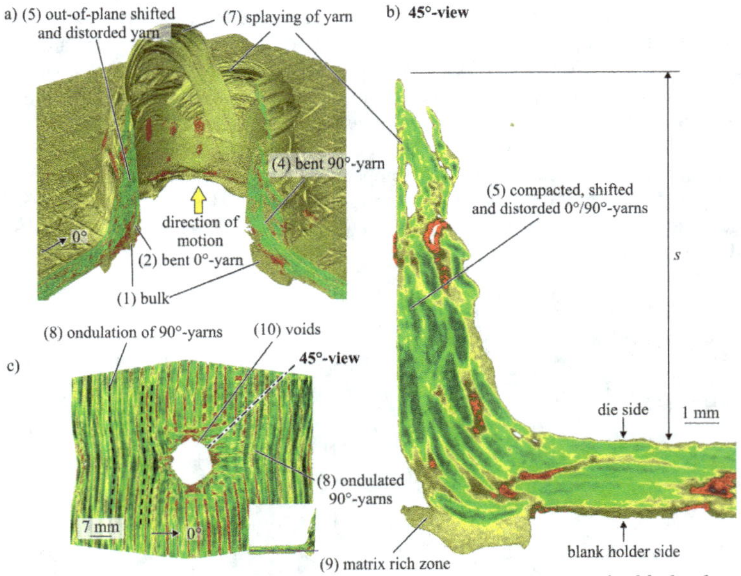

Fig. 5: CT-analysis of the resultant material structure for pin P2 (yarns highlighted in green, matrix in yellow and air volumes inside the TPC in red).

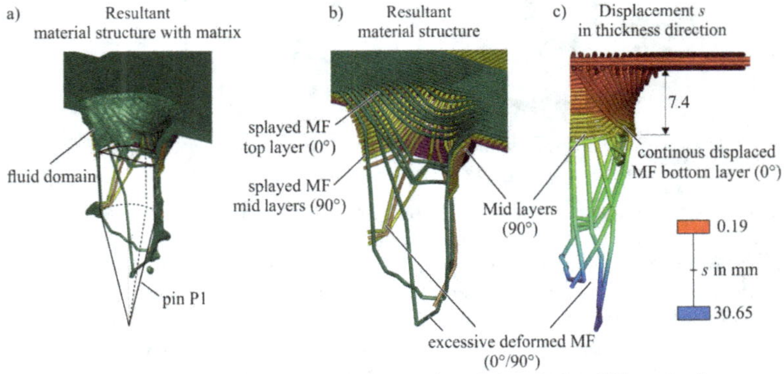

Fig. 6: Resultant material structure of pin P1 with the ALE method.

The resultant material structure for pin P2 shows a higher displacement of the MF in thickness direction in the piercing area. The splaying of the mid and bottom layer can be seen in (Fig. 7b and c). Additionally, an excessive deformation of the MF under the pin can be seen. It can be concluded that an unrealistic contact behaviour leads to the deformation. Due to the large displacement of the MF, the simulation exhibits areas with less matrix (Fig. 7a). Despite the non-accurate contact definition and the simplified material behaviour (constant viscosity, linear elastic), the results show main phenomena of the piercing process, which also occur in the experiment.

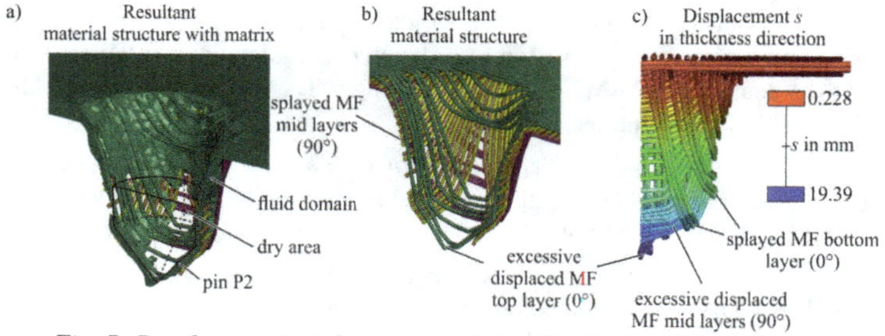

Fig. 7: Resultant material structure of pin P2 with the ALE method.

Summary and Outlook

The piercing forces and the resultant material structure of a piercing process are investigated with different pin shapes using isothermal process conditions. Based on the test setup, numerical simulations are carried out with the MF approach and ALE method. The numerical results represent the main phenomena of bending and displacement of MF in out-of-plane direction. Due to the limited contact definitions and the simplified assumptions used to determine the resultant stiffness of the MF, the shifting in fibre direction cannot be predicted accurately. For a more elaborated comparison, experimental and numerical investigations have to be carried out with focus on bending stiffness of yarns under process conditions. Additionally, the initial inhomogeneous textile architecture of the TPC sheets has to be modelled more accurately.

Funding

This research was funded by the Deutsche Forschungsgemeinschaft (DFG, German Research Foundation) -TRR 285-Project-ID 418701707 sub-project A03 and C04.

Acknowledgement

The authors are grateful to the Center for Information Services and High Performance Computing [Zentrum für Informationsdienste und Hochleistungsrechnen (ZIH)] at TU Dresden for providing its facilities for high throughput calculations.

References

[1] A. Galińska, C. Galiński. Mechanical Joining of Fibre Reinforced Polymer Composites to Metals-A Review. Part II: Riveting, Clinching, Non-Adhesive Form-Locked Joints, Pin and Loop Joining. Polymers (Basel) 2020;12(8). https://doi.org/10.3390/polym12081681

[2] F. Lambiase, S.I.Scipioni, C.-J. Lee, D.-C. Ko, Liu F. A State-of-the-Art Review on Advanced Joining Processes for Metal-Composite and Metal-Polymer Hybrid Structures. Materials (Basel) 2021;14(8). https://doi.org/10.3390/ma14081890

[3] B. Gröger, J. Troschitz, J. Vorderbrüggen, C. Vogel, R. Kupfer, G. Meschut et al. Clinching of Thermoplastic Composites and Metals-A Comparison of Three Novel Joining Technologies. Materials (Basel) 2021;14(9). https://doi.org/10.3390/ma14092286

[4] J. Troschitz, R. Füßel, R. Kupfer, M. Gude. Damage Analysis of Thermoplastic Composites with Embedded Metal Inserts Using In Situ Computed Tomography. J. Compos. Sci. 2022;6(10):287. https://doi.org/10.3390/jcs6100287

[5] H. Seidlitz, C. Gerstenberger, T. Osiecki, S. Simon, L. Kroll. High-performance lightweight structures with Fiber Reinforced Thermoplastics and Structured Metal Thin Sheets. JMSR 2014;4(1). https://doi.org/10.5539/jmsr.v4n1p28

[6] F. Hirsch, S. Müller, M. Machens, R. Staschko, N. Fuchs, M. Kästner. Simulation of self-piercing rivetting processes in fibre reinforced polymers: Material modelling and parameter identification. Journal of Materials Processing Technology 2017; 241(1):164–77. https://doi.org/10.1016/j.jmatprotec.2016.10.010

[7] J. Troschitz, B. Gröger, V. Würfel, R. Kupfer, M. Gude. Joining Processes for Fibre-Reinforced Thermoplastics: Phenomena and Characterisation. Materials (Basel) 2022;15(15). https://doi.org/10.3390/ma15155454

[8] N.W.A. Brown, C.M. Worrall, S.L. Ogin, P.A. Smith. Investigation into the mechanical properties of thermoplastic composites containing holes machined by a thermally-assisted piercing (TAP) process. Advanced Manufacturing: Polymer & Composites Science 2015;1(4):199–209. https://doi.org/10.1080/20550340.2015.1117748

[9] J. Popp, T. Kleffel, D. Römisch, T. Papke, M. Merklein, D. Drummer. Fiber Orientation Mechanism of Continuous Fiber Reinforced Thermoplastics Hybrid Parts Joined with Metallic Pins. Appl Compos Mater 2021;28(4):951–72. https://doi.org/10.1007/s10443-021-09892-0.

[10] M. Kraus, P. Frey, T. Kleffel, D. Drummer, M. Merklein. Mechanical joining without auxiliary element by cold formed pins for multi-material-systems. In: Proceedings of the 22nd international Esaform conference on material forming: Esaform 2019. AIP Publishing; 2019, p. 50006.

[11] S.T. Amancio-Filho, L.-A. Blaga (eds.). Joining of Polymer-Metal Hybrid Structures. Hoboken, NJ: John Wiley & Sons, Inc; 2018.

[12] E.E. Feistauer, S.T. Amancio-Filho. Ultrasonic Joining of Lightweight Alloy/Fiber-Reinforced Polymer Hybrid Structures. In: Amancio-Filho ST, Blaga L-A, editors. Joining of Polymer-Metal Hybrid Structures. Hoboken, NJ: John Wiley & Sons, Inc; 2018, p. 307–333.

[13] B. Gröger, A. Hornig, A. Hoog, M. Gude. Modelling of thermally supported clinching of fibre-reinforced thermoplastics: Approaches on mesoscale considering large deformations and fibre failure. ESAFORM 2021. https://doi.org/10.25518/esaform21.4293

[14] S. Roth, F. Pracisnore, S. Coutandin, J. Fleischer. A new approach for modelling the fibre path in bolted joints of continuous fibre reinforced composites. Composite Structures 2020;243(2):112184. https://doi.org/10.1016/j.compstruct.2020.112184

[15] J. Donea, A. Huerta, J.-P. Ponthot, Rodriguez-Ferran A. Arbitrary Lagrangian-Eulerian Methods. In: Stein E, Borst René de, Hughes TJR, editors. Encyclopedia of Computational Mechanics. Chichester, UK: John Wiley & Sons, Ltd; 2004, p. 657.

[16] S.D. Green, A.C. Long, B.S.F. El Said, S.R. Hallett. Numerical modelling of 3D woven preform deformations. Com. Struct. 2014; 108:747–56. https://doi.org/10.1016/j.compstruct.2013.10.015.

[17] Y. Wang, X. Sun. Digital-element simulation of textile processes. Composites Science and Technology 2001;61(2):311–9. https://doi.org/10.1016/S0266-3538(00)00223-2

[18] B. Gröger, V. Würfel, A. Hornig, M. Gude. Forming process induced material structure of fibre-reinforced thermoplastics - Experimental and numerical investigation of a bladder-assisted moulding process. Journal of Advanced Joining Processes 2022;5(7):100100. https://doi.org/10.1016/j.jajp.2022.100100

Methodical approach for the design and dimensioning of mechanical clinched assemblies

Christoph Zirngibl[1,a *], Sven Martin[2,b], Christian Steinfelder[3,c], Benjamin Schleich[4,d], Thomas Tröster[2,e], Alexander Brosius[3,f] and Sandro Wartzack[1,g]

[1]Friedrich-Alexander Universität Erlangen-Nürnberg, Engineering Design, 91058 Erlangen, Germany

[2]Paderborn University, Chair of Automotive Lightweight Design, 33098 Paderborn, Germany

[3]Technische Universität Dresden, Chair of Forming and Machining Processes, 01069 Dresden, Germany

[4]Technical University of Darmstadt, Product Life Cycle Management, 64287 Darmstadt, Germany

[a]zirngibl@mfk.fau.de, [b]sven.martin@uni-paderborn.de, [c]christian.steinfelder@tu-dresden.de, [d]schleich@plcm.tu-darmstadt.de, [e]thomas.troester@uni-paderborn.de, [f]alexander.brosius@tu-dresden.de, [g]wartzack@mfk.fau.de

Keywords: Joining, Structural Analysis, Machine Learning

Abstract. The focus towards multi-material and lightweight assemblies, driven by legal requirements on reducing emissions and energy consumptions, reveals important drawbacks and disadvantages of established joining processes, such as welding. In this context, mechanical joining technologies, such as clinching, are becoming more and more relevant especially in the automotive industry. However, the availability of only few standards and almost none systematic design methods causes a still very time- and cost-intensive assembly development process considering mainly expert knowledge and a considerable amount of experimental studies. Motivated by this, the presented work introduces a novel approach for the methodical design and dimensioning of mechanically clinched assemblies. Therefore, the utilization of regression models, such as machine learning algorithms, combined with manufacturing knowledge ensures a reliable estimation of individual clinched joint characteristics. In addition, the implementation of an engineering workbench enables the following data-driven and knowledge-based generation of high-quality initial assembly designs already in early product development phases. In a subsequent analysis and adjustment, these designs are being improved while guaranteeing joining safety and loading conformity. The presented results indicate that the methodological approach can pave the way to a more systematic design process of mechanical joining assemblies, which can significantly shorten the required number of iteration loops and therefore the product development time.

Introduction

Given their ability to join coated and dissimilar sheet metals, mechanical joining technologies provide a high applicability for generating multi-material connections and thus great potentials for the realization of novel lightweight designs, such as in car body components [1, 2]. In this context, novel developments are available in the field of mechanical joining, such as in [3], whereby this contribution concentrates on clinching as an example. Therefore, the technology offers a fast as well as environmentally friendly procedure to join overlapping sheets, profiles or tubes entirely based on cold forming without additional joining components [4, 5]. However, for the realization of reliable joined assemblies, it is often required to involve both expert knowledge and

Content from this work may be used under the terms of the Creative Commons Attribution 3.0 license. Any further distribution of this work must maintain attribution to the author(s) and the title of the work, journal citation and DOI. Published under license by Materials Research Forum LLC.

experimental studies. This results in time-intensive trial-and-error development iterations and therefore high costs until a final design is identified that satisfies all product requirements (see Fig. 1). Thus, it is crucial to not only focus on the selection of feasible and suitable joining technologies, but also on the requirement- as well as the manufacturing-oriented design of the entire joint connection. Especially, the matching of the loading conformity and the achievable individual joint properties are highly important to guarantee a satisfying joining safety of joined components. For the latter, already a few contributions and standards demonstrated the methodical design of clinched joint assemblies based on geometrical characteristics (neck, interlock and bottom thickness) as well as resistances against shear and tensile loading. For instance, in [6] the analytical dimensioning of joint connections by means of stresses is demonstrated. In this regard, the calculation of suitable joint positions enables the load-balanced design of assemblies. Furthermore, to support the analysis of component and joint loadings, the so-called load path analysis can be used for the adjustment of joining designs. Therefore, common methods base on results of a finite element analysis (FEA) calculating the load distribution in a component or an overall assembly. The visualization of load paths in components allows the assessment of the design and structural integrity and can qualitatively show how the loads are redistributed in case of effects of geometric defects [7]. The methods known from the literature depend on the orientation of the component in space and the selected direction of investigation [8]. Usually, a separate load path plot is generated for each coordinate axis of the model, which results in three individual plots for the load path analysis in a 3D simulation. In this regard, the combined evaluation of all plots is necessary, which can be influenced by a lack of information within the individual plots and thus potentially leads to inaccurate results. For this purpose, the method of the effective force developed by Steinfelder and Brosius results in a single plot, which, as a second advantage, is invariant with respect to the orientation in space [9]. The concept was applied to a 2D rotational symmetric analysis of a clinched joint [10]. Nevertheless, a methodical approach that combines the data-driven and knowledge-based initial design of the assembly with the subsequent analysis and adjustment of individual components regarding their energy, stress and loading conformity is not available yet.

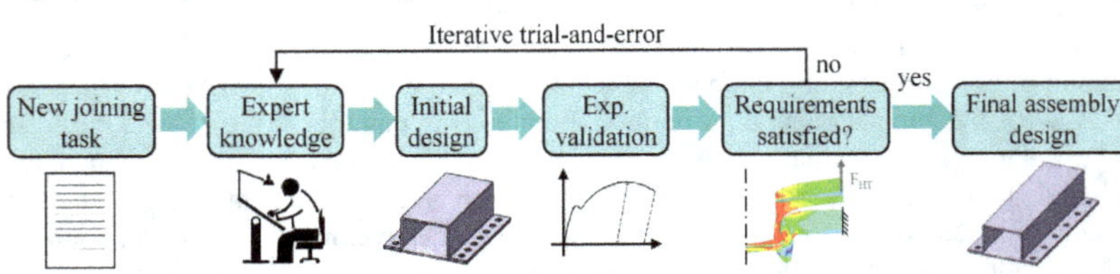

Fig. 1. Overview of the established approach for the design of mechanical joining assemblies

Overview of methodical approach

Motivated by this, the presented contribution introduces a methodical approach (see Fig. 2) that supports the product engineer regarding the comprehensive realization of reliable clinched assemblies by structuring, automating and simplifying the product development process. Especially, the involvement of data-driven and knowledge-based methods targets the efficient generation of high-quality initial product designs. Therefore, the implementation of an engineering workbench, described in [11], offers an efficient and applicable support for product engineers to configure individual joining elements by simultaneously guaranteeing a sufficient joining safety already in the early development phases. In addition to the transfer of expert and manufacturing knowledge into a machine-readable format, this also includes the utilization of predictive regression models for the accurate prediction of joint properties. Moreover, the setup of an interface between the engineering workbench and an available computer-aided design (CAD)

Sheet Metal 2023 Materials Research Forum LLC
Materials Research Proceedings **25** (2023) 179-186 https://doi.org/10.21741/9781644902417-23

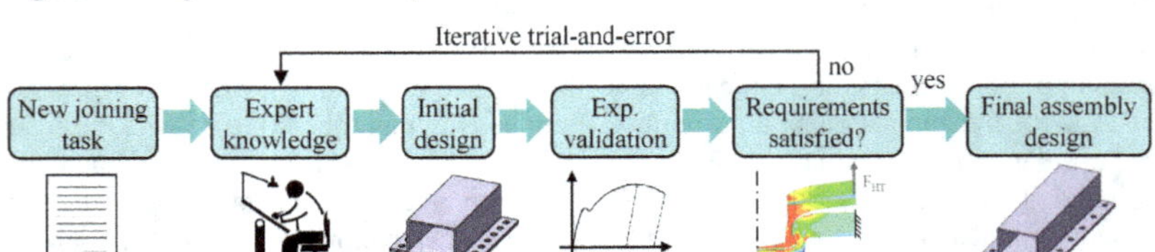

Software (here: Dassault Systèmes CATIA V5-6) enables both the automated generation of assembly visualizations and the following evaluation of the geometries regarding their manufacturing plausibility. Thus, the generated assembly design provides already a high-quality fundamental for the subsequent analysis of the dimensioning and the altering of the individual joining components. Therefore, the initial CAD models of the particular parts are meshed and a FE-model in LS-Dyna is generated. Moreover, previously calculated clinched joint stiffnesses are implemented in the simulation model and stand for the available elastic joint strength capabilities. Then, the definition of the loading combined with the bearing of the components enables the following evaluation and adjustment of the initial product design regarding a meaningful force distribution. In this context, a method analyzes the existing loads at joints, the load path and the effective force and, based on this, redistributes the force curve in the components by an altering product design. This can be achieved by changing either the position or property of the joints, the component geometry in the proximity of the joints or by changing the included number of joints. After the method has been executed, the final component and joining design can be determined and dimensioned regarding the given joining task requirements.

To get a deeper understanding of the approach, the next sections explain the methodical procedure for the generation of initial assemblies and the following analytical adjustment and dimensioning of these designs in more detail. Therefore, the joining task of a hat-profile and a blank considering a required minimum resistance against shear loading of 23000 N and given process characteristics (material: EN AW-6014-T4; sheet thickness: 2.0 mm) is used.

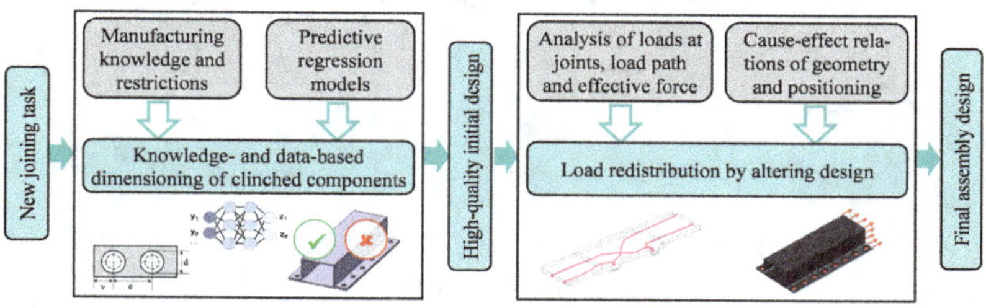

Fig. 2. Novel methodical approach for the design of mechanical joining assemblies

Data-driven and knowledge-based initial design of assemblies

Facing highly cost- and time-intensive development cycles, the following approach, illustrated in Fig. 3, enables the systematical generation of high-quality initial assembly designs already in the early product engineering phases. Therefore, the aim is to ensure both a requirement- and manufacturing-oriented dimensioning of joining connections and components. In this context, it is required to define relevant input information regarding a new joining task at the beginning. This involves for instance process-related characteristics, such as the applied joining tools or material properties, as well as requirements on the joint connection (e.g. shear or tensile load capacity). Based on this, the use of pre-trained predictive regression models enables the following estimation of individual clinched joint properties (e.g. geometrical characteristics). For this purpose and since this is among the most crucial step to guarantee a high reliability of the initial product design, linear and polynomial algorithms as well as artificial neural networks are implemented based on the results in [12]. Furthermore, the development of an engineering workbench [11] ensures both a high applicability of the approach and a consistent and automated data transfer between the product engineer and the involved software (CAD- or FE-environment). Therefore, the system offers estimated joint properties and the opportunity to configure a suitable assembly for the particular joining task. Based on the following design check regarding the compliance with manufacturing constraints and the fulfillment of requirements, an appropriate number of joints and

their positioning can be provided. In this context, the method proceeds according to the first-time-right principle and paves the way to a significant shortening of the product development cycle.
In the case of the chosen joining scenario, the product design involves a minimum of 16 clinched joints evenly distributed on both flanks of the hat-profile. For this purpose, the required minimum resistance against shear loading (23000 N) will initially be multiplied with a safety value of 1.3 in order to cover uncertainties within the joining process. Then, based on the estimation of achievable single joint properties, the minimum number of joints can be calculated. Since currently only less design principles are available in the field of mechanical joining, the selection of the safety value bases on few investigations regarding the process robustness of clinching, such as in [13]. Following, the application of plausibility checks ensures the compliance of the initial product design with implemented manufacturing specifications and limitations. Thus, the presented approach enables the generation of high-quality initial assembly designs already in the early product development phases considering data-driven and knowledge-based methods. As a summery, Fig. 3 shows an overview of the introduced steps.

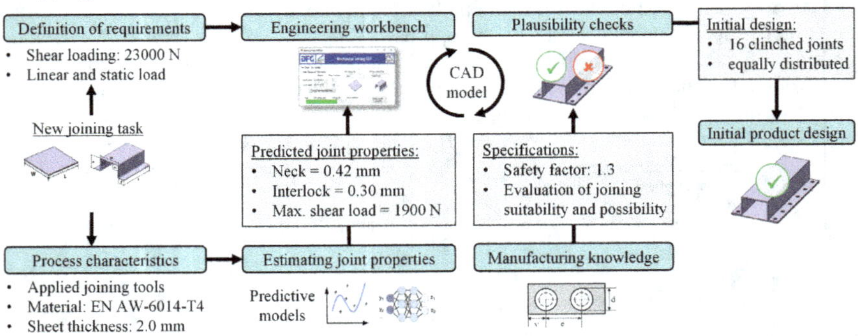

Fig. 3. Overview of the initial knowledge- and data-based design of an assembly

Analysis and load-adjustment of initial assembly design
The initial design of the assembly is then transferred as a CAD file combined with force-displacement information from the numerical experiments to the elastic dimensioning method shown in Fig. 2. The objective of the dimensioning is to ensure that the elastic joint safety is given at each clinched joint and that the sheet metal components are loaded elastically only. In this context, the method starts with the calculation of the maximum permissible force, which can be applied to the design components. For this purpose, the mechanical characteristics in normal and shear direction are determined (see Table 1) from the transferred force-displacement curves of the numerical characterization tests. In this regard, the stiffness is represented by the slope of a linear regression model (R^2 of 0.998, see Fig. 5 a), which is fitted to the provided data. Based on this, the components can be loaded elastically with a shear force up to 16000 N for the given use-case.

Table 1: Maximum force, maximum elastic force, stiffness in head-tension-test and shear-test

	Max. force in N	Max. elastic force in N	Stiffness in N/mm
Head tension test	~1700	~850	~1650
Lap shear test	~1900	~1300	~16000

For the check and dimensioning of the assembly, the components from the CAD file (Fig. 4 a) are meshed with shell elements on the mid surface for the finite element analysis. Here, the joints are mapped with a beam element and connected with a kinematic coupling by rigid spiders to the shells, which is depicted in Fig. 4 b) and 4 d). Therefore, the equivalent stiffness of the beam elements depends on the diameter of the rigid body spiders and thus on the joint's diameter and the edge length of the shell elements. In the presented use-case, the average shell length is 2 mm and the diameter of the rigid spider is 10 mm. The conversion of the experimentally determined stiffness into the stiffness of the equivalent beam is performed by curve fitting. The equivalent normal stiffness is approx. 1E6 N/mm and the equivalent shear stiffness is approx. 3E4 N/mm. The clinched joint is assumed unbendable, so the bending stiffness is set to 1E10 N/mm and the torsional stiffness is set to zero. In addition, Fig. 4 b) depicts the particular bearing conditions.

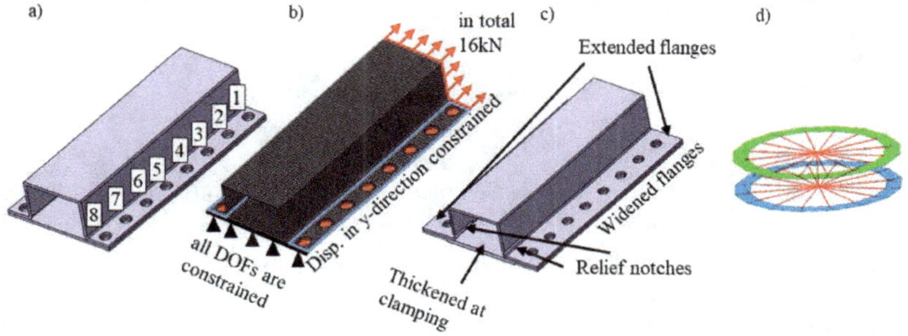

Fig. 4. a) Initial product design b) Load case the given joining task c) Final product design after improvements d) Beam-rigid-spider element

The evaluation of whether a joint is purely elastically loaded is based on a limit curve (Fig. 5 c). Here, this is represented in a simplified form via an ellipse equation in the first quadrant with the semi-axis maximum elastic shear force and maximum elastic normal force. In the fourth quadrant, the curve is parallel to the ordinate based on the equations in Fig. 5 d). The additionally drawn area by two lines with a slope of -1/3 and 1/3 illustrates the joints in which shear force is predominant. The influence of bending on the clinched joints is neglected. Due to symmetry, the subsequent evaluation is only carried out on the front flange with the joints 1 to 8 (see Fig. 4 a). Therefore, the individual loads at the joints are depicted in red (see Fig. 5 c). In that diagram, one can see that joint 8 (CJ8) is outside the limit curve. Moreover, the flanges and the walls of the U-section are tending to bent (see Fig. 5 b). To prevent this effect, the flange width is increased for 10 mm and the clinched joints are pushed 5 mm further out. In addition, the shear force can usually be reduced at the outermost joints if the distance to bearings is increased. For this reason, the profile is extended by 5 mm on the right and left. In order to reduce the shear force at joint 8 (CJ8), the left clamping is concentrated to the center. This enables an increased force flow to the other joints (CJ2 to CJ7). Furthermore, to prevent the aluminum sheet from being plastically deformed, the sheet thickness is increased locally from 2 mm to 3 mm at the left clamping. The changes are depicted in Fig. 4 c) and the obtained new loads at the joints are shown in black in Fig. 5 c). One can see that joint 8 (CJ8) is still outside the limit curve, which mainly bases on the high normal force share. Therefore, relief notches are provided on the side of the hat-profile. As a result, the joints 5 to 7 absorb more axial force based on a load redistribution (see Fig. 5 c green and black dots). However, joint 8 still lies very slightly above the limit curve. This can be changed by adjusting either the geometry again or the mechanical properties of joint 8 and thus the involved limit curve.

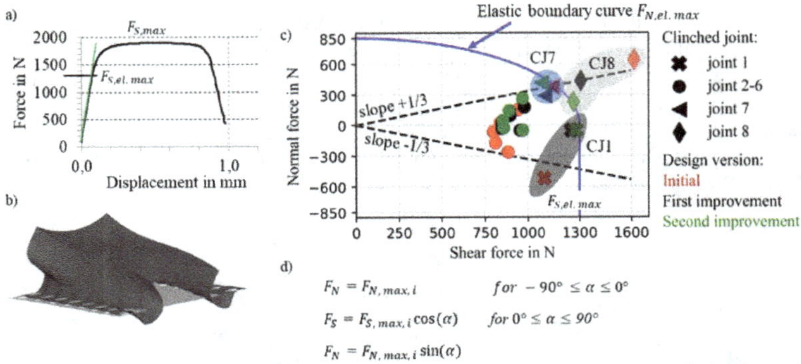

Fig. 5. a) Maximum forces and stiffness b) Deformation of the component with displacement scaled by factor 200 c) Loads at the clinched joints d) Equation of the elastic boundary curve

Besides this, the additional analysis of the normal and shear forces in the joints through the effective force method can identify the load path and the load direction in the assembly. This allows an evaluation of the areas that contribute more or less to the load transfer and, if necessary, how a load application in the component should be adapted. The method can currently also be applied to components with detailed modeling of the joints in 3D. The analysis of joined structures with a large number of joints modeled in detail is not efficient and feasible in terms of computing time. For an analysis of structures with shell element formulations and substitute models of joints, an adaptation of the calculation processes is necessary in further steps. Therefore, in the following, an exemplary investigation based on a 3D shear tension model with two clinched joints is considered. The following simplified example clearly shows why the change of the restraint has significantly relieved joint 8 in the first iteration step. The two clinched joints are arranged directly in series in the direction of loading, as in the flange area of the considered component, and are loaded to shear. The input data for the analysis are the results from a finite element analysis. The setup of the used model is shown in Fig. 6 a). For the calculation of the vector field, the principal stresses λ_i with the associated eigenvector $E_{i,j}$ are evaluated for each integration point per finite element in the model and the effective force $\overrightarrow{\boldsymbol{F}_{\text{eff}}}$, which refers to an infinitesimal area element A_{inf}, is calculated according to Eq. 1. Complementary to the normally used equivalent stress concept, this approach provides not only the information on the location of the highest load, but also the direction of the load in each case. Especially by the use of the direction information the versatility in the process chain can be increased, since modifications to compensate for unexpected changes in the process or at component level can be made in a targeted manner.

$$\overrightarrow{\boldsymbol{F}_{\text{eff}}} = \begin{Bmatrix} \lambda_{\text{I}} \cdot E_{\text{I},1} + \lambda_{\text{II}} \cdot E_{\text{II},1} + \lambda_{\text{III}} \cdot E_{\text{III},1} \\ \lambda_{\text{I}} \cdot E_{\text{I},2} + \lambda_{\text{II}} \cdot E_{\text{II},2} + \lambda_{\text{III}} \cdot E_{\text{III},2} \\ \lambda_{\text{I}} \cdot E_{\text{I},3} + \lambda_{\text{II}} \cdot E_{\text{II},3} + \lambda_{\text{III}} \cdot E_{\text{III},3} \end{Bmatrix} \cdot A_{\text{inf}}. \tag{1}$$

The method of effective force can be used as a support and provides additional information for the design of components and joints. In order to show load paths in a clear way, only nodes are used that have bearings or load inputs as boundary conditions. Two resulting load paths are shown in Fig. 6 b). It can be seen and concluded that loads applied closer to the plane of symmetry (Starting point 1) are mainly transmitted via the first clinched joint. Loads introduced further from the plane of symmetry (Starting point 2), on the other hand, are transmitted via the second clinched joint. When evaluating the effective force as a scalar value, see Fig. 6 c), additional information about the loads in the structure and joints can be visualized. This showed that the neck region of the first clinched joint is more involved in load transfer than the second clinched joint. Thus, the effective force method makes an important contribution to the analysis of loads in joined structures and provides further information about regions of the components involved in the load transfer and internal load directions, which assisted to adapt the joining design to the loadings.

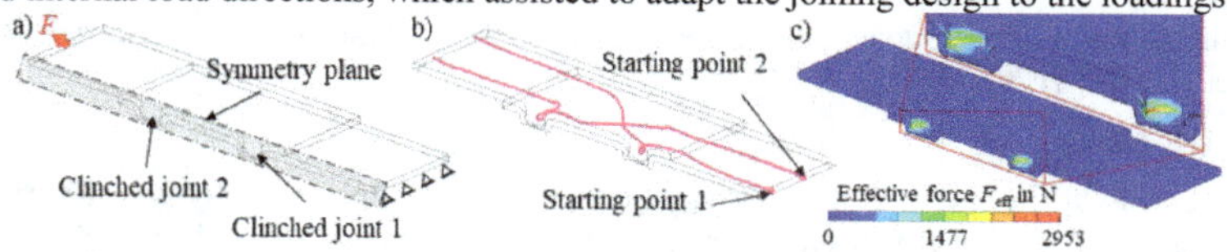

Fig. 6. a) Shear tensile specimen with two clinched joints b) Model visualization of load paths c) Effective force as a scalar value

Discussion

The demonstrated methodical approach provides the opportunity to assist the product engineer in the design and dimensioning of reliable and versatile mechanical clinched assemblies. In this regard, the involvement of knowledge-based and data-driven methods offers the manufacturing- and requirement-oriented generation of initial and high-quality designs already in the early product development phases. However, to guarantee a sufficient applicability of the method, it is crucial to not only focus on clinching but also on further mechanical joining technologies, such as pin joining or riveting. Moreover, currently the realization of complex joining tasks is limited on the available implemented parameterized CAD geometries, materials and sheet thicknesses. Thus, the generation of multi-material and -part combinations requires a deeper focus in future work. Furthermore, the functionality of the analysis and load path method has been shown for linear material and joint behavior. In order to take plasticity into account, it is required to consider a nonlinear behavior in the joint as well as in the material model. For the further analysis of the stress redistribution within a joint, it is recommended to replace the equivalent element by a detailed solid model for individual joints. In addition, the effective force method showed high potentials to visualize the load path in the component and joint graphically and thus to analyze, which regions of the component and which joints are involved in the load transfer. But the analysis of joined structures with a large number of joints currently requires an adaptation of the method with respect to the use of shell element formulations and substitute models for joints. The method is not dependent on a specific mechanical joining process and can also be applied to other processes.

Summary

Given costly and time-intensive development iterations, the presented contribution introduced an approach for the methodical design and dimensioning of mechanically clinched assemblies. Therefore, the integration of knowledge-based, data-driven, and simulation-based methods within an engineering workbench provide high potentials for the initial generation of high-quality designs already in the early product development phases. In this context, this requirement- and manufacturing-oriented configuration of components provides the fundamental for the following analytical adjustment of individual joining elements regarding a well-balanced energy, stress and load distribution. For this purpose, both the position and number of joints or the component's geometry can be systematically changed based on the introduced approach. After the method has been executed, the final component and the final joining design are determined and moreover, the joined component is dimensioned for its use. To demonstrate the entire methodical approach, the joining of a hat-profile and a blank considering a required resistance against shear loading of 23000 N and given process characteristics (material: EN AW-6014-T4; sheet thickness: 2.0 mm) was used as an exemplary joining task. In summary, the presented results can pave the way to a more methodical and simplified design and dimensioning process of mechanical clinched assemblies by providing a comprehensive support for the product engineer. This can also lead to a significant shortening of the required product development cycle and time.

Acknowledgement
Funded by the Deutsche Forschungsgemeinschaft (DFG, German Research Foundation) – TRR 285 – Project-ID 418701707.

Literature References

[1] J. Mucha, L. Kascak, E. Spisak, Joining the carbody sheets using clinching process with various thickness and mechanical property arrangements. Archives of Civil and Mechanical Engineering 11(1) (2011) 135-148. https://doi.org/10.1016/S1644-9665(12)60179-4

[2] R. Rzasinski, L. Kochanski, Joining methods in car body construction. IOP Conf. Ser.: Mater. Sci. Eng. 400022051 (2018). https://doi.org/10.1088/1757-899X/400/2/022051

[3] J.P.M. Pragana, R.F.V. Sampaio, I.M.F. Bragança, C.M.A. Silva, P.A.F. Martins, Injection Lap Riveting of Aluminum Busbars - A Thermo-Electro-Mechanical Investigation. J. Manuf. Mater. Process, 74(6) (2022). https://doi.org/10.3390/jmmp6040074

[4] M. Gude, G. Meschut et al., FOREL-Studie - Chancen und Herausforderungen im ressourcen-effizienten Leichtbau für die Elektromobilität. Dresden, (2015). ISBN 978-3-00-049681-3.

[5] L. Kascak, E. Spisak, J. Majernıkova, Clinching and Clinch-Riveting as a Green Alternative to Resistance Spot Welding. 2019 International Council on Technologies of Environmental Protection (ICTEP) (2019) 138-142. https://doi.org/10.1109/ICTEP48662.2019.8968973

[6] R. Neugebauer, F. Riedel, R. Marx, Entwicklung eines Konstruktionssystems für den rechnerischen Festigkeitsnachweis von punktförmig mechanisch gefügten Bauteilen. EFB-Forschungsbericht 323, EFB Hannover (2010). ISBN: 978-3-86776-359-2

[7] D. W. Kelly, M. Elsley, A procedure for determining load paths in elastic continua, Eng. Comput. 12 (1995) 415-424. https://doi.org/10.1108/02644409510799721

[8] K. Marhadi et al., Comparison of Quantitative and Qualitative Information Provided by Different Structural Load Path Definitions, Int. J. Simul. Multidiscip. Des. Optim. 3 (2009) 384-400.

[9] C. Steinfelder, A. Brosius, A New Approach for the Evaluation of Component and Joint Loads Based on Load Path Analysis, Lect. Notes Prod. Eng. (2020) 134-141. https://doi.org/10.1007/978-3-662-62138-7_14

[10] C. Steinfelder, S. Martin, et al., Load Path Transmission in Joining Elements, Key Eng. Mater. 883 (2021) 73-80. https://doi.org/10.4028/www.scientific.net/KEM.883.73

[11] C. Zirngibl, C. Sauer, et al., Knowledge and Data-Based Design and Dimensioning of Mechanical Joining Connections. ASME IDETC-CIE 2022, St. Louis Missouri (2022). https://doi.org/10.1115/DETC2022-89172

[12] C. Zirngibl et al., Application of reinforcement learning for the optimization of clinch joint characteristics. Prod. Eng. 16 (2022) 315-325. https://doi.org/10.1007/s11740-021-01098-4

[13] W.-G. Drossel, W. Israel, Sensitivitätsanalyse und Robustheitsbewertung beim mechanischen Fügen. EFB-Forschungsbericht 376, Hannover, (2013). ISBN: 978-3-86776-419-3

Influence of thermo-mechanical joining process on the microstructure of a hypoeutectic aluminium cast alloy

Thomas Borgert[1,a*], Moritz Neuser[2,b], Eugen Wiens[1,c], Olexandr Grydin[2,d], Werner Homberg[1,e] and Mirko Schaper[2,f]

[1]Forming and Machining Technology, Paderborn University, Warburger Straße 100, 33098 Paderborn, Germany

[2]Chair of Materials Science, Paderborn University, Mersinweg 7, 33100 Paderborn, Germany

[a]tb@luf.upb.de, [b]neuser@lwk.upb.de, [c]ew@luf.upb.de, [d]grydin@lwk.upb.de, [e]wh@luf.upb.de, [f]schaper@lwk.upb.de

Keywords: Joining, Casting, Microstructure

Abstract. Requirements of multi-material construction involve adjustments to standard joining techniques. Especially the growing importance of integral cast components poses additional engineering challenges for the industry. One approach to achieve these goals are adaptable joining elements formed by friction spinning. This approach uses friction-induced heat to form customisable joining elements to join sheets for different boundary conditions, even for brittle cast materials. It is possible to react immediately to adapt to the joining process inline and reduce the amount of different joining elements. As the joining partner serve casting plates of the aluminium casting alloy EN AC–AlSi9, which is processed in the sand casting. Joining hypoeutectic AlSi alloys constitutes a challenge because the brittle character of these cause cracks in the joint during conventional mechanical joining. Furthermore, the friction-induced heat of the novel joining process causes a finer microstructure in the hypoeutectic AlSi9 casting alloy. In particular, the eutectic Si is more fine-grained, resulting in higher joint ductility. This study indicates the joining suitability of a hypoeutectic aluminium casting alloy in combination with adaptive manufactured additional joining elements. Here, various mechanical and microstructural investigations validate the influence of the thermomechanical joining technique. In conclusion, the potential of this joining process is presented regarding the joinability of cast aluminium components.

Introduction

In the world of manufacturing technology, a constant state of change due to new requirements from government and society can be identified. The energy and resource efficiency of the production as well as the operation phase of products require an adaptation of conventional and well-known industrial processes [1]. In particular, the automobile industry, which produces many emissions, is continuously faced with new challenges [2]. In response, the industry is increasingly relying on a high-resolution multi-material mix to reduce the masses to be moved and thus the greenhouse gas emissions during the operation phase [3]. To increase productivity and to be able to produce application-adapted components, manufacturers rely on integral aluminium casting components. Aluminium casting alloys of the AlSi(Mg)–system are weldable to a limit, wherefore mechanical joining processes are a suitable alternative [4]. Due to the generally lower ductility and formability of these casting alloys, the connection of such components poses a challenge for joining technology, because the brittle properties, especially the low elongation at fracture, result in cracks in the joint [5].

Classical mechanical joining processes such as self-piercing riveting or clinching require a certain formability of the material, depending on the arrangement of the material [6]. Thermal joining processes such as the established resistance spot welding quickly reach their limits due to

Content from this work may be used under the terms of the Creative Commons Attribution 3.0 license. Any further distribution of this work must maintain attribution to the author(s) and the title of the work, journal citation and DOI. Published under license by Materials Research Forum LLC.

the conditional welding work of the cast alloys, the tendency to hot cracks and the resulting low dynamic fatigue strengths [4]. A way to overcome these limitations is to use alternative, ductile aluminium casting alloys that are adapted to the application due to targeted control of the solidification conditions [7]. These are more formable and thus better mechanically joinable than classic alloys. One such alloy is, for example, the AlSi9. Furthermore, there is the possibility to use refinement elements such as Sr or Na to modify the plate-like Si in the eutectic into a fine lamellar microstructure [8]. Hence, the brittle character of the microstructure can be improved into a more ductile, which allows better forming. In addition, to improve the ductility of the cast aluminium components, the (mechanical) joining process can also be adapted to the increased requirements of the joint. One possible group of processes are the thermomechanical joining processes. These processes use the friction-induced application of heat to reduce the materials flow stress locally (temperature is still below the melting temperature) in order to join them [9]. In recent years, friction stir welding in particular has been increasingly used to join aluminium sheets [10,11]. However, this method also reaches its limits with aluminium castings. On the one hand, thin sheets can only be processed to a limited extent, so that the potentially low weight cannot be sufficiently considered. Processes such as flow drilling can also be used with thinner materials, but depending on the arrangement of the materials to be joined, they also require ductile materials in which a viable thread can be formed [12].

A promising innovative thermomechanical joining process is found in the use of adaptive joining elements formed by friction spinning. This approach uses friction-induced heat to produce joining elements from a uniform starting material that are specially adapted to the boundary conditions of the joint. The elements are subsequently used to join sheets with each other. During the joining process, the joining element rotates around its own axis, so that heat is induced when penetrating the materials to be joined and forming the closing head. The friction-induced temperature input into the materials to be joined can be influenced by specific process parameterization in order to change its properties in a targeted manner. [13,14]

The aluminium casting alloy AlSi9 is particularly suitable for adapting the mechanical properties to the requirements of joining processes for this casting alloy by using a targeted cooling strategy. For this purpose, in this work, aluminium casting plates are manufactured by using the sandcasting process and are subsequently joined with a thermomechanical joining process based on adaptively produced friction elements. The input variables of the joining process in the form of the rotational speed and the feed rate are varied and the influence of the variation on the process parameters (temperature, process forces) on the joint (strength, microstructure) is described.

Process principle of the joining using adaptive friction elements

The basis of the versatile joining process with adaptive friction elements is a two-stage process chain with a wide variety of tools and setting variables, which are described in the following. In the first process stage, Friction Spun Joint Connectors (FSJC) are produced from a semi-finished product (d = 8 mm) of the steel C45e. These FSJCs are used in the second process stage to join two different metallic sheet materials. The basis of both process stages is the friction between the die and the semi-finished product or between the FSJC, the sheet metal and the die. The process temperatures of $T > 0.6 \cdot T_S$ within both process stages is set solely on the basis of friction without the use of external heating elements. The heat is required to reduce the flow stress of the auxiliary joining elements.

Process stage 1. Production of friction spun joint connectors (FSJC). In the following, the central requirements of the first process stage are described based on the four sub steps depicted in Fig. 1. A cylindrical bar section of a round rod with a diameter d = 8 mm and length l = 55 mm is first inserted into the collet of a spindle and set in rotation. To ensure continuous rotation, a horizontally arranged milling spindle from the manufacturer Weiss, Mobach Germany with a maximum rotational speed of n_{max} = 14,000 rpm is used. The first process step in the production of the FSJC starts with the positioning is the reaming and thus also the preheating of the rod by a specialised friction spinning tool made of the sintered carbide KX 40 (Co 9 %, WC 91 %). For the investigations described in this publication, the contact surface of the friction spinning tool consists of a cylinder with a diameter of $d = 3$ mm and a tip radius of $r = 1.5$ mm. The friction spinning tool is attached to a cross support, which allows the tool to be moved steplessly to the x and y coordinates shown in Fig. 1. Following positioning, the trajectory required to set a targeted FSJC geometry is executed at a constant rotational speed of $n = 12{,}000$ rpm and a feed rate of $f = 50$ mm/min. The last process step involves reworking the steel FSJC using a conventional turning process to set a uniform shaft diameter of $d_1 = 5.5$ mm.

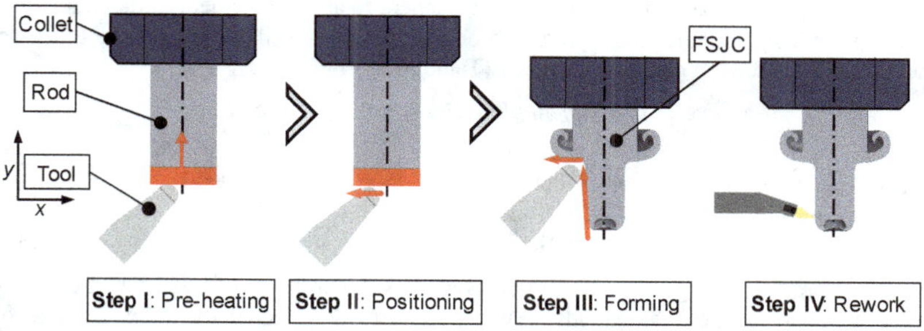

Figure 1: Principle of the first process stage to produce FSJC.

Process stage 2. Joining process. In the second process stage, the FSJC are used to join two sheets of different types (length in each case $l = 105$ mm and $b = 45$ mm). The sheets overlap by $\Delta l = 16$ mm and the joining point is placed centrally. In contrast to the sheet made of the aluminium casting alloy EN AC-AlSi9, the steel sheet of the alloy HCT590X has a pre-hole ($d = 6$ mm). In the joining process presented in this publication, the FSJC-side cast aluminium plate, which has no pre-hole (analogous to flow drilling), is driven-through (see Fig. 2). The FSJC produced in the previous process stage from the round bar of the steel material C45e has a significantly higher mechanical as well as temperature strength compared to the aluminium cast plate. This allows the aluminium cast plate to be driven-through, thus avoiding the insertion of a pre-hole. During the drive-through of the aluminium cast plate with the FSJC, heat is generated, which in this case is used to reduce the flow stress of the aluminium cast plate and thus to reduce the process forces as well as the tool loads. The FSJC heated by the drive-through is further fed in the negative y-direction, so that a frictional contact with the bottom of the die subsequently occurs, which leads to a further strong generation of frictional heat. The material volume of the FSJC, which is also reduced in strength due to this frictional contact, now flows laterally and completely fills the die, forming the interlock required for the high strength of the joint. After reaching the end position, both the feed and the rotational speed are stopped and the joint is removed.

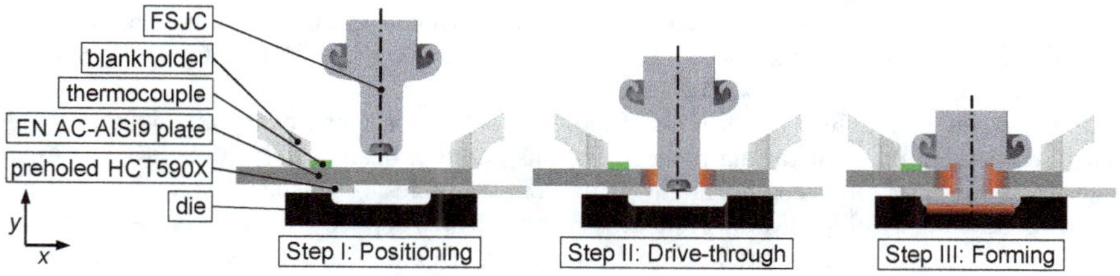

Figure 2: Principle of the second process stage, the joining of two sheets.

Materials, methods and processes

Central influencing variables of the joining and driving through process are the rotational speed as well as the feed of the auxiliary joining element [13], which have been varied in this paper in a wide interval in a two-stage test plan for the second process step of driving through (low/high). Four repetitions are carried out for each parameter combination. For the joining process itself, constant parameters of the rotational speed of n = 10,000 rpm and feed f = 30 mm/min are used. With regard to the relative movement that occurs between the rotating and fed auxiliary joining element and the stationary sheets, the rotational speed as well as the feed can be combined in one parameter, the related feed (feed per rotational speed). To investigate whether the rotational speed or feed as adjustable variables or the resulting dependent relative movement of the related feed is the main influencing variable on the driving through, the first as well as the fourth test series have an identical related feed of r_f = 0.015 mm/rev (Table 1).

Table 1: Experiment plan with resulting, related feed.

Test series	1	2	3	4
Rotational speed n [rpm]	2000	2000	10000	10000
Feed f [mm/min]	30	150	30	150
Related feed r_f [mm/rev]	0.015	0.075	0.003	0.015

To characterize the temperature and force as well as the properties of the joined component, different measuring and testing methods are used, which are described in the following. The forces occurring during the joining process (second process stage) are recorded by means of a strain gauging load cell (K6D175 50kN/5kNm/UP13) from ME-Meßsysteme GmbH Hennigsdorf, Germany, and subsequently processed with a measuring amplifier from Hottinger Brüel & Kjaer GmbH in Darmstadt, Germany (QuantumX Module MX440B). Due to the surface properties of the aluminium, a non-contact temperature measurement using a thermographic camera is here not possible. However, to record the different heat generation of the drive-through and the forming (depending on the rotational speed used and the feed rate), a type K thermocouple (G/G-24-KK-IEC) from Therma Thermofühler GmbH in Lindlar, Germany, is used at a distance of Δl = 20 mm from the joint. The measurement signal is also processed using the measurement amplifier characterized above. The previously characterised tooling system in combination with the position and alignment of the sheets is used to directly produce shear tensile specimens. These joints are then tested in accordance with the standard DVS/EFB-3480-1: Testing of Properties of mechanical and hybrid (mechanical/bonded) joints, which applies to the testing of steel and non-ferrous metals up to a thickness of $t \leq 4.5$ mm. In the course of the light microscopic (LM) examinations as well as the EBSD analyse, the joined samples were embedded in the embedding agent CEM1000 blue (Cloeren Technology) and then ground with a grain size of up to 4000. Subsequently, the specimens were polished for 24 h by an automatic polishing machine using the polishing suspension Silica suspension (Cloeren Technology) which includes nano particles with a size of 50 nm. Die LM images are taken by using the Keyence VHX5000 digital microscope. In addition, EBSD images are taken using the Zeiss Ultra Plus scanning electron microscope (SEM) to investigate the microstructure. For this application, a magnification of 500 is used at a working distance of 13.5 mm and a step size of 0.15.

The investigations involved the manufacture of cast plates consisting of the aluminium casting alloy AlSi9 using the sand casting process. For this purpose, quartz sand with an average grain size of 260 µm was used in combination with the cold-curing resin-hardener mixture Pentex L and the zircon coating 3139 KBV from Hüttenes Albertus. It was applied with a thickness of 250 µm and serves to prevent melt penetration into the moulding material. The quartz sand was mixed with the resin hardener mixture at a ratio of 100 : 0.6. Casting plates with a thickness of 2 mm and basic dimensions of 240 x 120 mm are produced.

Characteristic mechanical properties such as tensile strength, elongation at fracture and Brinell hardness of AlSi9 cast plates with a thickness of 2 mm can be taken from the previous study by Neuser et. al [15].

Characteristics and properties of the joining process

In this subchapter, the properties, and characteristics of the partial-pre-hole-free joining process are explained and the influence of processing the casting alloy as well as the influence of the different variables are described. First of all, an exemplary and characteristic force and temperature curve determined for partial-pre-hole-free joining is presented in Fig 3. and described for two different feed rates during driving through (in a) $f = 30$ mm/min, in b) $f = 150$ mm/min). In the overall evaluation, the results of four examinations for each parameter combination are considered. The figure shows that at the time $t = 0$ s a direct contact of the auxiliary joining element with the pre-hole-free cast plate occurs. For the low feed rate of $f = 30$ mm/min, a minor force increase followed by a force decrease can be determined in comparison to the high feed rate. The reason for the decrease in the force following the setting of the local maximum is the softening of the cast plate that occurs due to the frictional heat inserted. In the third process step (the forming of the closing head), a strong increase in the process force can initially be determined due to the significantly increased friction surfaces in the die with a diameter of $d = 14$ mm. After reaching the global maximum force in relation to the entire process, the process forces also decrease significantly here due to the softening of the auxiliary joining element made of C45e material.

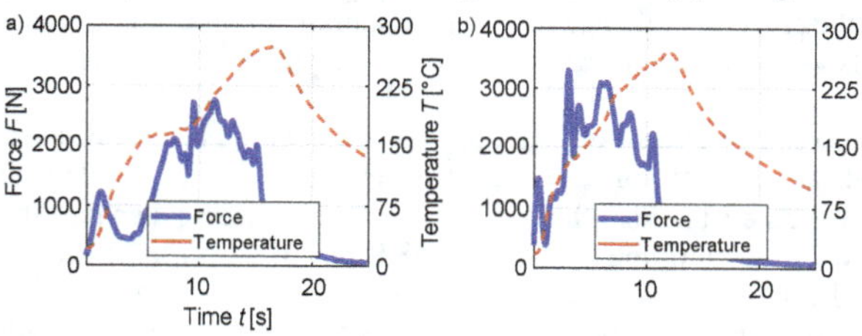

Sheet materials
EN AC-AlSi9, t = 2 mm
HCT590X, t = 1 mm
Joining element
C45e, d = 5.5 mm, l = 8 mm
Die geometry
Flat die, d = 14 mm, t = 1.5 mm

Figure 3: Variation of the process force and temperature during the driving through and forming of the closing head with a) n = 2000 rpm, f = 30 mm/min and b) n = 2000 rpm, f = 150 mm/min.

Besides the influence on the process force, a direct influence of the feed rate on the temporal course of the heat input can be shown. For the low feed rate of $f = 30$ mm/min, a constant temperature level in the range of $T \approx 155$ °C is established (for a short time) following the driving through of the cast plate. For the feed rate of $f = 150$ mm/min, an almost constant increase in temperature up to the maximum can be determined. An influence of the feed rate on the temperature maximum cannot be determined when considering the points of investigation presented here.

To illustrate the influence of the variation of the rotational speed and the feed (representable as the variation of the related feed) and to assess the robustness or repeatability of the process in the form of the setting of constant characteristic values of the force as well as temperature measurement, the results of the test points are compared in the following. No significant correlation can be determined between the rotational speed or the feed rate and the measured maximum temperatures. The reason for the insignificant correlation is the strong fluctuations of the maximum temperatures over a test series with constant rotational speed and feed. For example, the standard deviation $SD = 44$ °C of the test series with a rotational speed of $n = 10{,}000$ rpm and a feed rate $f = 30$ mm/min at an average maximum temperature of $T_{max} = 259$ °C, which corresponds to a percentage deviation of approx. 17 %. Although the results vary, a technological explanation can be seen in the deviations within the geometry of the auxiliary joining elements,

Materials Research Forum LLC
https://doi.org/10.21741/9781644902417-24

which are due to the wear of the sintered carbide tools. Despite the targeted, machining reduction of the shaft diameter to a uniform diameter, the edge radius as well as the diameter of the flange show deviations within a test series.

For the evaluation of the energy required to produce the joined joint, the measured process forces are considered. On the one hand, the maximum force during the joining process and, on the other hand, the average force during the forming of the closing head (process phase with the highest mechanical loads) are considered. For all investigation points, the maximum forces occurring during the process significantly exceed the average forces present during the forming of the closure head. In addition, it can be determined that, analogous to the observation of the temperatures, strong (standard) deviations also occur for the process forces within the individual test series with identical feed rates and rotational speeds. Nevertheless, the statistically non-significant tendency of increasing forces (maximum as well as average force) with increasing feed can be determined.

Characteristics of the joining connection

The mechanical testing of the joined tensile specimens shows that a high and suitable strength could be achieved for all specimens regardless of the process parameters used. The maximum forces in the shear tensile test before failure of the joints are between $F_{max} = 3.3$ kN and $F_{max} = 5.2$ kN. For all joints, it applies that due to the lower strength of the cast plate compared to the steel sheet, this cast plate breaks out and the joint fails at this point. Analogous to the results described above with regard to the process forces and temperatures during joining, the determined maximum forces of the joint also show no significant influence of the process parameters of the feed and the rotational speed.

In the scope of the investigations, LM images of the AlSi9 and HCT590X plates joined using frictional forces were obtained. An example of such an image is shown in Fig. 4. Here, the arrangement is as follows: on the die side, the steel sheet HCT590X is presented, and on the punch side, the AlSi9 cast plate is given. In the forming area of the AlSi9 cast plate, it can be seen that the dendritic microstructure in the die-side edge area (1) has formed a very fine Si eutectic as a result of partial remelting, as well as the dendritic α-aluminum has been formed by the joining process and has assumed a stem-like morphology. Just the latter can be seen in the LM image (2). During the joining process, no cracks were initiated in the cast plate, as is common, when using conventional mechanical joining processes. This can be avoided by modify the microstructure [15].

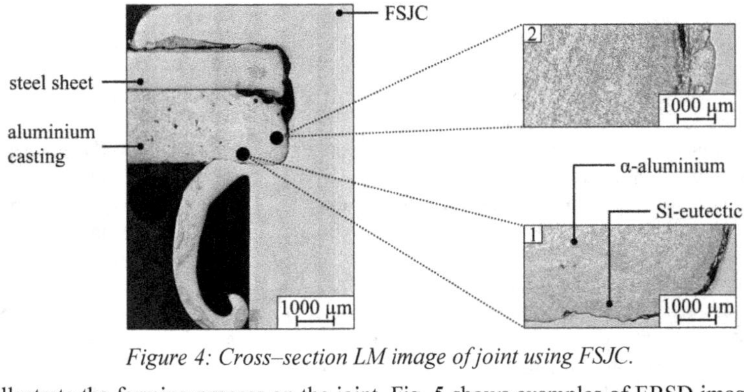

Figure 4: Cross–section LM image of joint using FSJC.

To illustrate the forming process on the joint, Fig. 5 shows examples of EBSD images of the forming zone and a comparison with the as-cast condition. In addition, the EBSD images of the

FSJC in the same conditions are shown. Due to the forming process as well as the partial melting, grain refinement can be observed in AlSi9. In the basic material (AlSi9), large grains are evident, which is typical for a sand-cast microstructure. Due to the low plate thickness of 2 mm, a solidification-related orientation of the grains is detectable. In comparison, in the heat-affected zone, a finer microstructure and, due to the forming process, an orientation alteration of the grains can be observed. In the case of the rivet material C45e, recrystallization takes place as a result of the previous forming and the thermal influence. The comparison of the two EBSD images shows the significantly smaller grains.

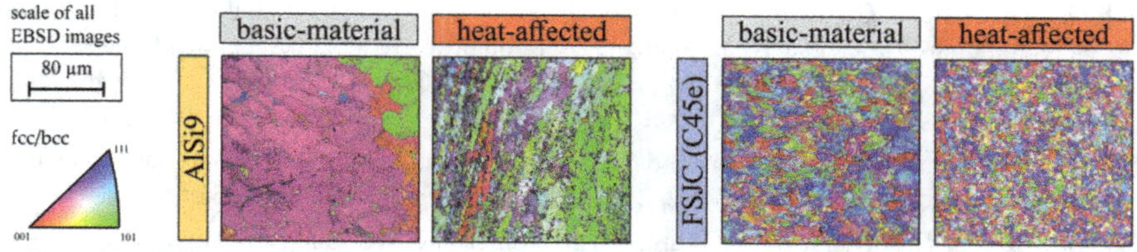

Figure 5: EBSD-analyse of aluminium casting (left); FSJC rivet consisting of C45e (right). In addition, a difference is made between images taken on the basic material and in the heat-affected zone.

Conclusion

The results of this study show that, regardless of the parameters used for the thermomechanical joining process, suitable strengths can be achieved for the joint between the hypoeutectic aluminium casting alloy plate and the steel sheet for all the joints investigated. The combination of joining processes and hypoeutectic aluminium casting plates used here can therefore overcome the challenges described in the prior state of the art with regard to joining aluminium casting alloys. An influence of the process parameters of the rotational speed and the feed rate on the force and temperature curves in the different process phases could be demonstrated. Due to the friction-induced application of heat and the associated partial melting, a grain refinement in the AlSi9 aluminium alloy due to recrystallisation can be determined. Furthermore, no cracks in the cast material can be detected at the joint, as is the case with conventional mechanical joining processes.

Funding

This research was funded by the Deutsche Forschungsgemeinschaft (DFG, German Research Foundation) – TRR 285 – Project-ID 418701707.

Acknowledgments:

The authors thank the German Research Foundation for their organizational and financial support.

References

[1] G. Pina Cipriano, A. Ahiya, J.F. dos Santos, P. Vilaça, S.T. Amancio-Filho, Single-phase friction riveting: metallic rivet deformation, temperature evolution, and joint mechanical performance, Weld World 64 (2020) 47–58. https://doi.org/10.1007/s40194-019-00803-3

[2] M. Chiaberge, New Trends and Developments in Automotive Industry, InTech, London, 2011. https://doi.org/10.5772/1821

[3] M. Haghshenas, A.P. Gerlich, Joining of automotive sheet materials by friction-based welding methods: A review, Engineering Science and Technology, an International Journal 21 (2018) 130–148. https://doi.org/10.1016/j.jestch.2018.02.008

[4] J. Machuta, I. Nová, P. Kejzlar, Structure and Mechanical Properties of Aluminium Alloys AlSi10 and AlSi5Mg, Manufacturing Technology 17 (2017) 772–777. https://doi.org/10.21062/ujep/x.2017/a/1213-2489/MT/17/5/772

[5] J.G. Kaufman, E.L. Rooy, Aluminum alloy castings: Properties, processes, and applications, ASM International, Novelty, 2004. https://doi.org/10.31399/asm.tb.aacppa.9781627083355

[6] K. Martinsen, S.J. Hu, B.E. Carlson, Joining of dissimilar materials, CIRP Annals 64 (2015) 679–699. https://doi.org/10.1016/j.cirp.2015.05.006

[7] G. Di Michele, P. Guglielmi, G. Palumbo, D. Sorgente, Investigation on the Strain Behaviour of a Precipitation-Hardenable Aluminium Alloy through a Temperature Gradient Based Heat Treatment, Key Eng. Mater. 639 (2015) 361–368. https://doi.org/10.4028/www.scientific.net/KEM.639.361

[8] H. Warlimont, W. Martienssen (Eds.), Springer Handbook of Materials Data, 2nd ed., Springer International Publishing, Cham, 2018. https://doi.org/10.1007/978-3-319-69743-7

[9] O. Siret, C. Desrayaud, M.A. Tourabi, Thermomechanical Joining of Aluminium Alloys: Effects of the Shear on the Quality of the Joining, MSF 638-642 (2010) 3716–3721. https://doi.org/10.4028/www.scientific.net/MSF.638-642.3716

[10] G. Çam, S. Mistikoglu, Recent Developments in Friction Stir Welding of Al-alloys, J. of Materi Eng and Perform 23 (2014) 1936–1953. https://doi.org/10.1007/s11665-014-0968-x

[11] J. Min, J. Li, Y. Li, B.E. Carlson, J. Lin, W.-M. Wang, Friction stir blind riveting for aluminum alloy sheets, J. Mater. Process. Technol. 215 (2015) 20–29. https://doi.org/10.1016/j.jmatprotec.2014.08.005

[12] M. Costas, D. Morin, J.K. Sønstabø, M. Langseth, On the effect of pilot holes on the mechanical behaviour of flow-drill screw joints. Experimental tests and mesoscale numerical simulations, J. Mater. Process. Technol. 294 (2021) 117133. https://doi.org/10.1016/j.jmatprotec.2021.117133

[13] C. Wischer, W. Homberg, A contribution on versatile process chains: joining with adaptive joining elements, formed by friction spinning, Prod. Eng. Res. Devel. 16 (2022) 379–388. https://doi.org/10.1007/s11740-021-01094-8

[14] C. Wischer, W. Homberg, Further Development of an Adaptive Joining Technique Based on Friction Spinning to Produce Pre-Hole-Free Joints, Key Eng. Mater. 926 (2022) 1468–1478. https://doi.org/10.4028/p-1n6741

[15] M. Neuser, O. Grydin, A. Andreiev, M. Schaper, Effect of Solidification Rates at Sand Casting on the Mechanical Joinability of a Cast Aluminium Alloy, Metals 11 (2021) 1304. https://doi.org/10.3390/met11081304

Sheet Metal 2023 Materials Research Forum LLC
Materials Research Proceedings **25** (2023) 195-202 https://doi.org/10.21741/9781644902417-25

Influence of distinct tool pin geometries on aluminum 8090 FSW joint properties

Harikrishna Rana[1,a*], Vivek Patel[2,b], Gianluca Buffa[1,c], Livan Fratini[1,d] and Rosa Di Lorenzo[1,e]

[1]Department of Engineering, University of Palermo, Viale Delle Scienze, 90128 Palermo, Italy

[2]Department of Engineering Science, University West, 46186, Trollhättan, Sweden

[a]harikrishnasinh.rana@unipa.it, [b]vivek.patel@hv.se, [c]gianluca.buffa@unipa.it, [d]livan.fratini@unipa.it, [e] rosa.dilorenzo@unipa.it

Keywords: Friction Stir Welding, Aluminum, Pin Profile

Abstract. Aluminum Lithium alloys are recuperating substantial interest from automotive and aerospace industries owing to their extraordinary specific strength as compared to conventional aluminum (2xxx, 6xxx, and 7xxx) alloys. The goal of the present investigation is to study AA 8090 joints produced with the unique solid-state welding technique friction stir welding (FSW). Tool pin profile induces remarkable influence on friction and further plastic deformation during FSW. Therefrom, the influences of three distinct but constant dynamic area conditioned tool pin geometries namely, square trapezoidal, hexagonal trapezoidal, and threaded taper on the resulting material flow patterns, mechanical properties, and the microstructure have been studied and discussed in detail. The FSW joint produced with hexagonal trapezoidal pin geometry delivered the highest joint resistance owing to grain refinement and almost flawless microstructure.

Introduction

The development of 2nd generation Al-Li alloys had stirred the aerospace and automotive industries by offering 2-8 % lesser density than most conventional 2000 and 7000 series aluminum alloys. This novel Al-Li alloy not only offers the potential for substantial weight saving in structural components but high specific strength and corrosion resistance [1]. Furthermore, these alloys possess excellent mechanical properties owing to a wide variety of precipitates present in the microstructure, depending on the heat treatment condition [2]. However, the joining of these alloys by conventional arc welding methods is problematic due to the defects associated viz., porosity, hot cracking, and evaporative loss of Li [3].

FSW is an innovative solid-state joining technology wherein such fusion-related defects can be eradicated up to a certain extent [4]. During FSW joining is carried out by the plastic material flow engendered by hot–shearing action engendered by rotating non-consumable tool. Until now, the majority of investigations in FSW of Al-Li alloy are focused on the influences of various process parameters like tool rotation (TR) and tool traverse (TT) speeds on microstructure and resulting mechanical properties [3, 5]. Lertora et al.[5] have discussed the role of revolutionary pitch (RP) on joint resistance and the fatigue strength of the prepared joint. They reported enhanced properties with lower RP values owing to defect-free microstructure obtained by higher specific thermal contribution (STC). As reported by Alam et al. [6], apart from the machine parameters like TR and TT, tool pin geometry plays a vital role in the material flow characteristics, STC and so the resulting microstructure. However, hardly a few researchers have reported the role of pin geometry on the microstructure and properties of FSW joints.

For a rotating body, eccentricity is related to the dynamic orbit generated due to the disparity between the two centers. As FSW is concerned, polygonal pin geometries such as square or hexagonal can be related to eccentricity. During tool stirring such eccentricity incorporates

Content from this work may be used under the terms of the Creative Commons Attribution 3.0 license. Any further distribution of this work must maintain attribution to the author(s) and the title of the work, journal citation and DOI. Published under license by Materials Research Forum LLC.

incompressible material to hover around the pin profile. As reported by Thomas et al. [7], the eccentricity of the rotating body can be associated with the dynamic volume of the pin profile owing to different centers. The volume of the engendered WZ at the end of FSW is basically similar to the dynamic volume of pin geometry. The trajectory of the material flow is majorly decided by the ratio of static to dynamic volume. In this regard, the present study is aimed at investigating material flow, microstructure, and mechanical properties resulting from various tool pin profiles possessing constant dynamic volume.

Experimental Design

Aluminum alloy 8090 (Li 2.7%, Mg 0.6-1.3%, Cu 1-1.6 %, Fe 0.6%, Si 0.2 %, Mn 0.1%, Cr 0.1, Al bal%) plates with dimensions:100mm×50mm×6mm were used as substrates (Supplier: Migliari Allumino srl). All the experiments were conducted on fully automatic ESAB (LEGIO) FSW machine. Butt joints were produced using three different M3 (Molybdenum enriched high-speed steel) tools as depicted with their geometries in Fig. 1. The milling center was used to manufacture polygonal pin profiles (Tool 1 & 2) whereas the turning center was used to manufacture taper threaded cylindrical pin profiles (Tool 3).

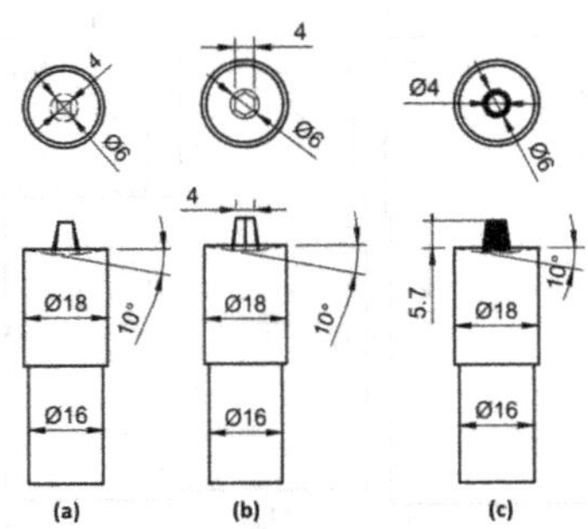

Figure 1: Tool geometries: (a) Square Trapezoidal pin (b) Hexagonal Trapezoidal pin (c) Tapered Threaded cylindrical pin[1]

All the experiments were carried out using the tool with the same shoulder diameter (SD) and pin length (PL) i.e. 18 mm and 5.9 mm, respectively. The static and dynamic conditions of distinct pin geometries are tabulated in Table 1. The faces of polygonal pins as well as conical pins were designed in accordance to have the same dynamic area conditions during FSW. Static dimensions are the actual dimensions of the pin which were used for the tool manufacturing. Dynamic dimensions are those which result during the stirring action (tool rotation). Dynamic volume decides the swept volume of the WZ. Apart, the TR, TT, and tilt angle were optimized with previous feasibility trials and kept constant for each tool pin geometry viz., 1700 rev. min^{-1}, 50 mm min^{-1}, and 3° respectively. Three FSW specimens were prepared with each welding condition as tabulated with their specimen IDs in Table 2. The temperature was also monitored during FSW for each specimen through mounting the K-type thermocouple 3mm beneath and 15 mm sideway from centre of weld line.

*Table 1: Details of tool pin design on static and dynamic area conditions**

Pin Profile	Pin Dimensions [mm]	V_s [mm³]	V_d [mm³]	V_d/ V_s	Static area condition	Dynamic area condition
Square Trapezoidal	PL: 5.9 a: 4	115.52		1.16		
Hexagonal Trapezoidal	PL: 5.9 a: 6	126.57	134.4	1.06		
Tapered Threaded cylindrical	PL: 5.9 Dr:6 Dt:4	130.24		1.03		

*PL: Pin Length, Dr: Root Diameter, Dt: Tip Diameter , a: No. of faces/sides of polygonal pin

Table 2: Specimen IDs with their FSW conditions

Specimen ID	Pin Geometry	Tool Rotation [rev minute^{-1}]	Tool Traverse [mm minute^{-1}]	Tilt Angle
A	Square Trapezoidal			
B	Hexagonal Trapezoidal	1700	50	3°
C	Tapered Threaded cylindrical			

Microstructural examination specimens were sectioned perpendicular to the welded joint line as exhibited in Fig. 2. They were further polished up to 1μm finish as per the standardized specimen preparation process. Each specimen was further etched with Keller's etchant (1 ml HF + 1.5 ml HCL + 2.5 ml HNO_3 + 95 ml DI water) [8]. The WZ area, grain morphologies, gross microstructure, and defects were analyzed with optical macro and microscopy.

Specimens prepared with each welding condition were sliced with a wire-cut electro-discharge machine (EDM) for extracting tensile specimens with dimensions according to ASTM E8M-04 guidelines.

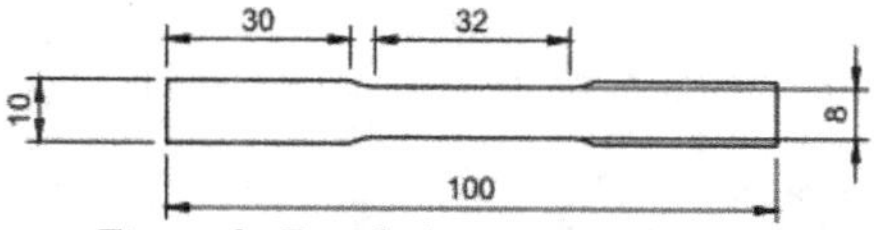

Figure 2: Tensile Specimen Schematic

Results And Discussion

Thermal Impact. Table 3 summarizes the effective pin and shoulder area, peak temperatures during FSW, and area of WZ resulting from different pin geometries. Therefrom, it is seen that specimen A, which was prepared with square trapezoidal pin geometry displayed the largest WZ area followed by specimen B and specimen C. This can be attributed to the different heat inputs generated by different pin geometries. The peak temperatures for the square trapezoidal pin geometry specimens are also the highest among the other geometry. The amount of frictional heat engendered during FSW can be attributed to two different interactions. First, the interaction between the tool pin and substrate material wherein the pin geometry makes the difference. Secondly, the shoulder-substrate interaction inducts a major amount of heat.

Table 3: Specimens Id wise tool geometry dimensions, peak temperature, and WZ area

Specimen Id	Pin Geometry	Effective pin surface area [mm²]	Effective shoulder surface area [mm²]	Peak Temp [°C]	Standard Deviation	Weld Zone Area [mm²]
A	Square Trapezoidal	119.56	236.16	393	+4, -3	47.12
B	Hexagonal Trapezoidal	124.49	231.27	384	+2, -3	38.58
C	Tapered Threaded cylindrical	128.57	229.68	363	+3, -4	29.75

It is important to notice that the effective shoulder area is higher than that of the pin for all the tools. Further, the effective shoulder area reduces as the number of faces is increased from 4 for square geometry to 6 for hexagonal geometry to infinite faces for cylindrical geometry. Conversely, the effective pin interaction area upsurges as the number of faces increases. Since the shoulder friction contribution to total heat input is much higher than the pin, the square pin geometry specimens exhibited the highest peak temperatures as compared to hexagonal and taper cylindrical pin geometry specimens. Accordingly, as the resulting WZ area is highly reliant on the peak temperature during FSW, the square trapezoidal pin geometry specimens exhibited the highest WZ area. Apparently, it is now well understood that WZ is the reflection of the tool pin geometry clubbed with the temperature history during FSW.

Micro and Macrostructural characterization. Fig. 3 exhibits macro and microstructural zones developed at distinct locations of the prepared FSW joints with distinct pin geometries. From the macrostructure, an interesting observation can be made about the material flow for all the samples. Therefrom, complete penetration of the tool pin was evident for all the specimens. Moreover, interesting observations can be made with distinct material flow patterns pertaining to each specimen. The tool substrate interaction generally happens in two areas namely, the shoulder interaction area and the pin interaction area. The material flow that happened at the top of WZ is attributable to shoulder interaction while for the middle and the bottom region, pin interaction. The microstructure of all the specimens has revealed the microstructural transition from WZ to thermos mechanically affected zone (TMAZ), and TMAZ to heat affected zone (HAZ). The WZ of all the specimens were characterized with finer grain structure which can be attributed to dynamic recrystallization while FSW[4, 9-11].

As reported by Elangovan et al. [12] polygonal pin profile produces the pulsating action whereas no such pulsation is produced with cylindrical or tapered cylindrical pin geometries. The number of pulses/second considerably contributes to the amount of plastic deformation and heat generation. The number of pluses can be calculated by multiplying the TR by the number of flat faces. Hence for a square pin, it is 113 pulses/second, while it is 170 pulses/second for the hexagonal trapezoidal pin. A higher number of pulses/second generates higher turbulence while stirring which ensues in finer grain structure along with well-distributed fine precipates [9]. The grain size comparison is portrayed in Fig. 3(d) for each sample.

The unprocessed substrate displayed a grain size of ~120 μm mostly shaped like a pancake as plates were obtained from a rolled stock of AA 8090. These grains were refined by almost 90-95% in the WZ of prepared FSW joint specimens. Such extreme grain refinement in WZ can be reportedly attributed to several DRX mechanisms [13]. Su et al. [13] have reported several grain sizes in different portions of WZ in multi-pass FSP ensuing from discontinuous DRX mechanism. As discussed earlier, since the WZ grains of joints prepared with polygonal pins experience multiple pulses engendering tremendous plastic deformation and turbulence, a discontinuous DRX mechanism can be opted in the present case. On the one hand, the joints prepared with polygonal pins exhibited more grain size variations in WZ as compared to threaded taper cylindrical pins.

This has resulted due to intermittent interaction of the pin's multiple faces and substrate during FSW using polygonal pins. This intermittent interaction further creates turbulence and leads to non-uniform grain deformation. On the other hand, the joints prepared with threaded taper cylindrical pin exhibited finely distributed grain structures with less variation.

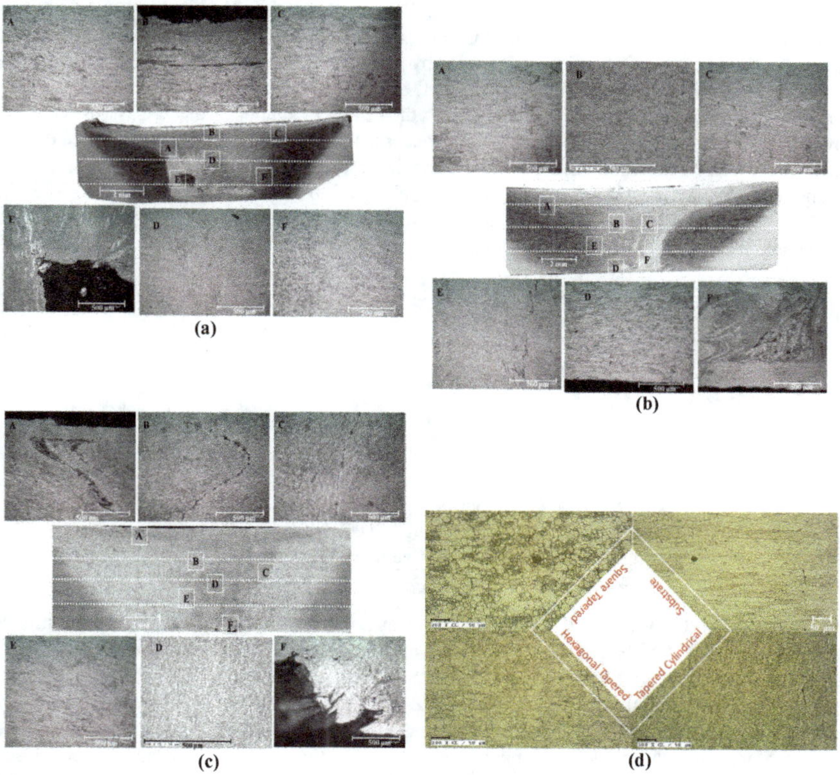

Figure 3: Macro and Microstructures of the specimens joined with different pin geometries: (a) Square Trapezoidal Pin (b) Hexagonal Trapezoidal Pin (c) Tapered Threaded cylindrical Pin (d) Grain structure comparison of all the specimens with parent metal

It is noteworthy that with an increase in the number of faces of the polygon the grain size variation diminishes as evident with the hexagonal pin geometry. As described in section 2.1 the dynamic volume of all the pin geometries was kept the same. For more regular shapes like cylindrical geometry, there is less or no disparity between the static and dynamic volume and so their ratio is unity. Specifically, in the case of polygonal pin geometries, this disparity increases with a reduction in the number of faces. For instance, the ratio is maximum in the case of square pin geometry, whereas it gets close to unity with hexagonal pin geometry (see Table 1). So, when the number of faces is increased from 4 to 6 the pin geometry gets closer to cylindrical geometry and creates less turbulence during FSW and deforms grain more homogeneously. Although the lowest grain sizes of 3-8μm were noticed with the hexagonal trapezoidal pin geometry, the threaded tapered cylindrical pin exhibited slightly larger (5-8μm) but more uniform grains in the WZ. The WZ of the square threaded specimen exhibited the largest grain size of 4-15μm.

Apart from that, the specimen processed with a hexagonal trapezoidal pin exhibited almost defect-free WZ (see Fig. 3(b)). Whereas the WZ of the square trapezoidal pin specimen was characterized with a large wormhole cavity, the WZ of the taper cylindrical pin specimen was identified with a small cavity at the bottom as exhibited in Fig. 3(a) and (c) respectively. As indicated earlier the pulsating action engendered by polygonal pin multiple faces fosters turbulence while stirring in FSW. As the number of faces in the pin decreases a larger material void is created intermittently around the pin which gives rise to the cavity or tunnel in the WZ. However, with increases in the number of faces, the geometry becomes alike circular and so in absence of such intermittent material flow and void generation, the defects are minimized.

Tensile test behavior. All the specimens were subjected to tensile testing as per ASTM E8M-04 guidelines. The joint resistance results along with joint efficiencies are compared for all the specimens through the chart portrayed in Fig. 4.

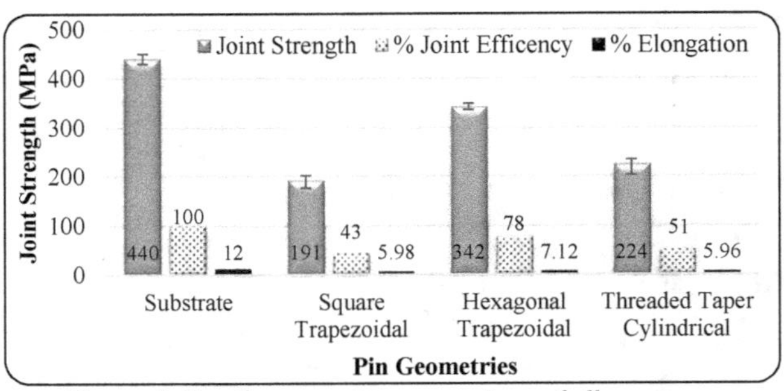

Figure 4: Tensile properties comparison of all specimens

Specimen B which was joined with hexagonal trapezoidal pin geometry delivered the highest joint resistance of 342 MPa and remarkable joint efficiency of ~78 %. It is noteworthy that the % elongation was almost 60% that of the substrate. Such remarkable joint properties were observed owing to highly refined grain structure in the WZ characterized by flawless microstructure. The highest grain refinement ensued from the pulsating action of hexagonal tool faces offers a higher grain boundary area which in due course hinders the dislocation motion. This hindrance not only increases joint resistance but also increases other properties like hardness and fatigue resistance [14]. Apart, precipitation strengthening offered by major precipitates like Al_2CuLi, Al_3Zr, and Al_3Li are also attributable for heightened properties [3]. Moreover, as WZ of specimen B possessed different sizes of grains, the larger grains aided in boosting the % elongation up to 60 %. As far as the other two cases are concerned, specimen A which was joined with square pin geometry delivered meager joint resistance owing to a large cavity in the center of the WZ, whereas specimen C produced with a threaded taper pin exhibited moderate joint resistance owing to the cavity at the bottom of the joint.

Apart from the joint resistance results, it is important to consider the location of the fracture for the tested specimens. The location of the fracture for each tensile specimen is depicted by white arrows in Table 5.

Materials Research Forum LLC
https://doi.org/10.21741/9781644902417-25

Table 5: Tensile test fracture specimens displaying fracture location for all the welding conditions

Specimen ID.	Fractured specimen: Location	Fractured Specimen: Side view
A	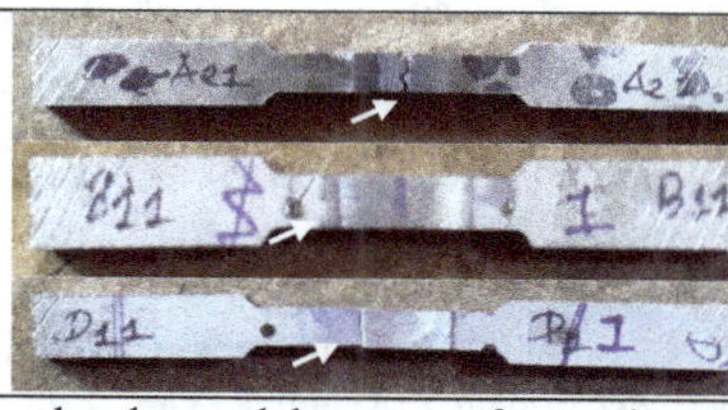	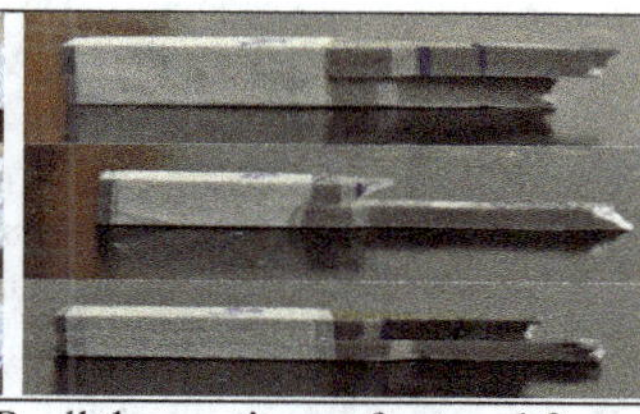
B		
C		

Therefrom, it can be observed that except for specimen B, all the specimens fractured from the center of the WZ. The WZ of specimens processed with a square trapezoidal pin and threaded taper cylindrical pin was identified with cavity and tunnel defects at the center/ bottom which may be the preferential site for the crack initiation and further growth leading to the fracture. However, it was interesting to notice that the specimens prepared with hexagonal pin fractured from the TMAZ/HAZ interface. This can be understood by the two circumstances. Firstly, the WZ of these specimens was characterized with immaculate microstructure. Secondly, as depicted in section E of Fig. 3 (ii) the transition zone of the TMAZ/HAZ was characterized with the transition of the fine grain to a larger grain structure. As the WZ possesses a highly refined grain structure, the HAZ region becomes the weak spot and a preferential site for crack initiation and further growth.

Conclusions

The present investigation described the influences of distinct pin geometries on thermal characteristics, microstructural characteristics, and mechanical properties of AA 8090 FSW joints. Further, the following conclusions were extracted from the investigation:

- The tool pin geometry significantly influences the heat generation, microstructural features like weld zone size, grain size, grain morphology, and mechanical properties like joint resistance and joint elongation.
- Square trapezoidal pin geometry generated the highest heat input with a peak temperature of 395 °C even though possessing the lowest pin face area. However, this geometry offers the highest effective shoulder interaction area and the shoulder's frictional heat contribution is greater than the pin's contribution. This pin geometry delivered the specimen with the highest WZ area owing to the highest peak temperature.
- The pulsating action of the polygonal pin characterized by the number of pulses/second contributes greatly to the grain refinement and distribution pattern during FSW. Specimen prepared with the hexagonal trapezoidal pin exhibited the finest grains owing to the turbulence created by the highest number of pulses/second. Simultaneously, the lower number of pulses creates intermittent material voids around the pin while tool stirring which is responsible for the greater non-uniformity in the grain sizes and the defects generation.
- The specimens joined with hexagonal trapezoidal pin exhibited the highest joint resistance along with the highest elongation in line with the grain size results. The rare combination of the high joint resistance and elongation was possible owing to a moderate disparity in the grain sizes in the weld zone of these specimens, wherein larger grains promoted the elongation. Only these specimens fractured from the TMAZ/HAZ interface owing to a large grain size gradient at the interface and a weak HAZ region characterized by larger grains.

References

[1] Y. Tao, D. Ni, B. Xiao, Z. Ma, W. Wu, R. Zhang, Y. Zeng, Origin of unusual fracture in stirred zone for friction stir welded 2198-T8 Al-Li alloy joints, Materials Science and Engineering: A 693 (2017) 1-13. https://doi.org/10.1016/j.msea.2017.03.079

[2] B. Decreus, A. Deschamps, F. De Geuser, P. Donnadieu, C. Sigli, M. Weyland, The influence of Cu/Li ratio on precipitation in Al–Cu–Li–x alloys, Acta Mater. 61(6) (2013) 2207-2218. https://doi.org/10.1016/j.actamat.2012.12.041

[3] H. Liu, Y. Hu, C. Dou, D.P. Sekulic, An effect of the rotation speed on microstructure and mechanical properties of the friction stir welded 2060-T8 Al-Li alloy, Mater. Charact. 123 (2017) 9-19. https://doi.org/10.1016/j.matchar.2016.11.011

[4] H. Rana, D. Campanella, G. Buffa, L. Fratini, Dissimilar titanium-aluminum skin-stringer joints by FSW: process mechanics and performance, Mater. Manuf. Process (2022) 1-14. https://doi.org/10.1080/10426914.2022.2116044

[5] E. Lertora, C. Gambaro, AA8090 Al-Li alloy FSW parameters to minimize defects and increase fatigue life, International Journal of Material Forming 3(1) (2010) 1003-1006. https://doi.org/10.1007/s12289-010-0939-1

[6] M.P. Alam, A. Sinha, Fabrication of third generation Al–Li alloy by friction stir welding: a review, Sādhanā 44(6) (2019) 1-13. https://doi.org/10.1007/s12046-019-1139-4

[7] W. Thomas, E. Nicholas, Friction stir welding for the transportation industries, Mater. Des 18(4-6) (1997) 269-273. https://doi.org/10.1016/S0261-3069(97)00062-9

[8] G.F. Vander Voort, S.R. Lampman, B.R. Sanders, G.J. Anton, C. Polakowski, J. Kinson, K. Muldoon, S.D. Henry, W.W. Scott Jr, ASM handbook, Metallography and microstructures 9 (2004) 44073-0002. https://doi.org/10.31399/asm.hb.v09.9781627081771

[9] V.V. Patel, V. Badheka, A. Kumar, Effect of polygonal pin profiles on friction stir processed superplasticity of AA7075 alloy, J. Mater. Process. Technol. 240 (2017) 68-76. https://doi.org/10.1016/j.jmatprotec.2016.09.009

[10] H. Rana, V. Badheka, Elucidation of the role of rotation speed and stirring direction on AA 7075-B4C surface composites formulated by friction stir processing, Proceedings of the Institution of Mechanical Engineers, Part L: Journal of Materials: Design and Applications (0) (2017) 1-18. https://doi.org/10.1177/1464420717736548

[11] H. Rana, V. Badheka, A. Kumar, A. Satyaprasad, Strategical parametric investigation on manufacturing of Al–Mg–Zn–Cu alloy surface composites using FSP, Mater. Manuf. Process 33(5) (2018) 534-545. https://doi.org/10.1080/10426914.2017.1364752

[12] K. Elangovan, V. Balasubramanian, Influences of pin profile and rotational speed of the tool on the formation of friction stir processing zone in AA2219 aluminium alloy, Materials Science and Engineering: A 459(1-2) (2007) 7-18. https://doi.org/10.1016/j.msea.2006.12.124

[13] J.-Q. Su, T.W. Nelson, C.J. Sterling, Friction stir processing of large-area bulk UFG aluminum alloys, Scr. Mater 52(2) (2005) 135-140. https://doi.org/10.1016/j.scriptamat.2004.09.014

[14] V. Patel, W. Li, A. Vairis, V. Badheka, Recent development in friction stir processing as a solid-state grain refinement technique: microstructural evolution and property enhancement, Critical Reviews in Solid State and Materials Sciences 44(5) (2019) 378-426. https://doi.org/10.1080/10408436.2018.1490251

Material characterization

Contact pressure-dependent friction compensation in upsetting tests with miniaturized specimens

Peter Hetz[1, a *], Marcel Rentz[1, b] and Marion Merklein[1, c]

[1]Institute of Manufacturing Technology, Friedrich-Alexander-Universität Erlangen-Nürnberg, Egerlandstraße 13, 91058 Erlangen, Germany

[a]peter.hetz@fau.de, [b]marcel.rentz@fau.de, [c]marion.merklein@fau.de

Keywords: Metal Forming, Friction, Upsetting Test

Abstract. Components produced by sheet metal forming can so far only be characterized locally to a limited extent using conventional tests, such as the tensile test or the hydraulic cupping test, due to the size of the specimen. A new approach based on miniaturized upsetting specimens taken from the sheet plane allows to determine the material properties of semi-finished products locally, for example in radii. However, in addition to this specific advantage, challenges such as friction between the specimen and the tool also appear with this testing method. For an accurate determination of the material parameters, it is therefore necessary to correct the test force by the friction coefficient. According to the state of the art, a variety of tests for friction coefficient characterization exist. By using a novel numerical method, friction coefficients for upsetting tests can be calculated contact pressure-dependent by means of an experimental single layer upsetting test. Therefore, in this paper, this method will be used for the first time to compensate the friction part in an upsetting test with miniaturized specimens as a function the contact pressure. This can improve the experimental mapping of the material behavior in the uniaxial compression stress state with increasing strain and hardening, since share of friction force increases with a higher deformation in the upsetting test. In contrast, a friction coefficient compensation based on an average value is provided in order to analyze, whether or in which case a contact pressure-dependent friction compensation is appropriate. In particular, for upsetting tests with miniaturized specimens, these results are relevant, since a major advantage of this test compared to conventional tests besides the local component characterization is the determination of material properties at high strains.

Introduction

The upsetting test with cylindrical specimens is used in bulk metal forming mainly to determine the flow behavior. In this test, a cylindrical specimen is upset between two flat, parallel tool surfaces. Assuming homogeneous deformation, the material behavior in the uniaxial compression stress state can thus be determined, for example, to create material models. The standard for this test is DIN 50106 and, in accordance with this norm, the specimen size may be scaled as required. However, it must be taken into account that size effects can occur in specimens with two dimensions in the submillimeter range [1]. Miniaturized upsetting specimens can be extracted directly from the sheet plane and thus the material behavior of sheet materials in the uniaxial compression stress state can be also characterized [2]. Compared with conventional test methods such as the tensile test, this test method provides the advantages that material properties can be analyzed locally in sheet metal products (dimensions of the specimen < 2 mm) and considerably higher degrees of deformation can be achieved. However, in addition to complicated specimen manufacturing, the test method also faces the challenge of friction influence. During the test, frictional forces occur due to the strain of the cylindrical specimen parallel to the tool surface, which affects the test force. Thus, for a correct calculation of the yield stress, the test force must be compensated to eliminate the friction influence. In forming technology, either the Coulomb

Content from this work may be used under the terms of the Creative Commons Attribution 3.0 license. Any further distribution of this work must maintain attribution to the author(s) and the title of the work, journal citation and DOI. Published under license by Materials Research Forum LLC.

friction coefficient model [3] or, for especially high contact pressures, the friction factor model [4] is used to describe the friction behavior. According to [5], the friction coefficient model is valid for friction coefficients to Tresca up to 0.5 or to Mises up to 0.577, if the normal stress σ_N is equal to the yield stress σ. For friction compensation of the test force in the upsetting test, there is the possibility to combine the friction coefficient model and the friction factor model [6]. But usually the Coulomb friction coefficient model is suitable for the upsetting test, because the normal stress in the upsetting test corresponds approximately to the yield stress and the friction coefficients are lower than 0.5 due to the application of lubrication. Consequently, the approach according to Siebel [7], which is based on these model, can be used for friction compensation. Assuming an isotropic deformation of the specimen, the related formula from Siebel allows to calculate the yield stress σ as a function of the specimen diameter d, the specimen height h, the punch force F_P, the specimen area A and the friction coefficient μ:

$$F_P = A \cdot \sigma \cdot \left[1 + \frac{1}{3} \cdot \mu \cdot \frac{d}{h}\right]. \quad (1)$$

Except for the friction coefficient and yield stress to be calculated, all parameters can be obtained by the measured test load and the strain in the loading direction based on the volume constancy under the assumption of homogeneous deformation. Thus, an exact experimental characterization of the friction influence is required to evaluate the yield stress applied in the material of the specimen. Various tests exist for this purpose, such as the strip drawing test or the pin on disk test [8]. But these tests are only suitable to a limited extent for characterizing the friction at high contact pressure or as a function of the contact pressure. In particular, however, it is necessary for upsetting tests, since the contact pressure increases steadily during compression. For example, a tribological system can be modified by the smoothing of lubrication pockets due to an increasing contact pressure. However so far only publications are available that investigated friction compensation with a constant friction coefficient. In this way, Christiansen et al. [6] reduced the friction influence with a numeric approach and Obiko [9] used an analytical approach for that. By means of a new combined numerical-experimental approach according to Kraus et al. [10], the friction can be determined dependent on the contact pressure using a single layer compression test as an improved alternative to the ring test method [11]. This method is adopted in this contribution in order to obtain a non-friction influenced flow curve based on the formula according to Siebel for the upsetting test with miniaturized specimens. Due to the friction compensation, the experimental mapping accuracy of the local material behavior from sheet metal products can be improved. As test material, the conventional deep-drawing steel DC04 with a sheet thickness of 2.0 mm together with the high-strength aluminum alloy AA7075-T6 with a sheet thickness of 1.5 mm are investigated.

Contact pressure-dependent friction coefficient characterization

In order to reduce a flow curve from an upsetting test about the frictional influence, the friction coefficient must first be determined experimentally with a comparable tribological system. The method according to Kraus et al. [10] is based on a single layer upsetting test and therefore excellently suited for characterizing a friction coefficient for an upsetting test. In addition, the method provides the advantage that the coefficient of friction can be evaluated as a function of the contact pressure. The associated methodological procedure is shown in a schematic illustration in Fig. 1. According to that, a single layer compression test must first be carried out experimentally. For this purpose, a round specimen with a diameter of 10 mm is cut from a sheet with the TruLaser Cell 7020 laser (Trumpf GmbH + Co. KG) and subsequently compressed in a tool with two parallel and flat surfaces on the walter + bai testing machine (ZwickRoell AG). A spherical cap in the lower part of the tool ensures parallelism. The specimen is upset by the total strain of $\varepsilon_{TS} = 0.5$ and the nominal strain rate during the test is $\dot{\varepsilon} = 0.005$ 1/s. Three repetition experiments are performed for the two materials DC04 and AA7075-T6. The sample height corresponds to the respective sheet thickness. Both the force and the displacement are measured by the machine throughout the test. Due to the stiffness of the machine, the force and displacement data must be corrected. Therefore a correction curve is generated by closing the upper and lower tool without a specimen, so that the system modulus of the machine and the tool is determined. Using this correction curve, the force and displacement data are corrected accordingly. Since conventional upsetting tests are carried out with a lubricant to reduce friction, the deep-drawing oil Dionol ST V 1725-2 (MKU-Chemie GmbH) is applied before the test. Force-strain curves are calculated for both materials after testing procedure.

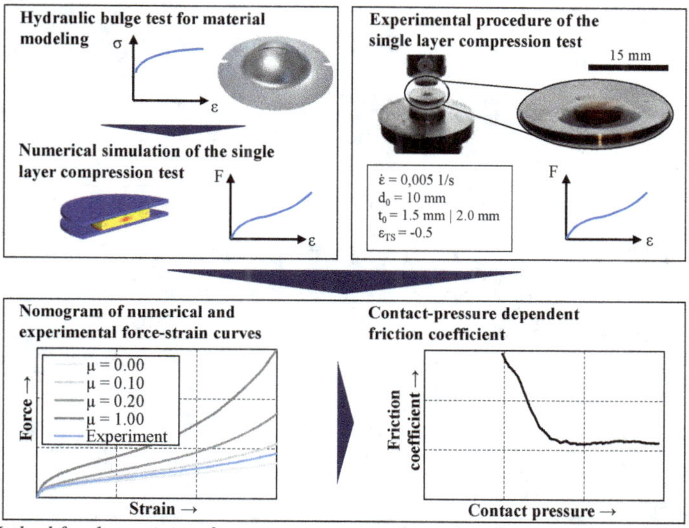

Fig. 1 Method for determining the contact pressure-dependent friction influence according to Kraus et al. [10]

In addition to the experimental performance of the single layer compression test, the experimental setup will also be numerically modeled by the simulation program Ls Dyna (Dynamore GmbH) in order to generate a nomogram for friction coefficient determination. The simulation model is based on the test conditions of the experimental setup. The material model used is Mat_24, which is based on yield locus curve of Mises. The flow curves are extrapolated according to Hocket-Sherby. It is particularly relevant in this method that the deposited flow curve for the material model is not influenced by friction. Furthermore, the yield curve should represent the same stress state as the single layer compression test. In this way, the flow curve from the numerical single layer compression test with a friction coefficient $\mu = 0.0$ is equivalent to the flow curve from the experimental single layer compression test without a friction component. According to Camberg and Tröster [12], the uniaxial compression stress state in the layer compression test with an isotropic deformation is identical to the biaxial tension stress state in the hydraulic bulge test. For that reason, the flow curve from a hydraulic bulge test based on DIN EN ISO 16808 is used to represent the two test materials in the respective simulation model, which is not influenced by friction. Subsequently, the simulation is explicitly calculated for the friction coefficients from $\mu = 0.0$ to $\mu = 1.0$ in the gradations as shown in Fig. 2. The 10 numerical force-strain curves are plotted then with the experimentally determined force-strain curve into a nomogram. By interpolation, a friction coefficient can be assigned to each strain value of the experimental force-strain curve. Since the contact pressure in the compression test corresponds approximately to the true stress and each strain value can be assigned to a stress value, the friction coefficient can then be depicted as a function of the contact pressure. It should be noted that the friction factor model is to be preferred if the determined friction coefficient is above 0.5. The respective nomograms for the two materials are shown in Fig. 2.

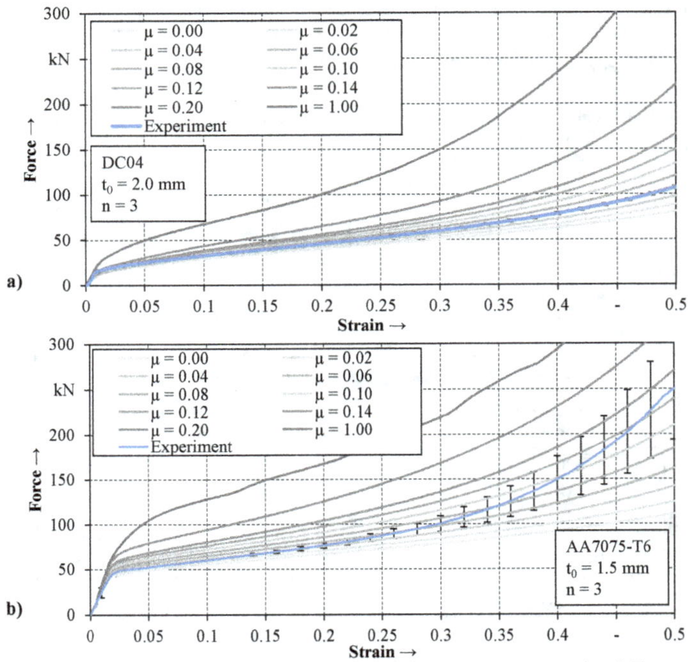

Fig. 2 Nomogram of the numerically determined force-strain curves for different friction coefficients with experimentally measured force-strain curve for the materials DC04 (a) and AA7075-T6 (b)

As expected, the upsetting force increases with higher friction coefficients for both materials. The standard deviation of the experimental force-strain curve is consistently low for DC04. However, for AA7075-T6, the standard deviation increases clearly at a strain of 0.2. After testing procedure, small amounts of the aluminum alloy remained visible on the tool surfaces. Therefore, the higher standard deviation can be attributed to the tendency of aluminum to adhesion. Moreover, adhesion leads to a considerable rise in the force, since the frictional force increases. It should be noted that at low strains, the simulated curves are located close to each other, because in this range the frictional influence is less pronounced. Therefore, it may be necessary to define a limit that determines the range of validity for the specified friction coefficients.

The friction coefficients as a function of strain can subsequently be determined from the nomograms by interpolation. Following Coulomb's friction law [3], the frictional force F_F is a function of the normal force F_N. By extending this formula about the contact area A, the friction stress τ_F can also be described as a function of the contact pressure σ_N and the friction coefficient:

Materials Research Forum LLC
https://doi.org/10.21741/9781644902417-26

$$F_F = \mu \cdot F_N \xrightarrow{\cdot A} \tau_F = \mu \cdot \sigma_N \quad (2)$$

Based on the friction law used, the friction influence is dependent on the normal force or contact pressure. Depending on the specimen geometry, for a certain strain there can be different contact pressures. Therefore, the friction coefficient must be shown as a function of the contact pressure and not as a function of the strain.

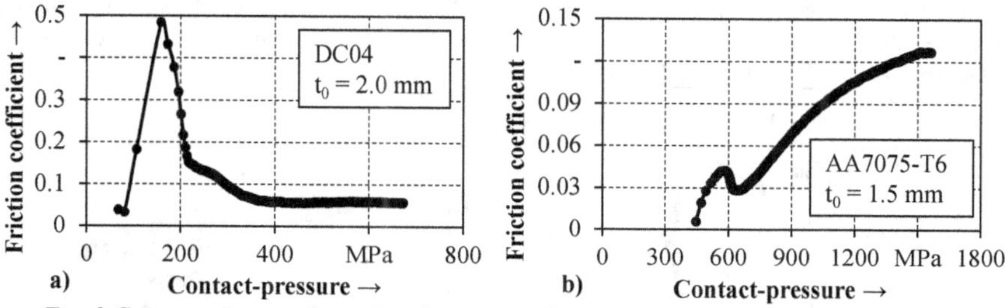

Fig. 3 Contact pressure-dependent friction coefficient for DC04 (a) and AA7075-T6 (b)

By assigning the corresponding true stress from the experimental single layer compression test to the strain, the friction coefficient is obtained in dependence on the contact pressure. In Fig. 3, this functional relationship is shown for DC04 and AA7075-T6. In both diagrams, the curves do not start at a contact pressure of 0 MPa, instead they start at approximately 80 MPa for DC04 as well as at 440 MPa for AA7075-T6. The reason for this is that in this range the experimental force-strain curve is outside the numerically determined curves in the nomogram and thus an interpolation of the friction coefficients is not possible. Initially, the friction coefficient increases for both materials up to a value of 0.48 for DC04 and 0.04 for AA7075-T6. For DC04, thereafter, the friction coefficient decreases continuously with increasing stress, while for AA7075-T6 it increases again at a pressure of 660 MPa. The cause of the increase in friction coefficient is due to adhesion and the resulting raise in friction influence with AA7075-T6. The previous decrease in friction coefficient might be because roughness peaks are smoothed by the increasing contact pressure and thus the friction influence decreases. Since no adhesion occurs for DC04, the friction influence decreases continuously for this material. However, because the friction coefficients for both materials initially increase from the first data point onward with values below 0.01, which are clearly below conventional friction coefficients, it can be assumed that these data points lie outside the previously mentioned range of validity of the nomogram. In this range, the numerically determined force-strain curves have such a close distance to each other in the nomogram that no reasonable determination of the friction coefficient is possible. Therefore, for DC04 the limit for the first valid friction coefficient is set to the maximum at $\mu = 0.48$ at a pressure of 160 MPa, and for AA7075-T6 the limit is set to the first maximum of $\mu = 0.04$ at 590 MPa. By considering a contact pressure-dependent modeling, effects such as adhesion or the smoothing of the surface during the upsetting test are taken into account.

Friction compensation in the upsetting test with miniaturized specimens

Therefore, the results from the method for contact pressure-dependent friction coefficient determination are now applied to the upsetting test with miniaturized specimens. The procedure for characterization based on Hetz et al. [2] of sheet metal using the miniaturized cylindrical upsetting specimens is shown in Fig. 4. First, sheet metal strips are taken from a sheet by wire electrical discharge machining using the UT2000 (GF machining Solution), from which in turn upsetting samples are separated parallel to the sheet plane by micro electrical discharge machining on the X-200-HPM (Sarix SA). The specimen diameter d_0 corresponds to the respective sheet thickness (DC04 = 2.0 mm, AA7075-T6 = 1.5 mm).

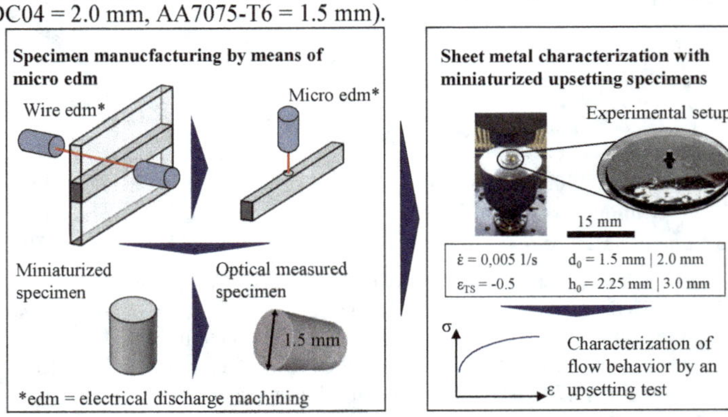

Fig. 4 Methodical procedure for the characterization of sheet metal using the upsetting test with miniaturized specimens according to [2]

Based on a specified upsetting ratio of 1.5, the height h_0 is therefore 3.00 mm for DC04 and 2.25 mm for AA7075-T6. A total of three repetition tests are carried out for each material. All specimens are measured with the Infinite Focus G5 optical measuring system (Alicona GmbH) to ensure dimensional accuracy. After specimen manufacturing, the samples are tested on the Z10 universal testing machine (ZwickRoell AG). The same upsetting tools are used as in the single layer compression test, as well as the same lubricant. Consequently, the tribological system of the upsetting test with miniaturized specimens is comparable to that of the single layer compression test. Also, the specimens are tested by $\varepsilon_{TS} = 0.5$ with a strain rate of 0.005 1/s. The parallelism of the upsetting tools is again ensured by a spherical cap in the lower part of the tool. Both force and displacement data, which are adjusted by a correction cure for the stiffness of the testing machine, are measured. Assuming isotropic deformation, the true cross-section and thus the true stress in the material during the test can be calculated via the volume constancy obtained by the strain in the loading direction. In this way, flow curves are at first created for DC04 and AA7075-T6 without compensation of the friction influence. These curves are shown in Fig. 5. Subsequently, the flow curves are compensated by their friction component according to formula (1) of Siebel. In the two diagrams, the curves are corrected for the respective contact pressure-dependent friction coefficient and for the mean value of the friction coefficients occurring in the respective stress range. Since limits for the validity of the nomogram were previously established, friction coefficients for contact pressures below 160 MPa of DC04 and below 590 MPa of AA7075-T6 are missing. Therefore, the first valid value of $\mu = 0.48$ for DC04 and $\mu = 0.04$ for AA7075-T6 are used for these contact pressure ranges. The mean values of the friction coefficients are $\mu = 0.091$ for DC04 and $\mu = 0.033$ for AA7075-T6.

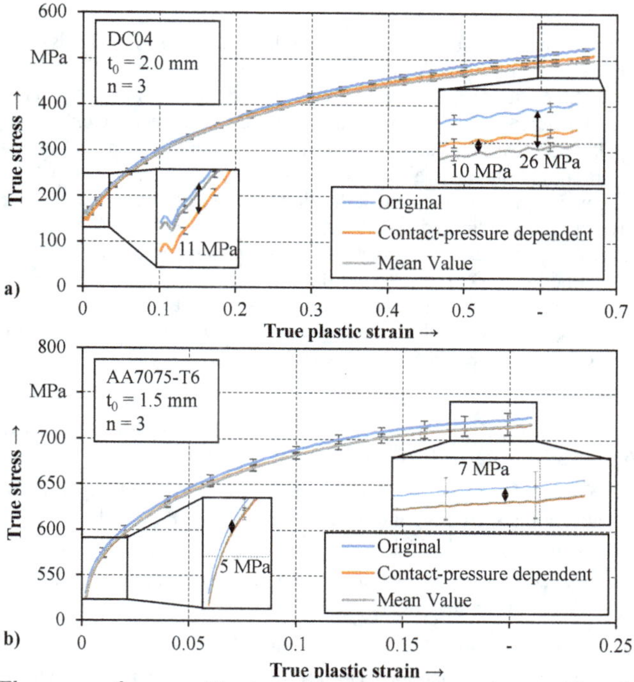

Fig. 5 Flow curves from upsetting test with miniaturized specimens with and without compensation of frictional influence for DC04 (a) and AA7075-T6 (b)

For AA7075-T6, the difference between contact pressure-dependent and mean value compensation is marginal and can be neglected. This is due to the fact that the friction coefficients in the contact pressure range, which is equivalent to the stress range of the flow curve, are quite small and change only slightly compared to the mean value. However, it must be noted that the influence of the friction coefficient would increase substantially at higher contact pressures due to adhesion. But the flow curves of AA7075-T6 are only available up to a contact pressure of 750 MPa, because due to microcracks in the miniaturized upsetting specimen about a true strain of 0.21. In the case of DC04, the contact pressure-dependent friction coefficients deviate clearly from the mean value to a certain extent. Therefore, this leads to distinct differences in the two types of friction compensation used at the yield strength and is increasing with higher strains. The difference is about 10 MPa at the yield strength, which corresponds to about 5 percent of the yield stress at this point. At higher strains, the difference is much smaller with about 2 percent of the yield stress resulting in an absolute value of 10 MPa as well. For both materials, it can be recognized that the share of friction force increases with higher strain. This is due to the fact that a compression of the specimen decreases the height and increases the diameter. It can be conducted in accordance with the formula of Siebel (1), that a reduction in the ratio of height to diameter increases the amount of friction and also leads to a change in surface properties. For this reason, friction influence at high strains should generally be compensated either as a function of contact pressure or independently of contact pressure.

Summary
Consequently, a contact pressure-dependent friction compensation leads to a substantial improvement in the mapping of material behavior in the characterization with upsetting tests. It should be noted that the friction coefficient as a function of contact pressure is clearly dependent on the material and differs significantly from an averaged friction coefficient at the yield strength in the case of DC04. For AA7075-T6, the difference is considerably smaller, but contact pressure-dependent friction compensation is also required here if stresses above 800 MPa are reached. Despite that, a contact pressure-dependent friction coefficient determination for low contact pressures is only possible to a limited extent with presented the method, so there is still a need for research for this case. In conclusion, it is recommended to compensate the friction influence in the upsetting test as a function of contact pressure. In this way, for example, the yield strength of the material in sheet metal formed parts can be measured much better by the upsetting test with miniaturized upsetting specimens, which has a particularly large influence in material models and thus in the numerical imaging accuracy.

References

[1] M. Geiger, M. Kleiner, R. Eckstein, N. Tiesler, U. Engel, Microforming, CIRP Annals 50 (2001) 445–462. https://doi.org/10.1016/S0007-8506(07)62991-6

[2] P. Hetz, M. Kraus, M. Merklein, Characterization of sheet metal components by using an upsetting test with miniaturized cylindrical specimen, CIRP Annals 71 (2022) 233–236. https://doi.org/10.1016/j.cirp.2022.03.010

[3] C.A. Coulomb, Théorie des machines simples en ayant égard au frottement de leurs parties et à la roideur des cordages, Bachelier, 1821.

[4] J.A. Schey, Metal Deformation processes/friction and lubrication, 1970.

[5] E. Doege, B.-A. Behrens, Handbuch Umformtechnik, Springer Berlin Heidelberg, Berlin, Heidelberg, 2010. https://doi.org/10.1007/978-3-642-04249-2

[6] P. Christiansen, P.A.F. Martins, N. Bay, Friction Compensation in the Upsetting of Cylindrical Test Specimens, Exp Mech 56 (2016) 1271–1279. https://doi.org/10.1007/s11340-016-0164-z

[7] Siebel E, Die Formgebung im bildsamen Zustand, Stahleisen, Düsseldorf, 1932.

[8] T. Trzepiecinski, H.G. Lemu, Recent Developments and Trends in the Friction Testing for Conventional Sheet Metal Forming and Incremental Sheet Forming, Metals 10 (2020) 47. https://doi.org/10.3390/met10010047

[9] J. Obiko, Friction correction of flow stress-strain curve in the upsetting process, IOPSciNotes 2 (2021) 14401. https://doi.org/10.1088/2633-1357/abdd96

[10] M. Kraus, M. Lenzen, M. Merklein, Contact pressure-dependent friction characterization by using a single sheet metal compression test, Wear 476 (2021) 203679. https://doi.org/10.1016/j.wear.2021.203679

[11] J. Danckert, T. Wanheim, Analysis of the Ring Test Method for the Evaluation of Frictional Stresses in Bulk Metal Forming Processes, CIRP Annals 37 (1988) 217–220. https://doi.org/10.1016/S0007-8506(07)61621-7

[12] A.A. Camberg, T. Tröster, A simplified method for the evaluation of the layer compression test using one 3D digital image correlation system and considering the material anisotropy by the equibiaxial Lankford parameter, IOP Conf. Ser.: Mater. Sci. Eng. 967 (2020) 12077. https://doi.org/10.1088/1757-899X/967/1/012077

Sheet Metal 2023 Materials Research Forum LLC
Materials Research Proceedings **25** (2023) 213-220 https://doi.org/10.21741/9781644902417-27

Preliminary study of the mechanical characteristics implementation of friction stir welded AA2024 joints by adding pure copper

Sara Bocchi[1, a *], Gianluca D'Urso[1,b] and Claudio Giardini[1,c]

[1] University of Bergamo - Department of Management, Information and Production Engineering, Via Pasubio 7b, 24044, Dalmine (BG), Italy

[a]sara.bocchi@unibg.it, [b]gianluca.d-urso@unibg.it, [c]claudio.giardini@unibg.it

Keywords: Friction Stir Welding, Aluminium, Solid State Diffusivity

Abstract. In the present paper, the mechanical properties of a AA2024 welded by Friction Stir Welding (FSW) were examined, both homogeneously and using a commercial pure copper sheet positioned between the edges to be welded in two configurations, T and C. The temperature trends reached during the FSW process were extrapolated through the development of a simulative model, to select the best combination of parameters to use in the experimental phase of the campaign. After that, the FSWed homogeneous and heterogeneous joints were executed and analysed. From the mechanical point of view, Rockwell B hardness tests and tensile tests were performed. It was possible to evidence a good relationship between the hardness distribution and the presence of the copper, especially at the nugget. On the contrary, the tensile tests executed orthogonally to the welding direction, showed a reduction of the tensile strength and of the real elongation percentage in the aluminium-copper heterogeneous FSWed joints.

Introduction

The Friction Stir Welding (FSW) is one of the most significant technological developments in metal joining of the last thirty years and it is considered a sustainable technology thanks to its energy efficiency. Friction Stir Welding is a welding technology developed by The Welding Institute (TWI) [1]. This solid-state mechanical joining technique is used to make joints between similar or dissimilar materials that are difficult to weld with conventional fusion techniques, such as sintered materials, magnesium, copper, Inconel, titanium, metal matrix composites and thermoplastics [2]. It is also used for high-strength aluminium alloys, which are difficult to join using traditional techniques due to their typical microstructural alteration during ageing hardening. In fact, the precipitates that are created thanks to the thermal treatment, interact with the dislocations within the matrix, influencing their mobility, and, consequently, the mechanical behaviour of the material. The morphology of the precipitates, the orientation, the distance between them and the degree of their coherence with the matrix contribute to the effectiveness of the precipitates themselves as a barrier to the movement of dislocations [3]. In particular, the interactions between dislocations and precipitates determine the mechanical behaviour of aluminium-copper alloys and, for the Al-Cu-(Mg) system, the precipitation sequence is as follows [4]:

$$\text{Solid supersaturated solution (SSSS)} \rightarrow \text{GP Zone} \rightarrow \theta'' \rightarrow \theta' \rightarrow \theta\ (Al_2Cu).$$

The precipitates represent a common microstructural feature of high strength aluminium-copper alloys, they are formed on the faces of the face-centred cubic structure of the matrix, and they provide an anisotropic contribution to the performance of the material stresses. At temperatures immediately below the solubilization, the Guinner-Preston (GP) zones begin to nuclear very rapidly at an increasing rate as the elapsed time decreases. These zones grow in such a way as to minimise the rate of increase in the deformation energy of the precipitates and the matrix.

Content from this work may be used under the terms of the Creative Commons Attribution 3.0 license. Any further distribution of this work must maintain attribution to the author(s) and the title of the work, journal citation and DOI. Published under license by Materials Research Forum LLC.

The second intermediate phase, called θ”, is still coherent with the matrix, resides on its planes and has a tetragonal structure. This phase is wider than the GP zones and maintains the same chemical composition.

As the ageing process continues, a final metastable precipitate (θ’) arises, with a tetragonal structure composed of Al_2Cu. These precipitates are semi-coherent to the aluminium matrix and swell with increasing ageing due to the surface reduction force.

The final equilibrium precipitate (θ) is mainly inconsistent with the matrix and can be nucleated directly from the supersaturated solution if the ageing temperature is high enough (about 300°C), or even at lower temperatures if the kinetic and thermodynamic conditions in the material are favourable. However, it was found that the phase diagram cannot explicitly explain what happens during the FSW process because it is constructed when the phases are in equilibrium. Knowing that the FSW process involves significant thermal changes, it is safe to assume that there is not enough time to have an equilibrium chemical reaction. The nucleation of the first metastable phases, which have a lower thermodynamic driving force, but a lower nucleation barrier due to greater coherence with the matrix, is therefore kinetically favourable [5].

One of the advantages of dissimilar soldering of copper to aluminium is that it can reduce the formation and growth of the layer formed by intermetallic compounds, introducing strong plastic deformation, and thus increasing the mechanical properties of the joint. The main advantages of diffusion are the obtainment of high-quality bonds and the possibility of producing joints with low porosity and good continuity along the weld [6].

Through a specific study on the solid-state diffusion of copper in aluminium, W. Bedjaoui et al. has shown that commercial 2xxx series aluminium and pure copper can be successfully joined using a solid-state diffusion welding process [7]. Furthermore, the increase in temperature and time during the diffusion bond leads to the formation of different types of intermetallic compounds, the dimensions of which vary as the parameters used vary.

The goal of this research is to improve the mechanical properties of a welding joint, operated by Friction Stir Welding between two AA2024 aluminium sheets, using a copper sheet positioned between the edges to be welded. By exploiting the solid-state diffusivity of copper in aluminium, the arrangement of copper within the aluminium alloy, through the passage of copper precipitates into the alloy which improve its mechanical characteristics, was ensured. The pair of parameters, rotational speed and advancing feed of the tool, which allows it to stay within the area of the aluminium-copper diagram in which the hardening precipitates are formed θ', is identified through finite element simulations. After that, an experimental campaign was conducted and subsequent mechanical tests, Rockwell hardness and tensile tests, were executed.

Materials and Methods

A 3D FEM Lagrangian model was set up using the commercial software DEFORM 3D™. The tool was modelled as a rigid object using AISI-1045 steel as reference material (defined between 20°C and 1100°C). The plates to be welded were modelled as a single plastic workpiece made of Aluminium-2024 (defined between 300°C and 500°C). Both steel and aluminium with their material flow stress data were chosen among those available in the Deform library database.

The assumption to consider the tool as a rigid object and the workpiece as plastic is reasonable because of the significantly higher yield strength of the steel tool with respect to the yield strength of the aluminium workpiece. Finally, the workpiece was discretized using 80000 tetrahedral elements, as shown in Fig. 1. Regarding the thermal part of the simulation, the considered parameters are reported in Tab. 1, as they were optimised in a previous work [8].

Materials Research Forum LLC
https://doi.org/10.21741/9781644902417-27

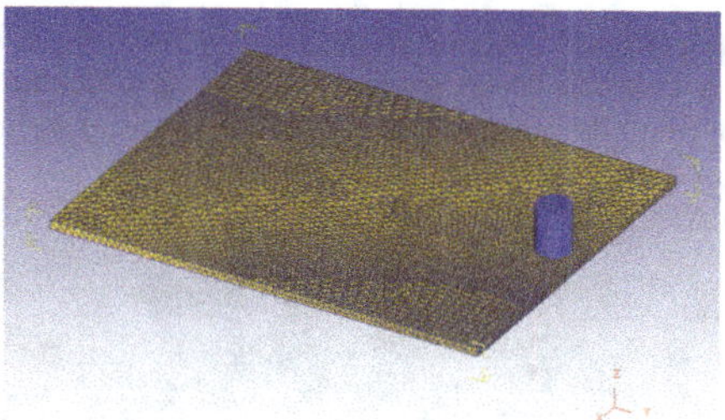

Fig. 1: Mesh of the aluminium workpiece.

Table 1: Thermal and friction parameters considered for the simulations.

Parameter	**Value**
Heat exchange with the environment	0.02 N/(mm·s·°C)
Friction coefficient aluminium-tool	1
Thermal conductivity	450 N/(mm·s·°C)
Aluminium emissivity	0.7
Heat transfer coefficient aluminium-tool	10000 N/(mm·s·°C)

As regards the process parameters, it was identified from a literature review a range of acceptable values to be considered for FSW welding of an AA2024 joint [9]–[13]. The chosen parameters are reported in Tab. 2.

Table 2: Process parameters used for the simulations.

Rotational speed (rpm)	**Advancing feed (mm/min)**
600-800-1000	40-120-200

Each combination of parameters was simulated. At the end of each simulation, the temperature trends at the nugget (0 and 5 mm from the centre of the weld), the TMAZ and the HAZ (20 mm and 35 mm from the welding line respectively) were evaluated.

The sheets were experimentally welded using the best combination of parameters resulting from the simulations performed: in order to obtain the right interaction between copper and aluminium.

The experimental campaign was performed using a CNC machine. Butt joints were carried out on AA2024-T3 sheets having a thickness equal to 4 mm (chemical composition and mechanical properties declared by the supplier in Tab. 3 and Tab. 4).

Table 3: AA2024-T3 composition.

Al	Si	Fe	Cu	Mn	Mg	Zn	Ti	Cr
bulk	0.06	0.09	4.50	0.46	1.40	0.08	0.04	0.03

Table 4: AA2024-T3 mechanical properties.

Yield Strength (YS) [MPa]	Ultimate Tensile Strength (UTS) [MPa]	Max strain [%]
344	464	16

A simple tool with a frustum of cone pin shape (pin minimum and maximum diameters equal to 4 and 6 mm, pin height equal to 3.9 mm) and flat shoulder (diameter equal to 16 mm) was considered. For all the welding, the tool inclination at 3° and the tool penetration depth into the sheets at 3.99 mm were fixed.

A homogeneous welding was carried out between two sheets of AA2024-T3 and then two welds were performed in which a commercially pure copper sheet (thickness equal to 0.1 mm) was interposed between the aluminium edges: one in a T- and one in a double C-configuration, as shown in Fig. 2. The thin copper sheet was ultrasonically cleaned in an acetone bath and then manually bent into the desired shape.

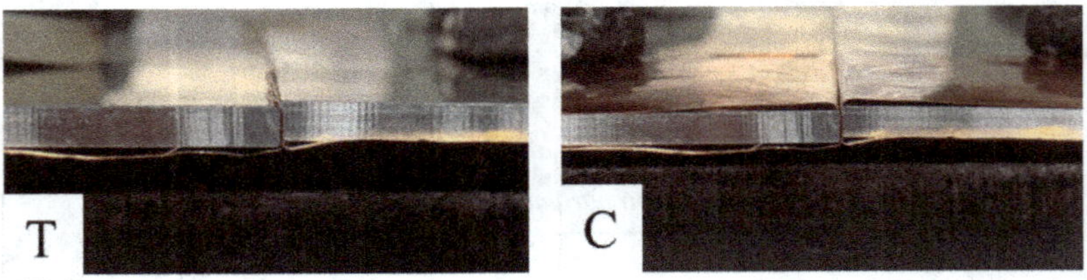

Fig. 2: T and double C copper configurations.

Rockwell B hardness tests (HRB) were performed following a 5 mm spaced grid in the central zone of the top of the specimen to avoid border effects, according to ISO 6508. Three profiles of indentations were carried out moving from the joint axis to the base material until the hardness of the base material was reached.

Tensile tests were executed orthogonally to the welding direction on the specimens having the welding nugget placed in the middle of gage length, according to UNI EN ISO 6892-1:2019. For the tests, a testing machine Galdabini with a load cell of 50 kN was used. The tests were carried out under speed control (1.2 mm/min). For the calculation of the engineering strain ε, strain gauges were applied to each sample.

Results and Discussion

From the simulations conducted in Deform, the temperature trends obtained in the four characteristic zones described in the methodology section were extrapolated: the peak temperatures reached were considered, to choose the best conditions to ensure the diffusivity of the copper within the aluminium (Tab. 5).

At the end of the experimental campaign, it was possible to note the absence of macroscopic defects in the homogeneous welding of AA2024, and in both welds made with copper interposed between the aluminium sheets (Fig. 3).

Table 5: Maximum temperatures obtained with the simulations. The pair of parameters chosen to conduct the experimental campaign is highlighted in orange.

Parameters		Temperatures [°C]			
[rpm]	[mm/min]	Centre [0 mm]	Nugget [5 mm]	TMAZ [20 mm]	HAZ [35 mm]
600	40	482.53	410.94	404.20	399.52
600	120	421.91	329.40	253.14	241.32
600	200	370.69	263.19	171.39	156.85
800	40	521.32	465.21	457.46	452.04
800	120	467.03	349.85	270.46	258.00
800	200	392.74	287.70	190.59	168.25
1000	40	485.46	409.89	403.48	398.69
1000	120	520.44	426.27	326.44	312.03
1000	200	463.77	360.46	219.94	199.23

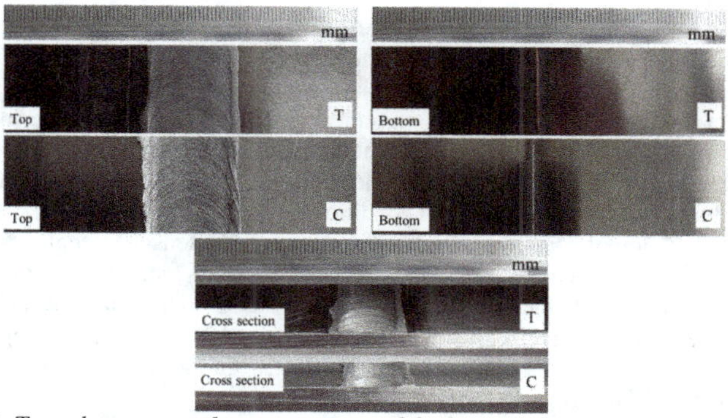

Fig. 3: Tops, bottoms, and cross sections of the heterogeneous joints AA2024-Cu.

Fig. 4 shows the graphs of the Rockwell B hardness tests performed both on the homogeneous welding AA2024-AA2024 and on the welds obtained with the addition of copper.

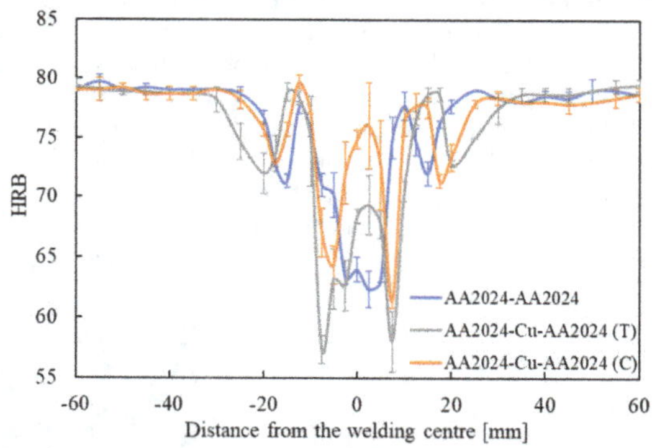

Fig. 4: HRB of the homogeneous and heterogeneous Friction Stir Welded joints.

With respect to homogeneous welding, the addition of copper has led to changes in the trend of the hardness curves. It can be observed that the minimum points of hardness have lower values in both the heterogeneous joints than in homogeneous welding, especially in the T-configuration. This behaviour could be attributable to the formation of embrittlement precipitates linked to the thermo-mechanical history of the area involved, that it is exactly on the nugget boundaries. The lowest hardness point in the homogeneous alloy is at the nugget, whilst in both heterogeneous configurations, these lower peaks are translated towards the base metal by about 5 mm to the right and about 5 mm to the left of the weld line.

Furthermore, on average there is an increase in hardness in the central area between -2.5 mm and +5 mm from the centre of the weld of about 4 HRB for the T-configuration and 11 HRB for the C-configuration. The increase in hardness of the C-configuration compared to the T-configuration is probably due to the greater quantity of copper used during processing, which is therefore able to form part of the brittle precipitates as well as the hardening precipitates. From the comparison it is also possible to observe a shift towards the base material of the minimum hardness points of the HAZ, of 5 mm for the T configuration and of 2.5 mm for the C configuration, with a less marked increase in hardness.

The addition of copper has determined an increase in hardness in the central area of the nugget, this is due to the passage in solid solution of the copper in the aluminium matrix, causing the formation of hardening precipitates, and the displacement of the minimum hardness values in the HAZ area and in the area between the nugget and the TMAZ towards the base metal causing a widening of the curves with their consequent removal from the nugget.

From the graph proposed in Fig. 5, it is possible to see the trend of the stress as the deformation varies tested and the fracture points for each condition.

In general, the addition of copper resulted in a reduction in the tensile strength (UTS). The UTS was 350.5 MPa for homogeneous welding, while for the C- and T-configurations it was, respectively, 301.75 MPa and 285.75 MPa. Furthermore, thanks to the tensile tests it was possible to ascertain that the addition of copper also led to a reduction in the percentage elongation at break. This value was equal to 2.56% for homogeneous welding, while for the C- and T- configurations it was 1.27% and 1.21%, respectively. It is also evident that, for all conditions, the fracture occurred preferentially in the central area of the nugget.

The reduction in the UTS and the percentage elongation at break compared to homogeneous welding is probably due to the presence of small layers of hard and brittle intermetallic precipitates (IMC) that are formed during welding, and which cause a reduction in mechanical strength. The formation of these very brittle intermetallic precipitates in a dissimilar AA2024-copper was demonstrated also by R. Khajeh et al. [14]. From the graph it is also possible to note that in the C-configuration the UTS and the percentage elongation are greater than those of the T-configuration. This is probably due to the higher amount of copper conveyed during welding in the C-configuration with respect to the T one. In fact, as the quantity of copper increases, the presence of the same in the weld increases, therefore in addition to the formation of the hard and brittle intermetallic precipitates, there is a greater formation of hardening precipitates by diffusion in the solid state which counteract the reduction of mechanical resistance due to IMC.

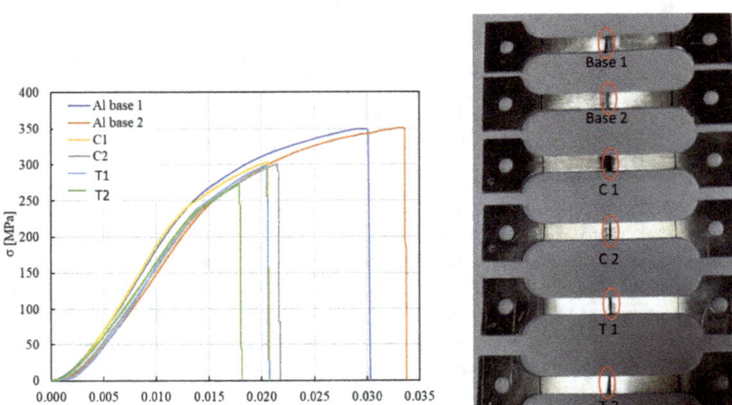

Fig. 5: On the left: tensile tests of the homogeneous and heterogeneous Friction Stir Welded joints (the number 1 and the number 2 are referred to the number of repetitions). On the right: fracture points of the homogeneous and heterogeneous joints.

Conclusions

The mechanical properties of the AA2024 welded by Friction Stir Welding were examined, both homogeneously and using a copper sheet positioned between the edges to be welded in two configurations, T and C.

Through an initial simulation phase and a subsequent experimental phase, it was possible to observe that:

- The parameters chosen from the simulations (S=800 rpm and f=40mm/min) proved to be suitable for all the welds, as they were free of defects and with good surface finishes.
- Rockwell hardness tests demonstrated that the addition of copper caused changes in hardness in the various characteristic areas of the joint. In particular, it increased the hardness in the central area of the nugget by 4 Rockwell points in the T configuration and 11 Rockwell points in the C configuration. Furthermore, it determined the shift of the hardness minima towards the base material.
- From the tensile tests it was found that the addition of copper resulted in a reduction in the tensile strength and the percentage elongation at break compared to homogeneous welding. However, the C-configuration has been observed to exhibit a higher tensile strength and percentage elongation at break than the T-configuration.

In conclusion, it can be stated that the improvement of the mechanical characteristics of a FSWed AA2024 by adding commercially pure copper to the interface of the aluminium edges to be welded, has been achieved partially but it is a good strategy that deserves to be studied.

Having observed that the mechanical characteristics of the C-configuration were superior to those of the T-configuration, to obtain better results it is suggested to orient the next studies in identifying copper configurations that allow the use of a greater quantity of copper in the welding. Moreover, new combinations of parameters had to be studied since it is not certain that the parameters optimised for the AA2024 alone are also optimal parameters for heterogeneous welds with copper, even if they do not show macroscopic defects.

References

[1] W. M. Thomas, E. D. Nicholas, J. C. Needham, M. G. Murch, P. Templesmith, and C. J. Dawes, "GB Patent application no. 9125978.8," 1991.

[2] M. Tabasi, M. Farahani, M. K. B. Givi, M. Farzami, and A. Moharami, "Dissimilar friction stir welding of 7075 aluminum alloy to AZ31 magnesium alloy using SiC nanoparticles," Int. J. Adv. Manuf. Technol., vol. 86, no. 1–4, pp. 705–715, Sep. 2016. https://doi.org/10.1007/s00170-015-8211-y

[3] H. Sehitoglu, T. Foglesong, and H. J. Maier, "Precipitate effects on the mechanical behavior of aluminum copper alloys: Part I. Experiments," Metall. Mater. Trans. A, vol. 36, no. 3, pp. 749–761, Mar. 2005. https://doi.org/10.1007/s11661-005-0190-4

[4] C. Sigli, F. De Geuser, A. Deschamps, J. Lépinoux, and M. Perez, "Recent advances in the metallurgy of aluminum alloys. Part II: Age hardening," Comptes Rendus Physique, vol. 19, no. 8. No longer published by Elsevier, pp. 688–709, 01-Dec-2018. https://doi.org/10.1016/j.crhy.2018.10.012

[5] M. Jariyaboon et al., "The effect of welding parameters on the corrosion behaviour of friction stir welded AA2024–T351," Corros. Sci., vol. 49, no. 2, pp. 877–909, Feb. 2007. https://doi.org/10.1016/j.corsci.2006.05.038

[6] E. B. Hannech, N. Lamoudi, N. Benslim, and B. Makhloufi, "Intermetallic formation in the aluminum-copper system," Surf. Rev. Lett., vol. 10, no. 4, pp. 677–683, Apr. 2003. https://doi.org/10.1142/S0218625X03005396

[7] W. Bedjaoui, Z. Boumerzoug, and F. Delaunois, "Solid-State Diffusion Welding of Commercial Aluminum Alloy with Pure Copper," Int. J. Automot. Mech. Eng., vol. 19, no. 2, pp. 9734–9746, Jun. 2022. https://doi.org/10.15282/ijame.19.2.2022.09.0751

[8] M. Quarto, S. Bocchi, G. D', N. A. Urso, and C. Giardini, "Hybrid finite elements method-artificial neural network approach for hardness prediction of AA6082 friction stir welded joints," Int. J. Mechatronics Manuf. Syst., vol. 15, no. 2/3, p. 149, 2022. https://doi.org/10.1504/IJMMS.2022.124919

[9] T. Rajkumar, K. Raja, K. Lingadurai, S. D. Vetrivel, and A. Godwin Antony, "Interfacial microstructure analysis of AA2024 welded joints by friction stir welding," J. New Mater. Electrochem. Syst., vol. 23, no. 2, pp. 123–132, Apr. 2020. https://doi.org/10.14447/jnmes.v23i2.a09

[10] N. S. Abtan, A. H. Jassim, and M. S. Marmoos, "Study on the effects of rotational and transverse speed on temperature distribution through friction stir welding of AA2024-T3 aluminium alloy," J. Adv. Res. Fluid Mech. Therm. Sci., vol. 53, no. 2, pp. 234–248, 2019.

[11] M. Cabrini et al., "Stress corrosion cracking of friction stir-welded AA-2024 T3 alloy," Materials (Basel)., vol. 13, no. 11, Jun. 2020. https://doi.org/10.3390/ma13112610

[12] A. W. El-Morsy, M. Ghanem, and H. Bahaitham, "Effect of Friction Stir Welding Parameters on the Microstructure and Mechanical Properties of AA2024-T4 Aluminum Alloy," Eng. Technol. Appl. Sci. Res., vol. 8, no. 1, pp. 2493–2498, Feb. 2018. https://doi.org/10.48084/etasr.1704

[13] S. Bocchi, M. Cabrini, G. D'Urso, C. Giardini, S. Lorenzi, and T. Pastore, "Stress enhanced intergranular corrosion of friction stir welded AA2024-T3," Eng. Fail. Anal., vol. 111, Apr. 2020. https://doi.org/10.1016/j.engfailanal.2020.104483

[14] R. Khajeh et al., "Microstructure, mechanical and electrical properties of dissimilar friction stir welded 2024 aluminum alloy and copper joints," J. Mater. Res. Technol., vol. 14, pp. 1945–1957, Sep. 2021. https://doi.org/10.1016/j.jmrt.2021.07.058

Sheet Metal 2023 Materials Research Forum LLC
Materials Research Proceedings **25** (2023) 221-228 https://doi.org/10.21741/9781644902417-28

Evaluation of the properties of AA7020 tubes generated by a heat treatment based hydroforming process

Jonas Reblitz[1,a *], Ricardo Trân[2,b], Verena Kräusel[2,c] and Marion Merklein[1,d]

[1]Institute of Manufacturing Technology, Friedrich-Alexander-Universität Erlangen-Nürnberg, Egerlandstraße 13, 91058 Erlangen, Germany

[2]Fraunhofer Institute for Machine Tools and Forming Technology, Reichenhainer Straße 88, 09126 Chemnitz, Germany

[a]jonas.reblitz@fau.de, [b]ricardo.tran@iwu.fraunhofer.de, [c]verena.kraeusel@mb.tu-chemnitz.de, [d]marion.merklein@fau.de

* Corresponding author

Keywords: Heat Treatment, Hydroforming, Tubes

Abstract. Energy efficiency and sustainability are getting more and more relevant in society and industry. Especially energy-intensive sectors offer a high potential for power savings. Thus, different lightweight strategies are applied in the automotive sector. In this context, high strength profile components are used as crash structures. As an example, these profiles are deployed to protect the battery of electrically powered vehicles. Therefore, high strength aluminum alloys of the 7xxx series are suitable for compensating the high battery weight partly. Compared to steel, this material is characterized by a high strength-to-weight ratio as well as an increased thermal conductivity. However, one challenge in forming high strength aluminum alloys is the limited formability at room temperature. For this purpose, heat treatment based process routes for tube hydroforming are investigated within this research work. By the combination of W-temper forming and hydroforming complex part geometries can be generated. A subsequent artificial ageing process realizes the required high strength of the profiles. Using AA7020 tubes, the properties of a demonstrator geometry are compared to the initial state T6 of the semi-finished parts. For this purpose, the wall thickness distribution and the grain structure are evaluated. Thereby, the influence of forming and heat treatment on the geometrical and microstructural properties is to be analyzed. For the evaluation of the achievable hardening of the aluminum alloy, the mechanical properties are investigated by tensile tests. This investigations provide an assessment of the process route under industry-related conditions.

Introduction

The reduction of energy consumption and CO_2 emissions is one of the most important aims in society. Therefore, the EU plans to reduce the emission of greenhouse gases up to 55% in 2030 compared to 2005 [1]. A high potential for reductions offers the transporting sector, which is responsible for 26% of total emissions within the EU [2]. Besides digitization of the traffic and the expansion of public transport [3], the electrification of vehicles can contribute to reduce CO_2 emissions. However, electrically powered vehicles have a high weight due to the battery, which moreover has to be protected against mechanical loads and crash impacts. On the other hand, a reduction of car weight leads to further savings in energy consumption. 100 kg additional weight leads to a reduction of range of 3.5% and an increase of the required engine power of 6% [4]. Nevertheless, a high percentage of aluminum is already used for the battery enclosure in current vehicles [5]. The application of tailored profiles made out of high-strength aluminum alloys of the 7xxx series offers a high potential for further weight reductions, which is particularly due to the high specific strength. One challenge is to reach both, a sufficient formability as well as the

Content from this work may be used under the terms of the Creative Commons Attribution 3.0 license. Any further distribution of this work must maintain attribution to the author(s) and the title of the work, journal citation and DOI. Published under license by Materials Research Forum LLC.

Materials Research Proceedings **25** (2023) 221-228 https://doi.org/10.21741/9781644902417-28

maximum strength. Current approaches for forming high strength aluminum alloys are W-temper forming and quench forming [6]. For sheets, investigations regarding the solution heat treatment for W-temper forming [7] and the influence on the artificial ageing [8] were already carried out. To transfer these results on the forming of high strength profiles a heat treatment-based hydroforming process is investigated within this work. Thus, the formability is enhanced to generate complex components. In a subsequent artificial ageing treatment, the mechanical properties are adjusted. In preliminary work, the forming behavior of tubular semi-finished parts was analyzed for different process routes [9]. Consequently, the properties of a more complex demonstrator geometry are evaluated for the heat treatment-based hydroforming process within this research work. In this context, the wall thickness distribution is compared between the initial semi-finished tube and the demonstrator geometry with the aid of an optical measurement system. Afterwards, the microstructure and the average grain size is evaluated for both states. To investigate the influence of the analyzed process route on the mechanical properties, tensile tests are performed on the different demonstrator components by using a thermo-mechanical simulator of type Gleeble 3500. Thereby, different heat treatment parameters are applied to the artificial ageing. Consequently, it can be evaluated if an industry-related process chain for the production of complex profile components can achieve the initial condition T6 of the extruded tubes.

Methodology

Experimental setup of the hydroforming tests. As part of the investigations, W-temper forming was used to manufacture the demonstrator components. For this purpose, the tubes were first solution-annealed and then quenched in a water bath. The set W-temper condition significantly increases the cold formability of the material. The tubes were then formed in a conventional hydroforming process without axial material feeding with a water-oil emulsion as the forming medium. A tubular ton geometry with a homogenously widened middle area of 95 mm length is used as demonstrator. In comparison to conventional tube bulge tests there is no free forming area. The tube is formed against a defined tool geometry. Thereby tensile specimens with homogeneous strain distributions can be cut out. The experimental setup can be seen in Fig. 1 a), the demonstrator geometry is depicted in Fig. 2 a).

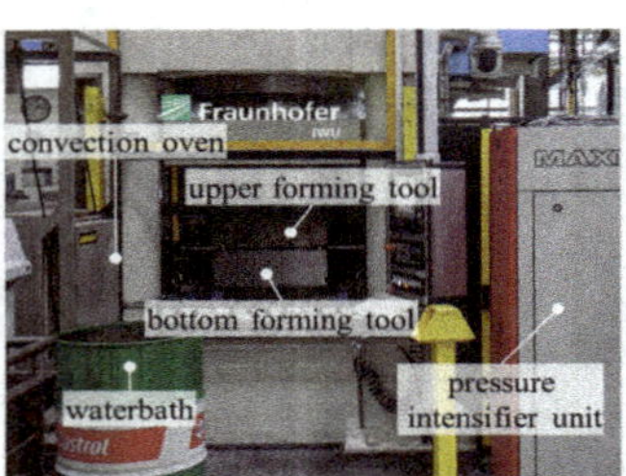

a)

		Process Route W-Temper-Forming
Semi-finished Tube	material	EN AW-7020 T6
	tube dimensions	Ø = 60 mm, t = 5 mm, l = 300 mm
	heat treatment	500 °C, convection oven, 60 min heating/holding
Forming Tool	temperature	20 °C room temperature
Process	max. forming pressure in MPa	34
	pressure build-up rate in MPa/s	5

b)

Fig. 1. a) Setup and b) parameters of the heat treatment-based hydroforming process

The tubes are heated to solution annealing temperature in a convection oven at 475 °C for 60 min. Afterwards the hot tubes can be quenched in the adjacent water bath. The experimental set-up can be used for different pressure build-up rates and maximum internal pressures. However, a pressure build-up rate of 5 MPa/s was used for the ton geometry. The maximum internal pressure was 34 MPa, as the tubes burst at an internal pressure of approximately 35 MPa. The test parameters that are based on common ageing temperatures and durations for AA7020 are summarized in Fig. 1 (b). Afterwards, the tubes are naturally aged for 24 h and then artificially aged in a two-stage ageing treatment with temperatures between 100 °C and 160 °C. The ageing times vary between 8 h and 24 h. The parameters of the ageing treatment are summarized in Table 1.

Table 1. Parameters for the artificial ageing process

Heat Treatment	Stage 1		Stage 2	
	Temperature	Time	Temperature	Time
CDC	185 °C	0.3 h	-	
Artificial ageing 1	100 °C	8 h	160 °C	16 h
Artificial ageing 2	100 °C	8 h	160 °C	24 h
Artificial ageing 3	100 °C	12 h	160 °C	16 h
Artificial ageing 4	100 °C	12 h	160 °C	24 h
Artificial ageing 5	125 °C	12 h	180 °C	4 h
Artificial ageing 6	125 °C	12 h	180 °C	6 h
Artificial ageing 7	125 °C	24 h	180 °C	6 h

Specimen preparation and experimental setup of the tensile test. The investigated demonstrator geometry is a barrel-shaped tube expanded by hydroforming, see Fig. 2 a). Due to the expansion on a length of 200 mm, a strain of 9% was reached. For the characterization of the mechanical properties, tensile specimens were cut out of the demonstrator by laser. Subsequently, the cutting edges were machined by milling. The corresponding tensile specimens are shown in Fig. 2 b). These are adapted A_{50} specimens, which have additional drillings in the clamping area to ensure a central positioning. Due to the preparation out of a tube, the specimens have a curvature. To avoid undesired influences of the clamping on the stress state, the clamping jaws have a curvature as well. These are deployed in the thermo-mechanical simulator Gleeble 3500, a testing machine that can also perform heat treatments and tests at elevated temperatures. The strains are measured by an optical strain measurement system of the Carl Zeiss GOM Metrology GmbH.

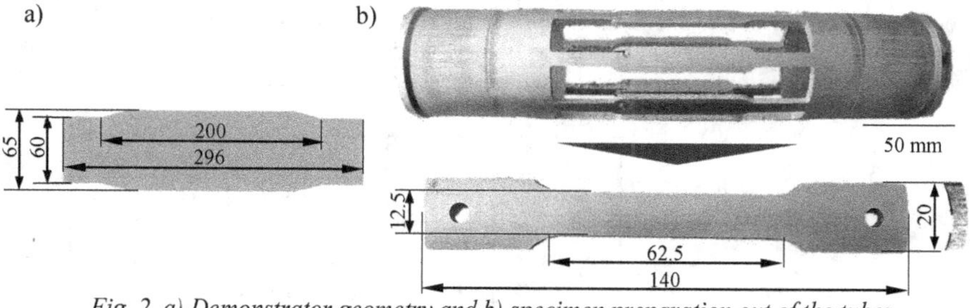

Fig. 2. a) Demonstrator geometry and b) specimen preparation out of the tubes

Results and discussion

Evaluation of the wall thickness distribution. At first, the wall thickness distribution of the initial state and the demonstrator geometry is analyzed by the optical measurement system ATOS from Carl Zeiss GOM Metrology GmbH. Thus, the influence of the hydroforming process on the properties of the tube component is investigated. Even in the as-delivered condition of the extruded tube, a wall thickness gradient in tangential direction was found. In the middle of the tubes and at the ends, the wall thickness varies continuous between 4.93 mm and 5.18 mm.. Consequently, the extrusion process causes a variation in the wall thickness of 0.25 mm. In axial direction, no gradient in wall thickness could be observed in the analyzed sections Due to the homogeneous expansion during the hydroforming process, failure is most likely at the thinnest point. For the analysis, a demonstrator of artificial ageing 7 according to Table 1 was used. For the optical measurement, the tube was cut in axial direction. Therefore, only one half is shown in Fig. 3 a). Hence, the curves in Fig. 3 b) do not match the lines for the minimum and maximum values.

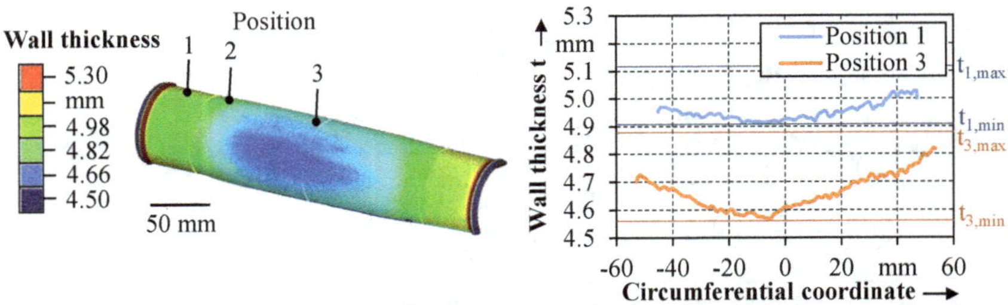

Fig. 3. a) Wall thickness distribution over the tube and b) wall thickness gradient in circumferential direction

For the evaluation of profile properties, the demonstrator geometry is split into three sections. Position 1 is located at the almost undeformed tube clampings. The maximum expanded section is represented by position 3 while position 2 describes the transition area. In this context, only the extrema in wall thickness are depicted in Fig. 3 b). As expected the smallest wall thickness can be found at position 3. The value is 4.59 mm whereas at position 1 the wall thickness is 4.90 mm. The wall thickness gradient in circumferential direction is shown in Fig. 3 b). There is a difference in the wall thickness of 0.21 mm at position 1 and 0.30 mm at position 3. Consequently, the thickness gradient increases during hydroforming. This has to be considered for the process design.

Influence of the heat treatment-based process route on the microstructure. The evaluation of the microstructure is performed for the initial state and the demonstrator geometry at position 1 and 3 in the middle of the wall, see Fig. 4. Probably due to the shear stresses during the extrusion, the micrographs in Fig. 4 a) and b) show a grain orientation in axial direction. The grain size and orientation increases due to the hydroforming process compared to the initial state T6 before annealing. One reason for this could be recrystallization during the extrusion process, which can lead to a small grain size in conjunction with high strain rates, as already shown for AA7075 [11].

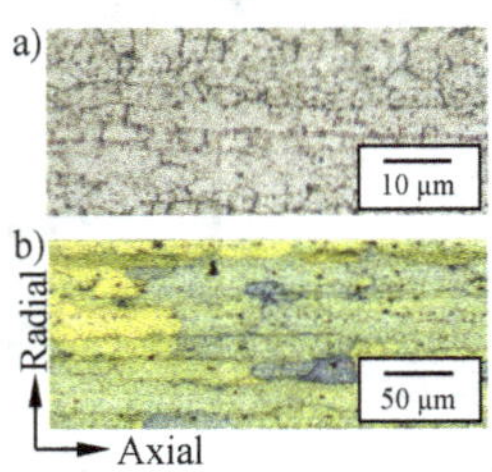

c)

		Initial state	Demonstrator
Average grain size in µm	Axial	7.00 ± 0.94	118.61 ± 6.58
	Radial	4.31 ± 0.29	25.70 ± 3.27
	Circum-ferential	9.39 ± 2.46	48.24 ± 4.21

Fig. 4. Microstructure of a) the initial state T6 before annealing and b) the demonstrator geometry at position 3 as well as c) the average grain size for both conditions

The average grain size for both states in different directions was determined by linear intercept method according to ASTM E 112; see Fig. 4 c). In the initial state the grain size is between 4.31 µm and 9.39 µm whereas the grain elongation is highest in the tangential direction. Significant higher grain sizes can be found in the demonstrator geometry with values between 25.70 µm and 118.61 µm, which might be caused by high solution annealing temperatures and durations. Due to low average heating rates in the convection oven of 0.14 K/s as well as the high wall thickness, dislocations are removed. Since dislocations are recrystallization nuclei, a coarse grain structure is generated [10]. Furthermore, Fujda et al. investigated an additional influence of the solution annealing temperature on the grain size for AA2024 [12]. For the demonstrator geometry, the grain structure is also analyzed for the different axial and radial positions. The average grain size is shown in Fig. 5. In preliminary investigations, different sections over the cross-section with varying mechanical properties could be found for the initial tubes [13].

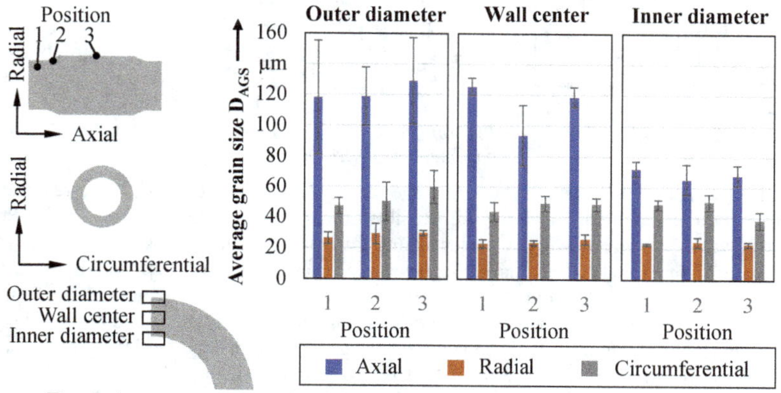

Fig. 5. Average grain size for different positions in axial and radial direction

In general, the highest grain size was found in axial direction due to the extrusion process. There are only small differences for the different positions in axial direction. In circumferential direction, the grain size grows with increasing pre-deformation at the outer diameter and in the center of the wall. One reason could be the expansion during the hydroforming process in circumferential direction. At inner diameter, no clear trend is visible. This section is characterized by a fine grain structure, which was also found in the initial state and can be caused by recrystallization [13]. In summary, there is a coarse-grained microstructure in the demonstrator profiles that shows similar gradients in the local grain size like the initial state. Thus, the reproduction of the initial mechanical properties despite the coarse microstructure has to be investigated.

Evaluation of the mechanical properties for different heat treatment parameters. First, common two-stage hot ageing processes are investigated for the AA7020 alloys, as well as cathodic dip coating CDC, which is mainly used in the automotive industry. The CDC generates the lowest strength values and the highest strains. In this case, probably no entire hardening was reached due to the short ageing duration. For the artificial ageing 3 and 4 with an ageing duration of 12 hours in the first stage, the highest hardening could be achieved. However, the ultimate tensile strength of 410 MPa in the initial state could not be reproduced by a value of 346 MPa. The uniform elongation is slightly lower than for the T6 state. Specimens with an ageing duration of eight hours at the first stage show lower strength values than for twelve hours. An increase in the ageing time in second stage tends to a further decrease in strength. The corresponding mechanical properties are shown in Fig. 6.

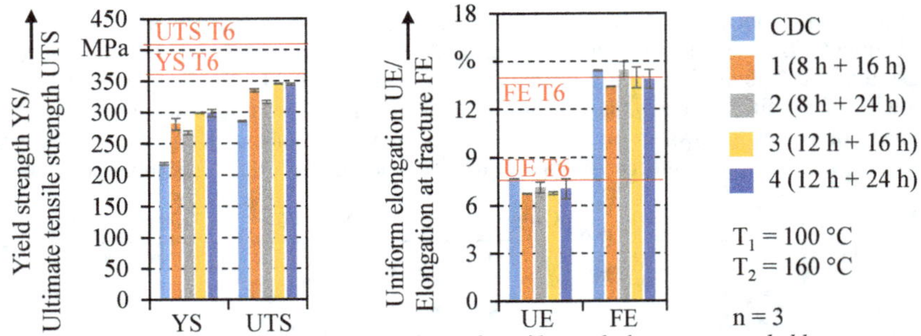

Fig. 6. Mechanical properties of the hydroformed profiles with the recommended heat treatment parameters for AA7020

For longer ageing durations in the first stage, an equilibrium state seems to be reached. The higher strength in heat treatment 3 and 4 is possibly due to a higher proportion of Guinier-Preston zones (GP zones) as a result of the longer ageing time in stage one. These act as nuclei for the formation of η'-precipitates which are crucial for maximum hardening [14]. Due to a lack of GP zones, a duration of eight hours in the first stage may induce an increased nucleation at dislocations. Therefore, the transformation of η'-precipitation into the stable and less solid η phase is accelerated. Thus, an enlargement of stage two might lead to overageing [15]. This effect was also observed by Krishnanunni et al. for AA7010 [16]. Based on the accelerated nucleation in the presence of dislocations, a reduction of the heat treatment time, especially in stage two, could be useful [14]. Hence, the ageing duration is reduced while the temperature is slightly increased, see Fig. 7. In general, the strengths achieved are lower than in the previous settings. It has already been shown for the alloy AA7020, that preforming has a positive effect on the strength, but that this effect is reversed if the ageing temperature is too high [15]. Dechamps et al. observed that in undeformed microstructures a higher proportion of homogeneously distributed η'-precipitates was present after an artificial ageing of one hour at 160 °C, while in predeformed microstructures η-precipitates were present at the dislocations [17]. The presence of dislocations hinders the nucleation of η'-precipitates and leads to coarse precipitate structures with lower strength. In addition, this is enforced by the annihilation of vacancies by the dislocations, which favor the nucleation of GP zones during cold aging [17]. Since an extension of stage two again reduces the strength in heat treatments 5 and 6, stage one is extended in parameter set 7. This results in higher strength values comparable to the tests in Fig. 6. Due to the longer heat treatment time at 125 °C, the influence of the pre-deformation is likely to be lower. In order to check whether a higher dislocation density leads to less hardening during artificial ageing, hardness measurements are carried out at the three previously defined positions, see Fig. 7 b). As expected, the Brinell hardness for the demonstrator component is lower than in the initial condition. At position 1, the highest hardness of 98.84 HB is obtained, while in the deformed areas it is slightly lower at about 95 HB. This tends to indicate that pre-deformation has led to a reduction in hardening in the tests performed, although the influence is quite slight for parameter combination 7. In general, an extension of stage 1 seems to have a positive effect on artificial ageing. On the other hand, high temperatures lead to lower hardening, as well as high ageing durations in stage 2 and the deformation preceded by the hydroforming process. Thus, the mechanical properties of the initial condition could not be achieved. An improvement of the properties might be reached by hot working to reduce the dislocation density as well as a further reduction of the ageing and solution annealing temperature.

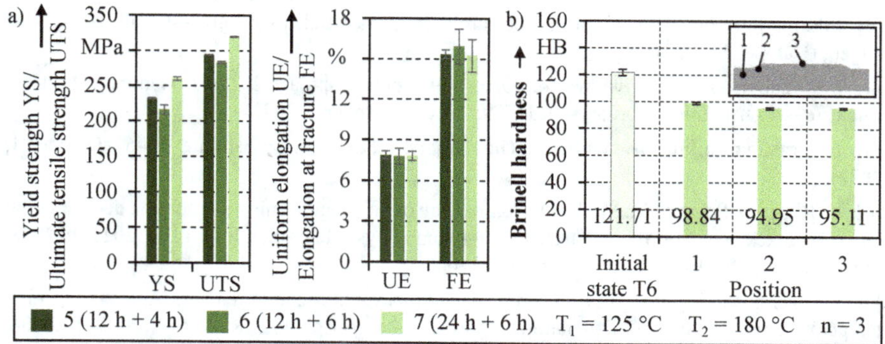

Fig. 7. a) Mechanical properties of the hydroformed profiles at higher ageing temperatures and b) hardness distribution in axial direction for heat treatment 7

Summary and outlook

Tube profiles made out of 7xxx aluminum alloys have a high specific strength. A challenge in forming technology is the low formability at room temperature in conjunction with the required high strengths. One approach is the combination of W-temper forming and hydroforming with a subsequent artificial ageing process. The achievable component properties were compared for a demonstrator geometry with the initial condition T6 of tubular semi-finished parts. In this context, a gradient in wall thickness was observed in circumferential tube direction for both states. The variation in wall thickness was increased by hydroforming so that the probability of failure is highest on the side with the smallest wall thickness. When investigating the microstructure, a fine-grained structure was observed in the initial state, while the grain size in the demonstrator component was up to 17 times higher. One possible reason for this could be the solution annealing, which results in grain growth. To determine the reachable mechanical properties, tensile specimens were extracted from the demonstrator components for different artificial ageing parameters. In accordance with common heat treatment processes for AA7020, the artificial aging was carried out in two stages. It was shown that an extension of stage one has a positive effect on the strength, while a longer second stage results in softening due to the higher temperature. This could be caused by the development of coarser precipitates and the increased formation of η-phases. A similar effect is expected with the increase of the artificial ageing temperature. Pre-deformation in the hydroforming process also causes an acceleration of the precipitation transformation and the growth of coarse precipitates, which tended to be confirmed with the aid of hardness measurements. Thus, the mechanical properties of the initial state could not be regained.

Further research needs to be carried out to increase the strength of the demonstrator components. Therefore, different heating rates during the solution annealing as well as different heat treatment temperatures have to be investigated. In addition, lower temperatures need to be considered for artificial ageing. An increase in strength could also be achieved by hot forming. In a further step, the achievable properties of a more complex component are to be analyzed.

References

[1] Information on https://www.eea.europa.eu/ims/total-greenhouse-gas-emission-trends, (October 1, 2022).

[2] Information on https: https://www.destatis.de/Europa/EN/Topic/Environment-energy/CarbonDioxideRoadTransport.html, (October 1, 2022).

[3] Information on https://transport.ec.europa.eu/news/efficient-and-green-mobility-2021-12-14_en, (October 1, 2022).

[4] Information on https://www.cargroup.org/wp-content/uploads/2021/02/Aluminum-Battery-Enclosures-Constellium-February-2021-FINAL.pdf, (October 1, 2022).

[5] Information on https://www.audi-technology-portal.de/en/download?file=1946, (October 1, 2022).

[6] N. Rigas, H. Schmid, M. Merklein, Comparison of different forming methods on deep drawing and springback behavior of high-strength aluminum alloys, Mater. Sci. Eng. 1157 (2021) 012048. https://doi.org/10.1088/1757-899X/1157/1/012048

[7] E. Sáenz de Argandoña, L. Galdos, R. Ortubay, J. Mendiguren, X. Agirretxe, Room temperature forming of AA7075 aluminum alloys: W-temper process, KEM 651 (2015) 199-204. https://doi.org/10.4028/www.scientific.net/KEM.651-653.199

[8] Hebbar, S., Kertsch, L., Butz, A., Optimizing Heat Treatment Parameters for the W-temper Forming of 7xxx Series Aluminum Alloys, Metals 10 (2020) 1361. https://doi.org/10.3390/met10101361

[9] R. Trân, J. Reblitz, R. Haase, V. Psyk, V. Kräusel, M. Merklein, Hydroforming of High-Strength Aluminum Tubes with Thermo-Mechanical Manufacturing Processes, Eng. Proc. 26 (2022) 13. https://doi.org/10.3390/engproc2022026013

[10]F. Ostermann, Anwendungstechnologie Aluminium. 3rd edn. Springer Vieweg, Berlin, Heidelberg (2014). https://doi.org/10.1007/978-3-662-43807-7

[11]Z.-C. Sun, L.-S. Zheng, H. Yang, Softening mechanism and microstructure evolution of as-extruded 7075 aluminum alloy during hot deformation, Mat. Charact. 90 (2014) 71-80. https://doi.org/10.1016/j.matchar.2014.01.019

[12]M. Fujda, R. Misicko, L. Rusnakova, M. Sojko, Effect of Solution Annealing Temperature on Structure and Mechanical Properties of EN AW 2024 Aluminium Alloy, Journal of Metals, Materials and Minerals 17 (2007) 35-40.

[13]J. Reblitz, S. Wiesenmayer, R. Trân, M. Merklein, Investigation of geometrical and microstructural influences on the mechanical properties of an extruded AA7020 tube, WGP 2022 LNPE (2023) 1-12.

[14]S. Lotz, E. Scharifi, U. Weidig, K. Steinhoff, Effect of Combined Forming and Aging Processes on the Mechanical Properties of the Precipitation-Hardenable High-Strength Aluminum Alloys AA6082 and AA7075, Metals 12 (2022) 1250. https://doi.org/10.3390/met12081250

[15]M. Merklein, J. Degner, Influence of pre-strain and simulated paint-bake on mechanical properties of high strength aluminum alloy AA7020, AMM 805 (2015) 115-122. https://doi.org/10.4028/www.scientific.net/AMM.805.115

[16]S. Krishnanunni, R. K. Gupta, G. Ajithkumar, V. Anil Kumar, Optimization of Heat Treatment Cycles and Characterization of Aluminum Alloy AA7010, JMEP 28 (2019) 776-787. https://doi.org/10.1007/s11665-019-3861-9

[17]A. Deschamps, F. Livet, Y. Bréchet, Influence of predeformation on ageing in an Al-Zn-Mg alloy-I. Microstructure evolution and mechanical properties, Acta mater. 47 (1999) 281-292. https://doi.org/10.1016/S1359-6454(98)00293-6

Sheet Metal 2023 Materials Research Forum LLC
Materials Research Proceedings **25** (2023) 229-236 https://doi.org/10.21741/9781644902417-29

On the utilization of radial extrusion to characterize fracture forming limits. Part I – methodology and tooling

Rui F.V. Sampaio[1,a], João P.M. Pragana[1,b], Ivo M.F. Bragança[2,c], Carlos M.A. Silva[1,d], Chris V. Nielsen[3,e] and Paulo A.F. Martins[1,f *]

[1]IDMEC, Instituto Superior Técnico, Universidade de Lisboa, Portugal

[2]CIMOSM, Instituto Superior de Engenharia de Lisboa, Instituto Politécnico de Lisboa, Portugal

[3]Department of Civil and Mechanical Engineering, Technical University of Denmark

[a]rui.f.sampaio@tecnico.ulisboa.pt, [b]joao.pragana@tecnico.ulisboa.pt, [c]ibraganca@dem.isel.ipl.pt, [d]carlos.alves.silva@tecnico.ulisboa.pt, [e]cvni@dtu.dk, [f]pmartins.tecnico.ulisboa.pt

Keywords: Forming, Failure, Stress State Transitions

Abstract. This paper and its second part introduce a new formability test based on double-action radial extrusion to characterize material formability in the bulk-to-sheet material flow transitions that are commonly found in metal forming. This first part draws from the presentation of a multidirectional tool, which was designed to convert the vertical press stroke into horizontal movement of the extrusion punches towards each other, to aspects of experimental strain determination, fractography and finite element analysis. The methodology and tooling that are introduced here pave the way for subsequent testing and modelling of the different modes of fracture in three-dimensional to plane-stress evolutions, typical of bulk-to-sheet material flow transitions, by means of the new proposed test.

Introduction

The fracture locus in bulk metal forming is commonly determined by means of upset compression tests performed on cylindrical and tapered specimens and is usually plotted in principal strain space $(\varepsilon_1, \varepsilon_2)$ as a straight line with slope '-1/2' falling from left to right [1]. This line is the graphical representation of the positive and negative strain values that a material withstands at the onset of cracking by out-of-plane shearing, as it was recently proved by Martins et al. [2], and earlier proposed by Kobayashi [3] after observing that vertical and inclined cracks found on the outer surface of these specimens do not run radially.

The proposal of a bilinear fracture locus in principal strain space by Erman et al. [4] resulting from combination of the fracture limit line of slope '-1/2' with a second fracture limit line of slope '-1', is essential to match the strains at failure of the cylindrical and tapered specimens with those of the tensile and upset compression tests performed on rod and cylindrical flanged specimens. This second fracture limit line of slope '-1' corresponds to failure by tension [5] and, in case of cylindrical flanged specimens, the morphology of the cracks is consistent with the experimental observation of vertical cracks running radially at the outer surfaces [6].

Although the bilinear fracture locus is consistent with the experimental results obtained with the different formability test specimens, it gives rise to an 'uncertainty region' (Fig. 1) within which failure occurs by mixed modes resulting from competition between cracking by tension (mode I) and by out-of-plane shearing (mode III). This was recently confirmed by Sampaio et al. [7], who designed a new test specimen to investigate the morphology of the cracks in the 'uncertainty region' of principal strain space and concluded that fracture was triggered by out-of-plane shearing (mode III) and propagate radially by tension (mode I).

Content from this work may be used under the terms of the Creative Commons Attribution 3.0 license. Any further distribution of this work must maintain attribution to the author(s) and the title of the work, journal citation and DOI. Published under license by Materials Research Forum LLC.

Materials Research Proceedings **25** (2023) 229-236 https://doi.org/10.21741/9781644902417-29

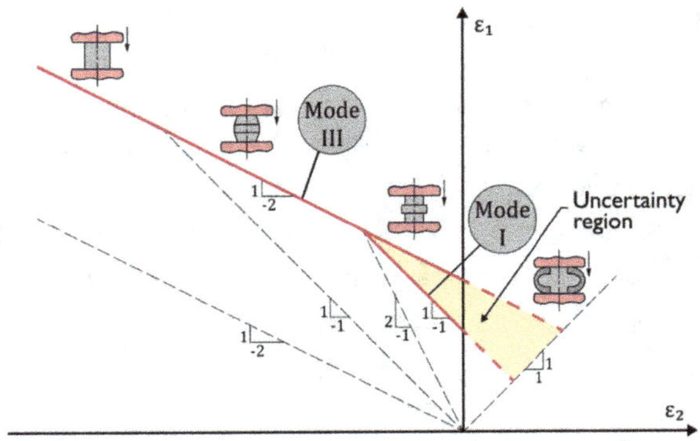

Fig. 1 Bulk formability tests and fracture limit lines in principal strain space.

Transitions from crack opening by mode I to propagation by mode III at the leftmost corner of the 'uncertainty region' were previously observed by the authors during upset compression of cylindrical flanged test specimens made from aluminum AA2030-T4 [8].

In view of the above, Sampaio et al. [7] proposed a new ductile damage criterion based on combination of the Cockcroft and Latham [9] criterion corresponding to crack opening by mode III with a modified version of the McClintock [10] criterion commonly used in crack opening by mode I to model ductile damage over the entire range of stress-triaxiality values, including the 'uncertainty region' of crack opening by mixed modes.

Although research on formability and crack opening modes has been directly, or indirectly, associated with stress triaxiality and, sometimes, also with the Lode parameter [11], there are aspects related to material flow transitions that are not commonly taken into consideration. Accountability of these aspects means, in practical terms, to include the three-dimensional to plane stress material flow transitions (i.e., bulk-to-sheet evolutions) in the list of parameters that are responsible for the competition between the different crack opening modes.

Fig. 2 shows examples of bulk formed parts where plane stress is likely to occur in regions subjected to extensive plastic material flow, (refer to the pink colored regions of the solid rod flange, connecting rod and tube flange, where the final thickness t is much smaller than that of the preforms). The connecting rod is a good example of a well-known occurrence of three-dimensional to plane stress transitions when excess material flows into the flash gutter during impression-die forging.

Under these circumstances, the main objective of this work is to investigate formability in three-dimensional to plane stress material flow transitions by means of a new formability test based on the working principle of double-action radial extrusion [12]. As will be shown, depending on the aspect ratio and overall ductility of the specimens, the diameter-to-thickness ratio of the radially extruded flange may reach typical plane stress values, thereby allowing to analyze strain path evolutions and failure in three-dimensional to plane stress material flow transitions by means of digital image correlation (DIC), scanning electron microscopy (SEM) and finite element modelling.

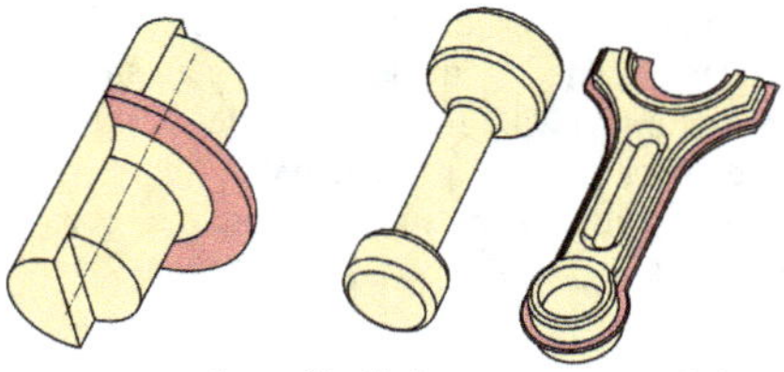

Fig. 2 Schematic representation of bulk forming parts exhibiting regions (refer to the pink color) where material flow takes place under plane stress conditions (solid rod flange, connecting rod and tube flange).

This paper is focused on the multidirectional tool where the double-action radial extrusion setup is installed and, on the methods, and procedures that are used in formability analysis and fractography. Special emphasis is given to a technique for determining the instant of cracking and the corresponding fracture strains by combination of the experimental force vs. time evolutions with the results obtained from digital image correlation.

Numerical simulation with finite elements is also addressed to introduce the fundamentals of exchanging data between two and three-dimensional models at the onset of diffuse necking when asymmetric plastic deformation of the outer flange begins.

Materials and Mechanical Characterization

The experimental work on the new formability test based on double-action radial extrusion made use of a soft aluminum AA1050-O and a medium strength aluminum-magnesium-silicon alloy AA6082 in the annealed (O) and solution heat-treated (T6) state. The materials were supplied in the form of solid rods and the flow curves were determined by means of compression tests carried out on cylindrical test specimens with 20 mm diameter and 20 mm height (Fig. 3).

The compression tests were performed at ambient temperature on an Instron SATEC 1200 kN hydraulic testing machine with a constant crosshead velocity of 5 mm/min. The specimens were lubricated with molybdenum disulphide (MoS2) to ensure near frictionless conditions.

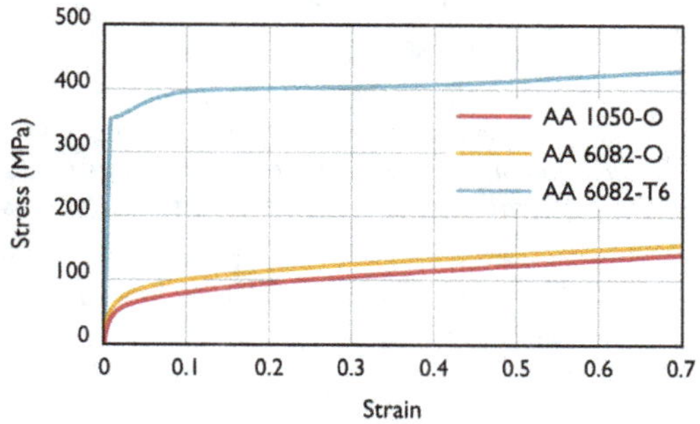

Fig. 3 Flow curves of the aluminum AA1050-O, AA6082-O and AA6082-T6.

Double-Action Radial Extrusion Formability Tests

The double-action radial extrusion formability tests were carried out in a multidirectional tool that was designed and constructed by the authors. The tool allows extruding specimens with different aspect ratios h/d_0 of the free gap height h to the initial diameter d_0 into the volume left between the two dies by compressing the material from two horizontal opposite sides in a single vertical press stroke.

Materials Research Proceedings **25** (2023) 229-236 https://doi.org/10.21741/9781644902417-29

Fig. 4a shows a detail of the tool set in which the movement of the extrusion punches (P) towards each other inside the dies (D) is obtained by means of two cam slide units consisting of a punch holder (H) and a die wedge actuator (A) with a working angle $\alpha = 30°$. Two guided pillars (G) ensure the alignment between the upper and lower tool halves.

The tool set was installed on the hydraulic testing machine that had been previously used to determine the material flow curves.

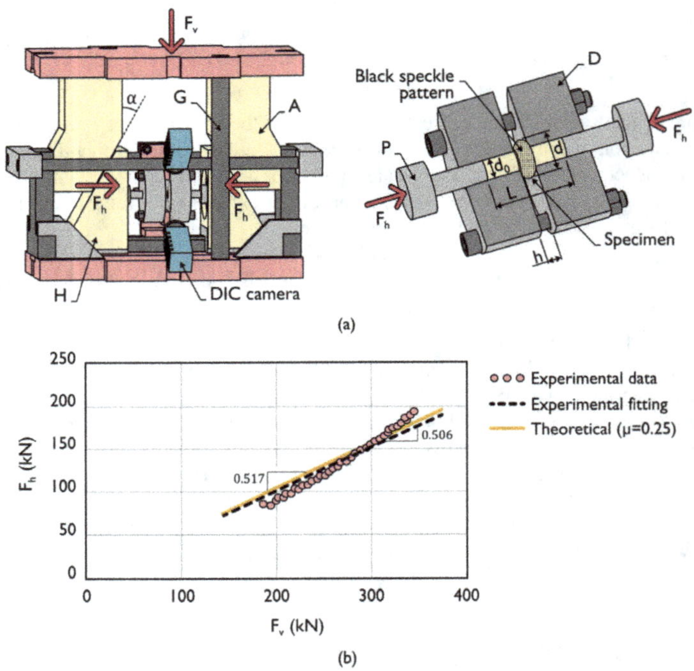

Fig. 4 Double-action radial extrusion formability test. (a) Schematic representation of the tool set with details of the dies and punches, and showing the cameras used by digital image correlation (DIC). (b) Experimental relation between the vertical F_v and horizontal F_h forces.

Force equilibrium across the cam slide units provides the following relation between the horizontal (F_h) and vertical (F_v) tool forces, where μ is the friction coefficient along the contact surfaces,

$$F_h = F_v \frac{\cos\alpha - \mu\sin\alpha}{2(\sin\alpha + \mu\cos\alpha)} = \begin{cases} 0.866\, F_v\,, & \mu = 0 \\ 0.517\, F_v\,, & \mu = 0.25 \end{cases} \tag{1}$$

Compression of cylindrical test specimens made from aluminum AA6082-T6 between straight parallel platens with the double-acting cam slide units and using a single-acting system in which the platens were fixed to the upper and lower bolsters of the tool set, allowed comparing Eq. 1 against experimental data. This is shown in Fig. 4b for $\mu = 0.25$ and the differences at the upper right end are attributed to tool stiffness, elastic deformation, and friction along the surfaces of the cam slide units. For this reason, a slope of 0.506 resulting from the black dashed linear trend line

 Materials Research Forum LLC
https://doi.org/10.21741/9781644902417-29

is used instead of the theoretical estimates with and without friction that are given in Eq. 1 to convert the vertical force F_v measured by the load cell into the horizontal applied force F_h.

Formability Analysis and Fractography

The evolution of the in-plane strains with time $(\varepsilon_1, \varepsilon_2) = f(t)$ on the outer surface of the radially extruded flanges was obtained with a commercial digital image correlation (DIC) system model Q-400 3D from Dantec Dynamics. For this purpose, the initial free gap heigh h of the specimens located in-between the two dies was painted in white and subsequently sprayed with a black speckle pattern (Fig. 4a).

The double-action working principle of the proposed radial extrusion formability test facilitates image acquisition because the measuring window of the DIC system, (equipped with two 6-megapixels resolution cameras with 50.2 focal lenses and f/8 aperture), is fixed and always centered with the vertical symmetry line of the tool set during the entire duration of the test. A frequency of 10 Hz corresponding to 10 images per second was used.

Transformation of the major and minor in-plane strain evolutions with time $(\varepsilon_1, \varepsilon_2) = f(t)$ obtained from DIC into the strain loading paths $\varepsilon_1 = f(\varepsilon_2)$ of principal strain space was carried out by combining and removing the time dependency from the individual evolutions of the major ε_1 and minor ε_2 in-plane strains. The procedure is schematically illustrated in Fig. 5a.

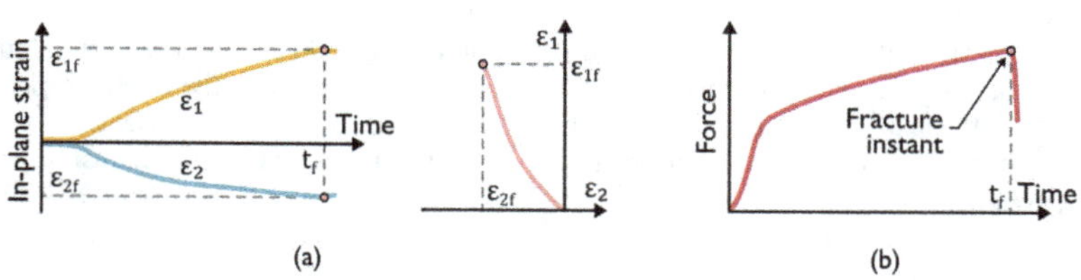

Fig. 5 Methods and procedures involving experimental data in formability analysis. (a) Individual in-plane strain vs. time evolutions determined by digital image correlation (left) and the result of their combination to obtain the strain loading path in principal strain-space (right). (b) Identification of the instant of time at the onset of fracture t_f from the force vs. time evolution.

The instant of time t_f at which cracks are triggered was obtained through identification of the sudden force drop in the experimental evolution of the radial extrusion force with time (Fig. 5b), as earlier proposed by Magrinho et al. [6]. Once t_f is determined, the strains at fracture $(\varepsilon_{1f}, \varepsilon_{2f})$ are easily retrieved from the individual in-plane strain evolutions with time obtained from DIC (Fig. 5a).

After testing, the fractured surfaces were cut out from the specimens for subsequent observation and analysis in a scanning electron microscope (SEM) S2400 from Hitachi. Results are presented in the second part of this paper and provide a correlation between the morphology of the cracks, their opening mechanisms and the corresponding strain loading paths obtained from DIC.

Numerical Simulation

The finite element computer program i-form was utilized to carry out the numerical simulation of the double-action radial extrusion formability test under different operating conditions. The computer program is built upon the finite element flow formulation and was developed by the authors [13] to account for the three major sources of nonlinearity (material nonlinearity, changes of static and kinematic boundary conditions and geometric nonlinearity) that occur in the numerical simulation of metal forming processes.

The numerical simulation strategy utilized by the authors took advantage of the facility that i-form offers to exchange data between two and three-dimensional models because experimental observations revealed that specimens undergo rotationally symmetric plastic deformation conditions up to approximately 30% of the total stroke. Only subsequently, with the occurrence of diffuse necking at the outer flange surface, there is evidence of asymmetric plastic deformation.

Fig. 6 illustrates the overall numerical simulation strategy for a test specimen made from AA1050-O with the different stages carried out by the computer program being illustrated by means of three distinct images:

- Fig. 6a refers to the initial two-dimensional simulation under rotationally symmetric conditions in which the test specimen is modelled as a deformable object and discretized by means of quadrilateral elements. The dies and punches are modelled as rigid objects and discretized by means of linear contact elements with friction,
- Once rotationally symmetric conditions can no longer be utilized, the quadrilateral mesh of the test specimen is automatically rotated counterclockwise about the z-axis to produce a temporary hexahedral mesh like that shown in Fig. 6b. Scalar field variables are rotated accordingly but second-order tensors, like for example the strain tensor ε_{ij}, must be properly transformed as follows,

$$\varepsilon_{ij}^{xyz} = R^T \varepsilon_{ij}^{r\theta z} R \tag{2}$$

In the above equation the superscripts xyz and $r\theta z$ refer to the Cartesian and rotationally symmetric coordinate frames, respectively. The symbol R is the rotation matrix and θ is the angle of rotation shown in Fig. 6,

$$R = \begin{bmatrix} \cos\theta & \sin\theta & 0 \\ -\sin\theta & \cos\theta & 0 \\ 0 & 0 & 1 \end{bmatrix} \tag{3}$$

- The temporary hexahedral mesh contains a significant number of wedge-shaped elements [14] along the z-axis. These irregular elements are automatically eliminated and replaced by regular hexahedral elements with field variables properly transferred between the two meshes (Fig. 6c). The resulting mesh is symmetric along the zx-plane because material was assumed as isotropic. The dies and punches resulting from the rotation of the axisymmetric finite element model continued to be assumed as rigid objects, but their contours were discretized by a mesh of spatial triangles with friction.

Sheet Metal 2023 Materials Research Forum LLC
Materials Research Proceedings **25** (2023) 237-244 https://doi.org/10.21741/9781644902417-30

On the utilization of radial extrusion to characterize fracture forming limits. Part II: testing and modelling

Rui F.V. Sampaio[1,a], João P.M. Pragana[1,b], Ivo M.F. Bragança[2,c], Carlos M.A. Silva[1,d], Chris V. Nielsen[3,e] and Paulo A.F. Martins[1,f] *

[1]IDMEC, Instituto Superior Técnico, Universidade de Lisboa, Portugal

[2]CIMOSM, Instituto Superior de Engenharia de Lisboa, Instituto Politécnico de Lisboa, Portugal

[3]Department of Civil and Mechanical Engineering, Technical University of Denmark

[a]rui.f.sampaio@tecnico.ulisboa.pt, [b]joao.pragana@tecnico.ulisboa.pt, [c]ibraganca@dem.isel.ipl.pt, [d]carlos.alves.silva@tecnico.ulisboa.pt, [e]cvni@dtu.dk, [f]pmartins.tecnico.ulisboa.pt

Keywords: Forming, Failure, Stress State Transitions

Abstract. This second part of the paper is focused on double-action radial extrusion testing and modelling to characterize material formability and failure in the bulk-to-sheet material flow transitions that are commonly found in metal forming. Results show that three-dimensional to plane-stress evolutions at the radially extruded flanges lead to different modes of fracture (by tension and by shear) that may or may not be preceded by necking. The use of double-action radial extrusion as a formability test also reveals adequate to characterize the failure limits of very ductile metallic materials, which cannot be easily determined by conventional upset compression tests, and to facilitate the identification of the instant of cracking and of the corresponding fracture strains by combination of the force vs. time evolutions with the in-plane strains obtained from digital image correlation.

Introduction

Competition between cracking opening modes by tension (mode I) and by out-of-plane shearing (mode III) in conventional bulk formability tests give rise to an 'uncertainty region' in principal strain space, as it was recently observed by Sampaio et al. [1]. Although research on formability and crack opening modes has been taking the competition between crack opening modes into account, there are aspects related to material flow transitions in regions of bulk metal forming parts subjected to extensive plastic material flow that are not commonly taken into consideration. Accountability of these aspects means, in practical terms, to include the three-dimensional to plane stress material flow transitions (i.e., bulk-to-sheet evolutions) in the list of parameters that are responsible for the competition between the different crack opening modes.

Under these circumstances, the main objective of this second part of the paper is to investigate formability in three-dimensional to plane stress material flow transitions by means of a new formability test based on the working principle of double-action radial extrusion [2]. It must, however, be said that the relevance of the new proposed formability test goes beyond the above-mentioned material flow transitions because it provides two significant advantages over the existing bulk formability tests.

Firstly, it allows carrying out tests with very slender cylindrical specimens, which would inevitably collapse by buckling during upset compression tests between flat parallel platens. This overcomes the difficulty of obtaining the experimental strains at fracture in very ductile materials, a topic that was also addressed by Bulzak et al. [3] when studying damage evolution in hot forming with rotary compression tests. However, the experimental evaluation of surface strains becomes very difficult for this test, due to the continuous movement of the specimen.

Content from this work may be used under the terms of the Creative Commons Attribution 3.0 license. Any further distribution of this work must maintain attribution to the author(s) and the title of the work, journal citation and DOI. Published under license by Materials Research Forum LLC.

Secondly, the fact that triggering and subsequent propagation of cracks in the double-action radial extrusion test will always lead to force drops in the force vs. time evolutions. This allows determining the instant of cracking and of the corresponding fracture strains by combination of the experimental force vs. time evolutions with the results obtained from digital image correlation (DIC). In fact, the force vs. time evolution of the double-action radial extrusion test is different from those found in conventional upset compression tests in which the combination of surface expansion with strain hardening commonly prevents the force from dropping at the onset of cracking.

Experimental testing conditions

The double-action radial extrusion formability tests were carried out in a multidirectional tool set that was installed in a hydraulic testing machine. Fig. 1 shows a detail of the tool set in which the movement of the extrusion punches (P) towards each other inside the dies (D) is obtained by means of two cam slide units consisting of a punch holder (H) and a die wedge actuator (A) with a working angle $\alpha = 30°$. Two guided pillars (G) ensure the alignment between the upper and lower tool halves. Further details on the tool set and on the conversion of the vertical force F_v of the hydraulic testing machine into the horizontal force F_h applied by the extrusion punches are given in the first part of this paper (Part I).

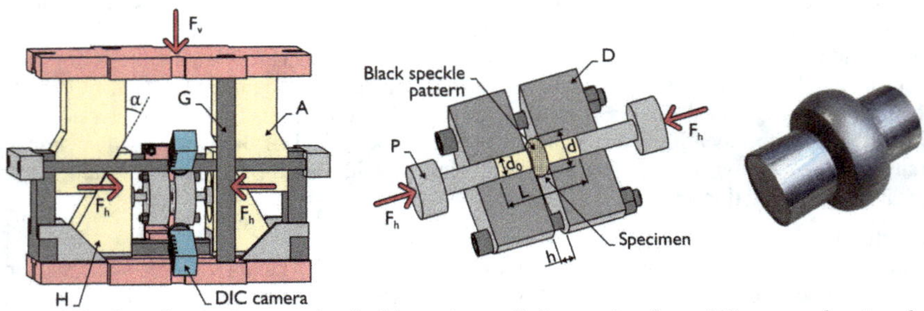

Fig. 1 Tool set for performing the double-action radial extrusion formability tests showing the cameras used by digital image correlation, the detail of the dies and punches and a test specimen.

The tests were carried out in a soft aluminum AA1050-O and a medium strength aluminum-magnesium-silicon alloy AA6082 in the annealed (O) and solution heat-treated (T6) state. The specimens were machined from the supplied rods and a molybdenum disulphide (MoS2) based lubricant was utilized in the operating testing conditions provided in Table 1.

Results and Discussion

Strain-Loading Paths in Principal Strain Space. Fig. 2 shows the experimental evolutions of the major and minor in-plane strains with time determined by DIC and of the radial extrusion force F_h with time after conversion of the vertical force F_v measured by the load cell. The in-plane strain evolutions were taken from the outer flange surface for the three different materials.

As seen, the instant of fracture t_f (refer to 'F' in Figs. 2a-c) is clearly recognized by a sudden drop in force and allows determining the major ε_1 and minor ε_2 in-plane strains at fracture from their evolutions with time $(\varepsilon_1, \varepsilon_2) = f(t)$ obtained from DIC. Point 'I' is the location in the outer flange surfaces where the strain evolutions were obtained.

Table 1 Operating conditions utilized in the double-action radial extrusion tests.

Dies	Gap height h	12 mm, 16mm, 32mm
	Fillet radius r	0.5 mm
Punches	Diameter	16 mm
	Velocity	5 mm/min
Specimens	Initial diameter d_0	16 mm
	Initial length l_0	50mm, 90mm
	Aspect ratio h/d_0	0.75, 1, 2

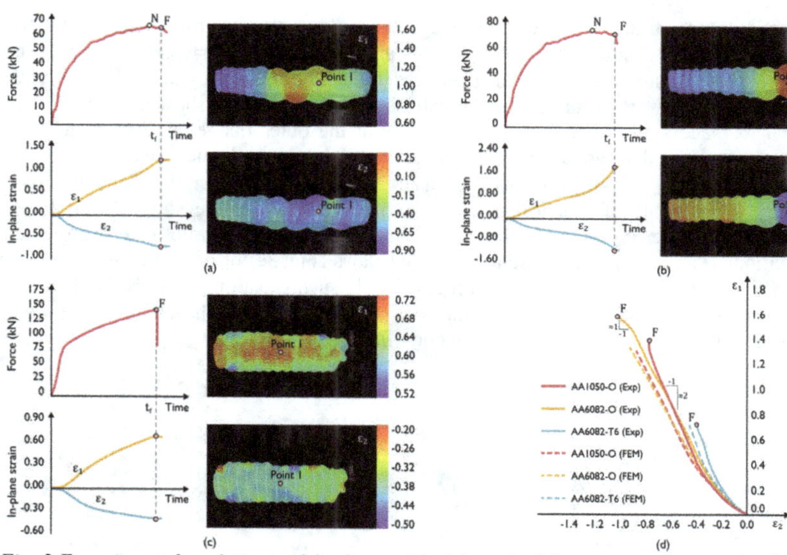

Fig. 2 Experimental evolutions of the force with time and of the major and minor in-plane strains with time for specimens with $h/d_0 = 0.75$ *made from aluminum (a) AA1050-O, (b) AA6082-O and (c) AA6082-T6. The images were taken from DIC immediately before fracture. (d) Experimental and finite element predicted strain loading paths in principal strain space.*

In case of the specimens made from aluminum AA1050 and AA6082 in the annealed (O) state, the instant of fracture is preceded by a change in the rate of force growth with time leading to the development of a neck on the outer flange surface (refer to 'N' in the force peaks of Figs. 2a, b). In contrast, the specimens made from aluminum AA6082-T6 only experience a sudden drop in force because cracking is not preceded by necking, as it was experimentally observed and will be shown later.

Application of the methodology that was described in the first part of this paper allows plotting the corresponding strain loading paths $\varepsilon_1 = f(\varepsilon_2)$ in principal strain space. The result is shown in Fig. 2d and allows distinguishing between monotonic and non-monotonic evolutions up to fracture thanks to the absence of necking (AA6082-T6) and to the sharp bends of the strain loading paths (in point 'F') towards plane strain (AA1050-O) and pure shear (AA6072-O) deformation conditions that happen in the specimens where fracture is preceded by necking.

Finite element computed evolutions of the strain loading paths (refer to the dashed lines in Fig. 2d) show a good agreement with DIC. However, the numerical estimates for the soft

Materials Research Proceedings **25** (2023) 237-244 https://doi.org/10.21741/9781644902417-30

aluminum AA1050 and the medium strength aluminum-magnesium-silicon alloy AA6082 in the annealed (O) state are only provided up to the onset of localized necking, which corresponds to approximately 80% of the total stroke. Beyond this point, the growth rate of the in-plane strains is very fast and the uncoupled finite element damage models utilized by the authors experience difficulties in handling the non-homogeneous plastic deformation and the elastic unloading caused by damage-induced softening.

However, because the onset of localized necking and point 'F' occur almost instantly, finite elements will continue to be used in the following sections of the paper to analyze the deformation mechanics and the different modes of failure that were observed in the double-action radial extrusion formability test.

Deformation Mechanics. Figs. 3a-c shows the initial and intermediate rotationally symmetric meshes at 30% of the total stroke for specimens with an aspect ratio $h/d_0 = 0.75$ made from the different materials and supplied conditions that were utilized in the investigation. As seen in the predicted distribution of vertical stresses, the radially extruded flanges start showing signs of three-dimensional to plane stress material flow transitions at the outer flange surfaces, which will become more pronounced as radial extrusion continues and the flange diameter increases.

The aspect ratio h/d_0 of the specimens plays a key role in the utilization of radial extrusion as a formability test capable of monitoring the three-dimensional to plane stress material flow transitions. In fact, larger aspect ratios than those shown in Figs. 3a-c, do not give rise to plane stress conditions (Fig. 3d) and may result in material flaws (Fig. 3e) or buckling, as reported in earlier investigations of the radial extrusion process [4]. In such situations not only the diameter-to-thickness ratio of the radially extruded flanges deviates from typical sheet forming values as plane stress conditions no longer prevail on the outer flange surfaces.

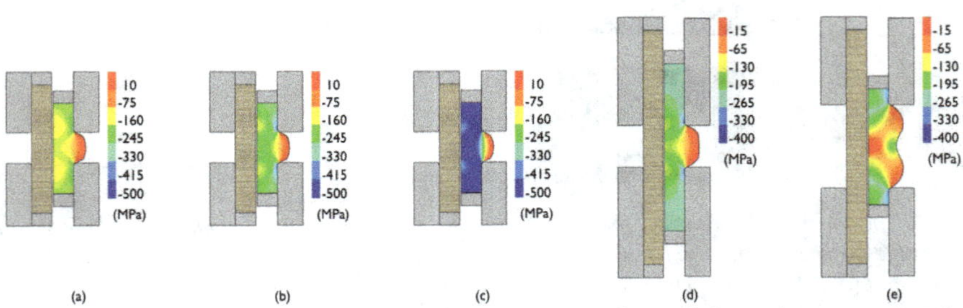

Fig. 3 Initial and intermediate rotationally symmetric meshes with finite element predicted distribution of vertical stress for specimens made from aluminum: (a) AA1050-O with $h/d_0 = 0.75$*, (b) AA6082-O with* $h/d_0 = 0.75$*, (c) AA6082-T6 with* $h/d_0 = 0.75$*, (d) AA1050-O with* $h/d_0 = 1$ *and (e) AA1050-O with* $h/d_0 = 2$.

The asymmetric deformation of the three specimens with an aspect ratio $h/d_0 = 0.75$ that progressively develops during the final stage of the double-action radial extrusion test was simulated with the three-dimensional finite element model built upon the interaction between two and three-dimensional meshes that was earlier described. For this purpose, the authors artificially imposed small gauge imperfections of approximately 0.2 mm on the outer flange diameter located at the vicinity of the xz-plane. These imperfections account for less than 1% of the flange diameter and correspond to localized material weaknesses that resemble those utilized by Marciniak and Kuczynski [5] in their theory of plastic instability in sheet metal forming.

The result of the above-mentioned strategy combined with intermediate remeshing operations aimed at replacing elements with excessive distortion at the center of the specimens made from

AA1050-O and AA6082-O is disclosed in Fig. 4. As seen, finite element computed geometries obtained shortly after the onset of cracking agree well with the experimental test specimens.

Incipient crack propagation was numerically modelled by means of an element deletion (removal) technique based on the critical value of a damage function D,

$$D = \int_0^{\bar{\varepsilon}} g \, d\bar{\varepsilon} \tag{1}$$

where g is a weight function and $\bar{\varepsilon}$ is the effective strain. Two different weight functions were used: (i) the McClintock [6] stress triaxiality ratio $g_{Mc} = \sigma_m/\bar{\sigma}$, corresponding to failure by tension (mode I), where σ_m is the hydrostatic stress and $\bar{\sigma}$ is the effective stress (Fig. 4a) and (ii) the Cockcroft-Latham [7] normalized principal stress ratio $g_{CL} = \sigma_1/\bar{\sigma}$, corresponding to failure by out-of-plane shear (mode III), where σ_1 is the major principal stress (Figs. 4b, c).

The choice between the proper weight function g to be used for each test case was based on the strain loading path evolutions disclosed in Fig. 2d, which clearly indicate failure by tension preceded by necking in AA1050-O, by out-of-plane shear preceded by necking in AA6082-O and failure without previous necking in AA6082-T6.

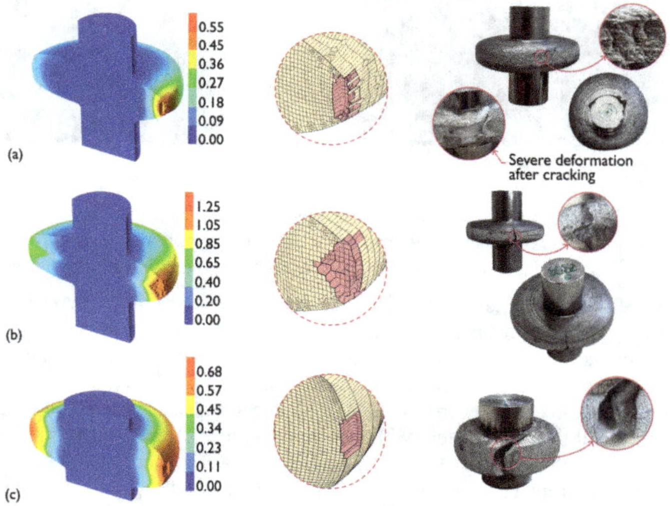

Fig. 4 Three-dimensional finite element prediction of ductile damage with mesh details showing cracks on the outer flange surfaces for three different test specimens with an aspect ratio $h/d_0 = 0.75$ *made from aluminum (a) AA1050-O, (b) AA6082-O and (c) AA6082-T6. Ductile damage was calculated according to (a) McClintock [6] and (b, c) Cockcroft-Latham [7] criteria and photographs of the specimens are enclosed for comparison purposes.*

Further observation of the two different types of necks combined with fractography analysis by means of scanning electron microscopy (SEM) to be presented in the next section provides a better identification and understanding of the different types of failure and modes of deformation that are likely to occur in three-dimensional to plane stress material flow transitions.

Necking and Fractography. Combination of the finite element computed meshes with the images obtained from DIC and SEM (Fig. 5) allows concluding that fracture in the outer surface flange of the AA1050-O test specimens is by tension (mode I). Results also show evidence of plastic instability in the form of diffuse (Fig. 5a) and localized necking (Fig. 5c) in good agreement with the deformation mechanics of sheet forming. In fact, the localization of the neck along a direction inclined at an angle of approximately 53.2° (Fig. 5c) to the circumferential direction value is very close to the theoretical estimate of 54.7° (Fig. 5b) obtained for a sheet (plane stress) specimen made from an isotropic material, subjected to uniaxial tension [8].

The morphology of the cracks consists of circular dimple-based structures (Fig. 5d), and further corroborates the experimental observation of cracks opening by tension (mode I) and running radially along the flange.

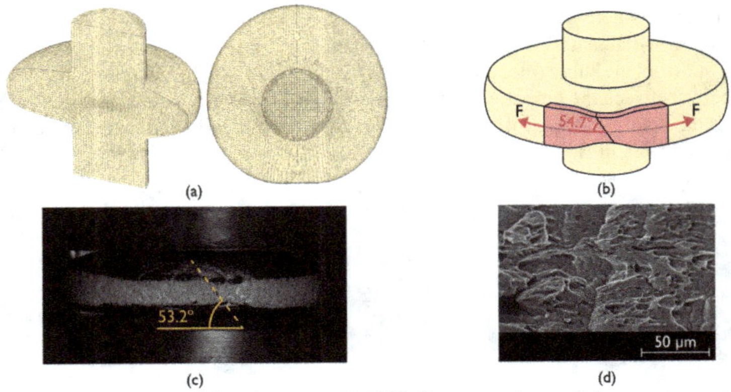

Fig. 5 Failure mode of the aluminum AA1050-O test specimen with an aspect ratio $h/d_0 = 0.75$ showing (a) the computed finite element mesh, (b) the analogy with localized necking in uniaxial tension of a sheet test specimen, (c) a DIC image showing localized necking at the outer flange surface and (d) a SEM picture showing the morphology of the cracks.

Application of the same methodology for the specimens made from AA6082-O (Fig. 6a) confirms that vertical cracks opened by shear stresses due to the occurrence of smooth parabolic dimple-based structures (Fig. 6e). Moreover, these cracks do not run radially along the flange (Fig. 4b), as it is commonly observed in the upset compression of cylindrical and tapered test specimens, due to significant distortion of outer flange surface placed within the neighborhood of the localized neck. All this is consistent with the strain path bend towards pure shear in principal strain space (Fig. 2d) before failing by through-thickness shearing (mode III).

The pure shear stress state acting at the outer flange surface (with $\varepsilon_\theta = -\varepsilon_z > 0$) gives rise to plane strain deformation along the radial direction ($\varepsilon_r = 0$) due to material incompressibility. Thus, assuming these conditions to hold in-between the top and bottom flange surfaces due to combination of small thickness and plane stress material flow conditions, an analogy with uniaxial tension of a sheet test specimen can also be made.

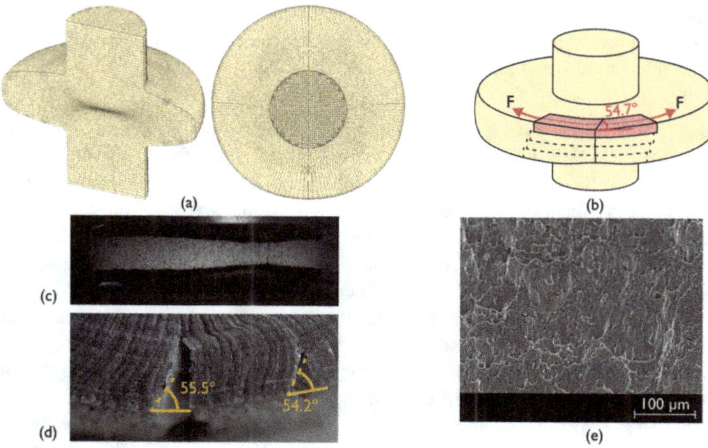

Fig. 6 Failure mode of the aluminum AA6082-O test specimen with an aspect ratio $h/d_0 = 0.75$ *showing (a) the computed finite element mesh, (b) the analogy with pure shear within the neighborhood of the localized neck, (c) a DIC image showing localized necking along the longitudinal direction at the outer flange surface, (d) top view detail of the flange near the neck showing crack opening on the* r-θ *plane with angles and (e) a SEM picture showing the morphology of the cracks.*

The analogy requires a similar approach to that taken for the specimen made from AA1050-O but now with diffuse necking taking place in the radial direction (Figs. 6b and 6c) and crack opening in the r-θ plane after the occurrence of localized necking with angles of approximately 54° ~ 55° (refer to the top flange view in Fig. 6d).

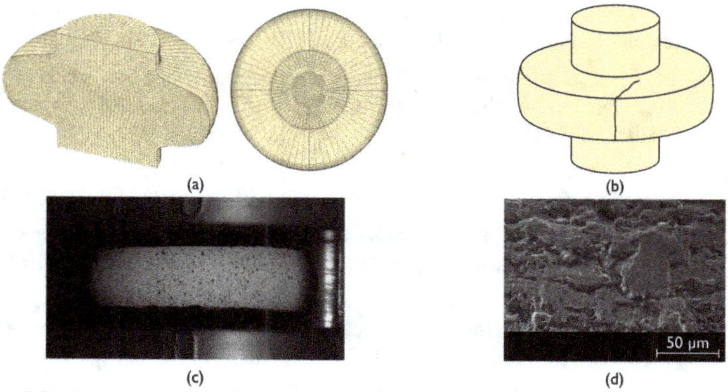

Fig. 7 Failure mode of the aluminum AA6082-T6 test specimen with an aspect ratio $h/d_0 = 0.75$ *showing (a) the computed finite element mesh, (b) a scheme showing vertical cracks that do not propagate radially, (c) a DIC image showing the absence of necking at the onset of cracking and (d) a SEM picture showing the morphology of the cracks.*

In what concerns the results obtained for the specimens made from AA6082-T6 (Fig. 7a) it is possible to conclude that vertical cracks open without previous necking (Fig. 7c). Crack propagation does not run radially, as it is schematically disclosed in Fig. 7b, due to the development of a mixed mode characterized by the simultaneous occurrence of circle and smooth parabolic dimple-based structures (Fig. 7d).

Conclusions

The work on the utilization of double-action radial extrusion as a new formability test to characterize failure in bulk-to-sheet material evolutions led to the following main conclusions:

- The test can be successfully used to characterize material formability in the three-dimensional to plane stress material flow transitions that are commonly found in bulk metal formed parts,
- Material flow transitions give rise to uniaxial tension states of stress that eventually lead to crack opening with or without previous localized necking,
- Cracks preceded by localized necking develop under plane strain or pure shear material flow conditions on the outer flange surface,
- The morphology of the cracks reveals crack opening by tension (mode I), by shear (mode III) and by a mixed mode resulting from combination of modes I and III,
- The new test is also adequate to characterize the formability limits of very ductile materials with large fracture strains, such as aluminum AA1050-O and AA6082-O, which cannot be easily determined by conventional upset compression tests.

Acknowledgements

The authors would like to acknowledge the support provided by Fundação para a Ciência e a Tecnologia of Portugal and IDMEC under LAETA- UIDB/50022/2020 and PTDC/EME-EME/0949/2020.

References

[1] R.F.V. Sampaio, J.P.M. Pragana, I.M.F. Bragança, C.M.A. Silva, P.A.F. Martins, Revisiting the fracture forming limits of bulk forming under biaxial tension, International Journal of Damage Mechanics 31 (2022) 882-900. https://doi.org/10.1177/10567895211072580

[2] R. Balendra, Process mechanics of injection upsetting, International Journal of Machine Tool Design and Research 25 (1985) 63-73. https://doi.org/10.1016/0020-7357(85)90058-7

[3] T. Bulzak, Z. Pater, J. Tomczak, Ł. Wójcik, A rotary compression test for determining the critical value of the Cockcroft–Latham criterion for R260 steel, International Journal of Damage Mechanics 29 (2020) 874–886. https://doi.org/10.1177/1056789519887527

[4] R. Balendra, Y. Qin, Injection forging: engineering and research, Journal of Materials Processing Technology 145 (2004) 189-206. https://doi.org/10.1016/S0924-0136(03)00670-8

[5] Z. Marciniak, K. Kuckzynski, Limit strains in the process of stretch-forming sheet metal, International Journal of Mechanical Sciences 9 (1967) 609–620. https://doi.org/10.1016/0020-7403(67)90066-5

[6] F.A. McClintock, A Criterion for ductile fracture by the growth of holes, Journal of Applied Mechanics 35 (1968) 363–371. https://doi.org/10.1115/1.3601204

[7] M.G. Cockroft, D.J. Latham, Ductility and the workability of metals, Journal of the Institute of Metals 96 (1968) 33–9.

[8] R. Hill, A theory of yielding and plastic flow of anisotropic metals, Proceedings of the Royal Society, London, Series A 193 (1948) 281–297. https://doi.org/10.1098/rspa.1948.0045

Sheet Metal 2023 Materials Research Forum LLC
Materials Research Proceedings **25** (2023) 245-252 https://doi.org/10.21741/9781644902417-31

Characterization of macroscopic local deformation behavior of thermo-mechanically graded high strength aluminum and steel alloys

Emad Scharifi[1, a], Sebastian Hirte[1, b], Steffen Lotz[1,2, c], Michael Mlynek[2, d], Ursula Weidig[1, e *], Agim Ademaj[2, f] and Kurt Steinhoff[1, g]

[1]University of Kassel, Metal Forming Technology, Kurt-Wolters-Straße 3, 34125 Kassel, Germany

[2]METAKUS Automotive GmbH, Department of Research and Development, Fehrenberger Straße 1a, 34225 Baunatal, Germany

[a]emad.scharifi@uni-kassel.de, [b]sebastian.hirte@student.uni-kassel.de, [c]steffen.lotz@uni-kassel.de, [d]michael.mlynek@metakus.com, [*e]ursula.weidig@uni-kassel.de, [f]agim.ademaj@metakus.com, [g]steinhof@uni-kassel.de

Keywords: Steel, Aluminium, Functional Gradation

Abstract. In this study, the local deformation behavior of the transition zone of thermo-mechanically graded boron steel 22MnB5 and high strength aluminum alloy AA7075 is investigated using an in-situ approach. For this aim, forming tool segments are differentially heated in order to obtain locally tailored microstructures and dissimilar mechanical properties based on the process-induced microstructural phenomena. Material strength distribution is examined by Vickers hardness measurements after thermo-mechanical processing. Tracking of the material- and geometry-dependent local deformation behavior of the resulted transition zone is performed by tensile tests coupled with digital image correlation technique. The experimental results reveal within the transition zone a characteristic reversed S-shape hardness distribution, similar to a sigmoidal curve. The local deformation characterization indicates the occurrence of plastic deformation mainly in the soft zones at the heated forming tools segments, while the harder zones at the cooled tool segments exhibit only an elastic deformation and remain undeformed. By comparing the macroscopic local deformation evolution shortly before failure, absolute strain values up to $\varepsilon_{von\ Mises} = 0.62$ and $\varepsilon_{von\ Mises} = 0.28$ and a distinct reduction in the area are obtained for 22MnB5 and AA7075, respectively. The occurred plastic strain pattern show the formation of shear bands after a homogenous localization at the necking point.

Introduction

In the field of metal forming technology, the adjustment of graded properties can be accomplished for hardenable steel materials by the combination of locally controlled thermal distribution and forming at elevated temperatures [1]. Therefore, in the framework of hot stamping of boron steel sheets different process variants based on thermo-mechanical approaches are proposed to gain dissimilar mechanical properties as well as microstructure distribution [2]. One of the promising techniques is the so-called differential cooling, which can be performed by differentially heated forming tools to provide locally varying cooling rates [2].

By applying this concept to boron steel 22MnB5, the local cooling rates above 30 °C/s at cooled forming tool segments result in a martensitic microstructure with a tensile strength of 1500 MPa, whereas lower local cooling rates lead to the formation of ductile zones, i.e., ferrite and pearlite with tensile strengths of 614 MPa [3]. This shows that the forming tool temperature plays a key role in determining specific microstructural properties for hardenable steel materials after austenitization. Using tool segment temperatures of 25 °C and 400 °C, a significant reduction of 52 % of hardness level from hard to soft is obtained within the transition zone, [4]. Further studies

Content from this work may be used under the terms of the Creative Commons Attribution 3.0 license. Any further distribution of this work must maintain attribution to the author(s) and the title of the work, journal citation and DOI. Published under license by Materials Research Forum LLC.

with tensile tests in combination with digital image correlation (DIC) technique gain insight into the plastic deformation behavior of graded components and visualize the failure location for the calibration of finite element (FE) simulations [3,5]. To ensure the occurrence of plastic deformation within the gauge length due to the dissimilar hardness distribution, tapered tensile sample geometries with different gauge section widths are used [3]. The experimental results reveal higher total displacement up to failure with increasing the gauge section widths. DIC measurements and FE-simulations show two competing zones for necking initiation at the tensile specimen surface, as a result of the strain hardening. However, a detailed analysis and visualization of local plastic deformation of the graded parts during tensile test is not considered. Therefore, tapered as well as standard tensile test specimen geometries are used in [6] to characterize the local deformation behavior of differentially cooled boron steel 22MnB5 after hot stamping. The DIC measurements reveal for the graded as well as for the martensitic structures higher local logarithmic strain within the necking zone compared to the soft ferritic pearlitic structures. In addition, using a standard tensile test specimen geometry reveals different types of strain localization and failure for the graded structure due to its complex hardening behavior and mixed microstructure leading to multiple necking sites.

Taking the ability of differential cooling to create tailored microstructures in boron steels within one single forming step into the account, this approach is recently proposed also for precipitation hardenable aluminum alloys to obtain dissimilar mechanical properties within one part [7]. Differing cooling rates generated by different tool segment temperatures lead to a characteristic sigmoidal hardness distribution along the longitudinal axes of the formed hat profile. The microstructural investigations based on the electron channeling contrast imaging (ECCI) technique within the transition zone reveal an increase in the precipitate's size with decreasing hardness level. Hence, in the zones of the cooled forming tool segments, a homogenous distribution of fine spherical particles is found, while in the zones of the heated forming tool segments, coarse lath-shaped particles are observed, explaining the obtained hardness distribution. However, despite the reported dissimilar material strength and the variation of microstructural features of the differentially cooled parts their local deformation behavior as well as their tensile deformation performance is not discussed in literature.

For this reason, the above reported open quotations, especially regarding the plastic deformation behavior of the transition zone, inspire the present study to characterize the macroscopic local deformation behavior of thermo-mechanically graded microstructures using DIC technique to highlight the logarithmic local strain fields of the transition zone. For this aim, dfferentially cooled ultra-high-strength boron steel 22MnB5 and high strength aluminum alloy AA7075 are compared to address aspects related to different materials and microstructural mechanisms like diffusionless transformation and precipitation hardening. Hence, relations between process, microstructure and deformation properties will be established with regard to the inhomogeneous microstructure and dissimilar mechanical properties.

Experimental procedure

Materials. The experimental investigations in this study are carried out on boron steel 22MnB5 and high strength aluminum alloy AA7075. The as received condition of 22MnB5 (ArcelorMittal) is Al–Si coated consisting of ferritic and pearlitic structure, an ultimate tensile strength (UTS) of 610 MPa and an elongation after failure of A = 20 %. The as received T6-condition (solution heat treated, quenched, and artificially aged to peak strength) of AA7075 (Al–Zn–Mg(–Cu)) shows a hardness of 190 HV and UTS of 580 MPa, respectively. Both materials are sectioned into blanks of 250 mm x 140 mm x 1.5 mm for the hot stamping experiments. A table of the detailed chemical composition of the selected materials is given in [6,7].

Thermo-mechanical forming process and heat treatment. The experimental set-up consists of a roller hearth furnace (type VKR 2800/e, Wärmebehandlungsanlagen für Industrie und

Umwelttechnik GmbH (WMU)), a manual transfer to a hydraulic press (type DS 400 Z 160, MAE Maschinen GmbH) equipped with a segmented hat-shaped forming tool and a chamber furnace with radiation heating (type KM70/13, ThermoConcept GmbH) for artificial aging treatment. The forming tool consists of two segments thermally separated by an air gap, in which one segment allows temperatures up to 700 °C whereas the other segment is cooled to a constant temperature of 24 °C by an integrated water-cooling circulation system. The chosen forming speed for all experiments was set to 48 mm^{-1} with a forming force of 1000 kN.

The first group of experiments, Tab. 1, aims to investigate the local deformation behavior of the transition zone of boron steel 22MnB5. For this purpose, the sectioned 22MnB5 sheets were first austenized at the soaking temperature of T_s = 870 °C and then soaked for t_s = 90 s to obtain a homogenous austenitic structure. The furnace temperature was set to T_F = 930 °C and the sheet temperature was recorded using thermocouples. Subsequently after the austenitization, the heated 22MnB5 sheets were transferred within 6 s to 8 s to the differentially tempered forming tool with T_{FT} = 24 °C and T_{FT} = 500 °C and then formed and held for $t_{Ht, FT}$ = 10 s within the closed forming tool.

The second group of experiments focuses on the local deformation behavior of high strength aluminum alloy AA7075 after solution heat treatment, differential cooling and artificial aging, Tab. 1. For this aim, the sectioned AA7075 sheets were heated to the soaking temperature of T_s = 480 °C to dissolve all alloying elements, formed and cooled within differentially tempered forming tools of T_{FT} = 24 °C and T_{FT} = 350 °C for $t_{Ht, FT}$ = 30 s after a manual transfer of 6 s to 8 s. Immediately after forming and cooling the parts were additionally quenched in water to obtain high cooling rates. Finally, the formed parts were artificially aged at T_{AA} = 120 °C for t_{AA} = 20 h to obtain full strength. The investigation and thermo-mechanical treatment of the materials is carried out separately and independently.

Table 1: Selected parameters for different thermo-mechanical processes to obtain tailored microstructure.

Material	Soaking temperature [°C]	Soaking time [s]	Tool temperature heated zone [°C]	Tool temperature cooled zone [°C]	Holding time within closed tools [s]	Aging
22MnB5	870	90	500	24	10	-
AA7075-T6	480	900	350	24	30	120 °C – 20 h

Property characterization and local strain measurement. Vickers hardness measurements (EMCO DuraScan) along the longitudinal axes of the formed parts were carried out at a test load of 98.07 N (HV10) for 22MnB5 and 49 N (HV5) for AA7075, respectively. Five repetitions were performed from the same measurement length to obtain the standard deviation. The measured hardness profiles were then used to highlight the property distribution and to select the position of the extraction of the tensile specimen at the top side of the formed hat profile, Fig. 1. Tensile test specimens, matched and aligned to the length of the transition zone were subsequently machined by electric discharge machining with a length 20 mm and width 7 mm, Fig. 1 [8]. The tensile properties were then determined with a screw driven uniaxial tensile testing machine (Hegewald & Peschke, max. press capacity: 100 kN) and an extensometer attached on the back side of the tensile specimen. In order to determine the geometrical changes as well as to determine the local strain field of the transition zone, 3D-DIC measurements coupled with tensile tests were carried out. The required surface pattern for DIC was applied using white spray to obtain a homogeneous uniform white background and subsequently black spray to adjust the spot pattern. In case of 22MnB5 the Al-Si coating was removed in advance. For the 3D-strain measurement, two CCD cameras (type Xenoplan 2.8/50-0511, Schneider Kreuznach) with a resolution of 1280 x 1024 pixel at the rate of 10 Hz were focused from different angles on the tensile specimen surface and

Materials Research Proceedings **25** (2023) 245-252 https://doi.org/10.21741/9781644902417-31

the measurement zone (transition zone). Finally, data analysis was carried out by ARAMIS software V.6.1 (GOM – Gesellschaft für Optische Messtechnik mbH, Germany).

Results

Effect of differential cooling on property distribution. The effect of differential cooling during hot stamping on ultra-high-strength boron steel 22MnB5 and high strength aluminum alloy AA7075 is presented in Fig. 1 (a – b). The hardness distribution along the longitudinal axes of both hat profile reveals a characteristic sigmoidal trend from hard to soft zone, as also reported in the literature [7,9]. In case of 22MnB5, a decrease in hardness level from the hard to soft zone of 480 HV10 to 200 HV10 is measured, revealing the fundamental influence of forming tool temperatures on the material strength, Fig. 1 (a). Hence, high cooling rates at cooled tool segments results in a martensitic structure, referring to the high material strength, while lower cooling rates in the heated tool segments causes a soft ferritic pearlitic structure.

Compared to 22MnB5, the same characteristic distribution with a lower hardness level is obtained for AA7075 after differential cooling and artificial aging, Fig. 1 (b). It reveals a clear transition and decrease of material hardness from 185 HV at the cooled tool segment to 100 HV at the heated tool segment. Thus, the creation of tailored microstructure distribution with a corresponding significant change in material strength within one part is also possible for precipitation hardenable aluminum alloys by means of locally differing contact cooling rates after solutionizing. Finally, a measured transition zone length of 20 mm is found for both 22MnB5 and AA7075.

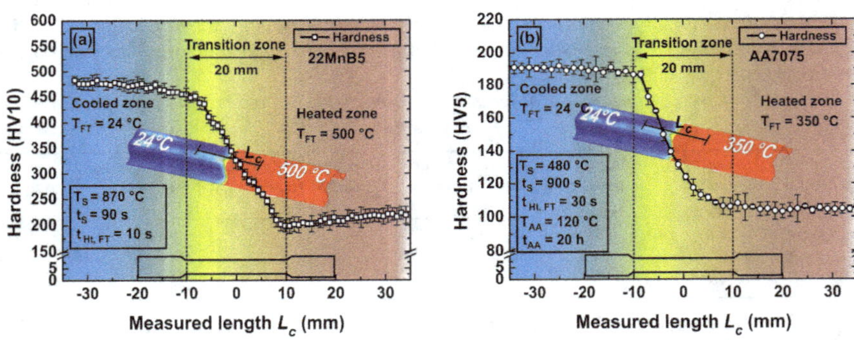

Figure 1: Hardness profile after differential cooling for (a) 22MnB5 and (b) AA7075.

Effect of differential cooling on tensile deformation performance. In addition to the hardness profile, tensile tests demonstrate the plastic deformation behavior of the transition zone of 22MnB5 and AA7075, Fig. 2 (a – b). For 22MnB5, after attaining the YS of 460 MPa, the tensile stress increases to an UTS of 685 MPa on average due to the high strain hardening of 22MnB5, Fig. 2 (a), whereas the elongation after failure, as indicator for ductility properties, reaches a maximum value of 7.4 %.

The tensile deformation performance of AA7075 after differential cooling and artificial aging shows Fig. 2 (b). The obtained YS of 375 MPa and UTS of 491 MPa are lower than the values of the as received T6-condition whilst the elongation after failure amounts to a maximum value of 6.5 %. Taking the obtained tensile curves for both, 22MnB5 and AA7075, into account the resulted strain hardening during plastic deformation at room temperature explain the difference of measured YS and UTS as well as the continuous increase of flow stress.

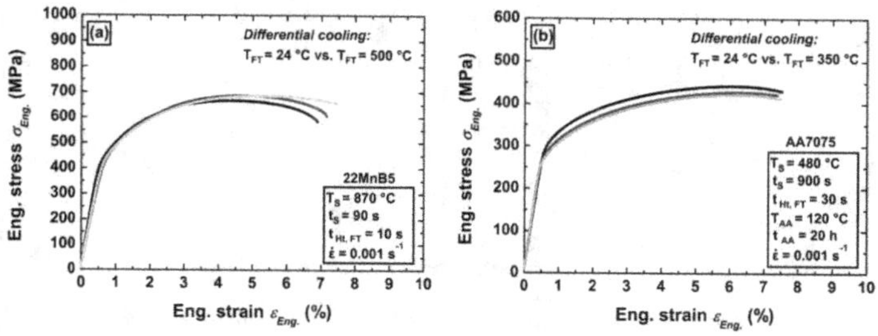

Figure 2: Tensile deformation of the transition zone of (a) 22MnB5 and (b) AA7075. Three repetitions are preformed from the same position.

Characterization of macroscopic local deformation behavior. The macroscopic local deformation evolution and the related logarithmic von Mises strain map during tensile tests of the transition zone of 22MnB5 and AA7075 after differential cooling is depicted in Fig. 3 (a – d). For both, plastic deformation appears mainly within the soft region, respectively heated forming tool segment zone. Fig. 3 (a) shows the logarithmic von Mises strain evolution for 22MnB5 within the transition zone during tensile deformation. At the beginning of plastic deformation and at the first stage of Eng. strain of $\varepsilon_{Eng.} = 2.0$ %, a slight difference in logarithmic von Mises strain in the cooled and heated zones along the measured length L_c can be seen. Accordingly, plastic deformation occurs mainly in the softer zones at the heated tool segments, while the harder zones at the cooled tool segments exhibit only an elastic deformation. With increasing the Eng. strain, plastic deformation occurs mainly in the heated tool zone, with this characteristic prevailing up to the necking point as well as to fracture. After the neck at $\varepsilon_{Eng.} = 4.0$ %, high localized logarithmic strain up to failure of $\varepsilon_{von\ Mises} = 0.62$ and a significant reduction in area is obtained for 22MnB5. The progression of the local logarithmic von Mises strain map in Fig. 3 (b) reveals and confirms the obtained local deformation evolution in Fig. 3 (a). It further can be deduced that the occurrence of plastic deformation is limited to the softer zones at the heated tool segments, which develop a strong strain localization leading to failure of the tensile specimen in the highly formed region with increasing plastic strain. Hence, within the transition zone of 22MnB5 the occurrence of plastic deformation during tensile tests underlies a well-defined delimitation. However, in contrast to the investigation in [6] multiple necking phenomena and a change of deformation locations after necking is not observed for the same used tensile specimen geometry.

Fig. 3 (c) shows the progression of the local strain until failure within the transition zone of AA7075 during tensile test. It reveals a similar logarithmic von Mises strain distribution compared to 22MnB5, Fig. 3 (a). From the initial stage of Eng. strain of $\varepsilon_{Eng.} = 2.0$ % and up to the necking point as well as to failure, local plastic strain is measured predominantly in the softer zone at the heated tool segment. After the neck at an Eng. strain of $\varepsilon_{Eng.} = 5.5$ %, a maximum localized logarithmic strain of $\varepsilon_{von\ Mises} = 0.28$ is obtained within the area of failure. It is worth to note that in case of AA7075, failure occurs subsequently and shortly after attaining the necking point and the load-supporting zone decreases rapidly, as shown in the local logarithmic von Mises strain map in Fig. 3 (d). However, plastic deformation also occurs in the zone at the cooled tool segment, which is comparatively lower to the measured logarithmic strain at the heated zone. This behavior shows also Fig. 3 (d), presenting the local logarithmic von Mises strain map. Nevertheless, this plastic deformation takes place only to a limited extent at areas of the cooled tool segment close to the heated tool segment, whilst the majority of this zone with higher hardness values remains undeformed up to failure.

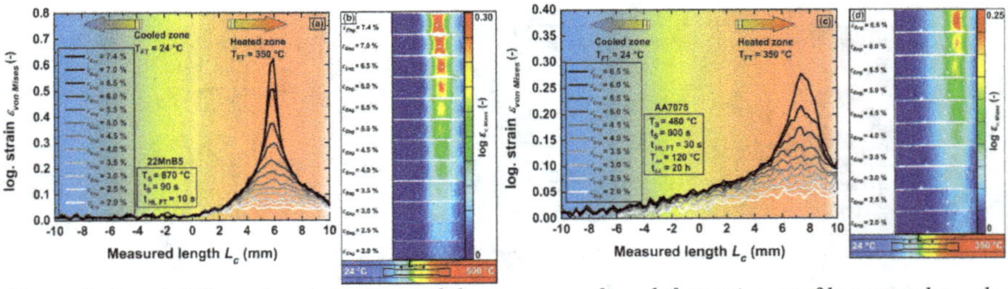

Figure 3: Local deformation behavior and the corresponding deformation profile maps along the tensile direction within the transition zone of (a – b) 22MnB5 and (c – d) AA7075 during tensile test.

The progression of the local strain at different specific locations on the tensile specimen of 22MnB5 and AA7075 and the corresponding logarithmic von Mises strain map until failure shows Fig. 4 (a – b). The first position (position 1) is located in the center of the specimen, where the highest logarithmic von Mises strain value occurs. The second position (position 2) is situated at the same height but left of position 1. The third position (position 3) is chosen at 1.2 mm above position 2. The tracking of the evolution of the local strain at 3 different positions of the transition zone intends to identify the type of failure and the nucleation of shear bands during the tensile test immediately before the occurrence of failure. The resulted logarithmic von Mises local strain of the different positions reveal for both, 22MnB5 and AA7075 that the highest strain appear in the center of the tensile specimen and exactly at the point where necking was initiated. In the course of the plastic deformation and after the neck initiation of 22MnB5, two crossed shear bands occur, which results in a higher logarithmic von Mises local strain for position 3 compared to position 2, Fig. 4 (a). This behavior is also observed for AA7075, Fig. 4 (b), although the resulting local strain is not as pronounced as for 22MnB5.

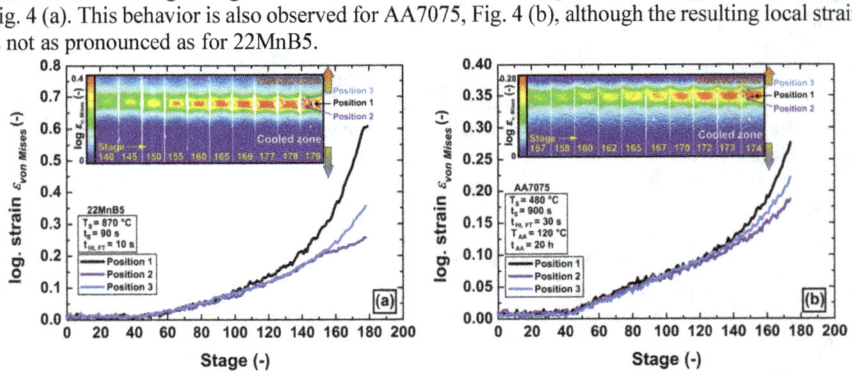

Figure 4: Progression of the local strain and the corresponding logarithmic von Mises strain map for (a) 22MnB5 and (b) AA7075.

Discussion and Conclusion

The present study investigates the local deformation behavior of the transition zone of thermo-mechanically graded boron steel 22MnB5 and high strength aluminum alloy AA7075 after differential cooling. After this thermo-mechanically gradation of 22MnB5 and AA7075 the obtained hardness distribution reveals the characteristic reversed S-shape. For 22MnB5, this property distribution is a result of the change in transformation mechanisms, i.e. from diffusionless lattice shearing to diffusion-controlled transformation, leading to a hard martensitic structure and a soft ferritic-pearlitic phase after differential cooling using segmented forming tools after uniform austenitization. Hence, from the hardness profile in Fig. 1 (a – b), a gradual change of present phases from hard martensite to soft ferrite–pearlite is derived. In case of high strength aluminum alloy AA7075, similar to 22MnB5, high cooling rates results in high material strength and low cooling rates to significantly lower material strength. This behavior is linked to a specific change in the microstructure, which is created by thermo-mechanically induced precipitation of particles or suppression of it during the differential cooling. At the low tool segment temperature of T_{FT} = 24 °C, the nucleation and growth of precipitates is suppressed leading to a high level of supersaturation in the solid solution. Hence, after artificial aging, the formation and growth of semi-coherent strengthening η′-phases explain the obtained high hardness level in the cooled zones. In contrast, forming at a higher forming tool segment temperature of T_{FT} = 350 °C together with an isothermal holding of 30 s leads to precipitation already during forming enabled by the sufficient thermal energy at 350 °C and results in a decreased level of supersaturation. During the following artificial aging the formation and growth of lath-shaped metastable η-phases is promoted. The resulted coarse particles hardly contribute to the yield strength increase and their interaction with moving dislocation is identified by the bypass mechanisms. Within the transition zone the gradual change of the supersaturation in the solid solution state possibly leads to mixed-precipitate structures and a gradual amount of growing fine precipitates. This may be the reason for the continuous decrease of hardness values from hard to soft.

The progression of the local strain and the related logarithmic von Mises strain map during tensile deformation show the occurrence of plastic deformation mainly in the zones at the heated forming tool segments, i.e. soft zones, while the hard zones at the cooled forming tool segments remain almost undeformed. The explanation for this behavior can be found in the specific microstructure of the soft zones, exhibiting lower yield strength compared to the harder zone, which enables and eases the initiation and accumulation of plastic deformation and movement of dislocations due to the high local stress concentration. These phenomena lead to higher formability of the soft zone and concomitantly to higher strain hardening rate. The high strain hardening can be found from the obtained stress strain curves showing in Fig. 2 by the continues increase of flow stress after attaining the yield point to UTS. This is gained by the mutual interactions of mobile dislocations as well as with other defects such as second phases only in limited zones. In this regard, the strain hardening rate increases, since in the hard zones the local stress is not sufficient to activate dislocations leading to enormous nucleation of dislocations in soft zones.

In case of AA7075, plastic deformation also occurs in the zone at the cooled tool segment, which is comparatively lower to the measured logarithmic strain at the heated zone, while the measured strain distribution in case of 22MnB5 is more localized. This behavior can be explained by the ratio of the hardness values from the hard and soft zones The ratio of the hardness values of 22MnB5 samples is higher and thus the gradient of transition zone is steeper and narrower. On top of that, due to the local deformation concentration in limited regions, the plastic instability after the necking point and the occurrence of crossed shear bands leads to a further increase of the amount of local plastic deformation. Therefore, local plastic strain as shown in Fig. 3 (a) for 22MnB5 accumulates as a result of plastic flow under plane stress domination and lead to significant high local logarithmic von Mises strain values at the necking zone short before failure. Due to the tailored mechanical properties by different microstructural features, the generation and propagation of necking as well as the failure can be controlled and limited to defined zones. Consequently, the mechanical performance of the mixed microstructure within the transition zone is governed by the macroscopic stress and strain partitioning as can be clearly seen by DIC analysis.

References

[1] K. Steinhoff, New process strategies and resulting product structures by locally induced thermo-mechanical interaction, in: M. Liewald, K. Siegert (Eds.), Neuere Entwicklungen Der Massivumformung, MAT INFO Werkstoff-Informationsgesellschaft mbH, Frankfurt, America, 2005, pp. 277–298.

[2] A. Ademaj, A. Donis, S. Prokoph, U. Weidig, K. Steinhoff, Hot stamping strategies for manufacturing of thermo-mechanically graded parts, in: Proc. 4th Int. Conf. Thermomechanical Process. Sheff., 2012.

[3] T.K. Eller, L. Greve, M.T. Andres, M. Medricky, A. Hatscher, V.T. Meinders, A.H. Van Den Boogaard, Plasticity and fracture modeling of quench-hardenable boron steel with tailored properties, J. Mater. Process. Technol. 214 (2014) 1211–1227. https://doi.org/10.1016/j.jmatprotec.2013.12.015

[4] R. George, A. Bardelcik, M.J. Worswick, Hot forming of boron steels using heated and cooled tooling for tailored properties, J. Mater. Process. Technol. 212 (2012) 2386–2399. https://doi.org/10.1016/j.jmatprotec.2012.06.028

[5] J. Min, J. Lin, Y. Min, Effect of thermo-mechanical process on the microstructure and secondary-deformation behavior of 22MnB5 steels, J. Mater. Process. Technol. 213 (2013) 818–825. https://doi.org/10.1016/j.jmatprotec.2012.12.012

[6] E. Scharifi, T. Schade, A. Ademaj, S.V. Sajadifar, U. Weidig, T. Niendorf, K. Steinhoff, Characterization of Mechanical Properties, Macroscopic Deformation Behavior, and Microstructure of Functionally Graded 22MnB5 Steel, Steel Res. Int. 92 (2021). https://doi.org/10.1002/srin.202000633

[7] E. Scharifi, M. Roscher, S. Lotz, U. Weidig, E. Jägle, K. Steinhoff, Functional Gradation in Precipitation Hardenable AA7075 Alloy by Differential Cooling Strategies, Key Eng. Mater. 883 (2021) 159–166. https://doi.org/10.4028/www.scientific.net/KEM.883.159

[8] E. Scharifi, S.V. Sajadifar, G. Moeini, U. Weidig, S. Böhm, T. Niendorf, K. Steinhoff, Dynamic Tensile Deformation of High Strength Aluminum Alloys Processed Following Novel Thermomechanical Treatment Strategies, Adv. Eng. Mater. 22 (2020) 2000193. https://doi.org/10.1002/adem.202000193

[9] A. Ademaj, A. Donis, U. Weidig, K. Steinhoff, Hot Stamping of Multi-Material Composites, in: Oldenburg, M.; Steinhoff, K.; Prakash, B. (Edtrs.): Proceedings of the 4th International Conference on Hot Sheet Metal forming of High-Performance Steel, Auerbach (D): Verlag Wissenschaftliche Scripten, 2013, pp. 481-488.

Modelling

Numerical investigation on the influence of hydroxide production on the high-cycle fatigue behavior of clinched joints

Sven Harzheim[1, a *], Martin Hofmann[1, b] and Thomas Wallmersperger[1, c]

[1]Institut für Festkörpermechanik, TU Dresden, Dresden, Germany

[a]sven.harzheim@tu-dresden.de, [b]martin.hofmann@tu-dresden.de, [c]thomas.wallmersperger@tu-dresden.de

* Corresponding author

Keywords: Finite Element Method (FEM), Fatigue, Galvanic Corrosion

Abstract. Corrosion usually has detrimental effects on the longevity of metallic structures. Several corrosion mechanisms exist, which affect the metallic structures differently. Galvanic corrosion is a major corrosion phenomenon that occurs when two dissimilar metals are in contact with each other and with an electrolyte. Due to galvanic corrosion, the less noble metal dissolves which can shorten the service life of the metallic component significantly. However, recent research has shown that corrosion can have a beneficial impact on metal joints. Brief exposure of clinched specimens to a salt spray environment extends the corroded specimen's fatigue life. The present work introduces a simulation procedure which can numerically predict the effects of corrosion product formation on the fatigue life of clinched joints.

Introduction

Mechanical clinching is an innovative joining method that is especially suitable for combining dissimilar metals [1]. In short, the joint is created by pushing two metal sheets into a die with a punch. Joining dissimilar materials has the advantage that lightweight structures can be formed, which reduce material cost while simultaneously providing the necessary structural integrity. However, combining dissimilar metals causes the risk of galvanic corrosion. Due to the different electrochemical potentials of the metals, electrons are transferred from the less noble metal to the noble metal, leading to oxidation of the less noble metal. Material or component integrity is often compromised by this process, meaning that metal components need to be replaced over time.

A common method to prevent material degradation by corrosion is to use a sacrificial anode or cathodic protection. An overview of the affection of clinched joints by corrosion is given in [2]. Recent literature [3] has shown that a brief exposure of clinched specimens to a corrosive environment can also have beneficial effects to the specimen's fatigue life. The EN AW-6014/HCT590X+Z specimens have been corroded for three weeks prior to the fatigue testing. The authors attributed the observed effect to the corrosion products building up in the gap area between the joined metal sheets. They provided a simple mechanical model that shows a change in the load bearing capabilities of the clinched specimen when the gap size is reduced. Previously, the influence of a salt spray environment on the load bearing capabilities of clinched joints has been experimentally studied in Ref. [4]. It was shown via shear tests that the load bearing capabilities of the joints decreased with the exposure to the salt spray.

Numerically, both corrosion and fatigue phenomena have been thoroughly studied in the literature. Mathematical models exist that describe corrosion and material fatigue very well. Modeling metal dissolution can be performed by using different approaches. Phase field models rely on solving the coupled Allen-Cahn and Cahn-Hilliard equations. The boundary between metal and electrolyte does not need to be moved by applying a distinct displacement. Thus, this approach

Content from this work may be used under the terms of the Creative Commons Attribution 3.0 license. Any further distribution of this work must maintain attribution to the author(s) and the title of the work, journal citation and DOI. Published under license by Materials Research Forum LLC.

is very suitable for modeling complex geometries or surfaces that combine during the simulation procedure.

In Ref. [5] a phase field model is used to study multiple corrosion phenomena, such as pitting corrosion in a hybrid joint. In Ref. [6], a phase field model is used to simulate stress corrosion cracking. In this computational model, the cracks initiate from pits. Pit growth under mechanical and environmental influence is studied. Another popular modeling approach for corrosion is the level set method, which is also very suitable for the simulation of merging surfaces. In Ref. [7], a level set model is presented, where among others, the lacy cover formation of pits is studied. The Arbitrary Lagrangian-Eulerian method to simulate the anodic dissolution is also widely used in the literature. A thorough overview of this method is given in Ref. [8].

High-cycle fatigue in metallic materials has been mathematically described and simulated in Ref. [9]. There is a research gap in the literature, where both concepts of galvanic corrosion and high-cycle fatigue are coupled and applied to clinched joints.

The present work expands on the idea that corrosion products improve the load bearing capabilities of clinched joints. First, a computational model is introduced, which simulates the formation of corrosion products and their influence on the fatigue life of clinched joints. Then, the test case is introduced, and the computational model is described. The results of the model are given and are discussed. Finally, a summary, as well as a short conclusion are given.

Test Case Description and Computational Model

In this section, the challenges in modeling of the corrosion phenomenon are described. The considered geometry for the test case is shown in Fig. 1. Modeling the positive effects of galvanic corrosion on the fatigue life of clinched joints requires a coupled electrochemical-mechanical model. The electrochemical part of the model describes the corrosion behavior of the clinched joint in the salt spray environment, while the mechanical part of the model describes the mechanical response of the joint under applied loads. In order to achieve the chemo-electro-mechanical coupling, three major steps are defined. (i) Simulate the hydroxide layer growth which is caused by galvanic corrosion. (ii) Simulate the load bearing during cyclic loading, which is affected by the hydroxide layer thickness, and (iii) numerically predict the fatigue life of the corroded clinched joint.

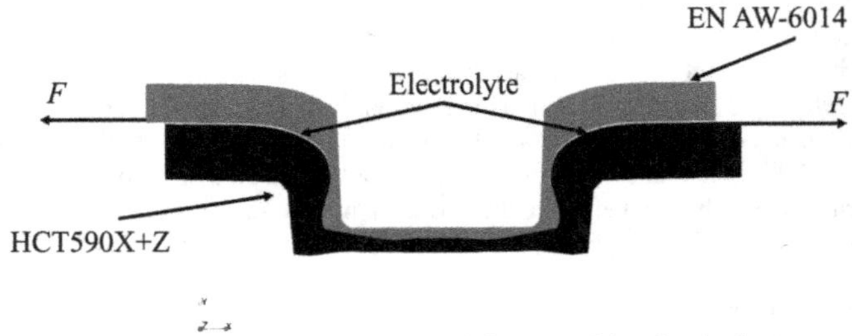

Fig.1: Two-dimensional computational domain of the clinched joint.

Performing step (i) usually requires a detailed model that accurately describes the corrosion behavior of the clinched joint. The corrosion behavior of EN AW-6014/HCT590X+Z/Electrolyte is reflected by the constituents' electric potentials and open circuit current densities. With those parameters, a corrosion velocity of the less noble metal's surface can be determined. During the dissolution of the less noble metal, metal ions are oxidized and dissolve into the electrolyte. This procedure is described by Eq. 1

$$\mathrm{Me} \rightarrow \mathrm{Me}^{n+} + n\,\mathrm{e}^{-}. \tag{1}$$

Depending on the pH of the electrolyte, hydroxide ions are produced at the cathode. This process is described by Eq. 2

$$2\,H_2O + O_2 + 4\,e^- \rightarrow 4\,OH^-. \quad (2)$$

Metal ions and hydroxide ions in the electrolyte react and form metal hydroxides, which usually have a low solubility and thus precipitate. The current work uses a simplified model to mimic the corrosion product formation. The corrosion behavior is assumed as follows. The zinc layer on the surface of HCT590X forms a galvanic couple with the aluminum sheet. As shown in [3], the zinc layer is less noble than the aluminum alloy and dissolves into the electrolyte. The zinc hydroxide production is modeled as a growing layer on top of the surface of HCT590X+Z. To achieve this, the geometry is divided into solid and liquid domains. Solid domains are the metal sheets and the metal hydroxide layer, and liquid domains are the electrolyte in the gap areas. This is implemented by using one concentration field for the solid (c_{solid}), and one concentration field for the liquid (c_{liquid}).

The normalized concentrations represent the phase fractions, i.e., they vary between 0 and 1, and thus reflect the different phases of the geometry, i.e., solid and liquid. In areas where c_{solid} is 1, c_{liquid} is 0, and vice versa. The transition between the liquid and the solid phases is resembled by a diffuse interface, see Fig. 2.

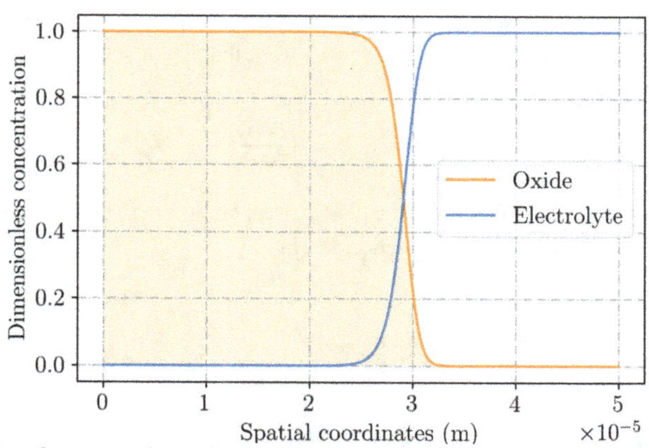

Fig. 2: One-dimensional interface between the solid phase (oxide) and the liquid phase (electrolyte) of the domain.

The conversion of the liquid domain to the solid domain in the gap area is realized by using a reaction term R_α. The system of equations is as follows. For the liquid and solid domains, a mass balance (Eq. 3) is used, respectively.

$$\frac{\partial c_\alpha}{\partial t} = D_\alpha \nabla^2 c_\alpha + R_\alpha \text{ with } R_\alpha = k\, c_{solid}\, c_{liquid}. \quad (3)$$

Here, c_α is the concentration of either liquid or solid, D_α is the diffusion coefficient, and R_α the reaction term. k is the rate constant. In this model, D_α and k define the size and the velocity of the moving interface during the phase transition. With the introduced chemical model, the hydroxide layer growth due to galvanic corrosion can be described.

In step (ii), the chemical field is coupled to the mechanical field. The mechanical model is based on linear elasticity. The coupling between the chemical and the mechanical field is achieved by multiplying the Young's modulus E with the solid phase concentration, resulting in distributions, which are shown in Fig. 3. A numerical problem arises where the solid phase is zero, since the resulting Young's modulus there would also be zero. This problem is solved by constraining the solid phase concentration in the liquid phase to very small values larger than zero.

In order to simulate the fatigue behavior, two load cases were applied to the clinched joint. One load, which represents the maximum force, and one load, which represents the minimum force during cyclic testing. The resulting stress fields can then be used in step (iii) to determine the fatigue life of the clinched joint.

In step (iii), the high cycle fatigue model, originally introduced by Lemaitre [9], is used. It assumes a representative volume element (RVE), which contains a micro inclusion. The RVE exhibits elastic behavior, while the micro inclusion behaves elastic plastic. For each loading cycle, a damage increment is numerically determined, which accumulates in a damage variable D. Upon reaching a critical value, i.e., $D = 1$, material failure occurs, and cracks can form.

Usually, the damage variable is calculated, e.g., for each node in the finite element mesh. However, computational costs can be reduced by limiting the high-cycle fatigue calculations to the location of the largest von Mises stress in the clinched joint. A similar approach has been used in [10]. A brief summary of the two-scale high-cycle fatigue damage model by Lemaitre is given in the following. The model considers two scales. The macro scale, which only considers elastic behavior, and the micro scale, where plastic strains occur. Variables on the micro scale are denoted with the superscript $()^{\mu}$. Stresses from the macro scale σ_{ij} are transformed to the micro scale σ_{ij}^{μ} by a localization parameter a (Eq. 4). The stresses from the macro scale are obtained by finite element simulations.

$$\sigma_{ij}^{\mu} = \sigma_{ij} - aE\varepsilon_{ij}^{\mu p} \quad \text{with } a = \frac{1-\beta}{1+\nu} \quad \text{where } \beta = \frac{2\,(4-5\nu)}{15\,(1-\nu)} \tag{4}$$

Here, E is the Young's modulus, $\varepsilon_{ij}^{\mu p}$ is the plastic strain on the micro scale, and ν is the Poisson ratio. During the static loading, a yield function f^{μ} (Eq. 5)

$$f^{\mu} = \left(\tilde{\sigma}_{ij}^{\mu} - X_{ij}^{\mu}\right)_{\mathrm{eq}} - \sigma_{\mathrm{f}} \tag{5}$$

is evaluated.

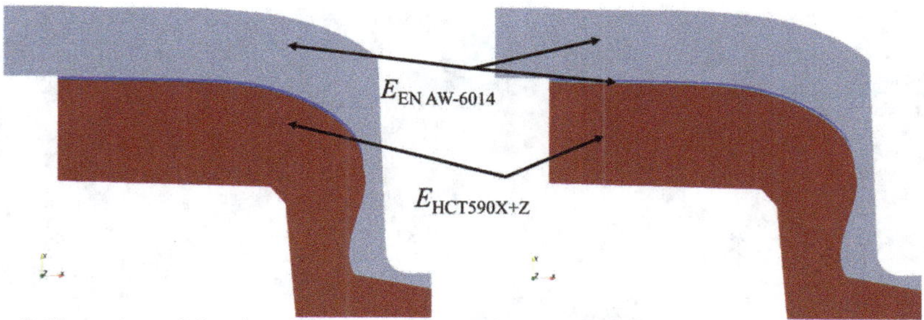

Fig. 3: Young's modulus distribution for the initial domain (left) and for the growing hydroxide layer (right).

$\tilde{\sigma}_{ij}^{\mu}$ is the effective stress on the micro scale, which is defined by $\tilde{\sigma}_{ij}^{\mu} = \sigma_{ij}^{\mu}/(1-D)$. X_{ij}^{μ} is the backstress tensor, which accounts for kinematic hardening on the micro scale, and σ_f is the fatigue limit. $(\,)_{eq}$ is the equivalent (von Mises) expression of the function inside the brackets. If the value of the yield function is zero or larger than zero, i.e., $f^{\mu} \geq 0$, plastic strains occur. The damage rate $\dot{D}$ of the current cycle can then be calculated by Eq. 6

$$\dot{D} = \left(\frac{Y^{\mu}}{S}\right)^{s} \dot{p}^{\mu}, \text{ where } \dot{p}^{\mu} = \frac{\dot{\lambda}^{\mu}}{(1-D)}. \tag{6}$$

Here, Y^{μ} is the strain energy density release function, S is the damage strength, s is the damage exponent, and $\dot{\lambda}^{\mu}$ is the plastic multiplier. Y^{μ} usually depends on a closure parameter h, which accounts for the closure of micro defects under compression. This means that the damage increments are smaller in areas where compression occurs.

In summary, the fatigue life of pre-corroded and non-corroded clinched joints can be numerically predicted by (i) simulating the hydroxide layer growth due to galvanic corrosion, (ii) simulating static loading during the fatigue tests and identifying the area where the largest stress occurs, and (iii) performing high-cycle fatigue calculations with the highest stress as an input. Both models for step (i) and (ii) are coupled and implemented within a finite element framework. Coupling means that the growing hydroxide layer is included in the static loading simulations. The high-cycle fatigue damage model is implemented via a python script, where the closed form solution shown in [11] is applied.

Influence of the Hydroxide Production on the Fatigue Life of Clinched Joints

In this section, the results of the simulation procedure are given and discussed. Fig. 4 shows the hydroxide layer thickness at different time stamps obtained from step (i). The small gap between both metal sheets is filled with corrosion products and thus shrinks in its size. Metal dissolution and corrosion product formation occur at the same time.

Materials Research Proceedings **25** (2023) 255-262 https://doi.org/10.21741/9781644902417-32

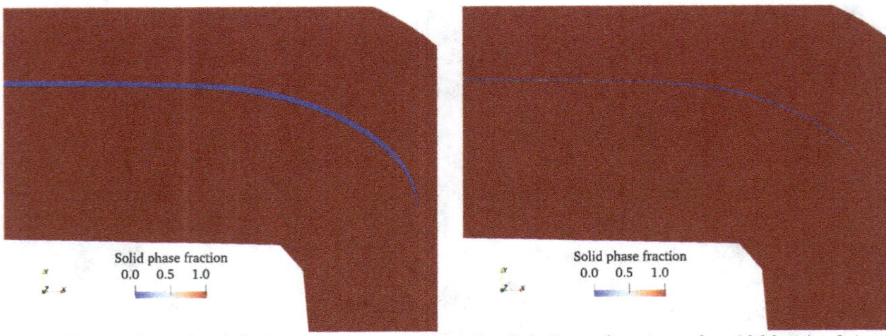

Fig. 4: Initial solid phase fraction (left) and solid phase fraction after 1200 s (right).

In the current model, metal dissolution is neglected. The specific volume of the corrosion product zinc hydroxide is much larger than the specific volume of the dissolving zinc, which is why only the hydroxide layer growth is considered. The time scales of the computational model do not necessarily reflect the real-world corrosion behavior of the clinched joint in the salt spray environment. For example, the clinched specimens were corroded for three weeks in the salt spray chamber in [3].

No details on the intermediate states of corrosion (e.g., hydroxide layer thickness) were given. It is thus not possible to precisely derive the growth rate of the hydroxide layer from the performed experiments. However, the parameters of the model (D, k) can be adapted to increase or slow down the hydroxide production rate to accurately reflect the actual corrosion behavior.

Furthermore, the presented results align very well with experimental findings, i.e., the formation of corrosion products leads to an overall better fatigue life. The introduced model shows the potential to accurately reflect the real-world behavior. Its parameters, however, have to be determined by further experimental investigations. For example, it is expected that mass transfer of the dissolving metal ions is slowed down by the growing hydroxide layer. This effect can be captured by a time dependent diffusion coefficient D. The hydroxide layer growth should thus slow down over time. In the current model, the growth rate is assumed to be constant, as shown in Table 1.

Table 1: Results of the high-cycle fatigue calculations with different hydroxide layer thicknesses.

Time t [s]		0	50	100	150
Hydroxide layer thickness [µm]		0	6	12	18
Number of cycles until failure	Force amplitude 700 N	1211	2397	4678	8602
	Force amplitude 500 N	21513	43338	84602	162120

Materials Research Proceedings **25** (2023) 255-262 https://doi.org/10.21741/9781644902417-32

The results obtained from step (ii) are shown in Fig. 5. In the left plot, no hydroxide layer is considered and consequently, large stresses occur in the left neck area. The right plot shows the von Mises stress for the same applied load. In this case however, the gap is fully closed with corrosion products. The overall occurring stresses in the final configuration are lower compared to the configuration without corrosion products. It can be concluded that the growing hydroxide layer influences the load bearing behavior of the clinched specimen. Near the tip of the left gap, very large stresses were observed. These large stresses originate from the fact that the clinched specimen is simulated as a continuous body with different material properties, i.e., no contact formulation was applied yet. For the rest of the proposed modeling procedure, these large stresses at the tip of the left gap were not considered.

In step (iii), the results from the static loading simulation are used in the high-cycle fatigue calculation. For this, two force amplitudes of 700 N and 500 N, respectively, were used in step (ii). Table 1 summarizes the calculated results. A clear trend is recognizable, a thicker hydroxide layer results in slower damage accumulation. This means that more loading cycles can be endured by the clinched specimen before material failure occurs. In 50 s simulation time of hydroxide layer growth, the number of cycles until failure roughly doubles.

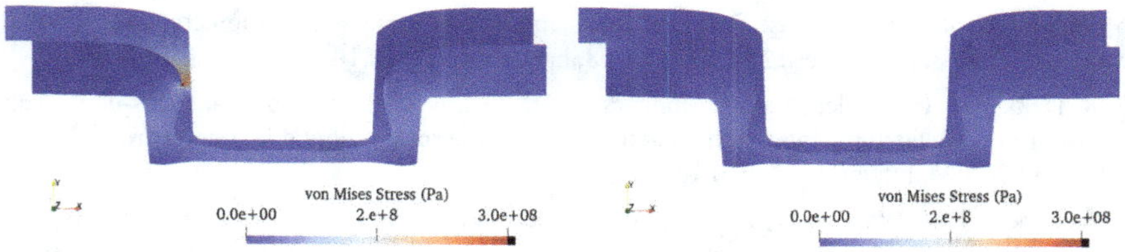

Fig. 5: von Mises stress of the initial configuration with no hydroxide layer (left) and von Mises stress of the final configuration with a fully closed gap (right).

Conclusion

In this work, a modeling procedure was introduced that can numerically simulate the impact of a hydroxide layer production on the fatigue life of clinched joints. It was shown that corrosion products enlarge the contact area of the metal sheets in the clinched joints, ultimately reducing the mechanical stresses. This stress reduction leads to an improved fatigue life. The present model has some shortcomings, which will be addressed in future research. The hydroxide layer production in the model does not accurately reflect the real-world time scale. This can be addressed by adjusting the reaction constant and the diffusion coefficients of the two species. Furthermore, the electrochemical behavior of the galvanic pair EN AW-6014 and HCT590X+Z is not fully described. This means that the hydroxide layer production is currently not influenced by the metals' electric potentials and is thus uniform along the metal surface.

Acknowledgement

We gratefully acknowledge the funding by the Deutsche Forschungsgemeinschaft (DFG, German Research Foundation) – TRR 285, subproject B03 – Project-ID 418701707.

References

[1] X. He, Clinching for sheet materials, Science and Technology of Advanced Materials. 18:1 (2017) 381-405. https://doi.org/10.1080/14686996.2017.1320930

[2] B. Schramm, S. Harzheim, D. Weiß, T. D. Joy, M. Hofmann, G. Meschut, J. Mergheim, T. Wallmersperger, A Review on the Modeling of the Clinching Process Chain – Part III: Operational Phase, Journal of Advanced Joining Processes. 6 (2022) 100135. https://doi.org/10.1016/j.jajp.2022.100135

[3] S. Harzheim, L. Ewenz, M. Zimmermann, T. Wallmersperger, Corrosion Phenomena and Fatigue Behavior of Clinched Joints, Numerical and Experimental Investigations. 6 (2022) 100130. https://doi.org/10.1016/j.jajp.2022.100130

[4] L. Calabrese, E. Proverbio, G. Galtieri, C. Borsellino, Effect of corrosion degradation on failure mechanisms of aluminium/steel clinched joints, Materials & Design. 87 (2015) 473-481. https://doi.org/10.1016/j.matdes.2015.08.053

[5] W. Mai, S. Soghrati, New phase field model for simulating galvanic and pitting corrosion processes, Electrochimica Acta. 260 (2018) 290-304. https://doi.org/10.1016/j.electacta.2017.12.086

[6] W. Mai, S. Soghrati, A phase field model for simulating the stress corrosion cracking initiated from pits, Corrosion Science. 125 (2017) 87-98. https://doi.org/10.1016/j.corsci.2017.06.006

[7] R. Dekker, F. P. van der Meer, J. Maljaars, L. J. Sluys, A level set model for stress-dependent corrosion pit propagation, International Journal for Numerical Methods in Engineering. 122(8) (2021) 2057-2074. https://doi.org/10.1002/nme.6614

[8] S. Sarkar, J. E. Warner, W. Aquino, A numerical framework for the modeling of corrosive dissolution, Corrosion science. 65 (2012) 502-511. https://doi.org/10.1016/j.corsci.2012.08.059

[9] J. Lemaitre, J. Sermage, R. Desmorat, A two scale damage concept applied to fatigue, International Journal of Fracture. 97 (1999) 67-81. https://doi.org/10.1023/A:1018641414428

[10] H. Argente dos Santos, F. Auricchio, M. Conti, Fatigue life assessment of cardiovascular balloon-expandable stents: a two-scale plasticity-damage model approach, Journal of The Mechanical Behavior of Biomedical Materials. 15 (2012) 78-92. https://doi.org/10.1016/j.jmbbm.2012.06.011

[11] R. Desmorat, A. Kane, M. Seyedi, J. P. Sermage, Two scale damage model and related numerical issues for thermo-mechanical High Cycle Fatigue, European Journal of Mechanics - A/Solids. 26 (2007) 909–935. https://doi.org/10.1016/j.euromechsol.2007.01.002

Sheet Metal 2023 | Materials Research Forum LLC
Materials Research Proceedings **25** (2023) 263-270 | https://doi.org/10.21741/9781644902417-33

Numerical analysis of failure modeling in clinching process chain simulation

Christian R. Bielak[1,a*], Max Böhnke[1,b], Johannes Friedlein[2,c], Mathias Bobbert[1,d], Julia Mergheim[2,e], Paul Steinmann[2,f] and Gerson Meschut[1,g]

[1]Laboratory for Material and Joining Technology (LWF), Paderborn University, Pohlweg 47-49, 33098 Paderborn, Germany

[2]Institute of Applied Mechanics, Friedrich-Alexander-Universität Erlangen-Nürnberg, Egerlandstrasse 5, 91058 Erlangen, Germany

[a]Christian.Bielak@lwf.upb.de, [b]Max.Boehnke@lwf.upb.de, [c]Johannes.Friedlein@fau.de, [d]Mathias.Bobbert@lwf.upb.de, [e]Julia.Mergheim@fau.de, [f]Paul.Steinmann@fau.de, [g]Meschut@lwf.upb.de

Keywords: Finite Element Method, Damage, Joining

Abstract. The application of the mechanical joining process clinching allows the assembly of different sheet metal materials with a wide range of material thickness configurations, which is of interest for lightweight multi-material structures. In order to be able to predict the clinched joint properties as a function of the individual manufacturing steps, current studies focus on numerical modeling of the entire clinching process chain. It is essential to be able to take into account the influence of the joining process-induced damage on the load-bearing capacity of the joint during the loading phase. This study presents a numerical damage accumulation in the clinching process based on an implemented Hosford-Coulomb failure model using a 3D clinching process model applied on the aluminum alloy EN AW-6014 in temper T4. A correspondence of the experimentally determined failure location with the element of the highest numerically determined damage accumulation is shown. Moreover, the experimentally determined failure behavior is predicted to be in agreement in the numerical loading simulation with transferred pre-damage from the process simulation.

Introduction

Clinching is a widely applied mechanical joining technique for sheet metal materials in which the joint is based on cold forming of the parts to be joined and no auxiliary element is required. It has proven particularly useful for the implementation of multi-material designs, which have become important in the context of lightweight construction in the automotive industry [1].

The material combination and clinching process parameters examined in this paper has been extensively investigated in [2]. In particular, the process chain of clinching aluminum EN AW-6014 T4 was described and experimental techniques were proposed to analyze the process-structure-property relationships from the sheet metals properties to the joined structure. 2D rotationally symmetric FE-models of this connection were presented in [3] and for 3D FE-models in [4]. With the FE models, results have already been obtained with regard to the joining point properties and the maximum transmissible force under load, which are in agreement with the experiments. The prediction of a neck crack in the joining process or the maximum strain under load could not be predicted yet due to a lack of damage and failure model in own previous studies. The plastic behavior and the pre-damage introduced into the material are relevant for the subsequent joining point properties. In [5] an isotropic elasto-plastic law coupled with a Lemaitre ductile damage model was used to investigate the clinching process. Subsequent strength tests and optimization of the process parameters through a Kriging metamodel were shown. In [6] the self-

Content from this work may be used under the terms of the Creative Commons Attribution 3.0 license. Any further distribution of this work must maintain attribution to the author(s) and the title of the work, journal citation and DOI. Published under license by Materials Research Forum LLC.

piercing riveting process (SPR) was examined and it was shown that the consideration of the damage behavior during the joining process itself is of major importance. From this, they conclude that it might also be relevant to investigate damage evolution for clinching, which is a similar joining technique.

In [7] a damage model was developed in order to analyse process-induced defects during SPR and clinching. 6xxx series aluminium sheets in the T6 condition were investigated with limited ductility and different damage criteria were compared to predict the influence of process-induced damage on the strength properties of the joints.

Failure modeling in mechanical joining processes was presented in [8] by applying the GISSMO damage model in LS-DYNA to the SPR process simulation to predict ductile damage in thin aluminum sheets. The parameters for this stress state dependent material model have been identified from tensile tests with various characteristic stress states. In [9], the complete process chain for self-pierce riveting was numerically investigated with respect to joint failure. The calibration of damage models and their application in strength analysis are integral parts of the present study. Furthermore, different damage models and the influence of the setting depth on the joint strength and the type of joint failure are compared.

In [10] a method for the calibration of failure models for the simulation of clinching processes was presented. Based on experimental tests and corresponding simulation models, failure-relevant data were transferred into three different failure models and calibrated. The Hosford-Coulomb model [11] was found to be highly predictive.

The purpose of this study is to investigate the numerical damage accumulation in the clinching process based on an implemented Hosford-Coulomb failure model using a 3D clinching process model and the influence of the transfered damage on the load-bearing behavior under shear tensile load.

Finite-Element-Models

The FE-model used in this study for the calculation of the clinch joining process was first presented in [12] and then applied for a numerical process chain in [4]. The tool and sheet geometries as well as boundary conditions were adapted. All simulations run with the LS-DYNA Solver ls-dyna__smp_d_R12.1. with implicit time integration. Due to the process time and low dynamic effects an implicit time integration was chosen.

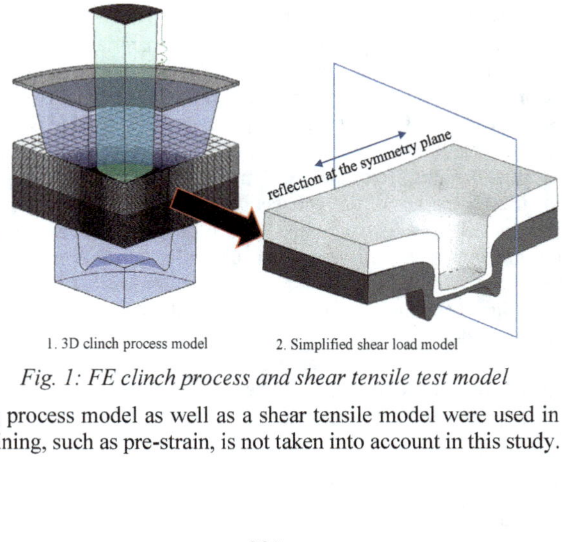

Fig. 1: FE clinch process and shear tensile test model

A clinch joining process model as well as a shear tensile model were used in this study. Pre-operation before joining, such as pre-strain, is not taken into account in this study. The models are modified with respect to the element formulation of the sheet materials and the material model used. Since the discretization of the parts to be joined is optimized for the process, there is no need for automatic adaptive remeshing during the simulation. The sheets are now discretized with hexahedral elements, as shown in Fig. 1.

The clinching process takes advantage of the symmetry of the process and is calculated using a quarter model of the real geometry. The shear tensile test is based on the geometry calculated in the clinching process and the material conditions. This is represented by a reflection at the symmetry plane as half a model. The shear tensile test is simplified in terms of geometry and does not reproduce the real specimen stiffness from the experiment. This must be taken into account in the subsequent evaluation.

Material model

A hyperelastic-based finite plasticity material model was implemented as user defined material model in the commercial FEM software LS-DYNA. At first, plastic anisotropy will not be considered such that the application of finite plasticity based on a multiplicative split of the total deformation is straightforward. The implementation is based on [13] , where the approach has already been extended to incorporate gradient-damage. The flexibility of a user material is necessary especially for our later extension to incorporate the inherent material softening of ductile damage and its physically motivated regularization by gradient-enhancement. In this study, the assumption was made that damage accumulation has no influence on material behavior and is only a failure criterion. An efficient strategy for the implementation of gradient-enhancement in LS-DYNA has been presented recently [14] utilizing the existing thermomechanical coupling.

The isotropic plastic hardening behavior is given by a flow curve presented in [15]. The extrapolated flow curve and the mechanical properties of EN AW-6014 T4 used in this study are shown in Fig. 2. The flow curve was extrapolated according to the Hockett-Sherby approach. The layer compression test is the experimental basis for this extrapolation.

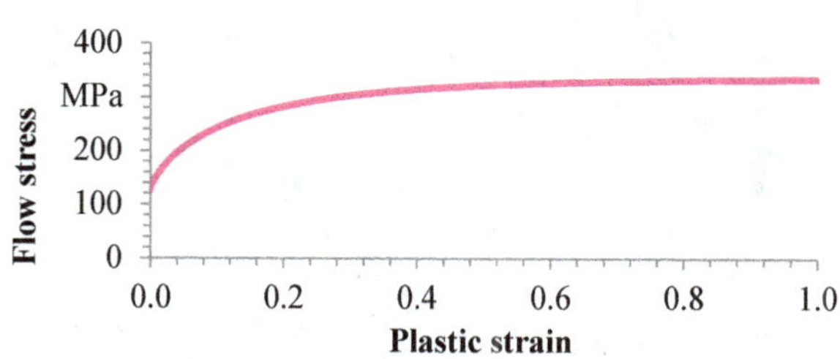

EN AW-6014 T4	
R_m [MPa]	231
$R_{p0.2}$ [MPa]	125
A_{50} [%]	25

Fig. 2: Mechanical properties of EN AW-6014 T4 [15]

Hosford-Coulomb failure model

The Hosford-Coulomb failure model was proposed in [11]. Therein, the von Mises equivalent stress was replaced by the Hosford equivalent stress in combination with the normal stress acting on the plane of maximum shear. The Hosford-Coulomb model is based on the extensive study on 3D unit cells in [16], thus it has a strong micromechanical foundation. In [11] fracture experiments on three different steels were conducted and used to calibrate the fracture criterion. When compared against the experiments, the simulations showed good agreement. Using and extending previous work [10], the failure prediction is based on the Hosford-Coulomb (HC) failure model [11]. The latter proved most suitable for the failure prediction of the also herein considered aluminum alloy EN-AW 6014. The HC model is the generalization of the Mohr-Coulomb model that was shown to agree well with localization analyses of various stress states [11], which is closely related to the failure prediction. Such stress-state dependent models are well suited to

Materials Research Proceedings **25** (2023) 263-270 https://doi.org/10.21741/9781644902417-33

describe the different damage processes taking place, ranging from void growth to shear banding. The HC failure model predicts the current failure strain $\bar{\varepsilon}_f$ as a function of the stress state characterized by the stress triaxiality η and the Lode angle $\bar{\theta}$. For a detailed description of the relations see [11]; [10]. The four parameters of the HC model are taken from [10] for the aluminium alloy EN-AW 6014 T4. Even though at this point, damage-related softening is not activated, a nonlinear damage accumulation law is utilized as proposed in [17]; [18] as

$$dD = n_d \left[\frac{\bar{\varepsilon}_p}{\bar{\varepsilon}_f(\eta,\bar{\theta})}\right]^{n_d-1} \frac{d\bar{\varepsilon}_p}{\bar{\varepsilon}_f(\eta,\bar{\theta})} \tag{1}$$

with the positive damage exponent n_d, here equal to 2. Such an incremental nonlinear approach is regarded as indispensable to predict the damage accumulation under non-proportional loading as present in experiments and distinct for the clinch joining and joint loading.

Results

The clinching process model is compared to the experiment for the connection EN-AW 6014 T4 t= 2mm in EN-AW 6014 T4 t= 2mm in order to show the influence of incorrect geometric properties or incorrect plastic deformations that would be calculated based on an incorrect plasticity model. For this purpose, the simulated contour of the clinched joint (blue) is compared with the experimental micro section, as shown in Fig. 3 (left). A good agreement can be observed. Furthermore, the experimental determined force-displacement curve is compared with numerical calculated values. A high agreement between experiment and simulation is evident. The clinched joint mesh calculated in the model is then transferred to the shear load model.

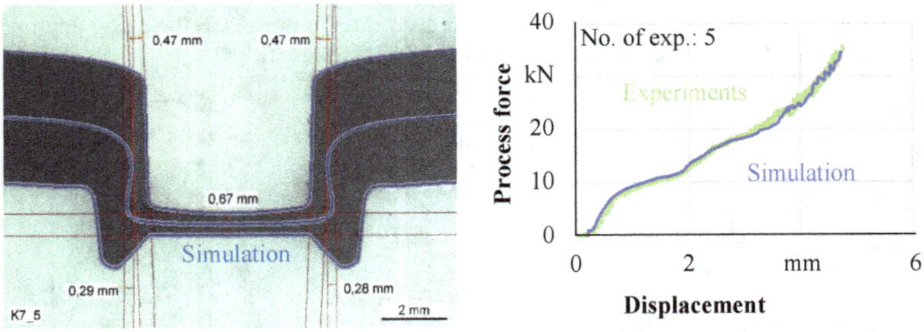

Fig. 3: Model comparison of the clinching process (simulation and experiment)

The element accumulating the highest damage was selected for the localization of the crack location due to the damage accumulation in the shear tensile test. This element was also evaluated in the clinching process. Afterwards, the influence of the joining process induced pre-damage on the shear tensile test was also taken into account (coupled model).

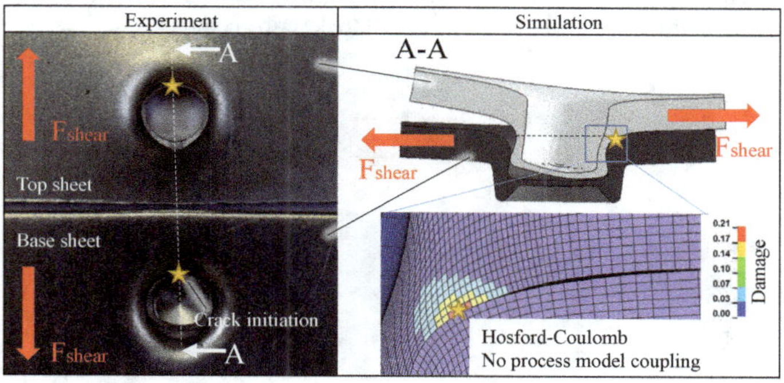

Fig. 4: Localization of the highest damage accumulation in the shear tensile test

The element is located in the top plate in the neck area and is shown in Fig. 4 on the right hand side (marked). The evaluation of the failure location was made at the beginning of the first occurring damage peak. The experimental clinched joint after shear tensile loading is shown on the left, with the fracture location corresponding to the simulated location of the highest damage accumulation. To achieve an understanding of the complex local stress state, the element with the highest final damage was selected and evaluated during the coupled process.

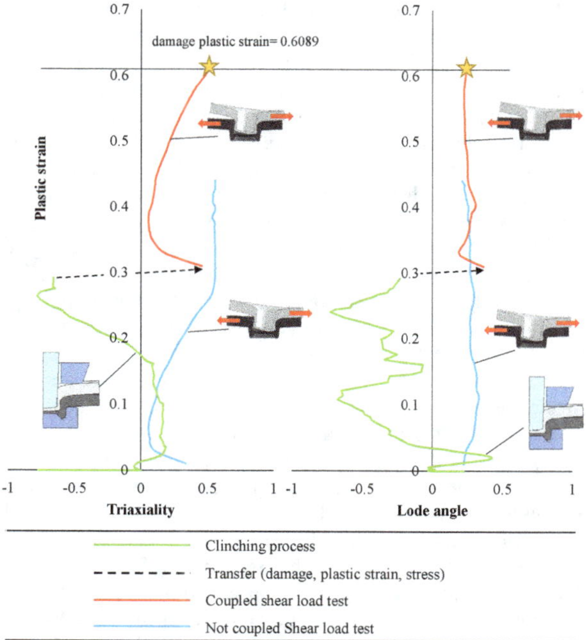

Fig. 5: Simulated stress state during the clinching process and shear tensile test

Fig. 5 shows the plasticity curves of the selected element during the process over the parameters stress triaxiality and the Lode angle, which are used to describe complex stress states.

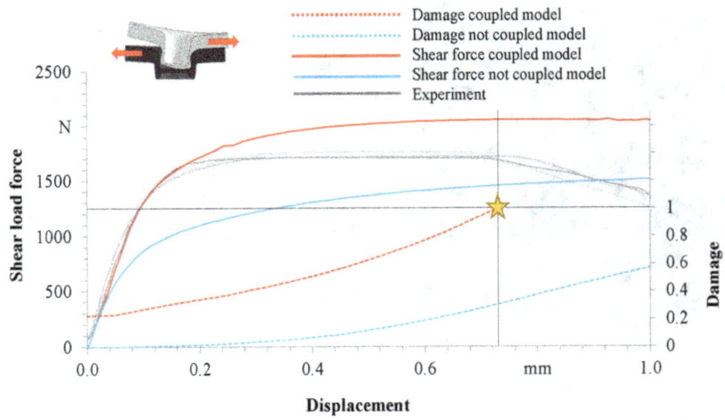

Fig. 6: Force curves during the shear load test and damage accumulation

The blue curve represents the progression of the uncoupled tensile shear test model. Therein, only the geometry (without strains, stresses, damage) was transferred from the clinching process model. The plasticity starts at a value of 0 and increases until the end of the simulation. Since the damage value of 1 is not reached, no failure will occur. The red curve describes the development of local plasticity with coupling to the clinching model, plastic strains, stresses, damage are transferred. Consequently, the red curve starts at an initial plasticity of approx. 0.3. Due to this pre-deformation, the damage accumulates higher, so that the element reaches a damage of 1 at a plasticity of 0.6089 and fails. The qualitative progressions of the uncoupled to the coupled model with regard to stress triaxiality and Lode angle are comparable. Initially, a tension-superimposed stress state is present in the element. In the further course of the test, the punch-sided sheet is pulled over the die-side sheet, which leads to a shear stress in the material, stress triaxiality around 0. The tilting of the specimen and an initial unbuttoning in the joining point (see failure mode in Fig. 4) again leads to a tensile superimposed state, stress triaxiality about 0.5. The green curve shows the plasticity curve of the element selected in the shear test model during the clinching process. The calculated properties (plastic strain, stress, damage) were then transferred to the coupled model (marked by black arrow). Fig. 6 shows the force-displacement curves of the shear tensile test of the coupled (red) and the uncoupled (blue) models. Furthermore, the damage accumulation of the models is also shown, at a value of D= 1, element deletion occurs and thus a failure of the joint. The uncoupled model underestimates the experimental force-displacement curve, since the strain hardening due to the plastic deformation that occurs in the clinching process is not taken into account. The damage starts at the value 0, since no pre-damage was transferred from the joining process. The damage accumulation in the uncoupled shear tensile test (blue) does not reach a damage value of 1 up to a displacement of 1.0 mm, thus no neck crack is predicted. The coupled model (red) overestimates the force curve of the experiment. Due to the simplified geometry of the FE shear tensile model shown here, a higher specimen stiffness results. Due to the missing total length of the specimen, the bending of the specimen parts characteristic for the specimen shape and the rotating clinch point are omitted, which leads to a simplified joint failure in the regular case. This simplification was adopted for this study, further investigation is required. The deviation of the maximum force can be attributed to the fact that the model does not show any material softening due to the damage. Furthermore, this can be attributed to the use of a quadratic von Mises yield criterion, which is known to overestimate the yield stress under shear for aluminum materials [19]. In the damage progression of the coupled shear strain model (red), an initial damage can be seen at the beginning. This increases due to the nonlinear Hosford-Coulomb damage accumulation during the shear tensile test until a damage value of 1 is reached. The element fails in the model, which subsequently leads to a neck crack. The simulated predicted failure displacement agrees with the experimental results.

Summary

In this study a modified FE clinching process model was presented, which does not require any remeshing and thus eliminates the influence of remeshing. The Hosford-Coulomb failure model identified in [10] by the modified punch test was implemented in the FE clinching process chain. The damage accumulation in the joining process as well as in the loading process could be determined. The influence of the pre-damage induced in the sheets during the joining process on the shear strength could also be shown. It is important to accumulate the material properties during the whole process chain in order to predict the failure location and time. Furthermore, it was shown that the calibrated Hosford-Coulomb model can predict the failure for the process shown. In further investigations, the material model will be extended to include a non-quadratic yield criterion as well as material softening due to ductile damage using gradient enhancement for appropriate regularization, as presented in [20]. Finally, the damage model is to be implemented in process chain models in order to predict the damage and failure behavior during a pre-operation, the joining process as well as more accurately during the loading phase.

Acknowledgement

This research was funded by the Deutsche Forschungsgemeinschaft, DFG, TRR 285 – project number 418701707 – TRR 285, subproject A01 and A05 is gratefully acknowledged.

References

[1] G. Meschut, V. Janzen, T. Olfermann, Innovative and Highly Productive Joining Technologies for Multi-Material Lightweight Car Body Structures, J. of Materi Eng and Perform 23 (2014) 1515–1523. https://doi.org/10.1007/s11665-014-0962-3

[2] R. Kupfer, D. Köhler, D. Römisch, S. Wituschek, L. Ewenz, J. Kalich, D. Weiß, B. Sadeghian, M. Busch, J. Krüger, M. Neuser, O. Grydin, M. Böhnke, C.-R. Bielak, J. Troschitz, Clinching of Aluminum Materials – Methods for the Continuous Characterization of Process, Microstructure and Properties, Journal of Advanced Joining Processes 5 (2022) 100108. https://doi.org/10.1016/j.jajp.2022.100108

[3] C.R. Bielak, M. Böhnke, R. Beck, M. Bobbert, G. Meschut, Numerical analysis of the robustness of clinching process considering the pre-forming of the parts, Journal of Advanced Joining Processes 3 (2021) 100038. https://doi.org/10.1016/j.jajp.2020.100038

[4] C.R. Bielak, M. Böhnke, M. Bobbert, G. Meschut, Development of a Numerical 3D Model for Analyzing Clinched Joints in Versatile Process Chains, in: K. Inal, J. Levesque, M. Worswick, C. Butcher (Eds.), NUMISHEET 2022, Springer International Publishing, Cham, 2022, pp. 165–172. https://doi.org/10.1007/978-3-031-06212-4_15

[5] E. Roux, P.-O. Bouchard, Kriging metamodel global optimization of clinching joining processes accounting for ductile damage, Journal of Materials Processing Technology 213 (2013) 1038–1047. https://doi.org/10.1016/j.jmatprotec.2013.01.018

[6] P.O. Bouchard, T. Laurent, L. Tollier, Numerical modeling of self-pierce riveting—From riveting process modeling down to structural analysis, Journal of Materials Processing Technology 202 (2008) 290–300. https://doi.org/10.1016/j.jmatprotec.2007.08.077

[7] M. Jäckel, T. Grimm, R. Niegsch, W.-G. Drossel, Overview of Current Challenges in Self-Pierce Riveting of Lightweight Materials, in: The 18th International Conference on Experimental Mechanics, MDPI, Basel Switzerland, p. 384.

[8] M. Otroshi, M. Rossel, G. Meschut, Stress state dependent damage modeling of self-pierce riveting process simulation using GISSMO damage model, Journal of Advanced Joining Processes 1 (2020) 100015. https://doi.org/10.1016/j.jajp.2020.100015

[9] A. Rusia, S. Weihe, Development of an end-to-end simulation process chain for prediction of self-piercing riveting joint geometry and strength, Journal of Manufacturing Processes 57 (2020) 519–532. https://doi.org/10.1016/j.jmapro.2020.07.004

[10]Böhnke M., Bielak C. R., Friedlein J. Bobbert M. Mergheim J., Meschut G., Steinmann P. (Ed.), A calibration method for damage modeling in clinching process simulations, The 20th International Conference on Sheet Metal, Nuremberg, 2023.

[11]D. Mohr, S.J. Marcadet, Micromechanically-motivated phenomenological Hosford–Coulomb model for predicting ductile fracture initiation at low stress triaxialities, International Journal of Solids and Structures 67-68 (2015) 40–55. https://doi.org/10.1016/j.ijsolstr.2015.02.024

[12]M. Böhnke, M. Rossel, C.R. Bielak, M. Bobbert, G. Meschut, Concept development of a method for identifying friction coefficients for the numerical simulation of clinching processes, Int J Adv Manuf Technol 118 (2022) 1627–1639. https://doi.org/10.1007/s00170-021-07986-4

[13]L. Sprave, A. Menzel, A large strain gradient-enhanced ductile damage model: finite element formulation, experiment and parameter identification, Acta Mech 231 (2020) 5159–5192. https://doi.org/10.1007/s00707-020-02786-5

[14]J. Friedlein, J. Mergheim, P. Steinmann, Efficient Gradient-Enhancement of Ductile Damage for Implicit Time Integration. Extended abstract of presentation, 16th LS-DYNA Forum 2022, Bamberg, Germany.

[15]M. Böhnke, F. Kappe, M. Bobbert, G. Meschut, Influence of various procedures for the determination of flow curves on the predictive accuracy of numerical simulations for mechanical joining processes, Materials Testing 63 (2021) 493–500. https://doi.org/10.1515/mt-2020-0082

[16]M. Dunand, D. Mohr, Effect of Lode parameter on plastic flow localization after proportional loading at low stress triaxialities, Journal of the Mechanics and Physics of Solids 66 (2014) 133–153. https://doi.org/10.1016/j.jmps.2014.01.008

[17]J. Papasidero, V. Doquet, D. Mohr, Ductile fracture of aluminum 2024-T351 under proportional and non-proportional multi-axial loading: Bao–Wierzbicki results revisited, International Journal of Solids and Structures 69-70 (2015) 459–474. https://doi.org/10.1016/j.ijsolstr.2015.05.006

[18]M. Nahrmann, A. Matzenmiller, Modelling of nonlocal damage and failure in ductile steel sheets under multiaxial loading, International Journal of Solids and Structures 232 (2021) 111166. https://doi.org/10.1016/j.ijsolstr.2021.111166

[19]A.M. Habraken, Modelling the plastic anisotropy of metals, ARCO 11 (2004) 3–96. https://doi.org/10.1007/BF02736210

[20]J. Friedlein, J. Mergheim, P. Steinmann, A Finite Plasticity Gradient-Damage Model for Sheet Metals during Forming and Clinching, KEM 883 (2021) 57–64. https://doi.org/10.4028/www.scientific.net/KEM.883.57

Sheet Metal 2023 | Materials Research Forum LLC
Materials Research Proceedings **25** (2023) 271-278 | https://doi.org/10.21741/9781644902417-34

A calibration method for failure modeling in clinching process simulations

Max Böhnke[1, a *], Christian Roman Bielak[1,b], Johannes Friedlein[2,c], Mathias Bobbert[1,d], Julia Mergheim[2,e], Gerson Meschut[1,f] and Paul Steinmann[2,g]

[1]Laboratory for Material and Joining Technology (LWF), Paderborn University, Pohlweg 47-49, 33098 Paderborn, Germany

[2]Institute of Applied Mechanics, Friedrich-Alexander-Universität Erlangen-Nürnberg, Egerlandstrasse 5, 91058 Erlangen, Germany

[a]max.boehnke@lwf.upb.de, [b]christian.bielak@lwf.upb.de, [c]johannes.friedlein@fau.de, [d]mathias.bobbert@lwf.upb.de, [e]julia.mergheim@fau.de, [f]meschut@lwf.upb.de, [g]paul.steinmann@fau.de

Keywords: Damage, Failure, Clinching

Abstract. In the numerical simulation of mechanical joining technologies such as clinching, the material modeling of the joining parts is of major importance. This includes modeling the damage and failure behavior of the materials in accordance with varying occurring stress states. This paper presents a calibration method of three different fracture models. The calibration of the models is done by use of experimental data from a modified punch test, tensile test and bulge test in order to map the occurring stress states from clinching processes and to precisely model the resulting failure behavior. Experimental investigations were carried out for an aluminum alloy EN AW-6014 in temper T4 and compared with the simulative results generated in LS-DYNA. The comparison of force-displacement curves and failure initiation shows that the Hosford–Coulomb model predicts the failure behavior for the material used and the tests applied with the best accuracy.

Introduction

Based on the multi-material design used in all industrial sectors today, the ongoing search for innovative construction methods puts high demands on joining technology. An efficient joining process for assembling different sheet metal materials is clinching, where the joint is created by solely cold forming the material without any auxiliary element [1]. In product development processes, numerical simulations can be used to reduce experimental effort and thus the product development cycles. In this context, accurate material modeling is of decisive importance. In addition to the plasticity and friction, the description of the damage and failure behavior of the material must also be taken into account. During the clinching process, the punch partially penetrates the sheet metal and causes a local metal flow of the material into the die. This generates a three-dimensional stress state in the parts to be joined [2].

The stress state dependent damaging and failure behavior of ductile metals in forming processes has been the research focus of many studies. In this regard, two mechanisms of damage have been identified as fundamental, depending on the stress state. Based on investigations of microscopic fracture surfaces in [3] it was found out that void growth takes place under hydrostatic tensile loading and void coalescence eventually leads to ductile failure. On the other hand, a different damage mechanism occurs for shear dominated loading, whereby voids are elongated and coalesced within a narrow band. In order to predict the failure behavior during deformation processes in ductile metal materials phenomenological failure models are widely used. Thereby, failure occurs if the equivalent plastic strain reaches a certain failure strain, which is derived from a predetermined relation as a function of the stress state. Widely known and frequently used stress state dependent failure models are the Johnson-Cook [4] and the Bao-Wierzbicki [5] model. Both

Content from this work may be used under the terms of the Creative Commons Attribution 3.0 license. Any further distribution of this work must maintain attribution to the author(s) and the title of the work, journal citation and DOI. Published under license by Materials Research Forum LLC.

are based on the stress triaxiality η, which is defined as the mean stress $\sigma_m = \frac{1}{3}(\sigma_1 + \sigma_2 + \sigma_3)$ divided by the von Mises equivalent stress $\sigma_{vM} = \sqrt{3J_2}$ leading to the dimensionless ratio expressed by Eq. 1, providing information on the stress state [6].

$$\eta = \frac{\sigma_m}{\sigma_{vM}} = \frac{\sigma_1 + \sigma_2 + \sigma_3}{3\sqrt{\frac{1}{2}[(\sigma_1-\sigma_2)^2+(\sigma_2-\sigma_3)^2+(\sigma_3-\sigma_1)^2]}} \text{ with } J_2 = \frac{1}{6}[(\sigma_1-\sigma_2)^2+(\sigma_2-\sigma_3)^2+(\sigma_3-\sigma_1)^2]. \quad (1)$$

With σ_1, σ_2 and σ_3 are the principal stresses with the assumption: $\sigma_1 \geq \sigma_2 \geq \sigma_3$. Developments and experimental observations in damage plasticity theory showed that the effect of stress state on fracture behavior could not be fully reflected by stress triaxiality alone [7]. Therefore, in addition to stress triaxiality η, the Lode angle parameter ξ must also be considered, which can be related to the third deviatoric stress invariant J_3 in following Eq. 2 [8]

$$\xi = \frac{27}{2}\frac{J_3}{\sigma_{vM}^3} = \frac{3\sqrt{3}}{2}\frac{J_3}{J_2^{3/2}} \text{ with } J_3 = (\sigma_1-\sigma_m)(\sigma_2-\sigma_m)(\sigma_3-\sigma_m). \quad (2)$$

The ongoing research on failure modeling has generated further models to describe the failure strain as a function of stress triaxiality and Lode angle. For example the Xue-Wierzbicki model, which is described in [7]. This model assumes that the fracture strain is always bounded by two limit cases corresponding to the axisymmetric stress state and the plane strain state. There is an abundance of experiential results showing a decrease of the failure strain with increasing stress triaxiality η. The most favorable condition for fracture, meaning the highest ductility is achieved in tension of round bars where the Lode angle parameter $\xi = 1$. This model is simple to calibrate and applied for aluminum material EN AA-2024 in [7].

Another well-known model is the modified Mohr-Coulomb model [9]. The model imposes a monotonic decrease in ductility for increasing stress triaxiality, which is coupled with an asymmetric function of the Lode angle. This model was calibrated for an EN AW-6016 aluminum alloy in [10]. In [11] a good predictability of the modified Mohr-Coulomb model was confirmed by comparing predictions to fracture experiments on TRIP780 steel.

A modification of this model named the Hosford-Coulomb model was proposed in [12]. Therein, the von Mises equivalent stress was replaced by the Hosford equivalent stress in combination with the normal stress acting on the plane of maximum shear. The Hosford-Coulomb model is based on the extensive study on 3D unit cells in [13], thus it has a strong micromechanical foundation.

Failure modeling in mechanical joining processes was presented in [14] by applying the GISSMO damage model in LS-DYNA to the self-piercing riveting (SPR) process simulation to predict ductile damage in thin aluminum sheets. The parameters for this stress state dependent material model have been identified from tensile tests with various characteristic stress triaxiality and Lode angle parameter values. Furthermore, the impact of different damage models on the accuracy of numerical joint strength prediction for the SPR process was studied systematically in [15]. An aluminum sheet – cast aluminum material combination was joined with SPR and the joint strengths were tested under various loading conditions. In [2] the stress states occurring during the clinching process of DC04 materials were investigated numerically. It was shown that in the failure critical necking stress triaxiality of $\eta = 0.3 - 0.7$ can occur, and Lode angle parameter $\xi = 0 - 1$.

For the experimental investigation of different stress states in order to identify damage parameters, a large variety of experimental tests and sample types exists. The punch test in several variants [16] has proven to be suitable, as it enables to generate different stress states. The experimental setup shows similarities to the clinching process, as it also consists of punch and die acting orthogonally to the sheet plane. However, the exact determination of the failure time or location has been problematic due to the lack of optical accessibility. Therefore, the development and functional proof of a modified punch test has been presented in [17]. This test rig allows the

Materials Research Proceedings **25** (2023) 271-278
https://doi.org/10.21741/9781644902417-34

application of defined stress conditions in different sheet materials using varying punch and die geometries. Due to the sectioned design and with the help of DIC, the deformation behavior as well as the crack initiation location and timing can be precisely determined. A corresponding numerical model has been explained and validated by comparison with experimental data.

In this study, a new method for calibrating failure models is presented. It mainly focuses on experimental data from modified punch tests and is intended to be useful for failure modeling in clinch simulations. A special focus is set on the modelling of the damage accumulation during the deformation of the material in order to be able to transfer it into a numerical process chain of clinching. For this purpose, three different damage models will be calibrated and simulated data will be compared with experimental results.

Materials and testing methods

The investigations in this study were carried out on an aluminum alloy EN AW-6014 T4 with a yield strength of 130 MPa and a minimum elongation A80 of 23 % [18]. The selected experimental tests for this study are shown in Fig. 1. It shows the modified punch test (MPT) in two different variants. One is a bar-shaped punch with blunt nose to induce shear stress states in the material. The second variant includes a punch with a hemispherical shape to achieve tensile superimposed stress states. The die used remained the same with a diameter of 11 mm and an edge radius of 0.3 mm. The detailed experimental setup and procedure can be found in [17]. The associated simulation models were validated using deformation behavior and force-displacement curves (without failure modeling). With the cross sectional design of the MPT, a biaxial stress state in the material cannot be achieved. However, this is essential for clinching processes. Therefore, data from a hydraulic bulge test was used as an additional interpolation point for the calibration. The bulge test is described in [19] and is characterized primarily by high plastic deformation with low frictional influence. The test generates an almost constant biaxial stress states with $\eta = 2/3$ and $\xi = -1$ in the material. Experimentally determined failure strains for the material under investigation were taken from [20] and amounts to $\overline{\varepsilon_f} = 0.71$. Furthermore, data from tensile tests were used for the calibration, since these represent the stressing of the material under uniaxial tension with $\eta = 1/3$ and $\xi = 1$ at the beginning and move over to biaxial stress states with $\eta = 2/3$ for increasing plastic deformation. These non-proportional load paths are also present in clinching processes [21]. Experimentally determined failure strains of tensile tests for the material were taken from [22].

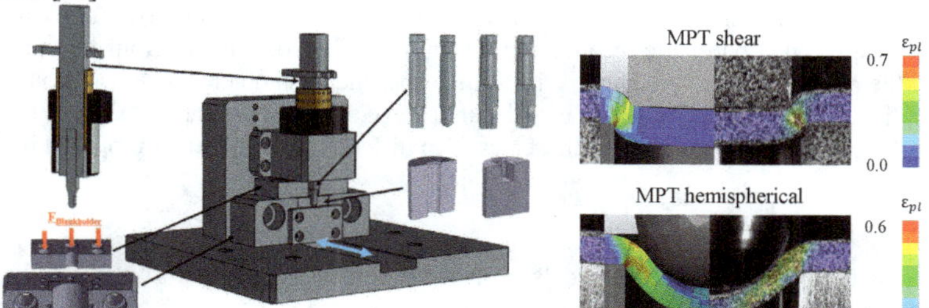

Fig. 1: Modified punch test for identifying damage parameters in sheet metal materials

Modelling and simulation

To model the failure behavior of the investigated material, three different models were calibrated in this study and their predictive ability was investigated. Firstly, the Xue-Wierzbicki model [7], which is described by Eq. 3.

$$\bar{\varepsilon}_f = c_1 e^{-c_2\eta} - (c_1 e^{-c_2\eta} - c_3 e^{-c_4\eta})(1-\xi^m)^{\frac{1}{m}}. \tag{3}$$

This model describes the failure strain $\bar{\varepsilon}_f$ as a function of stress triaxiality η and Lode angle parameter ξ as well as the four material parameters c_1 - c_4. The parameter m is an even integer closest to the value of *1/n*, which relates to the hardening exponent and is for the used material in this study taken as $m = 6.0$.

In addition, the modified Mohr-Coulomb model [9] was applied, which is described using Eq. 4.

$$\bar{\varepsilon}_f = \left\{\frac{A}{c_2}\left[\sqrt{\frac{1+c_1^2}{3}}\cos\left(\frac{\bar{\theta}\pi}{6}\right) + c_1\left(\eta + \frac{1}{3}sin\left(\frac{\bar{\theta}\pi}{6}\right)\right)\right]\right\}^{\frac{-1}{n}}. \tag{4}$$

Therein, the failure strain is described as a function of stress triaxiality η and an adapted Lode angle parameter $\bar{\theta}$, which derives from the Lode angle parameter ξ by Eq. 5.

$$\bar{\theta} = 1 - \frac{2}{\pi}\arccos\xi. \tag{5}$$

Furthermore, the model includes two material parameters c_1 - c_2 and the parameters A and n, which are related to the materials plasticity and are identified as $A = 355.15$ MPa and $n = 0.1548$. The third model to describe the failure behavior is the Hosford-Coulomb model [12]. It consists of the Lode angle dependent trigonometric functions $f_1 - f_3$ that are described by Eq. 6.

$$f_1 = \frac{2}{3}cos\left[\frac{\pi}{6}(1-\bar{\theta})\right] \; ; \; f_2 = \frac{2}{3}cos\left[\frac{\pi}{6}(3+\bar{\theta})\right] \; ; \; f_3 = -\frac{2}{3}cos\left[\frac{\pi}{6}(1+\bar{\theta})\right]. \tag{6}$$

The failure strain $\bar{\varepsilon}_f$ in the Hosford-Coulomb model is described by Eq. (7). The model includes three material parameters c_1 - c_3 and the strain hardening exponent $n = 0.1548$.

$$\bar{\varepsilon}_f = c_2\left(\frac{1+c_3}{g\,(\eta,\bar{\theta})}\right)^{\frac{1}{n}} \; ; \; g = \left[\frac{1}{2}(f_1-f_2)^{c_1} + \frac{1}{2}(f_2-f_3)^{c_1} + \frac{1}{2}(f_1-f_3)^{c_1}\right]^{\frac{1}{c_1}} + c_3(2\eta + f_1 + f_3). \tag{7}$$

For identification of the evaluation area and thus the calibration of the models, at first the most failure-critical element from each of the numerical models was evaluated in terms of plastic strain, stress triaxiality, and Lode angle according to the crack initiation determined from the experiments. To take the non-proportional loading paths of the experiments into account – also in view of the later application to clinching [21] – a non-linear damage accumulation law is applied, Eq. 8 with the damage exponent $n_d > 0$, herein chosen as 2. In differential form, this is proposed in [23].

$$dD = n_d\left[\frac{\bar{\varepsilon}_p}{\bar{\varepsilon}_f(\eta,\bar{\theta})}\right]^{n_d-1}\frac{d\bar{\varepsilon}_p}{\bar{\varepsilon}_f(\eta,\bar{\theta})}\;. \tag{8}$$

The material parameters of the models were adjusted using a nonlinear solver with respect to minimization of the mean squared error such that the damage value $D = 1$ occurs at exactly the time the crack initiation was determined in the experiment. An iterative approach was then applied for the calibration of the models. With a first failure model derived in this way, the numerical models of the MPT and tensile test were recalculated and the element with the highest accumulated damage selected and evaluated, in order to use the values determined from this recalculation again for the calibration. All simulations were run in LS-DYNA using implicit time integration. A hyperelastic-based finite plasticity formulation was implemented as user-defined material model including the presented stress-state dependent non-linear damage accumulation and failure by element deletion.

Results

The results of the models calibrated in this way are depicted in Fig. 2. This shows the curves for stress triaxiality and Lode angle over plastic strain in the evaluated element for the tensile test, MPT shear and MPT hemispherical. It is observable how the stress triaxiality in the tensile test starts at $\eta = 1/3$ but increases notably with increasing plastic deformation and necking of the specimen. The Lode angle starts at $\bar{\theta} = 1$ and progresses towards $\bar{\theta} = 0$. On the trajectories, the failure times ($D = 1$) predicted by the different models as well as experimentally determined are color coded. All models predict the failure time in the tensile test with high accuracy. The Xue-Wierzbicki model most closely matches the failure time for this experiment. The center plots show the stress triaxiality and Lode angle progressions for the MPT shear. The values run very consistently in the range $\eta = \bar{\theta} = 0$, which means that a pure shear stress state is well represented by this experiment. Again, all models agree precisely with the failure time from the experiments. The plots on the right in Fig. 2 depict the curves for the MPT hemispherical. In the evaluated element for this test, an inhomogeneous stress profile can be seen based on the stress triaxiality. The initially pressure-superimposed stress state changes to a tension-superimposed state as the plastic deformation progresses. This stress path is characteristic of the failure-relevant neck regions in clinched joints. The experimentally determined failure time is best predicted by the Hosford-Coulomb model. The discontinuous course is due to the finite time steps and discontinuously changing contact states in the simulation, which would require further refinement.

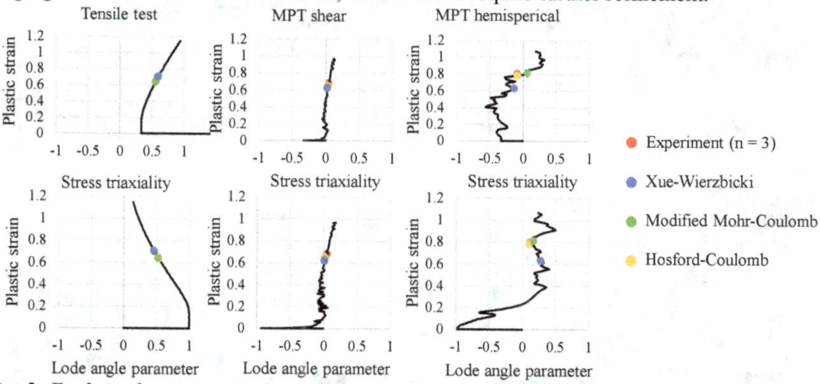

Fig. 2: Evaluated stress triaxiality and Lode angle for the experimental tests with failure times predicted using the different models

The averaged error over all tests is 4.3 % for the Xue-Wierzbicki model, 3.2 % for the modified Mohr-Coulomb model and 1.2 % for the Hosford-Coulomb model. The failure surfaces of the calibrated models with the associated material parameters can be seen in Fig. 3. It shows a dependence of the failure strain on the stress triaxiality for Lode angle values of -1 and 1, which flattens out in the region for Lode angle around 0. For all models, the failure strain decreases with increasing stress triaxiality. That means high failure strain for compression superimposed stress conditions and decreasing strain for tension superimposed stress conditions. The Xue-Wierzbicki model is symmetric with respect to Lode angle, whereas modified Mohr-Coulomb and Hosford-Coulomb are asymmetric, which renders a higher flexibility in Lode angle dependencies.

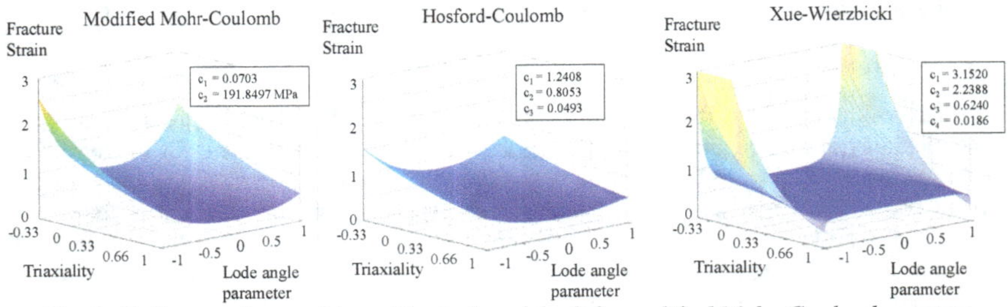

Fig. 3: Failure surfaces of the calibrated models: left: modified Mohr-Coulomb, center: Hosford-Coulomb and right: Xue-Wierzbicki

The calibrated Hosford-Coulomb model was implemented in the simulations of the MPT shear and MPT hemispherical experiments and compared with the experiments for verification. Fig. 4 shows the comparison of failure initiation and crack development in experiment and simulation. It shows that for both tests, crack initiation occurs at the same time at the identical location in experiment and simulation. Furthermore, the crack development over the further course towards complete fracture is already well predicted by the simulation as observed in the experiment.

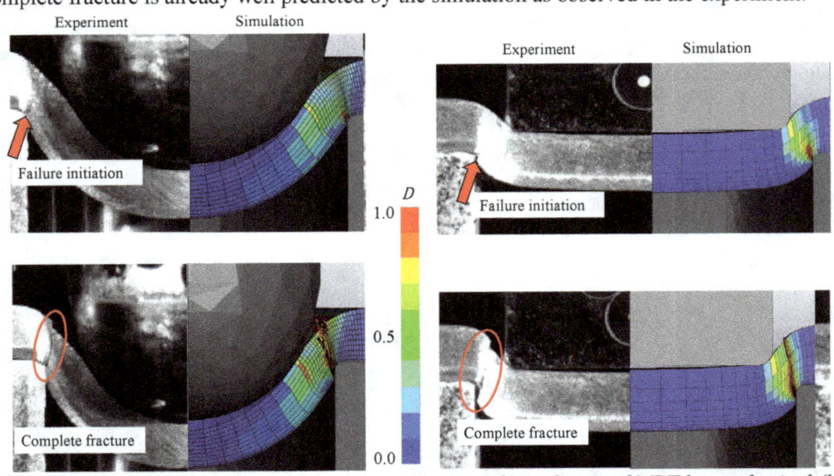

Fig. 4: Failure initiation and evolution in experiment and simulation of MPT hemispherical (left) and MPT shear (right) with implemented Hosford-Coulomb failure model

Fig. 5 shows the comparison of the simulated force-displacement curves with the experimentally determined values for the MPT shear and MPT hemispherical. For the MPT shear, the system response agrees well, especially the initial stiffness. The failure time is also predicted accurately. The discontinuous course can be optimized by decreasing the mesh size thus also stabilizing the contact. The deviation of the maximum force may be due to the lack of material softening effect in the model. Furthermore, this can be accounted to the use of a quadratic von Mises yield criteria, which overestimates the yield stress under shear for aluminum materials [24].

In the case of MPT hemispherical, the progression is essentially the same. However, a small offset of the force values can be seen, which is compensated for at the end of the curve.

The time of failure agrees well with the experimental curve. Based on these results, the calibrated and implemented failure model is verified and can be applied for further tests.

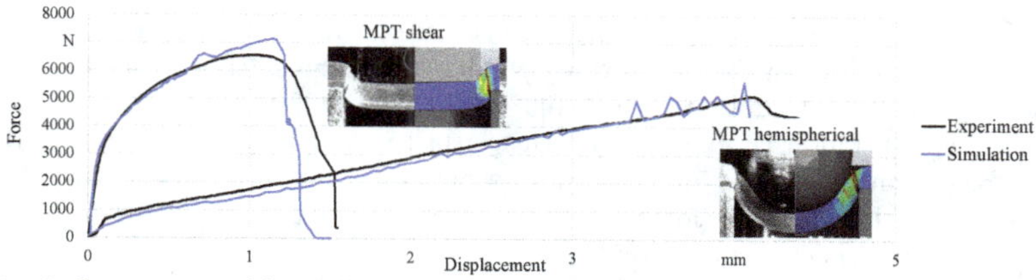

Fig. 5: Comparison of force-displacement curves between experiment and simulation of MPT shear and MPT hemispherical with implemented Hosford-Coulomb failure model

Summary and Outlook

In this study, a calibration method of failure models for the simulation of clinching was presented based on a modified punch test. Based on experimental tests and corresponding simulation models, failure-relevant data for an aluminum material were transferred into three different failure models and calibrated. The Hosford-Coulomb model was found to be highly predictive and is recommended for damage prognosis in clinching processes. The calibrated model was implemented in the simulations and verification was performed by comparing the failure behavior and force-displacement curves from experiment and simulation of the modified punch tests. The modified punch test proved helpful, because it allows a detailed observation of the failure initiation and allows the application of different relevant stress conditions in the sheet metal material.

For future investigations, even more variations of punch and die geometry of the punch test will be considered to cover additional stress conditions and to calibrate the model more accurately. Moreover, the material model will be extended by a non-quadratic yield criteria as well as material softening due to ductile damage using gradient-enhancement for proper regularization. Finally, the failure model will be implemented in clinching process models in order to accurately represent the damage and failure behavior during the joining process as well as during the loading phase.

Acknowledgement

Funded by the Deutsche Forschungsgemeinschaft (DFG, German Research Foundation) – TRR 285, subprojects A01 and A05 – Project-ID 418701707.

References

[1] DVS - Merkblatt DVS/EFB 3420: Clinching - basics (2018).

[2] M. Jäckel et al., Process-oriented Flow Curve Determination at Mechanical Joining, Procedia Manufacturing 47 (2020) 368–374. https://doi.org/10.1016/j.promfg.2020.04.289

[3] F. Bron, J. Besson, Simulation of the ductile tearing for two grades of 2024 aluminum alloy thin sheets, Engineering Fracture Mechanics 73 (2006) 1531–1552. https://doi.org/10.1016/j.engfracmech.2006.01.024

[4] G.R. Johnson, W.H. Cook, Fracture characteristics of three metals subjected to various strains, strain rates, temperatures and pressures, Eng. Frac. Mech. 21 (1985) 31–48. https://doi.org/10.1016/0013-7944(85)90052-9

[5] Y. Bao, T. Wierzbicki, On fracture locus in the equivalent strain and stress triaxiality space, International Journal of Mechanical Sciences 46 (2004) 81–98. https://doi.org/10.1016/j.ijmecsci.2004.02.006

[6] A.C. Mackenzie, J.W. Hancock, D.K. Brown, On the influence of state of stress on ductile failure initiation in high strength steels, Engineering Fracture Mechanics 9 (1977) 167–188. https://doi.org/10.1016/0013-7944(77)90062-5

[7] T. Wierzbicki, Y. Bao, Y.-W. Lee, Y. Bai, Calibration and evaluation of seven fracture models, International Journal of Mechanical Sciences 47 (2005) 719–743. https://doi.org/10.1016/j.ijmecsci.2005.03.003

[8] C. F. Guzmán, About the Lode angle infuence in ductile fracture, University of Liège, 2014.
[9] Y. Bai, T. Wierzbicki, Application of extended Mohr–Coulomb criterion to ductile fracture, Int J Fract 161 (2010) 1–20. https://doi.org/10.1007/s10704-009-9422-8
[10] H. Granum, D. Morin, T. Børvik, O.S. Hopperstad, Calibration of the modified Mohr-Coulomb fracture model by use of localization analyses for three tempers of an AA6016 aluminium alloy, International Journal of Mechanical Sciences 192 (2021) 106-122. https://doi.org/10.1016/j.ijmecsci.2020.106122
[11] M. Dunand, D. Mohr, On the predictive capabilities of the shear modified Gurson and the modified Mohr–Coulomb fracture models over a wide range of stress triaxialities and Lode angles, Journal of the Mechanics and Physics of Solids 59 (2011) 1374–1394. https://doi.org/10.1016/j.jmps.2011.04.006
[12] D. Mohr, S.J. Marcadet, Micromechanically-motivated phenomenological Hosford–Coulomb model for predicting ductile fracture initiation at low stress triaxialities, International Journal of Solids and Structures 67-68 (2015) 40–55. https://doi.org/10.1016/j.ijsolstr.2015.02.024
[13] M. Dunand, D. Mohr, Effect of Lode parameter on plastic flow localization after proportional loading at low stress triaxialities, J. Mech. Phy. Solids 66 (2014) 133–153. https://doi.org/10.1016/j.jmps.2014.01.008
[14] M. Otroshi, M. Rossel, G. Meschut, Stress state dependent damage modeling of self-pierce riveting process simulation using GISSMO damage model, J Adv Join Proc 1 (2020) 199-206. https://doi.org/10.1016/j.jajp.2020.100015
[15] A. Rusia, S. Weihe, Development of an end-to-end simulation process chain for prediction of self-piercing riveting joint geometry and strength, J. of Manu. Proc. 57 (2020) 519–532. https://doi.org/10.1016/j.jmapro.2020.07.004
[16] M. Buyuk, Development of A Tabulated Thermo-Viscoplastic Material Model with Regularized Failure for Dynamic Ductile Failure Prediction of Structures under Impact Loading: Thesis, 2013.
[17] M. Böhnke, C. Bielak, M. Bobbert, G. Meschut, Development of a Modified Punch Test for Investigating the Failure Behavior in Sheet Metal Materials, Proceedings of NUMISHEET 2022, pp. 575–584. https://doi.org/10.1007/978-3-031-06212-4_52
[18] Novelis Global Automotive, EN AW-6014 T4 - Novelis Advanz 6F - e170 (2019).
[19] R. Kupfer, D. Köhler, D. Römisch, S. Wituschek, L. Ewenz, J. Kalich et al., Clinching of Aluminum Materials – Methods for the Continuous Characterization of Process, Microstructure and Properties, Journal of Advanced Joining Processes 5 (2022) 100108. https://doi.org/10.1016/j.jajp.2022.100108
[20] S. Wituschek, M. Lechner, Material characterisation methods for a tumbling self-piercing riveting process, ESAFORM 2021 (2021). https://doi.org/10.25518/esaform21.398
[21] C.-R. Bielak, M. Böhnke, J. Friedlein, M. Bobbert, J. Mergheim, G. Meschut et al. (Eds.), Numerical analysis of damage modeling in clinching process chain simulation, 2023.
[22] M. Böhnke, F. Kappe, M. Bobbert, G. Meschut, Influence of various procedures for the determination of flow curves on the predictive accuracy of numerical simulations for mechanical joining processes, Materials Testing 63 (2021) 493–500. https://doi.org/10.1515/mt-2020-0082
[23] M. Nahrmann, A. Matzenmiller, Modelling of nonlocal damage and failure in ductile steel sheets under multiaxial loading, Int. J. Solids and Structures 232 (2021) 111-166. https://doi.org/10.1016/j.ijsolstr.2021.111166
[24] A.M. Habraken, Modelling the plastic anisotropy of metals, ARCO 11 (2004) 3–96. https://doi.org/10.1007/BF02736210

Presses and press tools

Friction behavior under magnetorheological lubricant in sheet metal forming process

Michele Brun[1,a], Enrico Simonetto[1,b*], Andrea Ghiotti[1, c] Stefania Bruschi[1,d]

[1]University of Padova, Dept. of Industrial Engineering, via Venezia 1, 35131, Padova, Italy

[a]michele.brun@phd.unipd.it, [b] enrico.simonetto.1@unipd.it, [c]andrea.ghiotti@unipd.it , [d]stefania.bruschi@unipd.it

*Corresponding Author: Dr. Enrico Simonetto

Keywords: Friction, Lubrication, Magneto-Rheological Fluid

Abstract. The increasingly high standards required in sheet metal forming industry, both geometrical and aesthetical, lead to continuous search of solutions to control the metal sheet flow during deformation. As alternative to traditional draw beads or hydraulically controlled segmented dies, the possibility of locally varying the material tangential speed by adapting the surface tribology at the interface between the blank and the blank holder represents a still unexplored scenario. The paper presents the feasibility analysis of the use of magneto-rheological (MR) fluids as lubricants in stamping, exploiting their ability to vary their rheological behaviour in response to external magnetic fields. To this aim, a new strip drawing test bench was developed to investigate the friction behaviour of MR fluids under different magnetic fields. The cold stamping of DC05 steel sheet was taken as reference case to investigate the influence of typical process parameters such as contact pressure, sliding speed and magnetic field.

Introduction

Magneto-rheological (MR) fluids are materials capable of reacting to applied magnetic fields by reversibly modifying their rheological behaviour and, typically, this change is manifested by an increase in the yield stress as an applied magnetic field increases [1]. MR fluids are basically made of three elements: the carrier fluid, iron particles and the stabilizing additives. The iron particles are dispersed in the carrier fluid and, if exposed to a magnetic field, they are aligned in chain structures that determine an increase in yield and a consequent greater resistance to wall-sliding in case of laminar flow. The additives are introduced to control the viscosity, the settling rate and the friction between the particles and avoid the thickening of the MR fluid after a defined number of off-duty cycles [2].

A review of the scientific literature reveals that MR fluids are mainly used in dampers for vibration control in civil structures [3], and automotive semi-active components, such as suspensions [4, 5], brakes [6], clutch [7] and differentials [8]. Despite their versatility and adaptability, MR fluids have seen few applications in metal forming, most of them related to the development of damping devices or actuations. In [9] MR fluids are applied to overcome the reverse load phenomena in blanking [9], while in [10] the same Authors demonstrate the possibility of enhancing the surface quality of the sheared surface in the final part. With regards to the capabilities of damping the dynamic phenomena, an MR device is implemented in turning of hard materials to suppress the cutting tool vibration, making the system stable and free from regenerative effect of the cutting process [11]. Moreover, MR dampers demonstrated their performances in reducing chatter phenomena during the cutting of thin-floor parts [12] and, consequently, improving the surface integrity of the final surfaces. With regards to deep drawing, a commercial MR damper was implemented in [13] and tested experimentally to retain constant the blank holder force during the process, defining the control system and his performance

Content from this work may be used under the terms of the Creative Commons Attribution 3.0 license. Any further distribution of this work must maintain attribution to the author(s) and the title of the work, journal citation and DOI. Published under license by Materials Research Forum LLC.

characteristics, in terms of response time and force errors. Recent studies proposed novel control algorithms [14, 15] capable to adjust the blank holder force with respect of the sheet draw-in to control the material flow.

The present paper aims at presenting a new approach for the control of the material flow in deep drawing, by using the MR fluid as lubricant and controlling the friction behaviour at the interface between the blank holder and the part. The effect of the tribological characteristics of the MR fluid was investigated by using a novel apparatus specifically designed and manufactured to exert different magnetic fields and evaluate the friction coefficient. The paper is organized into four parts: the former presents the characteristics of the selected MR fluid and the materials properties for both metal sheet and dies. The second part deals with the description of the magnetic tool design and numerical analysis to evaluate the magnetic field exerted to the MR fluid. The third part describes the overall experimental apparatus and the experimental plan. Finally, the last section presents the results of the experiments.

Material

The MR fluid selected for the test is the ARUS AMT-SMARTEC+®. It is a synthetic-based fluid with high yield strength to take up heavy loads and dissipate the energy with minimal consumption of power. The principal nominal characteristics are reported in Table 1.

Table 1. Properties of ARUS AMT-SMARTEC+®.

Viscosity (Pa*s) at 40 °C, calculated as sloop between 0-1200 s^{-1}	0.37
Density (g/cm^{3})	2.90
Solid content by weight (%)	81
Yield shear stress (kPa) at 140 kA/m	69
Operating temperature range (°C)	-20 - 150
Flash point (°C)	> 180

Table 2. Nominal chemical composition of AISI 1040 steel.

Chemical compositions in wt%							
C	Si (max)	Mn	P (max)	S (max)	Cr (max)	Ni (max)	Mo (max)
0.37 - 0.44	0.40	0.50 - 0.80	0.045	0.045	0.40	0.40	0.10

The punch was manufactured by using quenched AISI 1040 steel, which has a nominal surface hardness of 210 HB. Table 2 report the nominal chemical composition of the AISI 1040 steel while Fig 1a) and Fig. 1b) show the punch geometry. During the tests, the punch was in contact with the specimen through the active area, whose has an extension of 500 mm^2 (20 x 25 mm^2 surface). The active area was measured by means of an 3D surface profilometer to quantify the surface roughness. The profilometry of the measured area is reported in Fig. 2a), giving an average surface roughness S_a of 0.100(±0.012) µm.

The material of the specimen was the DC05 grade steel provided in 1 (±0.05) mm thick sheet, which is a low carbon steel commonly used in deep drawing processes. The chemical composition of the DC05 grade steel is summarized in Table 3, while Fig. 1c) shows the specimen geometry. The punch was made to slide in the central part of the testing zone, which was measured by a 3D surface profilometer, returning an average roughness S_a equal to 1.256 (±0.008) µm. The profilometry of the sheet surfaces is reported in Fig. 2b).

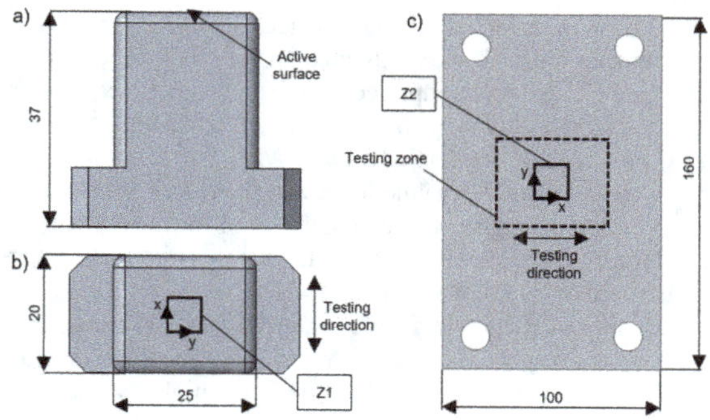

Fig. 1. Punch and specimen geometries: a) punch front view; b) punch plant view; c) specimen.

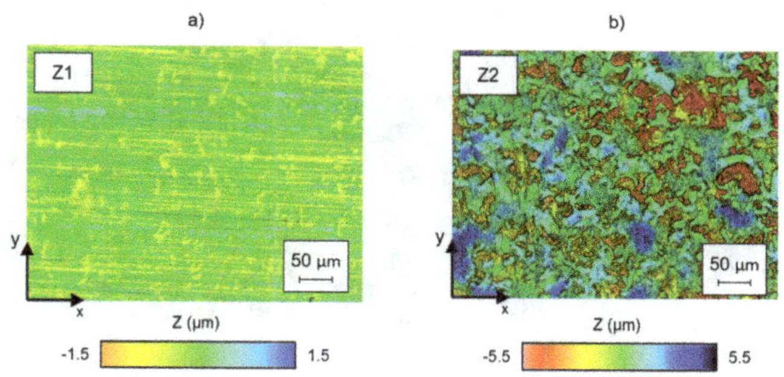

Fig. 2. Punch and specimen surface topographies: a) punch topography; b) specimen topography.

Table 3. Nominal chemical composition of the DC05 grade steel.

Chemical compositions in wt%			
Si (max)	Mn (max)	P (max)	S (max)
0.06	0.35	0.025	0.025

Table 4. Nominal chemical composition of the AISI 1015 steel.

Chemical compositions in wt%				
C	Si	Mn	S (max)	P (max)
0.13 – 0.18	0.15 – 0.35	0.30 – 0.60	0.045	0.045

Magnetic die

With the aim to expose the MR fluid to a controllable magnetic field, the die was designed and analyzed by means of ANSYS Maxwell software to evaluate the amplitude of the magnetic field *H*, which interests the MR fluid, see Fig. 3a).

Materials Research Proceedings **25** (2023) 281-288 https://doi.org/10.21741/9781644902417-35

It was made of a die base in annealed AISI 1040 steel on which a groove was manufactured. Tree magnetic cores with winding coils were mounted on the groove to produce the magnetic field when the winding coils are excited by an input current. The magnetic cores were manufactured in annealed AISI 1015 steel, whose chemical composition is reported in Table 4, while the winding coils were made of 70 turns each of enameled copper wire with a nominal diameter of 1.5 mm. The sheet specimen was mounted on the magnetic die and fixed with the blocker plate. The blocker plate used to fix the specimen during the test acted as a pool for the MR fluid. The blocker plate was manufactured in non-magnetic AISI 304 steel, to concentrate the magnetic field in the magnetorheological fluid volume.

Fig. 3b) and c) show the magnetic field distribution in the MR fluid along the longitudinal direction, for an input current respectively of 2.5 A and 5 A. The contour plots reveal an almost horizontal iso-line distribution of the magnetic field, with the higher values near to the surface of the specimen.

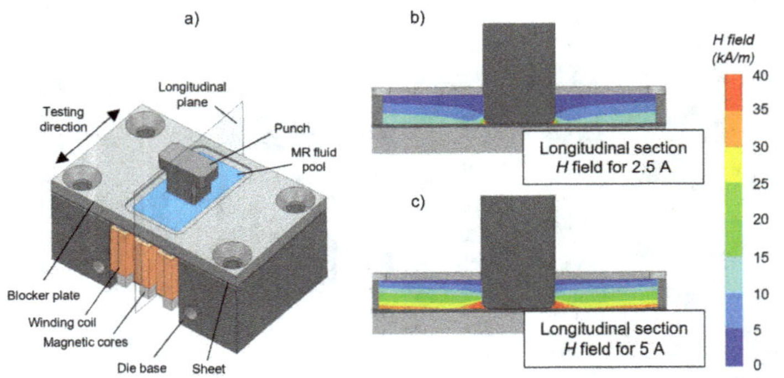

Fig. 3. Numerical model of the magnetic die and result: a) geometry implemented in ANSYS Maxwell; b) contour plot for an input current of 2.5 A; b) contour plot for an input current of 5 A.

Experimental

Fig. 4a) shows the overall experimental equipment, while Fig. 4b) reports a magnification of the magnetic die mounted on the equipment and connected to the power supply. The normal force F_N exerted by the punch was applied with a screw mechanism that embedded a piezoelectric load cell to monitor the normal pressure acting on the specimen.

During the test, the strip-sheet was fixed on the magnetic die and is made to slide against the punch. The sliding movement was transmitted to the carriage by using an electric actuator. The carriage is mounted on a linear roller guide with a negligible rolling friction coefficient, to avoid any disturbance in the tangential force F_T measurements. A load cell was mounted between the electric motor and the carriage to measure the tangential force F_T. The magnetic die was clamped on the carriage by means of a non-magnetic steel table, to avoid leakage of the magnetic field lines in the carriage material, hence enhancing the magnetic field applied to the MR fluid.

A dedicated power supply excited the winding coils of the magnetic dies with a maximum current of 5 A, which correspond 7.5 V.

Table 5 shows the experimental plan and the ranges of experimental parameters, namely normal pressure, sliding velocity and input current.

Materials Research Proceedings **25** (2023) 281-288
https://doi.org/10.21741/9781644902417-35

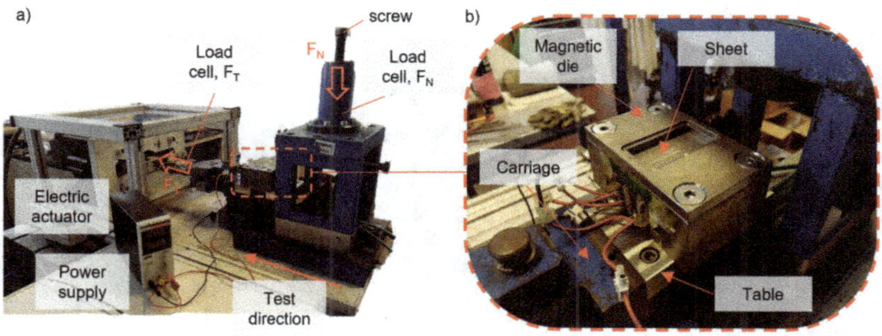

Fig. 4. Experimental equipment: a) overall view; b) detailed view of the magnetic die.

Table 5. Experimental plan for the strip drawing test

Normal pressure (MPa)	Sliding velocity (mm/s)	Input current (A)	Repeatability (-)
1, 5	5, 20	2.5, 5	3

Result and discussion

The coefficient of friction was calculated by using the Coulomb equation $\mu = F_T/F_N$ where F_T and F_N are respectively the tangential and normal forces measured by the embed piezoelectrical sensors.

Fig. 5 shows the experimental values of the friction coefficient with respect to the measured stroke for the investigated values of current, normal pressure and sliding velocity. The diagrams show an initial transit behavior for the first part of the stroke, then the coefficient of friction became stable after 10 mm. The exception is given in Fig. 5b), whereas the transit zone exceeds the first part of the stroke and the values of friction coefficient stabilize in the final part of the stroke. When the winding coils are powered with the input current, and the magnetic field is applied to the magnetorheological fluid, the coefficient of friction decreases from the initial values.

Fig. 6 shows the results of the friction coefficient according to the test parameters investigated in the experimental campaign. In the case of a sliding speed equal 5 mm/s and a normal pressure of 1 MPa the coefficients of friction are larger than the corresponding values investigated for a contact pressure of 5 MPa. For a normal pressure of 1 MPa, the coefficients of friction were respectively 0.294 (±0.021) at 0 A, 0.273 (±0.020) at 2.5 A, and 0.238 (±0.015) at 5 A. Rising the contact pressure up to 5 MPa, the coefficients of friction decrease to 0.280 (±0.017) at 0 A, 0.245 (±0.013) at 2.5 A and 0.215 (±0.016) at 5 A.

Regarding the test conducted with a speed of 20 mm/s, the tests reveal an opposite trend, with the highest friction coefficients in the case of contact pressure of 5 MPa. For a normal pressure of 1 MPa, the coefficient of friction ranges from 0.258 (±0.016) for a null input current, 0.220 (±0.019) in the case of an input current of 2.5 A and 0.198 (±0.019) for an input current of 5 A, while, in the case of a contact pressure of 5 MPa, the coefficient of friction is equal to 0.266 (±0.01) for a null current, 0.243 (±0.015) for a current of 2.5 A and 0.230 (±0.018) for a maximum current of 5 A. In these cases, the increase in the coefficient of friction is sharper for the contact pressure of 1 MPa, while, arising the contact pressure to 5 MPa, the coefficient of friction remains almost constant for 0 and 2.5 A, then decrease with a maximum input current of 5 A.

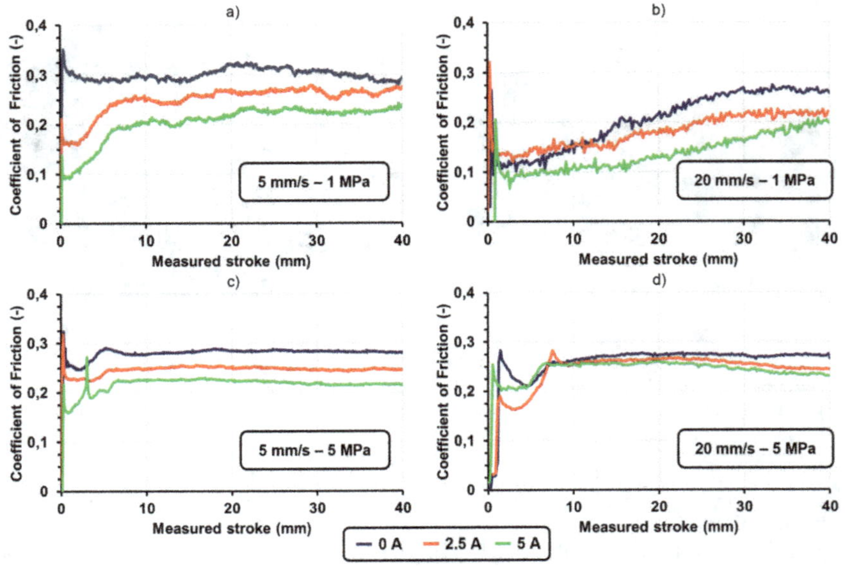

Fig. 5. Coefficient of friction vs measured stroke diagrams: a) 5 mm/s – 1 MPa; b) 20 mm/s – 1 MPa; c) 5 mm/s – 5 MPa; d) 20 mm/s – 5 MPa.

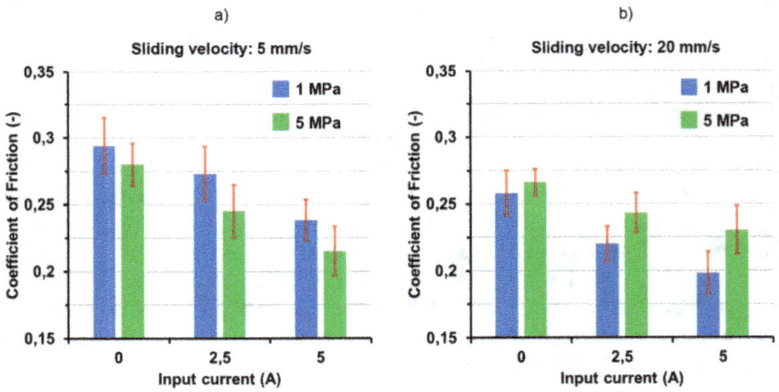

Fig. 6. Coefficient of friction values: a) at sliding velocity of 5 mm/s; b) at sliding velocity of 20 mm/s.

Finally, it is worth to notice how in both cases, higher values of applied DC led to lower COF. This can be related to the greater intensity of the magnetic field generated by higher DC, aligning the ferromagnetic particles dispersed in the ferrofluid with the magnetic lines, thus hindering the fluid flow between the tool and the sheet when the normal pressure is applied. In this way, the greater amount of lubricant at the interface can decrease the relative COF.

Conclusions

The present study aims at investigating the feasibility of using MR fluids as lubricants to vary the coefficient of friction in deep drawing. Moreover, by controlling the coefficient of friction becomes possible to adjust the restrain force on the blank during the draw-in. With this objective, a magnetic die was manufactured and installed in a strip drawing test equipment to apply controllable magnetic fields to the magnetorheological fluid.

To quantify the effects of the magnetic field on the MR fluid, the magnetic die was modeled in ANSYS Maxwell software. Two values of input current – respectively of 2.5 A and 5 A – were considered in the numerical analysis, to reproduce the test parameters.

From the experimental activities it was observed the dependency of the friction coefficient on the sliding speed and contact pressure. For the tests conducted at low speed, namely 5 mm/s, the coefficients of friction values were higher in the case of contact pressure of 1 MPa. The higher value equal to 0.294 was measured for a null current, and then it decreased as an input current was applied to the magnetic circuit, ranging to 0.273 and 0.238 respectively for an input current of 2.5 A and 5 A. By increasing the contact pressure up to 5 MPa, the values of the coefficient of friction became smaller. The coefficient of friction changed from 0.280 for a null current to 0.245 and 0.215 when a current of 2.5 A and 5 A was applied to the circuit. For the tests conducted at higher speed of 20 mm/s, the trend was opposite, with the lowest value in the case of contact pressure of 1 MPa. In this case, the coefficient of friction was equal to 0.258 for a null current, then increase to 0.220 and 0.2198 for input current of 2.5 A and 5 A. As the contact pressure was raised up to 5 MPa, the coefficient of friction was equal to 0.266 for a null current, 0.243 for an input current of 2.5 A and then decrease to 0.230 with a maximum input current of 5 A.

The results highlight a positive effect of a more intense magnetic field applied by increasing the coils supplied DC. A greater intensity of the applied magnetic field seems to hinder the lubricant flow under the dies improving the lubrication effects. Such results seem to open new possibilities for the in-process control of the sheet draw in applied to cold stamping processes.

References

[1] Mark R Jolly, Jonathan W. Bender and J. David Carlson, Properties and Applications of commercial Magnetorheological fluids, Journal of Intelligent Material Systems and Structures 10: 5 (1999). https://doi.org/10.1177/1045389X9901000102

[2] A. G. Olabi, A. Grunwald, Design and application of magneto-rheological fluid, Materials and Design, 28(2007) 2658-2664. https://doi.org/10.1016/j.matdes.2006.10.009

[3] G. Yang, B. F. Spencer Jr. J. D. Carlson, M. K. Sain, Large-scale MR fluid dampers: modelling and dynamic performance considerations, Engineering Structures 24 (2002) 309-323. https://doi.org/10.1016/S0141-0296(01)00097-9

[4] Jorge de Jesus Lozoya-Santos, Olivier Sename, Luc Dugard, Ruben Morales-Menendez, Ricardo Ramirez-Mendoza, A Semi-active Control-oriented Damper Model for an Automotive Suspension, IFAC Proceedings Volumes, 43: 7 (2010) 336-341. https://doi.org/10.3182/20100712-3-DE-2013.00158

[5] R. N. Yerrawar, R. R. Arakerimath, Development of Methodology for Semi Active Suspension System Using MR Damper, Materials Today: Proceedings 4 (2017) 9294-9303. https://doi.org/10.1016/j.matpr.2017.07.289

[6] Edward J. Park, Luis Falcao da Luz, Afzal Suleman, Multidisciplinary design optimization of an automotive magnetorheological brake design, computers and Structures 86 (2008) 207-216. https://doi.org/10.1016/j.compstruc.2007.01.035

[7] K. Hema Latha, P. Usha Sri, N. Seetharamaiah, Design and Manufacturing Aspects of Magneto-rheological Fluid (MRF) Clutch, Materials Today: Proceedings 4 (2017) 1525-1534. https://doi.org/10.1016/j.matpr.2017.01.175

[8] Antonio Lanzotti, Fabrizio Renno, Michele Russo, Riccardo Russo, Mario Terzo, Design and development of an automotive magnetorheological semi-active differential, Mechatronics 24 (2014) 426-435. https://doi.org/10.1016/j.mechatronics.2014.04.002

[9] A. Ghiotti, P. Regazzo, S. Bruschi, P.F. Bariani, Reduction of vibrations in blanking by MR dampers, CIRP Annals – Manufacturing Technology 59 (2010) 275-278. https://doi.org/10.1016/j.cirp.2010.03.111

[10] Andrea Ghiotti, Stefania Bruschi, Paolo Regazzo, Shear surface control in blanking by adaptronic systems, Procedia Engineering 81 (2014) 2512-2517. https://doi.org/10.1016/j.proeng.2014.10.359

[11] Rahul Kishore, Sounak K. Choudhury, Kashfull Orra. On-line control of machine tool vibration in turning operation using electro-magneto rheological damper, Journal of Manufacturing Processes 31 (2018) 187-198. https://doi.org/10.1016/j.jmapro.2017.11.015

[12] E. Diaz-Tena, L. N. Lopez de Lacalle Marcaide, F.J. Campa Gomez, D.L. Chaires Bocanegra, Use of Magnetorheological Fluids for Vibration Reduction on the Milling of Thin Floor Parts, Procedia Engineering 63 (2013) 835-842. https://doi.org/10.1016/j.proeng.2013.08.252

[13] Young-Won Yun, Jyung Sub Bae, Myeong-Kwan Park, A Study of the control of blank holding force using an MR damper in a drawing press, Journal of mechanical science and Technology, 24: 11 (2020) 2281-2288. https://doi.org/10.1007/s12206-010-0808-3

[14] Benny Endelt, Søren Tommerup, Joachim Danckert, A novel feedback control system – Controlling the material flow in deep drawing using distributed blank-holder force, Journal of Material Processing Technology 213 (2013) 36-50. https://doi.org/10.1016/j.jmatprotec.2012.08.003

[15] Martin Barthau, Mathias Liewald, New approach on controlling strain distribution manufactured in sheet metal components during deep drawing process, Procedia Engineering 207 (2017) 66-71. https://doi.org/10.1016/j.proeng.2017.10.740

Sheet Metal 2023
Materials Research Forum LLC
Materials Research Proceedings **25** (2023) 289-296
https://doi.org/10.21741/9781644902417-36

Numerical and experimental investigation on the applicability of elastomer tooling components for the manufacturing of undercut geometries by sheet metal forming

Michael Ott[1, a*], Yiran Li[1,b], Julian Krasselt[1,c], Florian Steinlehner[1,d], Patrick Haberkern[2,e], Albert Albers[2,f] and Wolfram Volk[1,g]

[1]Chair of Metal Forming and Casting, Technical University of Munich, Walther-Meissner-Strasse 4, 85748 Garching, Germany

[2]Institute of Product Engineering, Karlsruhe Institute of Technology, Kaiserstrasse 10, 76131 Karlsruhe, Germany

[a]michael.ott@utg.de, [b]yiran.li@tum.de, [c]julian.krasselt@tum.de, [d]florian.steinlehner@utg.de, [e]patrick.haberkern@kit.edu, [f]sekretariat@ipek.kit.edu, [g]wolfram.volk@utg.de

Keywords: Rapid Prototyping, Polymer, Metal Forming

Abstract. Due to their approximately hyperelastic properties, elastomers are suitable as a material for forming tools with extended forming capabilities. In this work, the use of two elastomer punches for manufacturing undercuts in sheet metal forming is performed experimentally for a demonstrator component. Since the manufacturing process does not require the use of cam slide units, it is aimed at rapid prototyping and small batch applications with the goal of reducing tooling cost and complexity. Additionally, the prediction accuracy of the elastomer tool deformation during forming was investigated in a finite element model of the manufacturing process. For comparison with the experimental implementation, punch force measurements and in-process optical strain measurements with a stereo camera system were carried out.

Introduction

The manufacturing of components with undercuts by sheet metal forming leads to challenges during the development of the tool concept. Forming operations that do not coincide with the direction of movement of the press ram are required. To perform such operations in industrial applications, cam slide units are used to deflect the direction of movement [1]. To enable the removability of the workpiece from the rigid tool after the forming step, they are commonly equipped with return springs [2]. As a design principle to keep costs and complexity low, the use of a minimum number of cam slide units for progressive forming and cutting operations is aimed at [2]. As an alternative approach to further reduce the number of cam slide units required, the formation of an undercut by reversible deformation of a punch made of an elastomer material is analysed. To prove the functionality of this approach, forming of an undercut without cam slide units is numerically evaluated and experimentally shown on a demonstrator component. This procedure is particularly aimed at prototypes, small batches and multiple variant designs.

State of the art

Current applications of elastomer tools in sheet metal forming. As alternatives to deep drawing with a rigid punch and a rigid die, DIN 8584-3 classifies deep drawing with a flexible punch and deep drawing with a flexible pad [3]. Known advantages of both approaches compared to tools made of steel are a cost-efficient and fast production of the tool components, a lower effort for the tool machining as well as a high achievable surface quality with no tool marks [4]. The disadvantages are an increased force requirement due to the additional deformation of the elastomer as well as the higher wear of the elastomer components [4]. Due to these properties, combinations of elastomer and rigid components are already widespread in industrial applications

Content from this work may be used under the terms of the Creative Commons Attribution 3.0 license. Any further distribution of this work must maintain attribution to the author(s) and the title of the work, journal citation and DOI. Published under license by Materials Research Forum LLC.

for the production of small batches. Elastomer materials commonly used for this purpose are often polyurethane-based due to their chemical resistance to the lubricants used in forming as well as their mechanical wear resistance, e.g. by [5]. For the Shore A hardness, values in the range 50 to 90 are common, e.g. by [5], [6] and [7]. The high achievable elastic elongation of polymer materials can be used to realize extended shaping possibilities. For example by [8], the design of two tool concepts is shown in which an elastomer punch is used to bulge a tubular part geometry. Another recent field of research is the development of elastomer forming tool components for the production of bipolar plates [9, 10].

Numerical simulation of elastomers under mechanical load. In order to be able to predict the material flow of the elastomer components during the development of the tools, the achievable accuracy of the numerical process simulation models is essential. Hyperelastic material models based on a strain energy density function are usually applied to model the material behaviour of elastomers. Examples of approaches implemented in widely used numerical simulation programs are the models according to Mooney-Rivlin [11], Valanis-Landel [12], Ogden [13] and Arruda–Boyce [14]. Furthermore, detailed modelling approaches for the friction in the contact between elastomer tool and sheet metal have been developed in [15] and [16]. For the comparison between numerical process simulation and real process, the distribution of the strains in the manufactured sheet metal components can be carried out by an optical deformation analysis [17].

Approach

A transfer of the principle used in tube bulging [8] to sheet metal forming for the locally defined formation of undercut features in the manufactured part is investigated. The approach is implemented experimentally for a demonstrator geometry. For this purpose, a symmetric undercut is formed in a pre-bent hat-shaped profile. With rigid tools, numerical models for the design of forming processes have already reached a high degree of maturity. In order to similarly test the usability for the design of the presented manufacturing of undercuts with elastic tools, a numerical model of the manufacturing process was created using the finite element method. The comparison of simulation and experimental implementation is performed by an optical stereo measurement of one side of the punch surface during the forming of the undercut. Complementary to [17], it is not the plastic strains in the finished sheet metal component that are observed here, but the elastic strains in the elastomer punch during the process.

Structure of the forming tool. To implement the forming of the demonstrator geometry, the tool concept shown in Fig. 1 was designed and manufactured. This forming tool was integrated into a tensile-compression testing machine, AllRound Line Z 150 Zwick Roell, Ulm, Germany, with a maximum force of 150 kN. As a semi-finished product, a hat-shaped profile ① is placed in the active element of the die ②. Subsequently the elastomer punch ③ is inserted. By screwing the blank holder ④ to the die cavity ⑤, a restraining force can be applied to the flange of the sheet during subsequent forming. The vertical downward movement of the upper steel punch component ⑥ results in a compressive load on the elastomer punch. A constant value of 5 mm/min was set as the speed of the upper steel punch in the experiments. The design of the manufacturing process is based on the fact that the Poisson's ratio of the applied elastomer material has a value close to 0.5. Thus the volume of the elastomer remains approximately constant under compressive load. Therefore, compression leads to lateral deflection of the elastomer, which results in the formation of an undercut in the vertical section of the hat-shaped profile. The springback of the elastomer to the initial configuration occurring during the subsequent upward movement of the punch enables the removal of the component. The open front side can optionally be closed with a transparent pane ⑦, which ensures that observability of the process is maintained.

Materials Research Proceedings **25** (2023) 289-296 https://doi.org/10.21741/9781644902417-36

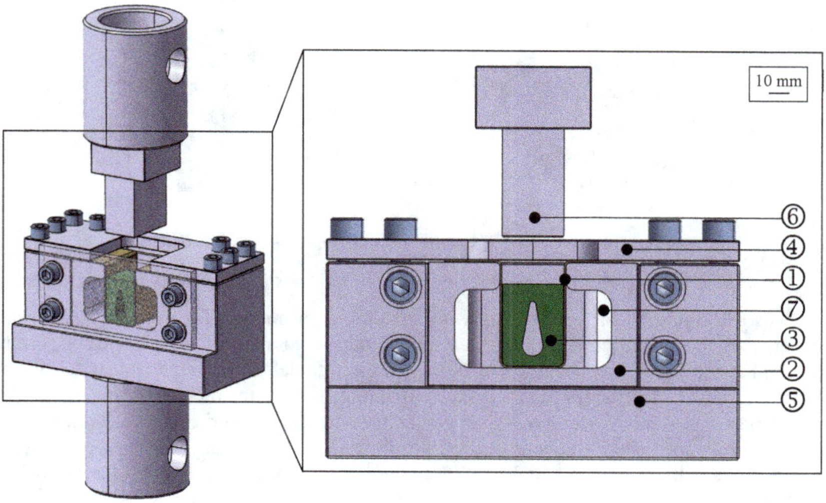

Fig. 1: Forming tool for forming of hat-shaped undercuts.

Preparation of the hat-shaped profile and the elastomer punch. The semi-finished geometry of the hat-shaped profile was bent on a Blech-Tec, Sauerlach, Germany, BT500 machine. DX56+Z100 with a sheet thickness of 0.8 mm and a non-grain oriented electrical steel sheet with a sheet thickness of 0.27 mm were used to test the suitability of the forming process for different materials and sheet thicknesses. Two different approaches were chosen for the creation of the elastomer punches. In the first approach, polyurethane with a Shore A hardness of 90 is cured through polyaddition of liquid two-component precursors. These are mixed and filled into a two-part mold of the punch. The mold was previously manufactured using the fused deposition modeling (FDM) process from polylactide (PLA) material. To reduce air inclusions in the elastomer, the polyurethane rubber is placed in a vacuum desiccator before curing and evacuated to 0.4 bar. In the second approach, additive manufacturing of the punch consisting of a rubber-like resin was carried out by photopolymer-based stereolithography (SLA) on an Anycubic Photon Mono X. Cleaning with isopropanol and subsequent curing with UV light were necessary as final manufacturing steps for the component generated by this method. As shown in Fig. 2, this punch contains a central opening to demonstrate the expanded shape capabilities of the direct additive manufacturing approach. Partial support structures, which could easily be removed manually afterwards, were required for creating the opening.

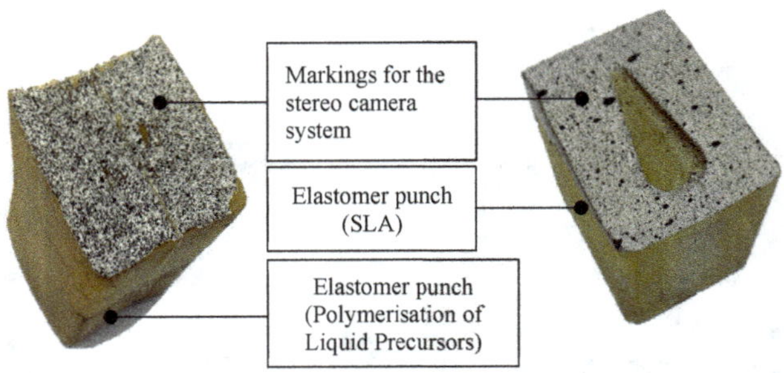

Fig. 2: Elastomer punches after preparation for the stereo camera measurements.

Sensor technology for comparison with the simulation model. The force acting on the steel punch and the displacements in the elastomer punch component are defined as criteria for evaluating the prediction accuracy of the simulation model. For comparison with the physical process, a load cell in the testing machine is used to determine the punch force versus the vertical punch position. A stereo camera system can be used to record the displacements during forming on the openly accessible front side of the elastomer. The GOM Aramis SRX system was used for this purpose with a frame rate of 1 Hz. For the preparation of the elastomers, a white colour base layer and a black stochastic dot pattern are coated on the front side of the punch, as shown in Fig. 2. From the acquired images, the strains in the marked area are calculated using Digital Image Correlation (DIC).

Structure of the simulation model. A numerical model of the manufacturing process was created in LS-DYNA using a Lagrangian finite element formulation as shown in Fig. 3. Following an explicit forming step, an implicit step was performed to calculate the springback of the sheet. Both the DX56 sheet and the electrical sheet were discretized by shell elements. For the modelling of the plastic material behaviour with the Hill48 model, data from previous tensile tests were available. For the non-linear stress-strain relation of the two elastomers, the hyperelastic material model according to Mooney-Rivlin (MAT_027 [18] in LS-DYNA) was used. The vertical punch movement during forming results predominantly in compressive stresses in the elastomer punch. Thus, for parameterization of the material model compression tests were carried out with the two materials on a Zwick Roell tensile-compression testing machine. Cylindrical specimens, produced under the same manufacturing conditions as the two elastomer punches, were used for this purpose. The resulting parameter values for MAT_027 are shown in Tab. 1, including the two empirically determined material constants of the Mooney-Rivlin model referred to as A and B.

Tab. 1: Shore A hardness and parameter values for the hyperelastic material model.

	Shore A hardness	**Mass density [g/cm³]**	**Poisson's ratio**	**Constant A**	**Constant B**
SLA material	95	1.1	0.495	4.82	-0.59
Polyurethane	90	1.0	0.495	4.63	-0.72

Materials Research Proceedings **25** (2023) 289-296
https://doi.org/10.21741/9781644902417-36

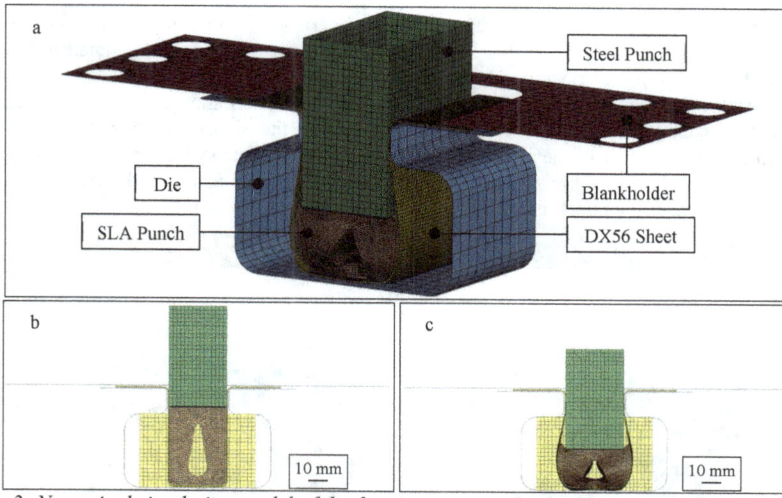

Fig. 3: Numerical simulation model of the forming process: Isometric view (a), front view with punch in top dead centre (b) and bottom dead centre position (c).

Results

Manufacturing of the demonstrator geometry with undercut. For both sheet materials investigated, the formation of a symmetrical undercut in the vertical section of the top hat profile was achieved with both of the punches. The resulting final part geometry after springback is shown in Fig. 4 for the DX56+Z100 sheet. A maximum overlap of 7 mm with the DX56 and 11 mm per side with the electrical sheet were achieved for the formed symmetrical undercut geometry. Compared to the initial width of the vertical sheet section of 30 mm, this corresponds to an increase in width of 47 % for the DX56 specimen and 73 % for the electrical sheet. In the bottom dead centre position, both punches were compressed by 20 mm. For the SLA-generated punch, this resulted in a maximum principal strain of 43.6 % measured by the stereo camera system. For the springback during the load release phase, both elastomers exhibit a viscoelastic behaviour. As shown in the image taken by one of the stereo cameras in Fig. 5, the elastomer punch does not follow the upward steel punch movement instantaneously. Instead it takes approximately 60 s for both punches to return to the initial geometry. To assess the durability of the tools, ten process repetitions were carried out. Neither significant wear on the two punches nor negative effects on the repeatability of the final sheet geometry were observed.

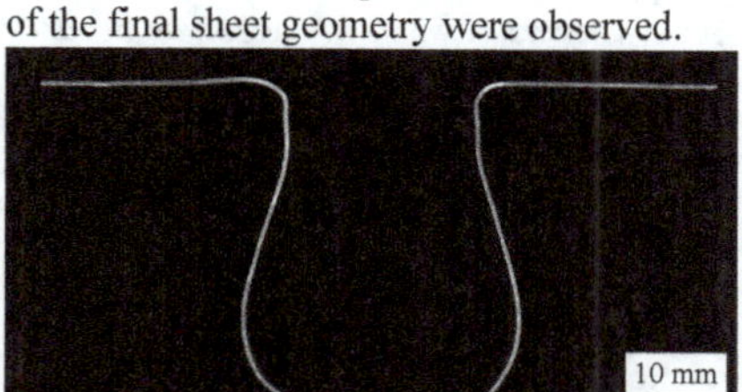

Fig. 4: Cross section of the resulting demonstrator geometry after springback (DX56+Z100).

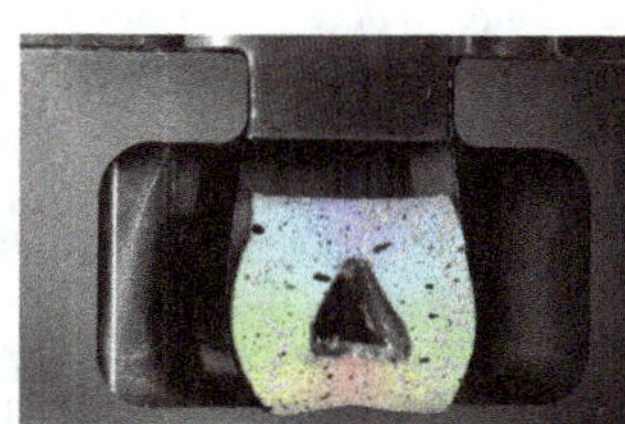

Fig. 5: Stereo camera image taken during the load release phase.

Comparison of the simulation model with the experimental implementation. Delamination of the colour coating occurred during forming at higher deformations for the punch produced by polymerisation of liquid precursors. Therefore, the comparison of simulation and measurements is performed for the SLA punch. For this configuration, the numerical model calculated stably up to reaching the lowest punch position examined in the experiment at a vertical displacement of -20 mm. As the first comparison criterion, the measured horizontal and vertical displacements in bottom dead centre position are shown in comparison to the simulation model in Fig. 6. An indication of the accuracy of the stereo camera measurements is the symmetry of the results. The numerical model shows the horizontal bulging of the elastomer. In addition, the formation of the upper two corners of the punch is predicted.

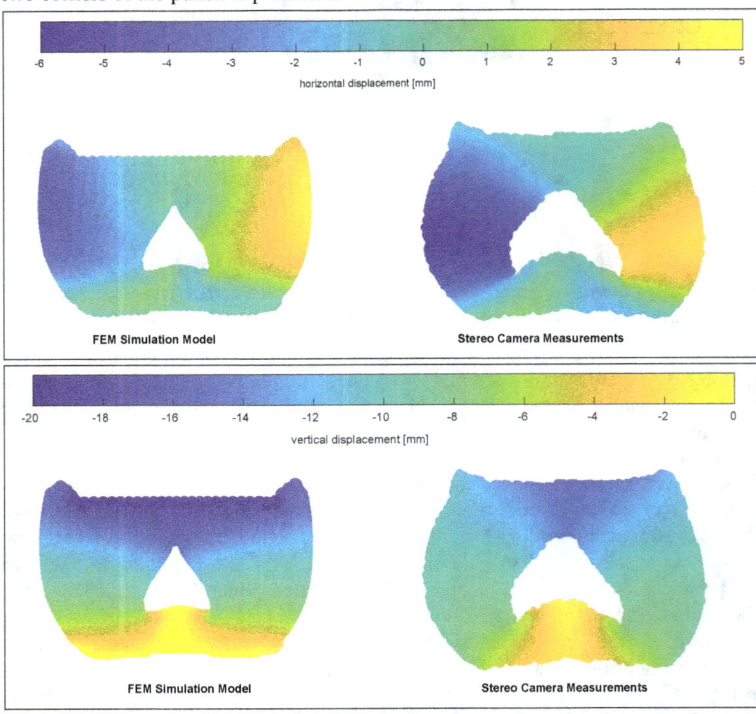

Fig. 6: Displacements (horizontal and vertical component) on the elastomer front side: calculation in the FEM simulation model and in-process measurements in the bottom dead centre position for the SLA-manufactured punch.

As the second comparison criterion, calculated and measured punch force - displacement curves are shown in Fig. 7. The initial height of the SLA punch was 40 mm. Up to a displacement of 10 mm, a good prediction accuracy of the simulation model is obtained. For larger deformations, the deviations increase. Further mesh refinement as a possible measure to increase the accuracy at high strains leads to a significant increase in the calculation time for the comparatively small elastomer punch. This would contradict the planned application goal of tool development for rapid prototyping applications. The main proportion of the force results from the deformation of the elastomer and only minor differences were measured when the experiment was performed without an inserted plate.

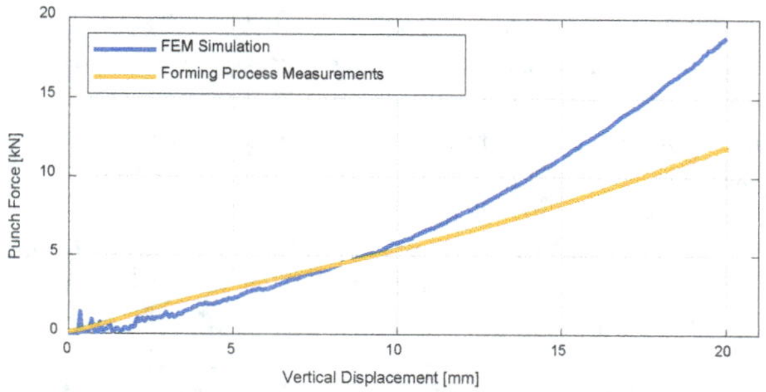

Fig. 7: Punch force - displacement curves: calculation in the FEM simulation model and measurement in the forming process.

Summary and Outlook

The experimental results obtained on the demonstrator geometry show the practical applicability of mechanical compression of an elastomer punch for local forming of undercuts. Overlaps up to 11 mm per side were achieved. Due to the almost complete springback of the punches to the initial configuration, removal of the parts from the forming tool set was made possible. The viscoelastic behaviour of the elastomer materials and the resulting time required for springback had a limiting effect on the achievable process times. However, this aspect often plays only a subordinate role for the intended application area of small batch production. Furthermore, the combination with the pre-bent hat-shaped profiles demonstrates the integration capability of the elastomer tools in multi-stage forming processes. Thus, the possibilities for prototyping can be expanded as in the manufacturing sequence shown in combination with swivel bending. Both the polymerisation of liquid precursor components in an additive manufactured mold and the direct additive manufacturing by the SLA process proved suitable for the production of elastomer tool components. Potentials of the second approach are an increased freedom of shape of the tool components and a further reduction of tool development times by eliminating the mold manufacturing.

The corresponding numerical process model shows stable behaviour despite the high strains in the elastomer during forming. Therefore it can be considered as an appropriate tool for simulative design and validation during the tool development. Both for the forces on the upper rigid punch component and for the in-process displacement measurements on the elastomer punch, there is good correlation up to punch displacements of 10 mm. For higher strains, the differences increased, especially for the predicted punch force. To further improve the agreement of the simulation and the experiment, more accurate models of the tribology between elastomer and sheet can be investigated. Space-fixed Eulerian meshing or the Element-Free Galerkin method can also be examined for applications with such high deformations of the elastomer in future work.

Acknowledgements

The authors gratefully acknowledge the German Research Foundation (Deutsche Forschungsgemeinschaft, DFG) for supporting this work carried out within the framework of the collaborative research project 431606085 (VO 1487/51-1).

References

[1] R. Canti, Analyse für eine belastungsgerechte Auslegung von Presswerkzeugen am Beispiel Niederhalter und Schiebersystem, PhD thesis, Technische Universität München, 2016.

[2] A. Birkert, S. Haage, M. Straub, Umformtechnische Herstellung komplexer Karosserieteile, Springer Berlin Heidelberg, Berlin, Heidelberg, 2013. https://doi.org/10.1007/978-3-662-46038-2

[3] DIN Deutsches Institut für Normung e.V., DIN 8584-3:2003-09, 2003.

[4] M. Ramezani, Z.M. Ripin, Rubber-pad forming processes: Technology and applications, Woodhead Publishing, 2012. https://doi.org/10.1533/9780857095497

[5] Y. Liu, L. Hua, J. Lan, X. Wei, Studies of the deformation styles of the rubber-pad forming process used for manufacturing metallic bipolar plates, J. Power Sources, 195 (2010) 8177–84. https://doi.org/10.1016/j.jpowsour.2010.06.078

[6] A. del Prete, G. Papadia, B. Manisi, Computer Aided Modelling of Rubber Pad Forming Process, KEM 473 (2011) 637–44. https://doi.org/10.4028/www.scientific.net/KEM.473.637

[7] L. Belhassen, L. Ben Said, S. Koubaa, M. Wali, Effects of Using Flexible Die Instead of Flexible Punch in Rubber Pad Forming Processm, in: M. Haddar, F. Chaari, A. Benamara, M. Chouchane, C. Karra, N. Aifaoui, (Eds.), Design and Modeling of Mechanical Systems-III, Springer International Publishing, Cham, 2018, pp. 259–267. https://doi.org/10.1007/978-3-319-66697-6_26

[8] Fibro GmbH, Hauptkatalog Update 2022, available on https://www.fibro.de/fileadmin/FIBRO/Blaetterkataloge/Hauptkatalog_DE/HTML5/1194/index.html, accessed 9 January 2023.

[9] Y. Liu, L. Hua, Fabrication of metallic bipolar plate for proton exchange membrane fuel cells by rubber pad forming, J. Power Sources 195 (2010) 3529–35. https://doi.org/10.1016/j.jpowsour.2009.12.046

[10] M. Elyasi, F.A. Khatir, M. Hosseinzadeh Manufacturing metallic bipolar plate fuel cells through rubber pad forming process, Int J Adv Manuf Technol 89 (2017) 3257–69. https://doi.org/10.1007/s00170-016-9297-6

[11] R.S. Rivlin, Large elastic deformations of isotropic materials IV. Further developments of the general theory, Phil. Trans. R. Soc. Lond. A 241 (1948) 379–97. https://doi.org/10.1098/rsta.1948.0024

[12] K.C. Valanis, R.F. Landel, The Strain-Energy Function of a Hyperelastic Material in Terms of the Extension Ratios, J. Appl. Phys. 38 (1967) 2997–3002. https://doi.org/10.1063/1.1710039

[13] R.W. Ogden, Large deformation isotropic elasticity – on the correlation of theory and experiment for incompressible rubberlike solids, Proc. R. Soc. Lond. A 326 (1972) 565–84. https://doi.org/10.1098/rspa.1972.0026

[14] E.M. Arruda, M.C. Boyce, A three-dimensional constitutive model for the large stretch behavior of rubber elastic materials, J. Mech. Phys. Solids, 41 (1993) 389–412. https://doi.org/10.1016/0022-5096(93)90013-6

[15] L. Deladi, Static friction in rubber-metal contacts with application to rubber pad forming processes, PhD thesis, University of Twente, 2006.

[16] M. Ramezani, Z.M. Ripin, R. Ahmad, Computer aided modelling of friction in rubber-pad forming process, J. Mater. Process. Technol. 209 (2009) 4925–34. https://doi.org/10.1016/j.jmatprotec.2009.01.015

[17] S. Kut, B. Niedziałek, Numerical And Experimental Analysis Of The Process Of Aviation Drawpiece Forming Using Rigid And Rubber Punch With Various Properties, Arch. Metall. Mater. 60 (2015) 1923–8. https://doi.org/10.1515/amm-2015-0327

[18] Livermore Software Technology, LS-DYNA Keyword User's Manual - Volume II Material Models (R13), available on https://www.dynasupport.com/manuals/ls-dyna-manuals, accessed 9 January 2023.

Sheet Metal 2023 Materials Research Forum LLC
Materials Research Proceedings **25** (2023) 297-304 https://doi.org/10.21741/9781644902417-37

Development of progressive tool system for ultrasonic vibration assisted microforming

Gandjar Kiswanto[1, a *], Hans Thiery Tjong[1, b], Siska Titik Dwiyati[1,c], Sugeng Supriadi[1,d], Wildan Zulfa Abdurrohman[1,e], Edward Joshua Patrianus Mendrofa[1,f] and Raditya Aryaputra Adityawarman[1,g]

[1]Department of Mechanical Engineering, Universitas Indonesia, Depok 16424, Indonesia

[a]gandjar_kiswanto@ui.ac.id, [b]hansthieryt@gmail.com, [c]siska.titik@ui.ac.id, [d]sugeng@eng.ui.ac.id, [e]wildan.zulfa@ui.ac.id, [f]emendrofa@gmail.com, [g]radityaaryaputra10@gmail.com

Keywords: Micro Forming, Tool, Ultrasonic Vibration

Abstract. The demand for biodegradable implants is increasing as their efficiency in reducing the number of removal surgery after the healing process. However, the fabrication of magnesium miniplate by microforming is still a challenge due to its mechanical properties at room temperature and the micro-scale of the implant's product. In this research, ultrasonic vibration is used to improve the production quality of micro-forming magnesium. The fabrication of miniplate magnesium includes micro blanking, micro punching, and micro stamping processes. The progressive system is adapted into the tool due to the constraint of design complexity. The simulation was conducted to validate the tool design and microforming process. The result of this research is a complete set of tool designs that is capable to integrate with Ultrasonic Vibration Assisted Microforming (UVAM) for producing miniplate implants with desired quality and accuracy. Furthermore, micro punching and microstamping stress decreased by around 30% when assisted with Ultrasonic Vibration.

Introduction

Osteosynthesis with a miniplate is a recovery method for a fractured bone which is carried out by making a bond between the fractured parts to be strong enough to hold and help in rejoining the bones. Orthopedic implants usually use titanium as the common material for miniplate, however, magnesium is superior due to its dissolving ability, therefore no extra surgery to remove the implant after the healing process is needed [1]. On the contrary, the magnesium miniplate fabrication by microforming is hard to be carried out because the forces and pressures causing cracks and fractures, especially at room temperature. The microcracking can spread to the overall structure causing damage and influencing the implant properties.

Currently, there have been researched regarding the ultrasonic technology that potentially can prevent and overcome the obstacles that exist in the character of magnesium. Ultrasonic Vibration Assisted Microforming (UVAM) embedded in the production system can help prevent cracking and produce a smoother surface roughness so that a more refined product is obtained with decreasing fracture rate and forming force [2], smaller grain and roughness [3]. In this research, the design of the tool for the miniplate manufacturing is done. The designing process will take account of the miniplate design, the constraints and the ability to be integrated with the UVA system, moreover the result will be simulated and analyzed on the stress reduction as expected.

Design of Progressive Tools

The designing of the tools will be conducted in few steps, starting from defining constraints, conceptual design, embodiment design and determining material of the components.

Content from this work may be used under the terms of the Creative Commons Attribution 3.0 license. Any further distribution of this work must maintain attribution to the author(s) and the title of the work, journal citation and DOI. Published under license by Materials Research Forum LLC.

Defining constraints: Product Design (Miniplate) and Machine Capacity & Working Area. The design of the product that should be obtained is shown in Fig 1. By considering the product functional and micro dimension-tolerance size, the tolerance of the product design that should be achieved also very tight which also a major challenge to be considered on the tool design.

The product is made of magnesium, in which the Ultimate Tensile Strength (UTS) is the main bottleneck quantified factor of forming. Magnesium has a very low density which make the product lightweight, however on room temperature, it is corrosive [4] and very brittle [5] which made it hard to form perfectly, especially in micro-scale with the high tendency to crack. The magnesium material structure features the Hexagonal Close-Packed (HCP) structure in which microcrack could happen with the mechanism discussed by Wang J et al. [6], Cauthen C et al. [7], Mordike et al. [8] that could be derived into more extensive crack in the structure overall. There are studies to optimize the forming process for magnesium, which includes the introduction of Ultrasonic Vibration Assistance to the microforming process. The assistance will help decrease the microcrack growth by less friction occurred and the thermal softening feature.

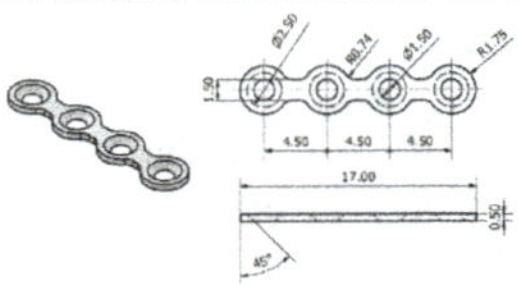

Figure 1 Miniplate design

The working envelope dimensions and capacity of the microforming machine should be taken into consideration when designing the tool with the ultrasonic vibration system integration. The microforming machine used in this study is the 5 kN microforming machine which was developed by the Manufacturing Laboratory, Department of Mechanical Engineering, Universitas Indonesia with an NX940MS-PS10-3 servo motor with a SCX10 motor controller as actuator. The tool together with UVA system will be mounted on the machine's bolster. The maximum area opening is 105 mm in diameter and 229 mm in height. [9]

Conceptual Design: Process Concept and the UVA System. Progressive tool system is chosen due to the complexity of the process. It is divided into two processes or strokes (illustrated in Fig. 2) to ensure the product quality and the process safety [9].

a. Blanking process: the making of main structure of the miniplate.
b. Punching with stamping: the making of four holes with chamfer 45°, resulting the product.

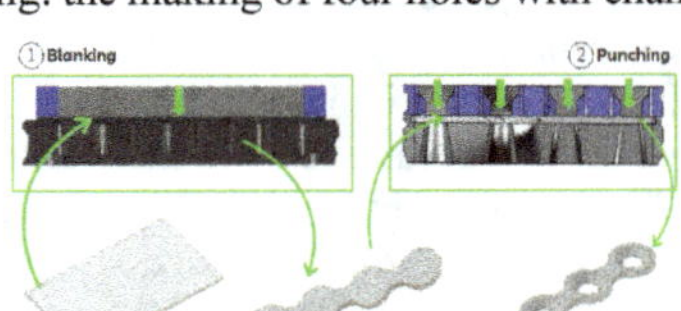

Figure 2 Step-by-step forming process of miniplate

The UVA research done by Hu et al. [2] and Liu et al. [3], analyzing the deformation of aluminum and copper material blanking process, shows that there is improvement on the surface roughness and smaller grain was achieved compared to the result without UVA, therefore similar UVA system is deployed in this research to analyze the differenced in quality and numerical effect. The same concept with the previous researches, including the determination of frequency and amplitude by the modal analysis to find the resonance frequency, moreover determine close-to-resonance frequency to achieve maximized longitudinal direction wave travel. The number will be

Materials Research Proceedings **25** (2023) 297-304 https://doi.org/10.21741/9781644902417-37

differed by the different shape, load and vibration mode. In this case, the longitudinal waves will be deployed on the UVA system which structured in such way (Fig. 3) that the vibration is inducted fully to the tool. Several studies have shown that the application of ultrasonic vibrations to the microforming process can increase the ratio of smooth zone with a polishing effect, reduce surface roughness [3] and improve product quality due to the cutting mechanism and thermal softening caused by temperature rise [10] as well as reduce forming forces because the friction between the workpiece and the tool is reduced [2]. Manufacturing on a micro scale is not the same as on a macro scale because the ratio between the dimensions and the microstructure changes when the scale is lowered. This reduction effect is called the size effect [11]. The size effect can be characterized as a comparison of the macro geometry to the microstructure, surface topography, and lubrication conditions, as well as parameters such as flow stress, forming limit, ultimate shearing strength, fracture, and roughness.[12]

Without the UVA, there will be more stress occurred during the process, especially on its peak which very potentially caused microcracking on magnesium material while will not be noticed happening during the process. The UVA system is expected not only to have impact on the forming force and stress or pressure, but also improve the product burr.

The UVA system deployed is the single longitudinal transducer with four stacks of piezoelectric ring clamped between the upper mass and lower mass which ensuring the vibration is channeled to the tool or punch that clamped to the lower mass, in this case, the Ultrasonic Vibration will work on the same axis of the forming process as seen illustrated on Fig. 3. The tool is attached to the UVA system perfectly, thus the vibration could be inducted completely without disturbance.

The designated UVA is integrated with the travel of tool set in the microforming machine, in which the wave occurred is the up and down travel on the main tool travel direction (moving down towards the die), therefore the UVA will move the tool set move back in negative and forward in positive in respect of the main travel axis along the process of machine movement in the main axis (tool set towards die), while without the UVA, the tool set will simply travel down towards the die according to the machine setting. Basically, with the UVA, by the ultrasonic wave, there will be enormous counts of tool-workpiece contacts, therefore causing heat for thermal softening, while also decrease the friction between the tool and workpiece, therefore the process become the multiple-micro-punch process, as if the accumulation of the sequence will spread the energy throughout the contacts and thus decrease the peak stress.

The use of UVA must follow the specified specifications, such as the frequency and amplitude determined by the simulation and trials so that the result should be obtained with no microcrack and burrs, which also being decreased in grain and roughness on the produced miniplate compared with without UVA, however the specific measurement is yet to be determined. The whole tool system with the UVA is illustrated in Fig. 3. The UVA will be activated throughout the whole process to ease difficulty in integrating ultrasonic timing and manufacturing procedure complexity.

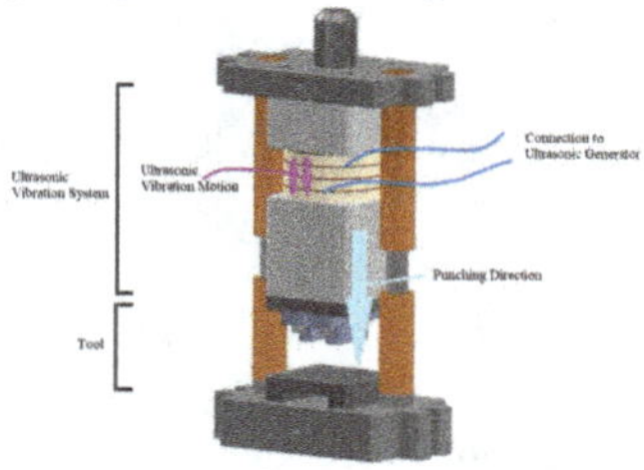

Figure 3 Tool System with UVA

Embodiment Design: Punches, Dies, Strippers and the materials. The detail design consists of components that build the whole tool system, which mainly including punch and dies [13]. The designing of the punch requires accuracy, functionality, and appropriate strength for each process. (illustration on Fig. 4)

a. Blanking process: the geometry is adapted from the main structure of the product design as it is the process of making the main structure, also including the dimension with subtraction adjustment of the clearance (around 10 μm).
b. Punching and stamping process: the geometry is adapted from the hole of product design with the slope in the stamping process, also including the dimension with summation adjustment of the clearance (around 10 μm).

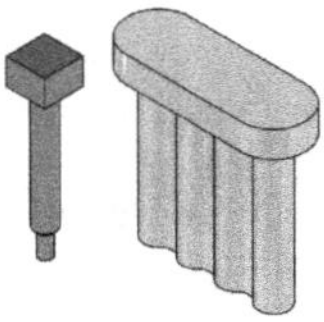

Figure 4 Punches Design, for punching (left) and blanking (right)

The punches also need to follow designing requirements, such as the maximum length (L_{max}) according to the Eq. 1 as follows [14]

$$L_{max} = \frac{\pi d}{8}\sqrt{\frac{Ed}{St}} \tag{1}$$

L_{max} : maximum length permitted [mm]
d : diameter [mm]
E : modulus elasticity of the workpiece material [MPa]
S : shear stress of the workpiece material [MPa]
t : thickness of workpiece material [mm]
$\frac{d}{t} = 1.10$ minimum

The design of the punches made is far from the limitation calculated from the maximum length, therefore fulfil the design fundamental.

The designing of the dies is done according to the system of the punch, as well as the strength for the functionality and the accessibility aspect, also including the adjustments of the clearance (around 10 μm). In this case the slot for the main structure to fall in the blanking process and the final product to be able to be taken from the dies slot. The thickness is also determined by the usability for handling the punching process and the material used for the punch and the dies itself. (illustration on Fig. 5)

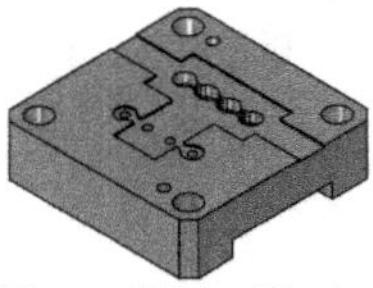

Figure 5 Dies Design

Polyurethane stripper is used for the design, aiming for releasing the scrap and workpiece from the punch, therefore the flexibility and forces are adjusted according to the constraints.

For the materials, plates, pins and the ring settings will be using S45C, while the punch and dies will be using SKD11 and stripper from polyurethane with respective properties as on Tab. 1.

Analysis

Calculation on Forces and Pressure. The shearing force for blanking and punching processes ($F_{punching}$) follows the Eq. 2 [14],

$$F_{punching} = L \times T \times \sigma = L \times T \times (0.7 \times UTS) \tag{2}$$

while for stamping process ($F_{stamping}$) follows the Eq. 3.

$$F_{stamping} = \frac{1}{\sqrt{5}} \times UTS \times A \tag{3}$$

L : punching perimeter [mm]
T : workpiece thickness [mm]
σ : workpiece material shear strength [MPa]
UTS : Ultimate Tensile Strength [MPa]
A : cross-section area [mm^2]

The resistant force by the stripper ($F_{stripper}$) also accounts for the total force: [14]

$$F_{stripper} = \frac{E.A.\Delta H}{H} \tag{4}$$

E : modulus elasticity of urethane [MPa]
A : cross-section area [mm^2]
ΔH : change of height [mm]
H : initial height [mm]

Following the equations, the blanking force including the striping force require around 3890 N, while punching and stamping of four holes approximately 1880 N and 2205 N respectively. The total forces for every process, blanking, punching and stamping with respective striping, have to be designed as progressive system as the compound system not only the complexity issue, but also the constraint of maximum force could be exerted by the machine (5000 N), while the compound system require at least 5770 N of force. Thus, the two-stroke process (progressive system) is chosen, besides the complexity of the design.

Stress Analysis and Buckling Simulation.The stress analysis and the buckling simulation is done using the Finite Element Analysis (FEA) software according to the maximum machine capacity force (5 kN) in actual dimension. The result (as seen in Fig. 6) is shown by color indicator for the stress distribution, from the lowest stress in blue and highest in red, in which on the punch (blanking on the left and punching on the right) have its highest stress on the red pointed spot and mostly focused on bottom part or the section in which will in contact with the workpiece. The same trend occurs to the die, in which the contact sections are the high stress area. The highest stress analyzed is around 1464 MPa (on the die) which is lower than the SKD11 material yield strength on the 1650 MPa. Therefore, with the maximum machine capacity of 5kN is still safe for usage in the SKD11 material (no buckling will occur).

Materials Research Proceedings **25** (2023) 297-304 https://doi.org/10.21741/9781644902417-37

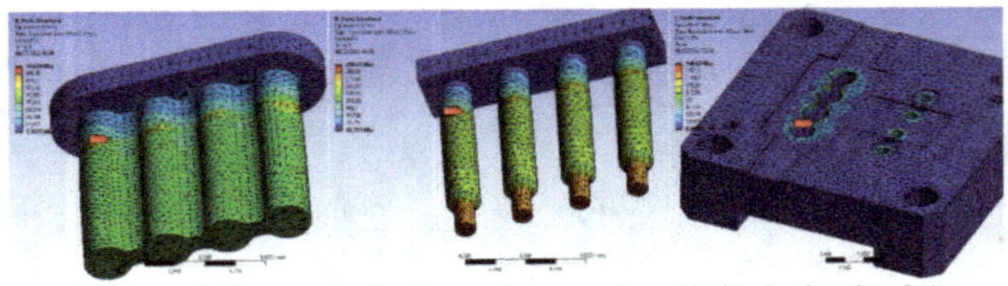

Figure 6 Stress analysis on the punches (left) & die (right)

Explicit Dynamic Simulation on the Tool. The explicit dynamic simulation is done using the Finite Element Analysis (FEA) software, simulating punching and stamping processes (stroke 2), done by the punch move towards the workpiece and die, with and without the Ultrasonic Vibration Assistance (UVA) at frequency of 37.8 kHz and amplitude of 2.5 μm, that were determined by the modal analysis and harmonic response of the whole system, with several adjustments strategy for the runtime such as the time scaling, meshing and erosion control. The material mode used in the simulation is as shown in Table 1. On the meshing strategy, it is determined by the smallest dimension tolerance or the smallest unit of the dimension for the focused part (die and punch) and rather large for only functional part (stripper), which is indicated on the simulated model (Fig. 7), while erosion control is retain default as it is not significantly effecting on the simulation process and time scaling is determined by 10 times faster. The simulation setting strategy is made for accuracy of the result as well as the simulation running time. The simulation was done only on one hole (the model shown as section view in Fig. 7) which already sufficient for substituting the whole process (half part equal substitutes) and the materials properties are shown in Table 1. The result will be shown in graph which value figures of one hole as it is similar in every hole, moreover the blanking process is also much more simple and lower in case of force and stress value, therefore is also considered substituted by the punching and stamping processes. In the simulation, the force and the stress data in the three processes that occurs on the miniplate is taken to be analyzed for both with and without assistance of Ultrasonic Vibration.

Figure 7 Section View of the simulated model with the meshing description

Table 1 Model Materials Properties for Simulation

SKD11 [15]				**Magnesium** [16]			
Properties		***Johnson-cook Strength***		***Properties***		***Steinberg-Guinan Strength***	
Density (ρ, kg/m^3)	8400	Initial yield stress (A, MPa)	1766	Density (ρ, kg/m^3)	1780	Initial yield stress (MPa)	190
Modulus of elasticity (E,GPa)	208	Hardening constant (B, MPa)	904	Specific heat (J/kg°C)	999	Maximum yield stress (MPa)	480
Poisson ratio	0.3	Hardening exponent (n)	0.39	Shear modulus (GPa)	16.5	Hardening constant (B)	1100
Bulk modulus (GPa)	173	Strain rate constant (C)	0.012			Hardening exponent	0.12
Shear modulus (GPa)	80	Thermal softening exponent	3.38			Derivative dG/dP G'P	1.6995
Thermal conductivity (W/m.k)	20.5	Melting temperature (K)	1733			Derivative dG/dT G'T	-8.4E+6
Thermal expansion (m/m.k)	11					Derivative dY/dP Y'P	0.0196
Specific heat (J/kg°C)	461					Melting temperature melt (°C)	876.85

Polyurethane [17]			
Properties		***Principal Stress Failure***	
Density (ρ, kg/m^3)	1265	Maximum Tensile Stress (MPa)	34,5
Bulk modulus (GPa)	2	Maximum Shear Stress (MPa)	1 E+17
Shear modulus (GPa)	5		

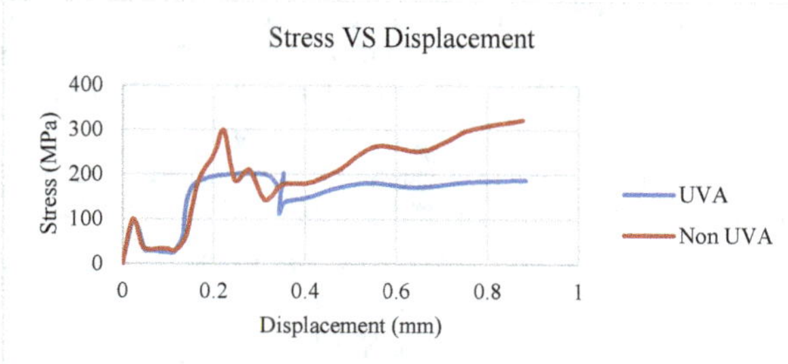

Figure 8 Graph of Tool Stress and Displacement without UVA and with UVA for one hole

It can be seen in Fig. 8 that according to the simulation, the use of UVA can reduce stress averagely around 30% on both punching and stamping (as seen on the graph) which is needed to carry out microforming processes, especially at its peak, however it may differ on the experiment as other factors also may take effect. From Fig. 8, it was found that the ultrasonic softening effect had an impact on the stress being lower, such in the punching and stamping processes as seen at the displacement of 0.2 mm until finish. The result achieved in this research also having the same trend of result done by Zhao et al. [18] which shows the effect on the stress-strain by having the ultrasonic vibration assisting the process. The UVA affect the stress as well as the force or the tool load which are lower than without UVA by respective average percentage. The cause of the lowering in stress and tool load is the mechanism of contacts accumulation between tool and workpiece, therefore the process is one with accumulated force and stress, while the magnesium material is also affected by the thermal softening that may take account on the material properties and characteristics.

The result also shows that the simulation and calculation by the equations match to each other with slight tolerated difference. On further research, experimental process can be executed and it is expected that by the use of UVA, there will be decrease in number of microcracks and burr occurred that could be monitored by visual control with visual magnifying device, throughout the whole microforming process, therefore producing more precise in tolerance product as well, however the specific number for the decrease is yet to be determined on the experiment result.

Summary

- The design of the punch and dies pass stress and buckling analysis on its UTS material (SKD11) and machinability (5kN), making it safe for use in its intended use and constraints.
- The effect of using UVA can be seen from the stress data which decreased by around 30%, as well as the quality of the resulting miniplate according to the simulation.
- This research will be continued for further analysis with the manufacturing and experimentation, therefore in this research is coped only up to the designing and simulating.

References

[1] S. Wiwanto, L.D. Sulistyani, F.D.E. Latief, S. Supriadi, B.P. Priosoeryanto, B.S. Latief, The experiment of magnesium ECAP miniplate as alternative biodegradable material (on male white New Zealand rabbits), in: 2018: p. 020013. https://doi.org/10.1063/1.5023947

[2] J. Hu, T. Shimizu, T. Yoshino, T. Shiratori, M. Yang, Ultrasonic dynamic impact effect on deformation of aluminum during micro-compression tests, J Mater Process Technol. 258 (2018) 144–154. https://doi.org/10.1016/j.jmatprotec.2018.03.021

[3] Y. Liu, C. Wang, H. Han, D. Shan, B. Guo, Investigation on effect of ultrasonic vibration on micro-blanking process of copper foil, The International Journal of Advanced Manufacturing Technology. 93 (2017) 2243–2249. https://doi.org/10.1007/s00170-017-0684-4

[4] S.V.S. Prasad, S.B. Prasad, K. Verma, R.K. Mishra, V. Kumar, S. Singh, The role and significance of Magnesium in modern day research-A review, Journal of Magnesium and Alloys. 10 (2022) 1–61. https://doi.org/10.1016/j.jma.2021.05.012

[5] M. Avedesian, H. Baker, eds., ASM Specialty Handbook: Magnesium and Magnesium Alloys, ASM International, n.d.

[6] J. Wang, N. Wang, X. Liu, J. Ding, X. Xia, X. Chen, W. Zhao, Compressive Deformation Behavior of Closed-Cell Micro-Pore Magnesium Composite Foam, Materials. 11 (2018) 731. https://doi.org/10.3390/ma11050731

[7] C. Cauthen, J. Pegues, J. Jr, M. Lugo, N. Shamsaei, Fatigue Micro-crack Growth and Modeling for Magnesium Alloys, 2015.

[8] B.L. Mordike, T. Ebert, Magnesium Properties-applications-potential, 2001. www.elsevier.com/locate/msea

[9] A. Mahmudah, G. Kiswanto, Development and Measurement of 5 kN μ-Forming Machine, Journal of Energy, Mechanical, Material and Manufacturing Engineering. 1 (2017). https://doi.org/10.22219/jemmme.v1i1.4474

[10] Y. Lou, X. Liu, J. He, M. Long, Ultrasonic-assisted extrusion of ZK60 Mg alloy micropins at room temperature, Ultrasonics. 83 (2018) 194–202. https://doi.org/10.1016/j.ultras.2017.03.012

[11] G. Kiswanto, S. Supriadi, S. Titik Dwiyati, Challenge in magnesium microforming, IOP Conf Ser Mater Sci Eng. 1070 (2021) 012121. https://doi.org/10.1088/1757-899X/1070/1/012121.

[12] A. Messner, U. Engel, R. Kals, F. Vollertsen, Size effect in the FE-simulation of micro-forming processes, J Mater Process Technol. 45 (1994) 371–376. https://doi.org/10.1016/0924-0136(94)90368-9

[13] A. Mahmudah, G. Kiswanto, D. Priadi, Fabrication of punch and die of micro-blanking tool, IOP Conf Ser Mater Sci Eng. 215 (2017) 012040. https://doi.org/10.1088/1757-899X/215/1/012040

[14] V. Boljanovic, Sheet Metal Forming Processes and Die Design Second Edition, 2019.

[15] J.L. Li, L.L. Jing, M. Chen, An FEM study on residual stresses induced by high-speed end-milling of hardened steel SKD11, J Mater Process Technol. 209 (2009) 4515–4520. https://doi.org/10.1016/j.jmatprotec.2008.10.042

[16] D.J. Steinberg, Equation of state and strength properties of selected materials, Lawrence Livermore National Laboratories, n.d.

[17] W. Toqueboeuf, B. Mortaigne, C. Cottenot, Dynamic Behaviour of Polycarbonate/Polyurethane Multi-Layer for Transparent Armor, Le Journal de Physique IV. 07 (1997) C3-499-C3-504. https://doi.org/10.1051/jp4:1997386

[18] J. Zhao, H. Su, C. Wu, The effect of ultrasonic vibration on stress-strain relations during compression tests of aluminum alloys, Journal of Materials Research and Technology. 9 (2020) 14895–14906. https://doi.org/10.1016/j.jmrt.2020.10.094

Processes

SheMet 2023

Sheet Metal 2023 Materials Research Forum LLC
Materials Research Proceedings **25** (2023) 307-314 https://doi.org/10.21741/9781644902417-38

Experimental study of the potential for extending the process limits in multi-stage forming of micro gears from sheet metal by a local short-term laser heat treatment

Miriam Leicht[1,a *], Martin Kraus[1,b,] Andreas Rohrmoser[1,c] and Marion Merklein[1,d]

[1]Friedrich-Alexander-Universität Erlangen-Nürnberg, Institute of Manufacturing Technology, Egerlandstraße 13, 91058 Erlangen, Germany

[a]miriam.leicht@fau.de, [b]martin.kraus@fau.de, [c]andreas.rohrmoser@fau.de, [d]marion.merklein@fau.de

Keywords: Micro Forming, Heat Treatment, Micro Gear

Abstract. Cold forming of micro gears is currently not possible for modules m < 0.2 mm due to the occurrence of high tool stresses, handling problems and size effects. A multi-stage process chain of micro bulk forming from sheet metal offers the potential to solve these challenges. However, the industrial applicability is limited due to material efficiency. Against this background, the aim of this research work is an improvement of the material utilization as well as a further reduction of the tool stresses. Therefore, the influence of different heat treatment strategies as well as the materials copper Cu-OFE and brass CuZn30 are analyzed. Thus, a local laser heat treatment and a global heat treatment of the blanks are investigated. The focus of this research work is on the first process stage, pin extrusion, which is used for forming of the wheel blank. Since this process setup provides the material for the further forming stages, it has a huge influence on the material efficiency and on the forming of the gearing in the second process stage. Therefore, the component and process properties are evaluated to assess the influence of the measures. The results prove higher material efficiency at a similar tool stress by using local short-term heat treatment.

Introduction

In times of scarcity of resources and energy, increasing global warming and the awareness of the necessity for reduction emissions, the efficiency of production methods and the performance of technical systems have gained even more importance. In order to solve these global challenges, a transformation in the industry is required. The product miniaturization as well as the increasing functional integration offer the potential for the saving of raw material, space, costs and energy. Due to a lower use of resources and benefits for applications, there is a rising demand for micro components [1].

Important micro parts for motion and power transmission are micro gears as components of micro driving systems, which are used in watch industry, medical applications and in aerospace technology [2]. Microdrive systems enable high-precision and backlash-free motion transmission in a small space [3]. Metallic micro gears are industrially produced by cutting processes such as milling, hobbing or electrical discharge machining [4]. Furthermore, lithographic manufacturing technology (LIGA) as well as micro metal powder injection molding (μ-MIM) are used for production of micro gears [4]. However, these technologies are limited due to productivity, material and energy efficiency, mechanical properties and producible gear geometries [2].

In view of the increasing demand for geared, metallic micro components, resource-efficient and productive manufacturing processes are gaining importance. In macro scale, cold forming of gears offers technological, economic and ecological benefits like high output volume, energy and material efficiency and improved component properties [5]. However, due to the occurrence of size effects [6], handling problems [7] and high tool load [4], the production of micro gears with modules m < 0.2 mm by cold forming is currently not industrially possible [2]. In Rohrmoser et

Content from this work may be used under the terms of the Creative Commons Attribution 3.0 license. Any further distribution of this work must maintain attribution to the author(s) and the title of the work, journal citation and DOI. Published under license by Materials Research Forum LLC.

al. [8], a new multi-stage process chain for forming micro gears from sheet metal was presented for the first time. In the first stage, a pin is extruded from sheet metal, which serves as a wheel blank for the further process stages. Subsequently, the pin is geared by lateral extrusion. In the last stage, the micro gear is separated from the sheet by shear cutting. By forming from sheet metal, which serves as a workpiece carrier, the handling and transport of the micro components is facilitated. Additionally, the multi-stage forming process causes grain refinement, which can reduce size effects [8]. In addition, the die stress is significantly lower compared to single full-forward extrusion [9]. Furthermore, the applicability of the process chain for the materials copper Cu-OFE and brass CuZn30 has already been demonstrated. However, a limited material utilization is one of the existing challenges, which currently limit an industrial establishment. Furthermore, the forming of cold-rolled materials is not possible due to high tool stress. An approach of sheet metal forming to extend these process limitations is the application of a local short-term heat treatment in the forming zone. The local modification of material properties is a promising method to control the material flow purposefully. In addition, the hardness gradient within the circular blank offers the potential to achieve an increase in material utilization, by limiting the lateral flow of material outwards.

Objectives and methodology

The objective of this study is to determine the influence of a local short-term heat treatment on a multi-stage bulk micro forming process from sheet metal for manufacturing micro gears with a module m = 0.1 mm. For this purpose, the type of heat treatment is varied by a global heat treatment of the semi-finished products as well as by a local laser short-term heat treatment in the forming zone. Furthermore, the influence of material using the single-phase copper Cu-OFE and the two-phase brass CuZn30 is evaluated. The focus is on the first process stage, the pin extrusion, which determines the material efficiency of the process chain and the quality of the wheel blank. For this purpose, the resulting component and process properties are analyzed. The evaluated component properties are the pin height and funnel depth, as well as the micro hardness distribution and grain structure. For the analysis of the process properties, the force-displacement curves are investigated. Additionally, the process results of both materials are compared to ensure their transferability.

Materials and experimental setup

Setup for local short-term heat treatment. For local modification of material properties, a short-term laser heat treatment is used. The process setup is depicted in Fig. 1. The blanks are heat treated with a diode laser Laserline LDM3000-100, which emits light at a wavelength of 900 to 1070 nm with a maximum power of 3 kW. The beam is focused with the optic OTS-5, which has a round spot geometry with a diameter of 4 mm. The laser unit is attached to a multi–axial robot arm from Kuka AG to enable precise positioning of the laser spot. To increase the absorption rate, the surface of the blanks is coated with a thin layer of graphite for a uniform emission coefficient of $\varepsilon = 0.95$ [10]. The rotationally symmetrical specimens are positioned with the help of a stencil plate. An infrared camera of type FLIR SC 7600 is installed to control and monitor the temperature distribution along the blank during the laser process.

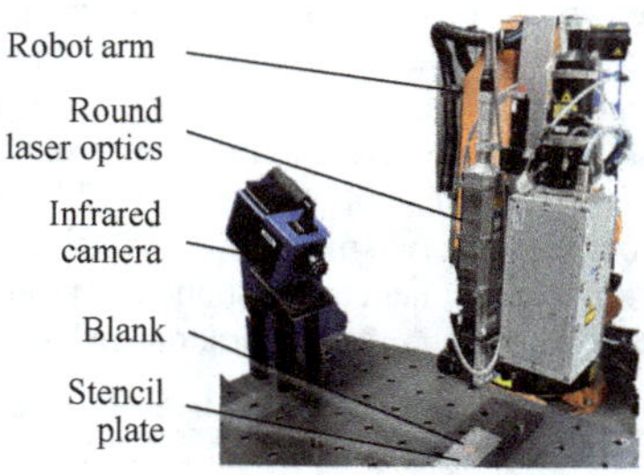

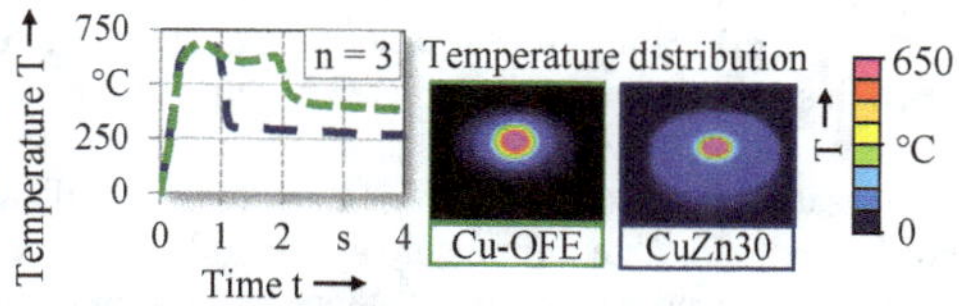

Laser parameters (power-controlled)		
	Cu-OFE	CuZn30
Laser power	0.9 kW (30 %)	0.6 kW (20 %)
Holding time	0.8 s	1.75 s

Figure 1: Setup for local short-term heat treatment.

Materials and setup for global heat treatment. The sheet materials used in this work are the single-phase copper Cu-OFE and the brass CuZn30. The oxygen-free Cu-OFE is frequently used in electronics and in vacuum technology. In addition, the high-purity copper is well suited as a model material for fundamental research of micro forming processes. Brass CuZn30 is mainly used in metal goods industry as well as in mechanical and electrical engineering. Both materials exhibit excellent cold formability. However, the high zinc content of the brass increases mechanical strength and decreases thermal conductivity. Thermal conductivity has a major influence on the local short-term laser heat treatment process. For copper, the thermal conductivity is 394 W/mK. Brass, at 121 W/mK, has only 30 % of the thermal conductivity of copper.

In this study, the materials are analyzed in global and local heat-treated condition. On delivery, the materials were in cold-rolled condition with a high pre-hardening of 86 % for Cu-OFE and 66 % for CuZn30. For global heat treatment of the entire blank, a tube furnace NR 40/11 from Nabertherm GmbH with inter gas atmosphere (argon) is used. The temperature of 650°C was held for one hour. Then the specimens were cooled slowly in the furnace. In Fig. 2, the grain structure in cold-rolled and global heat-treated state are shown. Through the heat treatment, a homogeneous grain structure with large grains is achieved over the entire blank. To characterize the mechanical properties and the flow behavior, uniaxial tensile tests were carried out in rolling direction according to DIN EN ISO 6892-1. The experimental flow curves are presented in Fig. 2.

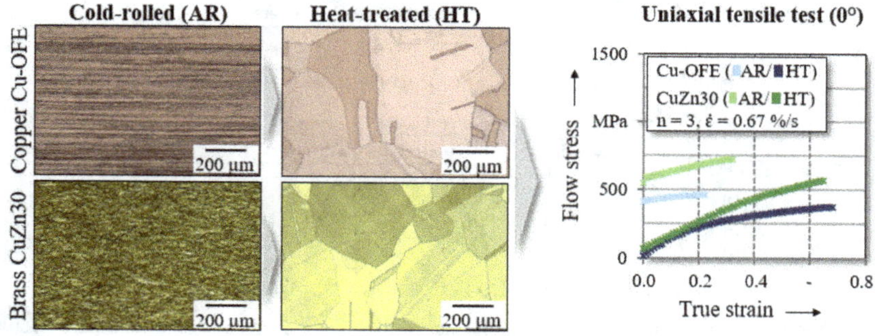

Figure 2: Grain structure and flow behavior of Cu-OFE and CuZn30 in cold-rolled and global heat-treated condition.

The flow curves show a behavior typical of the material, with the global heat treatment resulting in a reduction in strength and an increase in elongation at fracture. As a further heat treatment strategy, a short-term laser heat treatment is investigated in order to locally influence the material structure. The efficiency of this strategy is qualitatively evaluated based on the grain size change of the grain structure, which is depicted in Fig. 3. In the area of local heat treatment, uniform, small grains are located, which replace the grains deformed by the rolling process through recrystallization processes and grain growth. This indicates a decrease in strength in this area. In the border area elongated grains are still visible as in the cold-rolled condition, suggesting higher strength in this zone. Furthermore, there is a difference between the two materials in the size of the heat-treated zone. In case of copper, the heat-treated zone has a diameter of about 4 mm. For brass, a diameter of about 6 mm is identified, thus the heat-treated zone is more pronounced.

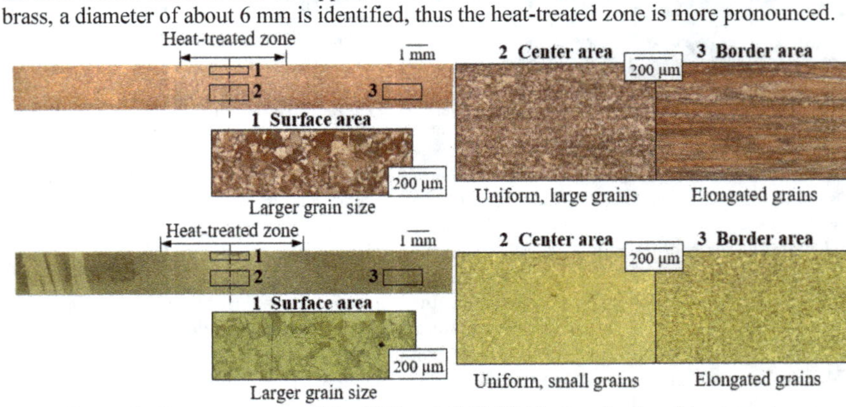

Figure 3: Grain structure of Cu-OFE and CuZn30 blanks after local heat treatment.

Process setup for pin extrusion. For the manufacture of micro gears, a multi-action tool system is required, which is installed in a universal testing machine Walter+Bai type FS 300 with a maximum testing force of 300 kN. The process scheme and the parameters are illustrated in Fig. 4. The extrusion punch and the ejector are driven by hydraulic cylinders. The blank holder is controlled by the universal testing machine, which prevents a lifting of the blanks and reduce the radial material flow outwards in the direction of the sheet metal. The blank holder pressure is set material-specifically in order to prevent plastic deformation due to the blank holder.

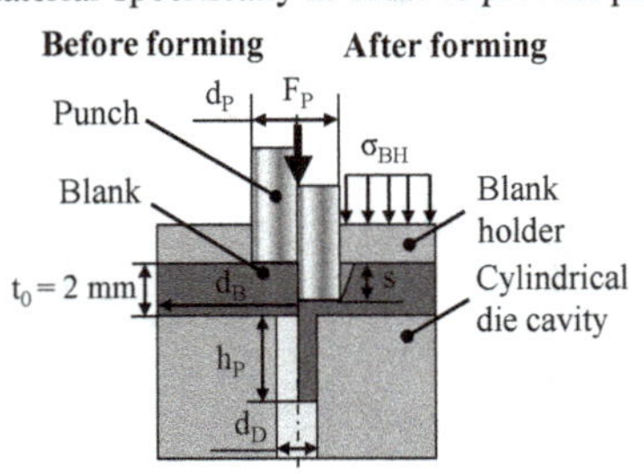

Process-related parameters	
Punch stroke s	1.8 mm
Blankholder pressure σ_{BH}	Global HT: 0.99 * $\sigma_{0, HT}$ Local HT: 0.60 * $\sigma_{0, AR}$
Forming speed	5.0 mm/min
Tool-related parameters	
Punch diameter d_P	3.0 mm
Die diameter d_D	1.5 mm
Blankholder/ Blank diameter d_B	20.0 mm

Figure 4: Process scheme of pin extrusion and process parameters.

At the beginning of the process, the blank holder moves downwards and constrains the circular blank on the die. The blank has a diameter of 20 mm and an initial sheet thickness of 2.0 mm. After applying the blank holder pressure, the extrusion punch moves axially downwards with a constant forming velocity of 5 mm/min and penetrates the blank. This movement causes an axially and a radially material flow. The material located above the cavity is displaced axially downwards into it. Furthermore, the material under the punch, which is not directly above the cavity flows laterally into the cavity and radially outwards into the sheet plane. Mechanical stops limit the punch penetration at a depth of 1.8 mm, which corresponds related to the sheet thickness to a relative penetration depth of 90 %. As lubricant, the extrusion oil Dionol ST V 1725-2 from MKU-Chemie GmbH is used in a quantity of more than 10 g/m².

Experimental results and discussion

In this chapter, the experimental results of pin extrusion are presented. As geometrical properties, the pin height as well as the funnel depth are compared. The funnel depth quantifies the resulting piping defect. This represents a material deficit inside the pin, which can occur during extruding pins from the sheet plane. Funnel formation occurs when the axial material flow velocity, which corresponds to the forming velocity of the punch, is lower than the lateral material flow velocity towards the die cavity. This velocity gradient inside the pin causes the funnel forming. Therefore, in order to improve the material efficiency of the process, the objective is to maximize the pin height and minimize the funnel depth. Furthermore, mechanical properties in form of the micro hardness distribution are analyzed in order to analyze the influence of the local heat treatment. By evaluating the grain structure of the components, the material flow should be confirmed.

Process properties. The force-displacement curves are analyzed to draw conclusions about the tool stresses. During the forming process, forces and displacements are recorded by sensors with a resolution of 1.5 N and 1.0 µm, which are installed in the upper tool unit. In Fig. 5 a the force-displacement curves of the pin extrusion processes are presented.

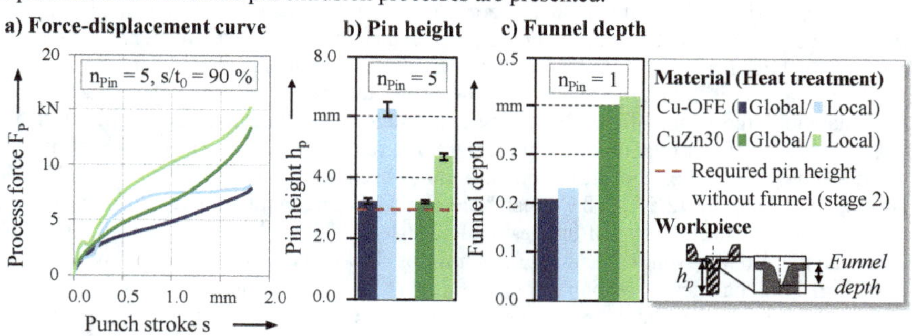

Figure 5: Force-displacement curve (a), pin height (b) and funnel depth (c).

The force-displacement-curves of the global heat-treated materials are similar to the curves during upsetting and corresponds to the described phases by Ghassemali [11]. Compared to Cu-OFE, CuZn30 shows a higher increase of the force curve and a 70 % higher maximum force, which can be explained by the higher yield stress and stronger work hardening of the brass. However, the force-displacement curve of the local heat-treated specimens displays differences from the typical progression for both materials. The process starts with a nearly linear increase of the process force until a displacement about 0.1 mm (CuZn30) respectively 0.07 mm (Cu-OFE). Subsequently, a slight decrease of force occurs, before increasing again. For Cu OFE the process force reaches a saturation value at 8.2 kN, while for CuZn30 it continuously rises to 15.2 kN. This can be explained by the previously described grain structure across the sheet thickness. Due to the larger, softer grains in the upper area of the sheet (Fig. 3), there is a slight drop in force or a stagnation of the increase in force after elastic deformation. Thus, a low force is required for the plastic deformation of material in the upper area. In the further process, the punch penetrates deeper into the sheet, forming the area with smaller, harder grains, which required a higher forming force. The force-displacement curves of the two types of heat treatment differ significantly during the force progression. The maximum forming forces are only 4 % higher for Cu-OFE and 14 % higher for

CuZn30 due to the local heat treatment, which does not significantly increase the tool stress. A forming of the cold-rolled materials was not feasible due to high force requirement and tool stress.

Component properties. The pin height and the funnel depth are displayed in Fig. 5 b, c in order to analyze the material efficiency of the process. A digital absolute dial indicator from Mitutoyo is used to measure tactile the pin height with an accuracy of ± 0.5 µm. The funnel depth is measured by using micrographs.

In order to completely fill the gearing in the second process stage, a minimum material volume of 5.1 mm³ is required. This corresponds to a height of 2.9 mm without funnel formation. In the globally heat-treated state, the pins formed from both materials only slightly exceed this specification, while the material deficit due to funnel formation (Fig. 5 c) lead to incomplete die filling of the gearing. The application of a local heat treatment results in a 95 % (Cu-OFE) increase in pin height up to 6.24 mm and a 47 % (CuZn30) increase up to 4.70 mm. This is attributed to the hardness gradient along the diameter of the blank and the higher hardness in the passive forming zone, which inhibits the lateral flow of material outward in the sheet metal, resulting in an increase in the lateral flow of material into the die cavity. In the global heat-treated condition, on the other hand, softer material is present in the outer area, resulting in less outward flow inhibition and thus more material being able to flow into the sheet metal. Thus, the use of local heat treatment is a suitable strategy to increase material efficiency during pin extrusion. Furthermore, a comparison of the materials shows a 33 % higher pin for Cu-OFE in the local heat-treated state. This can be explained by different sizes of heat-treated zones (Fig. 3) of the materials. Due to the larger heat-treated zone of the brass, there is a lower hardness gradient along the blank diameter compared to copper, which results in a higher material flow radially outward into the sheet plane.

Another parameter is the funnel depth, which quantifies the resulting piping defect (Fig. 5 c). For both materials, the depth of the funnel increases slightly by 0.02 mm due to the local heat treatment. However, in relation to the pin height, the relative proportion of the funnel depth is significantly reduced. For copper, the proportion has been minimized from 6.5 % to 3.7 %. For brass, a reduction from 12.5 % to 8.9 % can be identified. Furthermore, for CuZn30 a larger funnel depth is observed independent of the heat-treated state, which could be due to the higher yield stress and the stronger work hardening of the brass, as shown in the flow curves in Fig. 2.

In Fig. 6 the hardness distribution and the grain structure are shown. For micro hardness measurement, the Vickers hardness test method according to ISO 14577 is used. The measurement was carried out with a Fischerscope HM2000 (Helmut Fischer GmbH) with a test force of 100 mN (HV0.01) and a holding time of 10 s The initial micro hardness in heat-treated condition has a value of 66 ± 7 HV0.01 for copper and of 92 ± 5 HV0.01 for brass. The highest forming induced strain hardening for all specimen results in the residual sheet and in the upper zone of the pin. A maximum micro hardness of 170 ± 9 HV0.01 (Cu-OFE) and 294 ± 8 HV0.01 (CuZn30) is achieved in the pin of global heat-treated condition, which equals an increase of 158 % and 220 %, compared to the initial hardness. Furthermore, the micro hardness decreases in the direction of the pin head, with a hardness of 85 ± 4 HV0.01 (Cu-OFE) and 128 ± 10 HV0.01 (CuZn30). This is attributed to the forming of the pin head at the beginning of the process, where the material flows predominant axially downward into the cavity. Thus, low strain hardening is achieved. In the outer area of the blank, no significant hardening can be determined due to the limited blank holder force.

Materials Research Forum LLC
https://doi.org/10.21741/9781644902417-38

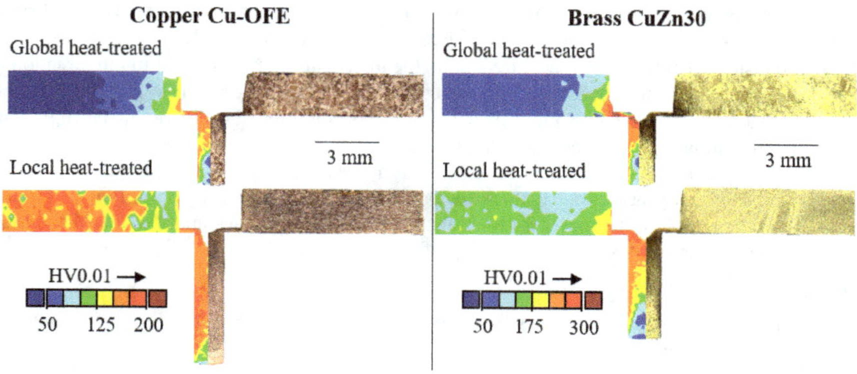

Figure 6: Micro hardness distribution and grain structure of the investigated components.

In initial local heat-treated condition, the micro hardness in the border area is 172 ± 11 HV0.01 for copper and 182 ± 15 HV0.01 for brass. In the forming zone, the micro hardness is 69 ± 4 HV0.01 (Cu-OFE) and 83 ± 3 HV0.01 (CuZn30). Due to the forming induced hardening, a maximum hardness in the upper area of the pin of 165 ± 1 HV0.01 for copper and 245 ± 5 HV0.01 for brass is achieved. In the head area, the micro hardness is 123 ± 7 HV0.01 for copper und 125 ± 3 HV0.01 for brass. Thus, a similar distribution of micro hardness and strain hardening in the pin area is present for brass for both heat treatment types. For copper, there is a 145 % increase in micro hardness in the pin, which is clearly visible from the larger red color details in the image. The differences between the materials can be explained by the different laser heat treatments.

In addition to the investigated micro hardness distribution, micrographs (Fig. 6) are analyzed to evaluate the expected material flow through an interpreting of the grain deformation. In the border area of the blank und in the area of the pin head of the investigated specimens, there is no significant deformation in grain structure in comparison with the initial grain structure visible. Furthermore, in residual sheet as well as in the upper part of the pin near the funnel, there is a significant deformation of the grains corresponding to the evaluated micro hardness. The comparison of the heat treatment strategy shows especially for copper significant differences in the grain structure in the pin area. In the local heat-treated condition, many small grains are visible in the pin area, whereas in global heat-treated condition large grains are present.

Summary and outlook

In this paper, the potential for extending the process limits during pin extrusion from sheet metal by a local short-term laser heat treatment was investigated. Therefore, the influence of the heat treatment strategy was evaluated. In order to analyze the transferability of the results, the investigations were carried out with the single-phase copper Cu-OFE and the application material brass CuZn30. The following main conclusions can be drawn:

- Using a laser as heat source, the heat treatment can be locally limited to the area of the forming for Cu-OFE and CuZn30. This type of heat treatment offers the potential to save process time and energy in comparison with heat treatment in a furnace.
- Through local heat treatment an improvement of the material flow and an increase in material utilization is achieved. The pin height was increased by 95% for Cu-OFE and 47 % in case of CuZn30.
- In addition, no significant influence on the funnel formation was identified. Only a small funnel depth occurs.

- The maximum force requirement of the two types of heat treatment and thus the tool stress is very similar, but the course of the force-displacement curves differs significantly.
- By using local short-term heat treatment, the micro hardness of the pin in the direction of the pin head is increased, especially for copper. In the case of brass, however, the hardness distribution is similar to the global heat-treated condition. Furthermore, the maximum micro hardness in the area of the funnel is in similar range.

Future research should focus on the investigation of the influence of heat treatment strategy on the forming of micro gearing by lateral extrusion. In addition, the investigation of different laser process parameters offers the potential to adjust the material properties and the size of the heat-affected zone to achieve desired material properties for pin extrusion.

References

[1] J. Zhao, Microforming Technology: Theory, Simulation and Practice, Elsevier Science & Technology, San Diego, 2017.

[2] N.K. Jain, S.K. Chaubey, Review of Miniature Gear Manufacturing, in: M Hashmi (Eds.), Comprehensive Materials Finishing, Elsevier Inc., Oxford, 2017, pp. 504-538. https://doi.org/10.1016/B978-0-12-803581-8.09159-1

[3] R. Slatter, Mikroantriebe für präzise Positionieranwendungen: Spielarme Getriebe, Antriebstechnik 42 (2003) 30-34.

[4] VDI-Standard, Microgears, VDI 2731, Beuth Verlag GmbH, Berlin, 2009.

[5] K. Gupta, R.F. Laubscher, J.P. Davim, N.K. Jain, Recent developments in sustainable manufacturing of gears: a review, J. Clean. Prod. 112 (2016) 3320-3330. https://doi.org/10.1016/j.jclepro.2015.09.133

[6] F. Vollertsen, Categories of size effects, Prod. Eng. Res. Devel. 2 (2008) 377-383. https://doi.org/10.1007/s11740-008-0127-z

[7] R. Mahshid, H.N. Hansen, M. Arentoft, Accuracy of Transferring Microparts in a Multi Stage Former, KEM 554-557 (2013) 900-907. https://doi.org/10.4028/www.scientific.net/KEM.554-557.900

[8] A. Rohrmoser, M. Kraus, M. Leicht, M. Merklein, Experimental Evaluation of the Feasibility and Reproducibility of a Novel Multi-step Forming Process for Cu-OFE Micro Gear Manufacturing Compared to Macro-scale Forming, Manuf. Lett. 33 (2022) 183-189. https://doi.org/10.1016/j.mfglet.2022.07.024

[9] M. Reck, A. Rohrmoser, A. Jobst, F. Pilz, M. Merklein, Forming of Complex Functional Elements on Sheet Metal, in: M. Merklein, A.E. Tekkaya, B.-A. Behrens (Eds.), Sheet Bulk Metal Forming, Springer International Publishing, Cham, 2021, pp. 30-52. https://doi.org/10.1007/978-3-030-61902-2_2

[10] M. Merklein, M. Johannes, M. Lechner, A. Kuppert, A review on tailored blanks—Production, applications and evaluation, J. Mater. Process. Technol. 214 (2014) 151-164. https://doi.org/10.1016/j.jmatprotec.2013.08.015

[11] E. Ghassemali, M.-J. Tan, A.E.W. Jarfors, S.C.V. Lim, Progressive microforming process: towards the mass production of micro-parts using sheet metal, Int. J. Adv. Manuf. Technol. 66 (2013) 611-621. https://doi.org/10.1007/s00170-012-4352-4

Sheet Metal 2023 Materials Research Forum LLC
Materials Research Proceedings **25** (2023) 315-322 https://doi.org/10.21741/9781644902417-39

High speed impact cutting of continuous fiber reinforced thermoset plastics

Fabian Dietrich[1,a*], Matthias Nestler[2,b], Verena Kräusel[2,c], Matthias Kolbe[1,d]

[1] Westsächsische Hochschule Zwickau, Kornmarkt 1, 08056 Zwickau, Germany

[2] Fraunhofer Institut für Werkzeugmaschinen und Umformtechnik, Reichenhainer Straße 88, 09126 Chemnitz, Germany

[a] Fabian.Dietrich@fh-zwickau.de, [b] matthias.nestler@iwu.fraunhofer.de, [c] verena.kraeusel@iwu.fraunhofer.de, [d] Matthias.Kolbe@fh-zwickau.de

Keywords: Fiber Reinforced Plastic, Quality, High-Speed Impact Cutting

Abstract. Endless fiber-reinforced plastics are being used to an increasing extent as alternative materials for highly stressed or lightweight components instead of metallic materials. In order to achieve the geometric requirements, peripheral machining of the raw parts is necessary. Instead of the currently mainly used cutting processes, which are not suitable for clocked production, high-speed impact cutting (HSIC) was examined in the presented experiments. This technology is known as adiabatic cutting from the processing of metallic materials. Due to the high process energy which is released in a very short time resulting in high punch speed, the prevailing separation mechanism changes. Instead of bending the fibers due to the shear force the high-speed cutting experiments with a punch speed of 10 m/s lead to a brittle shearing of the glass fibers and a locally very limited heating and hence softening of the matrix material resulting in a clean surface of the cut specimen. The inter fiber breakage, meaning the separation between fibers and matrix called delamination, can be avoided or at least be sealed at the surface due to heat induced smearing of the matrix material. The resulting surface quality of the cutting edge is exceptionally good. However, the technically necessary cutting clearance leads to a jump in diameter within the cut surface.

Introduction

Even today a significant proportion of the energy required by humans is obtained by the use of fossil fuels. One problem of this approach is the limited availability of oil and gas, another is the environmental impact caused by the combustion of these fuels that leads to damage of the environment and reduction in human quality of life. As a result, in addition to revolutionizing drive systems and power generation, consistent lightweight construction is necessary to achieve the environmental goals that have been set. This concerns especially the area of transport, which is responsible for around 20 % of global CO_2 emissions [1]. Lightweight design is a key driver to reduce vehicles energy consumption and is often applied in the automotive industry [2]. Fiber-reinforced plastics have been used as a substitute for the classic metallic materials to a steadily increasing extent for several decades [3]. Application fields range from aviation up to shipbuilding and vehicle construction. Endless fiber-reinforced duromers (EFRP) are of particular importance here because they are suitable for the production of components with the same function but significantly lower mass in comparison to conventional materials due to their up to five times higher specific strength compared to steel [4]. In contrast to components made of metallic materials, for components made of EFRP near netshape manufacturing and subsequent machining of holes and edges is the typical approach to comply with the required geometric dimensions. However, conventional processes such as milling are poorly suited due to the high hardness of the reinforcing fibers and the heterogeneity and anisotropy of the composite [5]. The milling tool

Content from this work may be used under the terms of the Creative Commons Attribution 3.0 license. Any further distribution of this work must maintain attribution to the author(s) and the title of the work, journal citation and DOI. Published under license by Materials Research Forum LLC.

experiences a high level of wear, which manifests itself in rounding of the cutting edges and in turn leads to damage to the workpiece which is being machined [6]. Alternative processes for producing the required dimensions are currently cut-off grinding, sawing, abrasive waterjet cutting and laser cutting. In addition to the respective process-specific advantages and disadvantages of these individual production methods, they all lack suitability for a clocked production process, since the entire contour of the component to be produced always has to be traversed in order to perform the separation process. An alternative to this is peripheral machining using shear cutting, in which one or even several components can be manufactured using only one press stroke. Shear cutting is widely used in the sheet metal processing industry [7] and it offers high cost-effectiveness [8]. Aiming for an improvement of the geometric accuracy of the cut surface and good surface quality, high speed impact cutting (HSIC) is a promising process variation [9]. In contrast to conventional shear cutting, this technology is characterized by the high speed (i. e. above 3 m/s) of the cutting tool (punch). This causes high and strongly localized strain rates and temperatures, leading to a nearly rectangular cut surface, a reduction of burr and a reduced rollover height for metallic materials [10]. The presented study aims to show the general feasibility of cutting EFRP with high cutting speed.

Experimental Methodology

Samples with a reinforcement made of glass fiber fabric (fiber alignment 0/90 °, fiber volume content $\varphi_F = 55$ %) with an epoxy resin matrix in thicknesses of 1, 2, 3 and 4 mm were used for the investigations. Three slugs with the same parameters were cut out from a flat workpiece with the dimensions of 175x50 mm^2 for each material thickness. The punch had a diameter of 20 mm and the cutting clearance was varied between 0.05 to 0.3 mm by using differently sized dies. After inserting the sample into the tool, the cutting punch was accelerated by the high-speed press Adiapress ADIAflex to a speed of 10 m/s. For this purpose, an energy of 1 kJ was applied. Fig. 1 shows the press on the left and the tool on the right side.

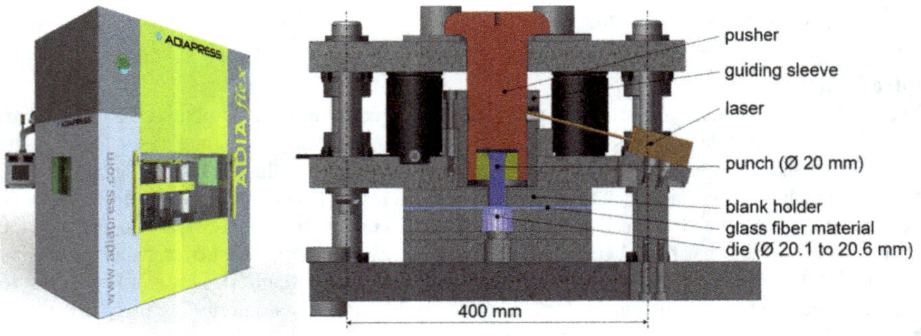

Fig. 1: Experimental setup for the HSIC tests: left) high-speed press ADIAflex; right) CAD drawing of the used tool

The cutting punch dipped into the sample for 2/3 of its thickness. The slug was then released from the workpiece by kinetic energy. After the HSIC tests the cut surfaces of the slugs and the workpiece were analyzed, as shown in Fig. 2. The percentage of breakage of the cut surface was measured manually using calipers at three positions distributed over the circumference of the slug and the perforated plate. This procedure allows fast and efficient determination of the general effects of changing parameters (DIN SPEC 25713-2017). On selected samples with typical appearance a conductive gold layer with a thickness of a few nanometers was applied using cathode sputtering (Cressington Sputter Coater 108auto). This is the only way to subsequently examine the non-electrically conductive compound by using a scanning electron microscope (Karl-Zeiss EVO MA15). The used procedure is state of the art for nonconductive sample materials and a significant influence on the surface topology is not to be expected. A systematic, manually controlled scanning and exact surface measurement of typical or conspicuous separating edge areas was performed.

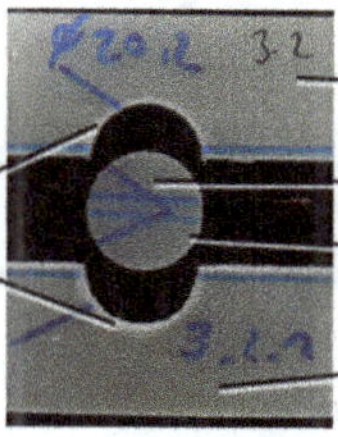

Fig. 2: Slug and divided workpiece prepared for measurements and visual inspection

Results and Discussion

Fig. 3 shows four slugs, cut by HSIC, with thicknesses of 1, 2, 3 and 4 mm (from left to right). The cutting clearance was 0.05 mm for the specimens with 1 mm and 2 mm thickness and 0.1 mm for the specimen with 3 mm and 4 mm thickness presented in the picture. With this variation of the absolute cutting clearance a relative cutting clearance between 2.5 % and 5 % was ensured, which is typically used in high-speed impact cutting of metallic materials.

Fig. 3: Cut slugs featuring thicknesses of 1, 2, 3 and 4 mm from left to right

Fig. 4 illustrates the dependency between the cutting clearance and the fraction of breakage for all examined material thicknesses comparing the workpiece (WP) and the slug (S). It is shown that the fraction of breakage increases with an increase of the cutting clearance. The slugs show a slightly higher (5-20 %) proportion of breakage, also referred as rough cut zone, than the workpiece. The minimum area of fracture occurs at a thickness of 1 mm and a cutting clearance of 0.05 mm. The relative deviation was due to the manual measuring, the heterogenous material structure and the resulting deviating cutting proportions rather high (up to 10 %). However, the influences of cutting clearance, high punch velocity and material thickness on the shear cutting process could still be identified in this study. It is clearly visible that smaller cutting clearances reduce the fraction of breakage for every tested specimen. However, the proportion of rough cut zone is not consistent through the increasing material thicknesses. Reasons for that could be different compositions of the bought EFRP material as well as the measurement inaccuracy.

Materials Research Proceedings **25** (2023) 315-322 https://doi.org/10.21741/9781644902417-39

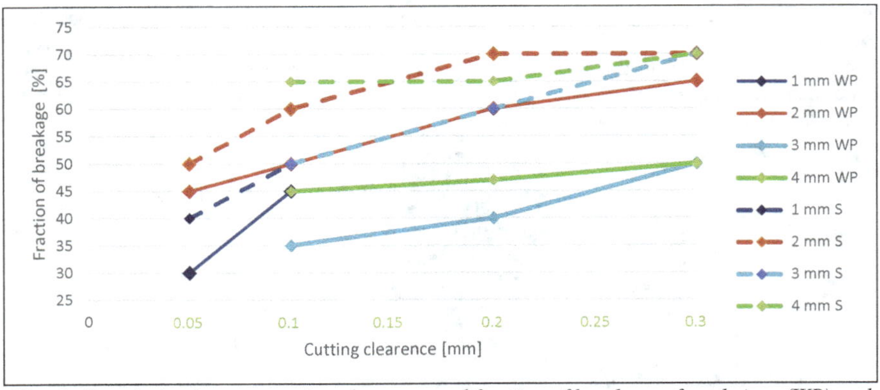

Fig. 4: Correlation between cutting clearance and fraction of breakage of workpiece (WP) and slug (S)

For one slug with a thickness of 4 mm and a cutting clearance of 0.3 mm, Fig. 5 shows the varying smooth and rough cut proportions. On the left side a photograph of the part is depicted. On the right side the result of a 3D measurement made with the scanning electron microscope is shown. Fig. 5 gives a better understanding on the resulting surface topology and shows that the proportions are irregular distributed over the perimeter which leads to a high scatter of the taken measurements. This could only be reduced by time consuming complete measurements of ever sample with the SEM for example.

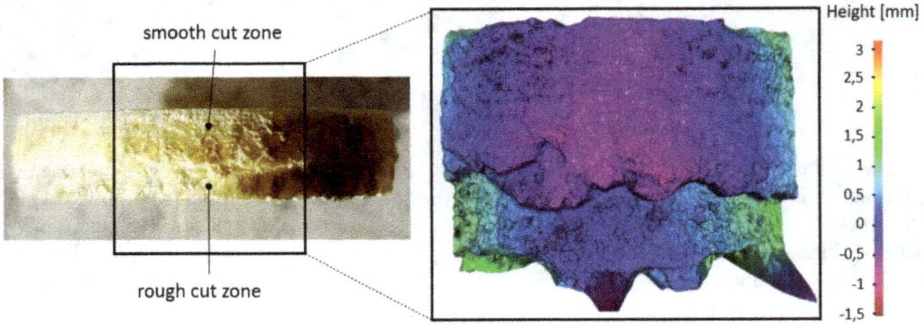

Fig. 5: Varying smooth and rough cut proportions on a slug, v_{Die} = 10 m/s, t = 4 mm, c = 0.3 mm

A more detailed analysis of the cut surface was performed by using the SEM for recording detailed images of the fracture surface, pictured in Fig. 6. In the area of the rough cut zone, numerous protruding fibers and fiber bundles can be seen in addition to an irregular surface topography, which is typical for shear cutting of EFRP (compare “punchside” in Fig. 6). The part of the workpiece facing the punch and the part of the slug facing away, both being classified as smooth cut zone, show an even and less rough surface. In this area, thermal damage to the matrix material can be recognized by a slight brown discoloration visually and by a significantly higher oxygen content measured by laser-induced plasma spectroscopy, indicating the temperature induced oxidation of the matrix material. This diffusion of oxygen into the matrix material has to be activated by heat according to [11]. The observed changes of the matrix material indicate heating above the glass transition temperature up to the temperature of thermal decomposition.

However, as it is already known from the adiabatic cutting of metallic materials, this damage is strongly localized [10]. This results in "blurring" of the matrix material in the direction of the slug ejection. As a result, the cutting edge consists exclusively of matrix material. Loose fiber ends or cracks cannot be seen in the smooth cut zone. This means that the material seals itself as a result of the heating during the shear cut, which significantly simplifies further processing of the component, such as painting. Traces resulting from the withdrawal of the stamp from the cutting zone cannot be seen, because the residual heat from the cutting process transferred into the poorly heat conductive composite material was not enough to keep the matrix material in a soft state until the punch withdrawal. This confirms the assumption of strong local limitation of the thermal influence.

The effect of matrix softening increases with increasing thickness of the workpiece. At t = 1 mm, only a very small thermal influence and unsealed surface with clearly visible fibers can be observed as shown in Fig. 7, while at t = 4 mm an extensive smooth-cut zone with closed surface can be seen (Fig. 6). This is caused by the energy necessary to cut the samples. While cutting thicker samples, a lot of kinetic energy from the punch is transformed into heat while cutting the slug. However, if thinner specimens are being processed only a small part of the provided energy is absorbed by the composite, which means that there is very little heat generated and no matrix softening takes place. Despite this, the high velocity of the punch still ensures a clean cut of the glass fibers, resulting in a good surface quality for the 1 mm specimen. It has to be noted nonetheless, that the matrix softening increases the smoothness of the cutting surface significantly.

Looking at the top surfaces of the samples white discoloration occur along the cutting edges. This suggests that the applied shear forces cause delamination due to fiber bending especially at high material thicknesses. Also, the transition from smooth cut surface to fracture surface is clearly marked by a step in the separation edge for thicknesses bigger 1 mm, where the diameter of the cut hole and the slug increases/decreases. This abrupt change in diameter, determined by the 3D measurement of a sample using SEM, is shown on the embedded diagram in Fig. 6. The cutting clearance for the sample pictured is 0.3 mm, which roughly corresponds to the change in radius visible in the diagram. This shows that the cutting clearance is too large and should be significantly smaller instead according to examination results of shear cutting metallic materials [12]. Since the cut begins at the punch and at the die side of the workpiece, the two cutting fronts are running almost parallel to each other at about the same distance as the set cutting clearance. When both separation zones in the material have reached the same level of the laminate consisting of matrix material (black, horizontal line in the diagram), a crack develops in almost horizontal direction through this matrix layer, which connects both separation fronts.

 https://doi.org/10.21741/9781644902417-39

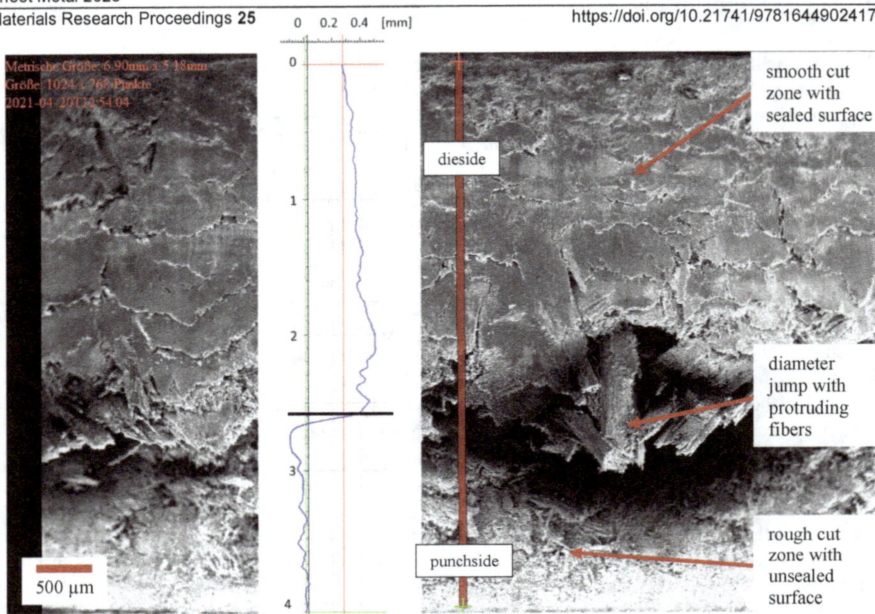

Fig. 6: Variance of the radius along the cut surface on a slug at v_{Die} = 10 m/s, t = 4 mm, c = 0.3 mm

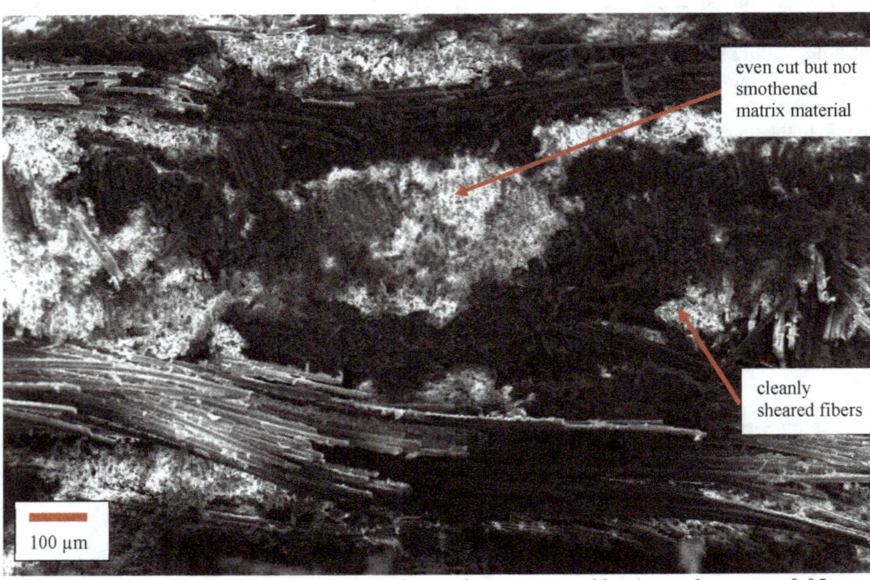

Fig. 7: Unsealed but even cut surface on the workpiece, v_{Die} = 10 m/s, t = 1 mm, c = 0.05 mm

Materials Research Proceedings **25** (2023) 315-322
https://doi.org/10.21741/9781644902417-39

The fracture surface on the die and on the punch side of the workpiece or slug respectively is characterized by fiber tear out and a rough surface, comparable to preliminary tests carried out with a conventional punch speed. Fig. 8 shows a 4 mm thick sample, analyzed in previous studies, for comparison. In contrast to the high-speed impact cutting this sample does not show any sealing effects in the smooth cut zone but pushed out, delaminated fibers at some places. Furthermore, there are loose matrix particles and a lot of protruding fibers in the rough cut zone. Also, severe delamination took place and on the resulted surface no finish could have been applied without extensive rework such as grinding.

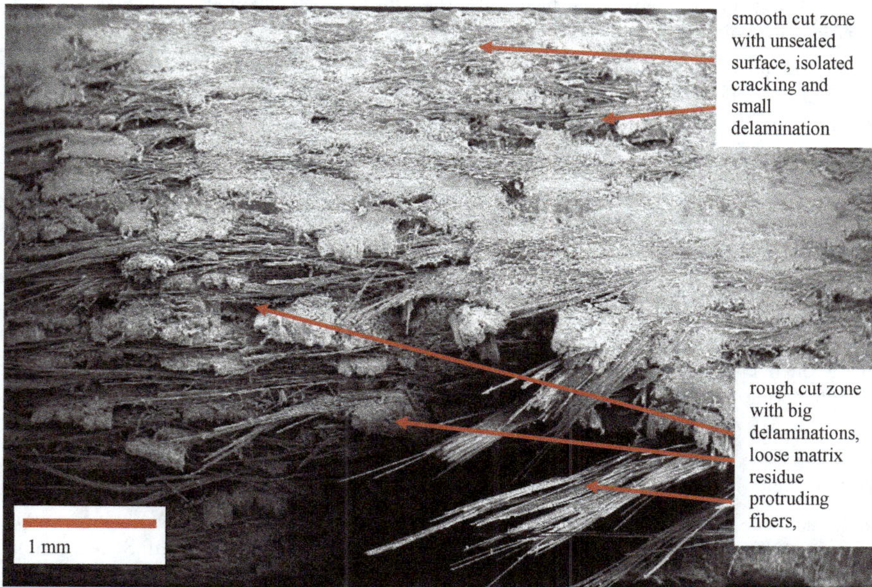

Fig. 8: Surface of a preliminary test specimen cut with low punch speed

Summary and outlook

The presented study shows that the high-speed impact cutting process, which is currently mainly used for cutting metallic materials, is fundamentally suitable for cutting components consisting of endless fiber-reinforced duromers (EFRP) even if the heterogenic material composition and the resulting crack formation properties are vastly different. The resulting cut surfaces show very little delamination in comparison to cutting tests at conventional punch speeds. The resulting finish of the edge is clean and uniform in the smooth cut areas, free of fibers and sealed by the matrix heated in the process. Moreover, thermal damage that is strongly localized takes place. However, depending on the thickness and the cutting clearance there is a jump in diameter and the fracture area shows numerous protruding fiber bundles and fiber tears. The resulting surface is still of better quality in regards of evenness, fiber protrusion and delamination than samples cut with a low velocity punch. Nonetheless further study is necessary in order to achieve a uniform, sealed cut surface over the whole thickness for all workpiece dimensions. This should include investigations of crack formation and propagation, the influence of different compositions of fibers and matrices as well as the relevance of every cutting parameter and their optimal value. The aimed goal should be a uniform temperature increase along the cutting surface to seal all fibers and voids which appear due to the occurring mechanical separation mechanism and the avoidance of the observed step formation for samples with higher thicknesses. The result would be a cut surface that requires no post-processing, is dimensionally accurate (± 0.1 mm) and can be painted immediately. Beneficial would also be the possibility to implement this process into clocked production for mass producing parts.

References

[1] Statista, Anteil der Verkehrsträger an den weltweiten CO2-Emissionen aus der Verbrennung fossiler Brennstoffe in den Jahren 2018 und 2019, BDL (2022). [https://de.statista.com/statistik/daten/studie/317683/umfrage/verkehrsttraeger-anteil-co2-emissionen-fossile-brennstoffe/] (accessed October 11, 2022).

[2] S. Rosenthal, F. Maaß, M. Kamaliev, H. Marlon, S. Gies, A. Erman Tekkaya, Lightweight, Automotive Components by Forming Technology, Automotive Innovation 3 (2020) 195-199. https://doi.org/10.1007/s42154-020-00103-3

[3] Carbon Fibers Technology – 9 Dimensional Assessment, Frost & Sullivan, (2014) 24

[4] F. Henning, E. Moeller, Handbuch Leichtbau, Methoden, Werkstoffe, Fertigung., 1. Aufl., Carl Hanser Verlag München, 2011. https://doi.org/10.3139/9783446428911.001

[5] M. Kreutzbruck, P. Fey, W. Essig, M. Rahammer, S. Joas, T. Erb, I. Solodov, Faserkunststoffverbunde – Hochleistungswerkstoffe in Leichtbausystemen als eine neue Herausforderung für die ZfP, ZfP-Zeitung 151 (2016) 31-39.

[6] Y. He, H. Qing, S. Zhang, D. Wang, S. Zhu, The cutting force and defect analysis in milling of carbon fiber-reinforced polymer (CFRP) composite, The International Journal of Advanced Manufacturing Technology 93 (2017) 1829-1842. https://doi.org/10.1007/s00170-017-0613-6

[7] F. Sergey. Golovashchenko, N. Wang, Q. Le, Trimming and sheared edge stretchability of automotive 6xxx aluminum alloys, Journal of Materials Processing Technology 264 (2019) 64–75. https://doi.org/10.1016/j.jmatprotec.2018.09.001

[8] I. Paetzold, F. Dittmann, M. Feistle, R. Golle, P. Haefele, H. Hoffmann, W. Volk, Influence of shear cutting parameters on the fatigue behavior of a dual-phase steel. IOP Conf. Series: Journal of Physics: Conf. Series 896 (2017) 012107. https://doi.org/10.1088/1742-6596/896/1/012107

[9] S. Winter, M. Nestler, E. Galiev, F. Hartmann, V. Psyk, V. Kräusel, M. Dix, Adiabatic Blanking: Influence of Clearance, Impact Energy, and Velocity on the Blanked Surface, Journal of Manufacturing and Materials Processing 5 Nr. 2 (2021) 35. https://doi.org/10.3390/jmmp5020035

[10] F. Schmitz, S. Winter, T. Clausmeyer, M.F.-X. Wagner, A. Erman Tekkaya, Adiabatic blanking of advanced high-strength steels, CIRP Annals 69 Nr. 1 (2020) 269–272. https://doi.org/10.1016/j.cirp.2020.03.007

[11] S. Moll, Untersuchung der Auswirkungen einer thermisch-oxidativen Beanspruchung auf das Emissionsverhalten von ABS und PP und der Korrelation mit dem Alterungsgrad, Friedrich-Alexander-Universität Erlangen-Nürnberg (2017) 18.

[12] P. Sachnik, Methodik für gratfreie Schnittflächen beim Scherschneiden, TU München (2016) 16.

Sheet Metal 2023
Materials Research Forum LLC
Materials Research Proceedings **25** (2023) 323-328
https://doi.org/10.21741/9781644902417-40

Definition and validation of a customized classification system for sheet metal bending parts

Daniel Schmid[1,a *] and Marion Merklein[2,b]

[1] Recaro Aircraft Seating GmbH, Daimlerstr. 21, 74523 Schwäbisch Hall, Germany

[2] Institute of Manufacturing Technology, Friedrich-Alexander-Universität Erlangen-Nürnberg, Egerlandstraße 13, 91058 Erlangen, Germany

[a] daniel.r.schmid@fau.de, [b] marion.merklein@fau.de

Keywords: Optimization, Bending, Sheet Metal

Abstract. The industry demands an increasing individualization, which leads to a rising variety of sheet metal parts. The consequence for sheet metal manufacturers are larger amounts of product data, smaller batch sizes and a larger quantity of different parts per year. The challenge for an efficient company is the availability and use of the corporate product data. Classification is a suitable tool for organizing corporate product data sets. In order to better reflect the individualization of customer demands and the resulting flexible manufacturing processes in a classification system, a method is needed to adapt the classification system based on the frequently used feature list. In the context of this paper, features are defined and parts are classified that are manufactured with the technology of air bending. It is a contribution to the creation of more precise approaches to the sheet metal part classification with simultaneous validation by a measurable characteristic. Focus is placed on the validation of the system with empirical measurement data. This allows every company to optimize and validate their own classification system.

Introduction

Over the last decades, the increasing individualization of customer demands led to a rising variety of sheet metal parts [1]. The result is a wide spectrum of sheet metal products, that ranges from simple articles of daily use to precision parts for the automotive or aviation industry [2]. Especially the aviation industry today is largely based on sheet metal forming technologies [3]. Due to individualization, product data is growing in all departments of a company like development, industrial engineering or production, as new products are constantly being developed or old products are being modified [4]. Manufacturers of sheet metal parts have adapted to the new situation. Associated with the larger variety of sheet metal parts, a significant range of flexible manufacturing processes for the production of those were developed [2]. The direct consequence for sheet metal manufacturers are smaller batch sizes and a larger quantity of different parts per year [5]. In addition, they have to handle larger amounts of product data for the higher variety of different parts.

In this environment it is important to find an efficient way to cope with the amount of data. The challenge for an efficient company is the availability and use of the corporate product data [6]. Classification is a suitable tool for organizing corporate product data sets and thus allows access to empirical knowledge [4]. Greska recognized that classification is a basic strategy for acquisition and processing of knowledge [6]. The necessity of a classification system becomes obvious when engineers of a manufacturing company spend on average about 45 % of their working time to search for existing parts and assemblies [6]. A review from Duflou et al. [7] has shown that a large number of publications contain proposed classification systems, that differentiate between certain features of design or manufacturing. Gupta et al. [8] use also features as they distinguish between deformation and cut features. In their work, they define sheet-metal parts as a combination of a

Content from this work may be used under the terms of the Creative Commons Attribution 3.0 license. Any further distribution of this work must maintain attribution to the author(s) and the title of the work, journal citation and DOI. Published under license by Materials Research Forum LLC.

number of individual features and each feature is related to another directly or indirectly. Furthermore, Greska et al. [6] give an overview of the three types of classification systems: 1. form order systems, 2. special feature list and 3. pattern recognition. The form order systems are almost irrelevant, as they are error-prone and time-consuming. An example of a frequently used feature list is a classification code. That code can be derived in form of a five-sign chain, based on the analysis of different features of parts [9]. The research is focused on automated processes and artificial intelligence methods to recognize patterns and features of parts and classify the parts into categories [7]. The methods cover a wide range of manufacturing processes [4]. The automated approach has yet been only partially successfully applied in most industrial companies.

To better reflect the individualization of customer demands and the flexible manufacturing processes, a method is needed to adapt the frequently used feature list. A classification system tailored to the specific manufacturing process of a company contributes to increasing this rationalization potential [10]. To achieve the target of an appropriate classification system, a process is required that adapts the classification system to the manufacturing process. The selected classification system must be validated against measurable data. A specific classification system supporting the different work processes could be created for each functional department of a company, e.g., purchasing, customer service, sales, or engineering [4]. This paper is a contribution to the creation of a more precise approach to the sheet metal part classification with simultaneous validation by a measurable characteristic. To overcome some limitations of general classification system, a method is shown for developing a classification system that fits the process.

Experimental Procedure

Feature list and calculation of the complexity score. In the context of this paper, parts are classified that are manufactured with the technology of air bending. The achievable dimensional accuracy in air bending is dependent on variations in sheet thickness and material properties, because these lead to variations in springback [2].

The production of tightly toleranced bending angles therefore requires a good execution of the process [2]. The major advantage of air bending over the other frequently used technology die bending is its much greater flexibility, because one tool can be used for many bending angles and radii [11]. In order to determine the relevant features for a feature list of the air bending process, the basic studies of Greska et al. [10] are used. With this set of features as a basis, the features fitting to the air bending process are selected. Additional features are combined with them based on expert knowledge from engineering and production. The initial selection includes 28 features, which are shown in Fig. 1. To facilitate data analysis and to be able to compare and rank parts, a complexity score (CS) is introduced. Once the complexity of parts can be quantified, re-use of the knowledge available for certain categories of parts is possible [7]. The CS is generated out of the selected features. To quantify the complexity of the parts, the weighting of the features is necessary using the method of pairwise comparison. In addition, a normalization of the feature values is done to the range 0 to 10. As a result the different feature values become comparable and the differentiation of the final CS score is improved. The CS is the sum of the normalized feature values multiplied by the weighting. The theoretical maximum range of the complexity score is up to 1000. The actual parts assessed are in the range from 51 to 468.

Validation of the features based on the measurement results. In order to validate the classification system for the air bending technology, a hypothesis is formulated which is verified using actual measurement data. The hypothesis is the higher the complexity score of the part, the larger is the deviation of the profile shape tolerance. To verify the hypothesis a selection of sample parts from the company's portfolio needs to be defined with a minimum of three parts per set. For this paper, sample parts of RECARO Aircraft Seating GmbH & Co. KG are used to demonstrate the method for the air bending technology. The selection is made on the basis of the geometry and the production quantity with an amount of selected parts less than 15 to be cost effective. The

Materials Research Forum LLC
https://doi.org/10.21741/9781644902417-40

selection of parts is optimized according to the most possible appearance variety and the largest production quantity. The selected parts are classified on basis of the developed complexity score, manufactured with a batch size of 20 and measured for an empirical data set. The measurement data are used to investigate the correlation of complexity score and the deviations of the parts based on the profile shape tolerance. The average values of the profile shape tolerances of a production lot are determined. The correlation between averaged profile shape tolerances and complexity score is calculated. Additionally, a scatterplot is created for the visual interpretation of the empirical data. Recognizing an association in the pattern does not guarantee that there is actually a cause-and-effect relationship between the variables [12]. The calculated correlation coefficient is a measure of the strength of the relationship between the two variables and does not imply causation. Based on this, it is important to conduct a substantive interpretation of the correlative relationship to justify causality.

Flat pattern – geometry and further features	material properties length [mm] width [mm] thickness [mm] symmetrie of outer contour [yes/no] symmetrie of inner contour [yes/no] boreholes/sinkholes in critical area [yes/no] number of bend lines [qty.]
Flat pattern - orientation of the bend lines	number of bend lines parallel to the length of the part [qty.] number of bend lines at 90° to the length of the part [qty.] number of bend lines at a "free" angle [qty.]

Finished part - dimensions	length [mm] width [mm] height [mm] price per part [€]
Finished part - geometry	number of bending segments [qty.] max. length of beding segment [mm] related height of segment[mm] related angle [deg] related radius [mm] number of L/90°-bends [qty.] number of U-bends [qty.] number of Z-bends [qty.] number of double-Z-bends [qty.] number of free angle [qty.] number of boxes [qty.] number of layers [qty.] appearance [closed / open]

Figure 1: Feature list for air bending

Reduction of features and verification of correlation. For an efficient classification system, it is evaluated whether the initial number of features can be reduced. To evaluate the reduction of the features, a linear optimization is performed based on the sample parts. Using linear optimization, the weighting factors of the score are modified to improve the correlation between averaged profile shape tolerances and the part's complexity score. With this approach, non-relevant features can be excluded from the calculation of the score. Further parts from the company's portfolio are added in order to verify the linear optimization performed. In most companies, measurement data from first article reports (FAI) of parts are available. All measurement data used in this paper are measured with the scanner system ATOS Compact Scan. The measurement data of these parts are used to check the correlation between complexity score and the profile shape tolerance.

Results and Discussion

The scatterplot in Fig. 2 shows the distribution of the parts based on the profile shape tolerance in relation to the complexity score. The position of the dots indicates values for an individual data point of the sample. The 62 blue dots represent parts with one measured value, during the part first article inspection. The eleven orange dots are average values ($n = 20$) of the profile shape tolerances of the selected sample parts. The sample correlation coefficient (r) shows a strong positive association between the computed complexity score and the measured profile shape tolerance. When looking at all the 73 parts, the correlation coefficient r is 0.75. Limiting the correlation coefficient to the 11 parts used to optimize the system with the linear optimization, the value increases to 0.93, which means a very strong positive association.

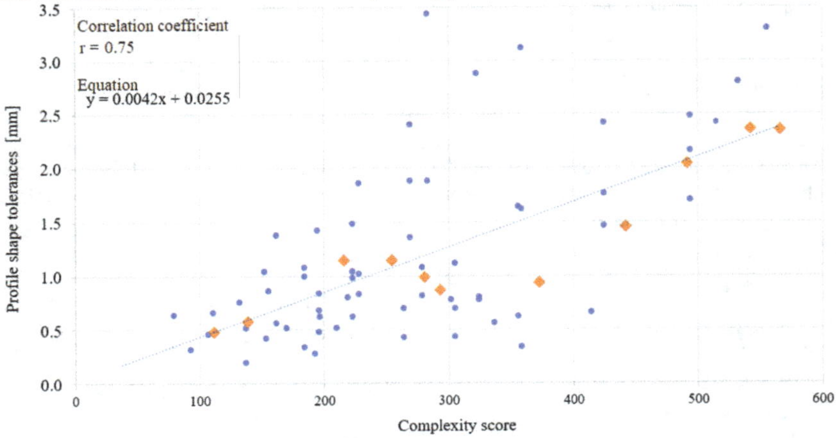

Figure 2: Correlation between complexity score and profile shape tolerances; orange: measurement series, averaged n = 20, blue: single parts n = 1

The hypothesis is the higher the complexity score of the part, the larger is the deviation of the profile shape tolerance. The association between the two variables is strong, therefore the hypothesis is verified. However, evidence of a correlative association does not guarantee that there is causation [12]. Interpreting the causal relationship, the feature list from Fig. 3 is taken into account. A causal relationship exists if the remaining features have an impact on the profile tolerance. The initially selected classification features are reduced using the linear optimization from 28 to 11. Only features of the finished part are used to compute the optimized complexity score. Features such as dimensions of the finished part or the number of determined bending lines, e.g., the number of U-bends, are also confirmed in the literature as influencing variables for part accuracy. Greska et al. [6] show the factors impacting the part accuracy and Söderberg et al. [13] describe the different geometrical sensitivities that lead to variation in individual part dimensions. The individual part dimensions are reflected in the profile shape tolerance as it is defined as a 3-D tolerance zone around a surface. Therefore, the relationship between CS and profile shape tolerance is considered causal. With the validation, the classification system is ready to use in all relevant departments of a company. Furthermore, a process is defined based on the data and experience gathered during the experiment.

The developed process works for all sheet metal bending technologies and is a more precise approach to the sheet metal part classification due to the simultaneous validation by a measurable characteristic. It overcomes the limitations of general classification system, by using features for classification which fits the specific process. Fig. 4 visualizes the process and describes the four steps to a customized classification system. It uses the available knowledge from literature and experts to derive features.

Sheet Metal 2023
Materials Research Proceedings **25** (2023) 323-328

Materials Research Forum LLC
https://doi.org/10.21741/9781644902417-40

Flat pattern – geometry and further features	0% material 0% length [mm] 0% width [mm] 0% thickness [mm] 0% symmetrie of outer contour [yes/no] 0% symmetrie of inner contour [yes/no] 0% boreholes/sinkholes in critical area [yes/no] 0% number of bend lines [qty.]
Flat pattern - orientation of the bend lines	0% number of bend lines parallel to the length of the part [qty.] 0% number of bend at 90° to the length of the part [qty.] 0% number of bend at a "free" angle [qty.]

Finished part - dimensions	20% length [mm] 10% width [mm] 40% height [mm] 0% price per part [€]
Finished part - geometry	50% number of bending segments [qty.] 20% max. length of beding segment [mm] 0% related height of segment[mm] 0% related angle [deg] 0% related radius [mm] 10% number of L/90°-bends [qty.] 10% number of U-bends [qty.] 10% number of Z-bends [qty.] 10% number of double-Z-bends [qty.] 10% number of free angle [qty.] 10% number of boxes [qty.] 0% number of layers [qty.] 0% appearance [closed / open]

Figure 3: Reduced feature list for air bending after linear optimization

Room for improvement lies in the area of digitalization and Industry 4.0. One challenge is to automatically collect the features in an end-to-end PLM system. This increases efficiency and contributes to a company-wide use of the classification system. After the developed classification system has been validated by empirical measurement values, it is still required to define what actions are required in the individual departments depending on the score. For example, at high values of the CS, a simulation of the part or the production and measurement may be necessary.

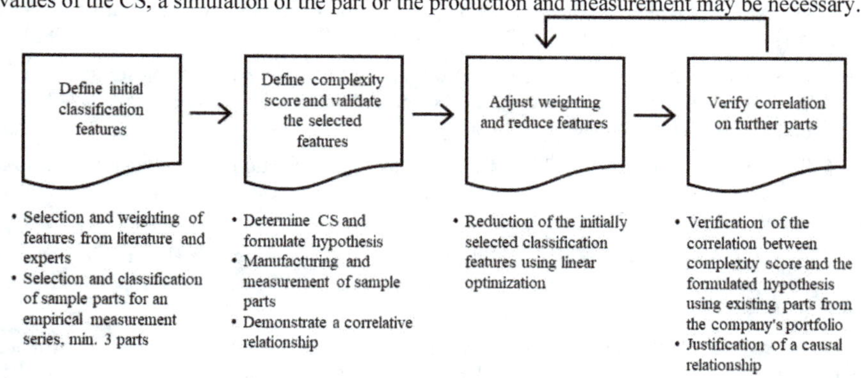

Figure 4: Technology-based approach to a customized sheet metal part classification

Summary

Within this paper a technology-based approach to a customized sheet metal part classification system is introduced. The individualization of customer demands and the flexible manufacturing processes require a precise classification. The customized classification system eliminates the restrictions of a general classification system. Because of the success in the industry, a feature-based classification is customized to one specific manufacturing process. A method is shown for developing a classification system that fits the process. The relevant features for the air bending technology are identified and weighted. A complexity score is computed based on the selected features. To quantify the complexity of the parts, weighting of the features is necessary using the method pairwise comparison. In addition a normalization of the feature values is done to the range 0 to 10. The complexity score is the sum of the normalized feature values multiplied by the weighting. The hypothesis defined at the beginning of the experiment is verified, which means that the developed classification system is relevant for the considered air bending technology. The sample correlation coefficient ($r = 0.75$) shows a strong positive association between the computed complexity score and the measured profile shape tolerance.

References

[1] D.T. Bowen, I.M. Russo, C.J. Cleaver et al. (2022) From art to part: Learning from the traditional smith in developing flexible sheet metal forming processes. Journal of Materials Processing Technology Volume 299(Article 117337):1–18. https://doi.org/10.1016/j.jmatprotec.2021.117337

[2] M. Kleiner, R. Schilling (1992) Umformtechnik. Ideen, Konzepte und Entwicklungen. Vieweg+Teubner Verlag, Wiesbaden. https://doi.org/10.1007/978-3-322-99878-1

[3] T. Trzepieciński (2020) Recent Developments and Trends in Sheet Metal Forming. Metals - Open Access Metallurgy Journal 10, 779(6):1–53. https://doi.org/10.3390/met10060779

[4] A. Guldi (2005) Unternehmensspezifisches Klassifikationssystem zur effizienten Datenverwaltung (mit Anwendungsszenarien aus der Praxis). Abschlußbericht des Verbundprojektes "Klassifikationssysteme automatisiert erstellen" (KLASTER). Univ.-Verl. Karlsruhe, Karlsruhe

[5] Z. Liu, J. Li, Y. Wang et al. (2004) Automatically extracting sheet-metal features from solid model. J Zhejiang Univ Sci 5(11):1456–1465. https://doi.org/10.1631/jzus.2004.1456

[6] W. Greska, V. Franke, M. Geiger (1997) Classification problems in manufacturing of sheet metal parts. Computers in Industry 33:17–30. https://doi.org/10.1016/S0166-3615(97)00008-0

[7] J.R. Duflou, J. Váncza, R. Aerens (2005) Computer aided process planning for sheet metal bending. A state of the art. Computers in Industry 56(7):747–771. https://doi.org/10.1016/j.compind.2005.04.001

[8] R.K. Gupta, B. Gurumoorthy (2013) Classification, representation, and automatic extraction of deformation features in sheet metal parts. Computer-Aided Design(45):1469–1484. https://doi.org/10.1016/j.cad.2013.06.010

[9] P. Šugár, J. Šugárová, M. Kolník (2011) Technology-Based Sheet Metal Classification and Coding System. Journal for Technology of Plasticity 36(1):1–8. https://doi.org/10.2478/v10211-011-0005-0

[10] W. Greska (1995) Wissensbasierte Analyse und Klassifizierung von Blechteilen. Bericht aus dem Lehrstuhl für Fertigungstechnologie, LFT. Zugl.: Erlangen, Nürnberg, Univ., Diss., 1995. Fertigungstechnik - Erlangen, vol 49. Hanser, München

[11] H. Hagenah (2003) Simulationsbasierte Bestimmung der zu erwartenden Maßhaltigkeit für das Blechbiegen. Zugl.: Erlangen-Nürnberg, Univ., Diss., 2002. Fertigungstechnik - Erlangen, vol 141. Meisenbach, Bamberg

[12] M.L. George, J. Maxey, M. Price et al. (2016) Das Lean Six Sigma Toolbook. Verlag Franz Vahlen GmbH. https://doi.org/10.15358/9783800648535

[13] R. Söderberg, L. Lindkvist, K. Wärmefjord et al. (2016) Virtual Geometry Assurance Process and Toolbox. Procedia CIRP 43:3–12. https://doi.org/10.1016/j.procir.2016.02.043

Sheet Metal 2023
Materials Research Forum LLC
Materials Research Proceedings **25** (2023) 329-336
https://doi.org/10.21741/9781644902417-41

Numerical analyses of the influence of a counter punch during deep drawing

Bernd-Arno Behrens[1,a], Daniel Rosenbusch[1,b], Hendrik Wester[1,c], Joshua Weichenhain[1,d] and Philipp Althaus[1,e,*]

[1]Institut für Umformtechnik und Umformmaschinen, Leibniz Universität Hannover, An der Universität 2, 30823 Garbsen, Germany

[a]behrens, [b]rosenbusch, [c]wester, [d]weichenhain, [e]althaus@ifum.uni-hannover.de

* Corresponding author

Keywords: Fracture Analysis, High Strength Steel, Process Extension

Abstract. In the automotive sector, the demand for high crash safety and lightweight construction has led to an increased use of steels with higher strengths. However, the rising number of varying materials with different strengths and ductilities lead to an increasing complexity in production, making it more challenging to ensure robust processes. Therefore, the focus of current researches still lays on the further development and extension of forming processes to enable high productivity and reliable production. A powerful tool for an efficient optimisation and extension of forming processes is the Finite Element Method (FEM), which offers time- and cost saving potentials in the design phase. In deep drawing, the use of a counter punch offers the possibility of extending the process limits. By superimposing compressive stresses on the workpiece, the initiation of cracks can be delayed, thus higher drawing ratios can be achieved. The aim of this research is therefore the numerical investigation of a deep drawing process with a counter punch to analyse the influence on the crack initiation and identify optimisation potentials for the process. For this cause, the applied force as well as the position and geometry of the counter punch are varied and the influence on fracture initiation is evaluated. It is found that the applied force on the counter punch is the major influencing factor for crack initiation. Furthermore, it was concluded that the contact between the counter punch and the workpiece should be applied as soon as the bottom of the cup is shaped. A further improvement can be achieved if the counter punch is geometrically adapted to the bottom of the workpiece.

Introduction

The process limits regarding fractures in deep drawing can be extended by reinforcing or relieving the force transfer zone. Reinforcement can be achieved either by increasing the initial sheet thickness or by using materials with higher strength [1]. Furthermore, locally adapted semi-finished products, such as tailored rolled blanks [2] can be used.

In order to relief the force transfer zone, the stress state can be influenced. Wu et al. [3] found that the formability of a sheet metal can be increased by superimposing a hydrostatic pressure on the material. In hydromechanical deep drawing, additional compressive stresses are realised by means of an active medium [4]. However, this is associated with significant higher tool costs and increased regulatory expenditure compared to conventional deep drawing, as it requires the integration of complex process technology into the tool set up.

Petzold and Otto [5] extended a deep drawing process by generating material flow from the outer part of the flange towards the developing cup. During forming, the material from the edge of the workpiece is pushed forward by a mechanical force, which is applied by a compression ring. In experimental tests, a significant increase in the applicable drawing ratio could be achieved. In

Content from this work may be used under the terms of the Creative Commons Attribution 3.0 license. Any further distribution of this work must maintain attribution to the author(s) and the title of the work, journal citation and DOI. Published under license by Materials Research Forum LLC.

addition, the reduction of the sheet thickness at the transition zone between the bottom and the wall could be reduced compared to conventional deep drawing.

Another possibility is deep drawing with a counter punch, where an additional punch is integrated into the conventional tool structure, which applies a force perpendicular to the bottom or both the bottom and the radius of the deep drawn cup. Morishita et al. [6] investigated the effect of a counter punch during deep drawing of tailored blanks, in which centrally welded blanks of different materials were deep-drawn into square cups. Experimental tests showed that the activation of a counterforce enables an increase in the achievable drawing height. While a cup height of less than 30 mm could be achieved in conventional deep drawing, a counterforce of 100 kN made it possible to produce cups with drawing heights of up to 60 mm without fracture.

A further investigation was carried out by Behrens et al. [7], where a counter punch was used during deep drawing of two high-strength steels HX340LAD and HCT600X. The counterforce was applied at the bottom and the radius of rectangular cups. It was found, that higher limiting drawing ratios can be achieved with the additional counter force, thus leading to an extension of the process window. Furthermore, in a recent study [8], deep drawing was performed with a flat counter punch which applies pressure only in the bottom of the cup. The extension of the process limits was achieved by higher applicable blank holder forces, which increased by 14.3% for HX340LAD and 17.9% for HCT600X. Additionally, the process was simulated and validated using the FE-software Abaqus to investigate the effect of the counter force, which was defined as a surface pressure field on the bottom of the cup. It was found that the stress state slightly shifts to lower stress triaxialities, thus allowing higher deformations to be reached.

Based on the aforementioned literature, this research investigates the influence of the counter punch numerically to evaluate the effect of the force on the stress state in the bottom of the cup and derive further optimisation potentials for industrial processes. For this purpose, deep drawing of two high-strength steels HX340LAD and HCT600X with and without the activated counter punch are simulated and the influence of varying process parameters on the fracture initiation is evaluated. In this study, the counter punch will be modelled as a deformable body instead of a pressure field, so that the friction to the workpiece is taken into account.

Materials and Methods

In order to investigate the influence of the superimposed force on fracture initiation, the deep drawing process was numerical depicted in Abaqus/Explicit. Fig. 1 a) shows the process set up. The tool system consists of a punch, a blank holder and the forming die, which were modelled as ideal rigid bodies and meshed with shell elements with an element edge length of 1.5 mm. For the application of the blank holder force, a uniform surface pressure was defined. With a thickness of 1 mm and initial dimensions of 281 mm x 213 mm, the sheet is formed to a final size of 160 mm x 80 mm, which corresponds to a deep-drawing ratio of 1.9. For the sheet, hexahedral elements with an element edge length of 1.5 mm were used, which were refined to 0.5 mm in the area of the bottom of the final cup. The forming punch moves with a constant velocity of 10 mm/s for a drawing depth of 55 mm. For the contact between the tools and the workpiece, a friction coefficient of 0.05 was defined on the basis of a previous study [9]. To minimize the computational effort, mass scaling with a factor of 100 and a time scaling factor of 10 were applied. Exemplary simulations with and without scaling were carried out for both materials to ensure that the application of the scaling factors has no significant influence on the results.

Materials Research Forum LLC
https://doi.org/10.21741/9781644902417-41

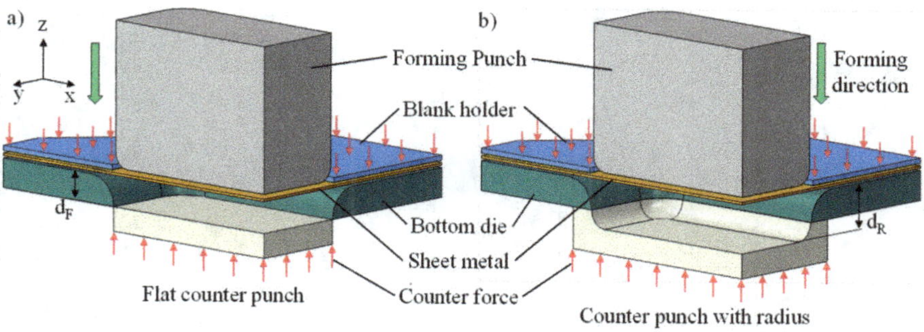

Fig. 1: Deep drawing model with a) flat counter punch and b) counter punch with radius

For the counter punch, which is located below the forming die, shell elements with an element edge length of 1.5 mm were used and an elastic material model with a Young's modulus of 210 GPa and a Poison's ratio of 0.3 was adopted. An ideal rigid behavior was not applicable, since numerical instabilities occurred when the sheet metal came in contact with the counter punch. The counterforce was induced by means of a surface pressure on the counter punch. In the experimental tool set up, the counterforce of 15 kN is applied by a gas pressure spring. Additional counterforces of 30 kN and 60 kN were defined in the numerical simulation to analyse the influence of the applied force. Moreover, the position of the counter punch was varied, so that it comes into contact with the sheet metal after a distance of d_f = 0 mm, 12 mm and 19 mm. At a distance of 12 mm, the flat counter punch is located directly under the die, which is shown in Fig. 1 a). Furthermore, a simulation was carried out in which the counter punch was designed with a radius to extend the contact area, which can be seen in Fig. 1 b). Due to the radius, however, the counter punch can only come into contact with the workpiece after a distance of at least d_R = 19 mm. Table 1 summarises the investigated process and geometrical parameters.

Table 1. Investigated parameters

Material	Blank holder force [kN]	Counter force [kN]	Distance [mm]	Counter punch geometry
HX340	320	15, 30, 60	12	Flat
		15	0, 12, 19	
			19	Flat, Radius
HCT600	230	15, 30, 60	12	Flat
		15	0, 12, 19	
			19	Flat, Radius

For the material modelling, elastic-plastic models of the high-strength steels HX340LAD and HCT600X were defined. The data was determined in a previous study by means of tensile and hydraulic bulge tests [9]. To model the flow curve, the approach of Gosh was chosen for HX340LAD and a combined approach of Swift and Hocket-Sherby for HCT600X, since these approaches ensured the best approximation of the experimental data. In addition, the flow criterion according to Hill'48 [10] was defined for both steels. The yield loci and flow curves with the corresponding model coefficients are shown in Fig. 2.

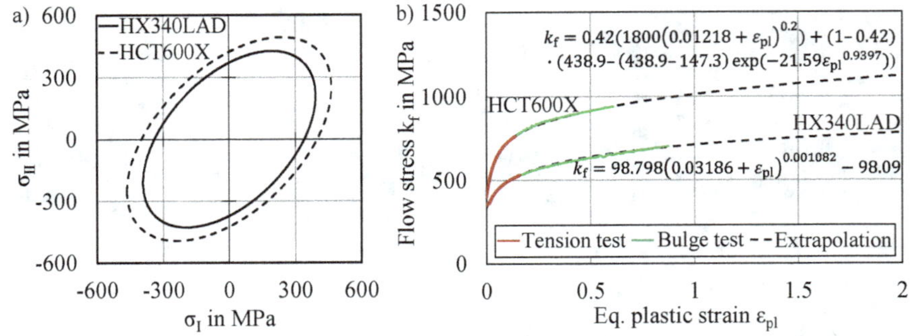

Fig. 2: Yield loci a) and flow curves b) for HX340LAD and HCT600X according to [10]

In order to predict the initiation of fracture during forming, the damage behaviour was modelled using the Modified Mohr-Coulomb (MMC) criterion [11], where the equivalent plastic strain at fracture $\varepsilon_{\mathrm{pl}}^{\mathrm{f}}$ is defined depending on the stress triaxiality η, normalised Lode angle $\bar{\theta}$, strain hardening exponent n and the material specific constants A, c_1, c_2, c_Θ^{s}, c_Θ^{ax}:

$$\varepsilon_{\mathrm{pl}}^{\mathrm{f}} = \left\{\frac{A}{c_2}\left[c_\Theta^{\mathrm{s}} + \frac{\sqrt{3}}{2-\sqrt{3}}\left(c_\Theta^{\mathrm{ax}} - c_\Theta^{\mathrm{s}}\right)\left(\sec\left(\frac{\bar{\theta}\pi}{6}\right) - 1\right)\right]\left[\sqrt{\frac{1+c_1^2}{3}}\cos\left(\frac{\bar{\theta}\pi}{6}\right) + c_1\left(\eta + \frac{1}{3}\sin\left(\frac{\bar{\theta}\pi}{6}\right)\right)\right]\right\}^{-1/n} \quad (1)$$

The MMC criterion was chosen because its applicability was proven for various ductile materials, such as aluminium alloys and high-strength steels [11]. Furthermore, the applicability for the materials considered in this research was verified in a previous study [12]. Therefore, tensile tests with butterfly specimens were carried out to determine the material specific constants. A special tooling system was used, which enables the variation of the force direction prior each test. This allowed different stress states ranging from shear to uniaxial tensile to be induced in the centre of the butterfly specimens. Afterwards, numerical simulations were used to evaluate the damage parameters $\varepsilon_{\mathrm{pl}}^{\mathrm{f}}$, η and $\bar{\theta}$ at the onset of fracture to calibrate the model parameters A, η, c_1 and c_2 by the least square method. For the parameters c_Θ^{s} and c_Θ^{ax}, the default values of one were assumed analogous to [11]. The resulting parameters are summarised in Table 2.

Table 2. Parameters of the used MMC criterion [13]

	A	n	c_1	c_2
HX340	713.0	0.165	0.208	401.934
HCT600	1018.6	0.143	0.166	560.033

In the simulation of deep drawing, the damage variable D is evaluated to predict the onset of fracture. D increases monotonically with the plastic strain $\varepsilon_{\mathrm{pl}}$ according to Eq. 2.

$$D = \int \frac{d\varepsilon_{\mathrm{pl}}}{\varepsilon_{\mathrm{pl}}^{\mathrm{f}}} \quad (2)$$

Material damage is present, when D reaches or exceeds the value of 1. The evolution of damage or the deletion of elements after failure is not taken into account in this study.

Results

In order to validate the location of crack initiation, experimental deep drawing tests were carried out. As mentioned earlier, the experimental procedure is presented in detail in [8]. Fig. 3 a) shows an experimental cup made of HX340LAD, which was formed with a blank holder force of 320 kN and without a counter punch. In b), the numerical distribution of the damage variable D is shown. In the simulation, material damage is visible in the corner of the cup, where the damage variable D exceeds the value of 1 and is marked grey. Since no elements are deleted after failure and no fracture evolution is taken into account, the crack does not spread as intensively as in the experiment over the side of the cup. Nevertheless, the damage criterion provides good agreement with the experimental results regarding the location of fracture initiation in the corner of the cup.

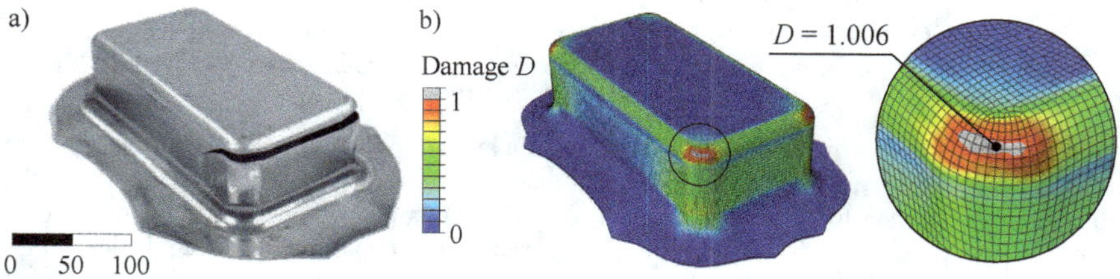

Fig. 3: Results without counter force: a) experimental cup made of HX340LAD with a blank holder force of 320 kN and b) distribution of damage from the simulation

In the next step, the experiment and simulation were carried out again with the flat counter punch at a distance of 12 mm and an applied counter force of 15 kN, while aforementioned parameters were kept constant. Fig. 4 compares an experimentally formed component with the numerical result of the simulation with a counter force of 15 kN.

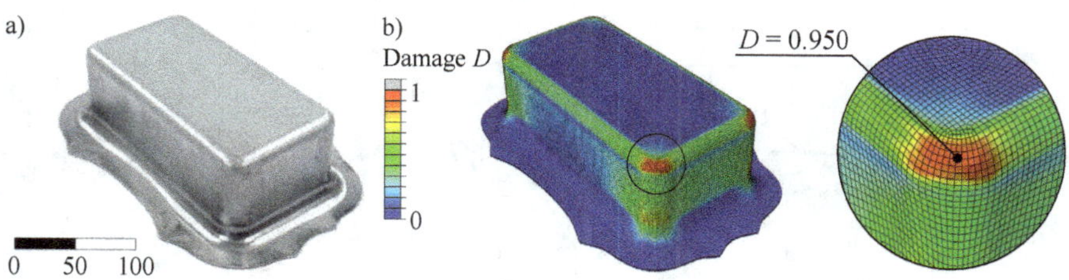

Fig. 4: Results with activated counter force of 15 kN: a) experimental cup made of HX340LAD with a blank holder force of 320 kN and b) distribution of damage from the simulation

It can be clearly seen, that the experimental component could be formed without any formation of cracks. Furthermore, no material damage is predicted by the simulation, which can be attributed to the influence of the counter punch. In order to evaluate the effect of the counter force more precisely, the course of the plastic strain over the stress triaxiality at the corner of the cup is shown in Fig. 5 a) for HX340LAD and in b) for HCT600X. Differing from the HX340LAD, a blank holder force of 230 kN is used for the HCT600X. Additional simulations with counter forces of 30 kN and 60 kN were carried out to analyse the influence of the applied force. The results represent the mean of 10 elements located in the fracture initiation area. Additionally, the mean damage D is indicated for each simulation.

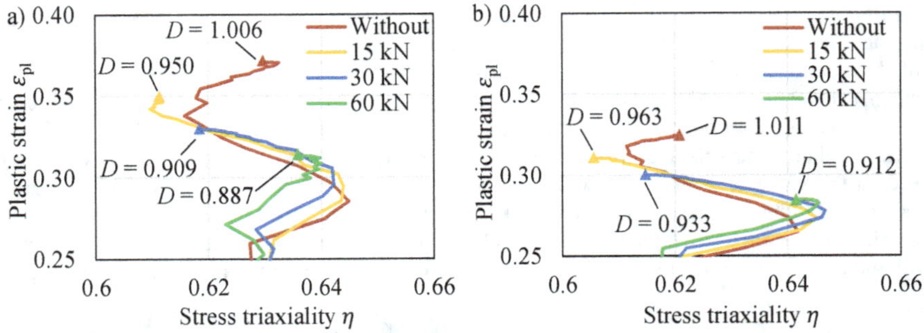

Fig. 5: Plastic strain over stress triaxiality depending on the counter force for a) HX340LAD and b) HCT600X

HX340LAD shows higher plastic strains due to a higher applied blank holder force. Thus, the effect of the counter punch is more significant than for HCT600X. However, for both materials, the increasing counter force leads to a reduction of the maximum plastic strain at the radius of the cup. This can presumably be attributed to the restraining frictional force between the sheet metal and the counter punch, which prevents the material from exceeding the forming limit and thus the formation of fractures. An increase of the counter force and therefore has a positive effect on fracture prevention. However, higher counter forces also increase the required press forces, which enhances the tool loads and potentially requires larger forming machines.

In the next step, the influence of the punch position was varied so that the sheet metal would come into contact with the counter punch after a distance of 0 mm, 12 mm and 19 mm. Here, the flat counter punch with constant force of 15 kN and blank holder forces of 320 kN for HX340LAD and 230 kN for HCT600X were applied. The results are shown in Fig. 6.

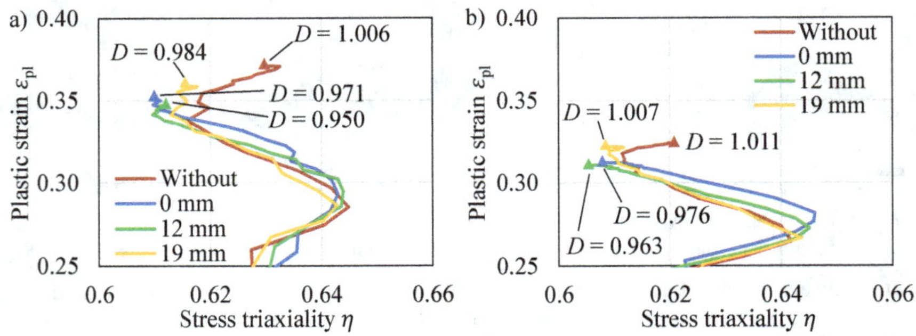

Fig. 6: Plastic strain over stress triaxiality depending on the punch distance for a) HX340LAD and b) HCT600X

For both materials, the best results are achieved with a distance of the counter punch of 12 mm. Shifting the counter punch to a distance of 0 mm does not reduce the resulting plastic strain compared to a distance of 12 mm and therefore does not diminish the risk of fractures. This concludes that the counter punch has no positive influence in the first phase of forming, when the radius of the cup is shaped around the radius of the forming punch.

Furthermore, if the sheet metal comes into contact with the counter punch after a distance of 19 mm, a higher increase in plastic strain can be seen, resulting in a closer approach to the forming capacity of the materials. It can even be seen that material failure is present for HCT600X, since the mean damage D exceeds the value 1. This leads to the conclusion that the sheet metal should come into contact when the bottom of the cup is shaped out, which is in the considered case a distance of 12 mm, to achieve an optimal effect on the fracture prevention.

Finally, the influence of the punch geometry was investigated. For this purpose, the flat counter punch was extended by a radius that merges tangentially into the edge of the forming die, which is shown in Fig. 1 b). The goal hereby was to increase the contact surface in the corner area of the workpiece. Fig. 7 shows the resulting plastic strain over the stress triaxiality and compares it to conventional deep drawing without counter force and deep drawing with a flat counter punch at a distance of 19 mm. Here, the counter force was again held constant at 15 kN and blank holder forces of 320 kN for HX340LAD and 230 kN for HCT600X were applied.

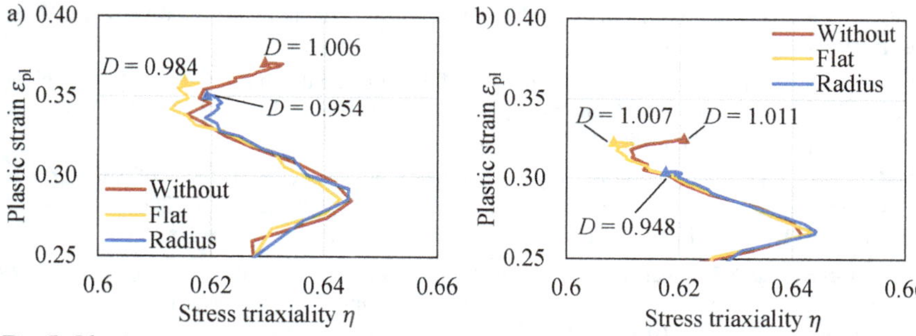

Fig. 7: Plastic strain over stress triaxiality depending on the punch geometry for a) HX340LAD and b) HCT600X

The use of a counter punch with a radius leads to a further reduction of the plastic strain compared to a flat counter punch for both materials. This can be attributed to the fact that the contact area was extended and the counter force could be applied in the most critical area for material damage. Therefore, the use of a counter punch that is geometrically adapted to the geometry of the workpiece has a positive effect on the prevention of fractures and thus on the extension of the process limits. However, the contact to the sheet metal can only be established after 19 mm due to the radius of the counter punch. Hence, similar results can be achieved with the flat counter punch, which comes in contact after a distance of 12 mm.

Summary and Outlook

In this work, the influence of a counter punch on the formation of cracks during deep drawing of two high-strength steels HX340LAD and HCT600X was investigated. It was found that the counterforce leads to a reduced plastic deformation in the bottom and radius of the cup and thus counteracts the initiation of fractures. As a result, higher blank holder forces and thus higher limiting drawing ratios can be achieved, which results in the extension of the process window. By means of numerical parameter studies, it was found that the highest influencing factor on fracture initiation is the force applied to the counter punch. As a result, the maximum force of 60 kN provided the best results in this research. Furthermore, it was found that the counter punch should come into contact with the workpiece as soon as the bottom of the cup is shaped, as this results in lower damage values compared to an earlier or later contact during the process. A further reduction of the damage could be achieved by a counter punch with a radius due to the increased contact area. But since the contact to the sheet can only be applied after a distance of 19 mm, only slight improvements could be achieved compared to a flat punch, which comes into contact after a distance of 12 mm. Therefore, the use of a flat counter punch is preferred in the considered process, as it less complex and simpler to implement into the tool system than a counter punch with radius.

In future investigations, the use of a counter punch should be applied in an industrial process to investigate the extension of the production limits. Furthermore, an economic feasibility study to weigh the benefits of the counter punch versus the additional costs due to a more complex tool design as well as higher required press forces would be beneficial.

Acknowledgements
The results presented were obtained in the project "Extension of the forming limits during deep drawing by additional force transmission" – 212270168. The authors thank the German Research Foundation (Deutsche Forschungsgemeinschaft, DFG) for their financial support.

References

[1] C. P. Singh, G. Agnihotri, Study of deep drawing process parameters: a review, Int. J. Sci. Res. Publ. 5 (2) (2015) 352-366.

[2] A. Meyer, B. Wietbrock, G. Hirt, Increasing of the drawing depth using tailored rolled blanks – Numerical and experimental analysis, Int. J. Mach. Tools Manuf. 48 (2007) 522-531. https://doi.org/10.1016/j.ijmachtools.2007.08.003

[3] P.D. Wu, J.D. Embury, D.J. Lloyd, Y. Huang, K.W. Neale, Effects of superimposed hydrostatic pressure on sheet metal formability, Int. J. Plast. 25 (9) (2009) 1711-1725. https://doi.org/10.1016/j.ijplas.2008.10.002

[4] A. Anil Kumar, S. Satapathy, D. Ravi Kumar, Effect of Sheet Thickness and Punch Roughness on Formability of Sheets in Hydromechanical Deep Drawing, J. Mater. Eng. Perform. 19 (2010) 1150–1160. https://doi.org/10.1007/s11665-010-9602-8

[5] W. Petzold, M. Otto, Increasing the LDR through additional radial compressive stress, J. Mater. Process. Technol. 170 (1–2) (2005) 151-155. https://doi.org/10.1016/j.jmatprotec.2005.04.114

[6] Y. Morishita, T. Kado, S. Abe, Y. Sakamoto, F. Yoshida, Role of counterpunch for square-cup drawing of tailored blank composed of thick/thin sheets, Jo J. Mater. Process. Technol. 212 (10) (2012), 2102-2108. https://doi.org/10.1016/j.jmatprotec.2012.05.011

[7] B.-A. Behrens, A. Bouguecha, C. Bonk, N. Grbic, M. Vucetic, Validation of the FEA of a deep drawing process with additional force transmission, AIP Conf Proc 1896 (2017) 080024. https://doi.org/10.1063/1.5008104

[8] P. Althaus, J. Weichenhain, S. Hübner, H. Wester, D. Rosenbusch, B. A. Behrens, Numerical and experimental failure analysis of deep drawing with additional force transmission, in: M. Liewald (Eds.), Production at the Leading Edge of Technology, Springer Nature, Switzerland, 2023, pp. 1-10.

[9] B.-A. Behrens, C. Bonk, N. Grbic, M. Vucetic, Numerical analysis of a deep drawing process with additional force transmission for an extension of the process limits, IOP Conf. Ser.: Mater. Sci. Eng. 179 (1) (2017) 012006. https://doi.org/10.1088/1757-899X/179/1/012006

[10] R. Hill, A theory of the yielding and plastic flow of anisotropic metals, Proc. R. Soc. Lond. 193 (1033) (1984) 281-297. https://doi.org/10.1098/rspa.1948.0045

[11] Y. Bai, T. Wierzbicki, Application of the extended Coulomb-Mohr model to ductile fracture, Int. J. Fract. 161 (1) (2010) 1-20. https://doi.org/10.1007/s10704-009-9422-8

[12] B.-A. Behrens, D. Rosenbusch, H. Wester, P. Althaus, Comparison of three different ductile damage models for deep drawing simulation of high-strength steels, IOP Conf. Ser.: Mater. Sci. Eng. 1238 (2022) 012021. https://doi.org/10.1088/1757-899X/1238/1/012021

Sheet Metal 2023 Materials Research Forum LLC
Materials Research Proceedings **25** (2023) 337-344 https://doi.org/10.21741/9781644902417-42

Stainless steel deep drawing with polymer punches produced with fused filament fabrication technology: effect of tool orientation on the printing plate

Luca Giorleo[1, a *] and Elisabetta Ceretti[1,b]

[1]Advanced Prototype Laboratory, University of Brescia, Italy

[a]luca.giorleo@unibs.it, elisabetta.ceretti@unibs.it[b]

Keywords: Additive Manufacturing, Deep Drawing, Polymer Tool

Abstract. In this research the potential of using polymer punches in a deep drawing process to realize stainless steel cups having diameter equal to 40 mm and drawing depth equal to 18 mm was investigated. Punches have been Additive Manufacturing printed following two different orientations (horizontal and vertical) and two different wire orientations with respect to the drawing direction (45° and 90°). Cups and punches radii, heights, roundness tolerances and linear profiles have been acquired to compare process performances. Results highlight benefits and problems of all tested punches.

Introduction

Nowadays manufacturing paradigm is changing into a production-oriented model to meet consumer expectations for a huge variety of products and allows the achievement of ad hoc product characteristics (mass customization) [1]. Highly customized production offers unique and very low-production-volume products; however, this causes an excessive increase of costs for the tooling, consequently customization it is not yet a competitive market.

In this scenario, the additive manufacturing (AM) process could be a good solution for producing tools for prototypes, small batches, or customized production. This technology includes a group of manufacturing techniques in which the shape of the physical part is generated by adding the material layer by layer [2]. AM technologies are characterized by short setup times, a wide range of printable materials for both plastics and metals, and a higher design complexity. Currently, there is a great demand for AM technologies to support product development of tooling or tooling inserts in different industrial applications such as casting [3, 4], injection moulding [5, 6] or forming [7, 8].

For deep drawing process a specific AM technology, the Fused Filament Fabrication (FFF), was tested to produce tools with polylactic acid (PLA) polymer and to deform cold-rolled steel sheets such as DC01 or DC04 with a thickness of 0.7–1 mm [9-12]. Results showed that PLA can be used to form steels sheets, but the process loads caused the punch flattening so affecting the cups drawing radius and height. The main limit is that all tested geometries have been produced with punch axis printed parallel to z axis that could be not the best solution for parts that must withstand to compression forces.

The present study focuses on the use of polymers reinforced with short carbon fiber to produce tooling for deep drawing applications with FFF technology; in particular the authors tested the influence of punches orientation on the print plate and measured punches performance in terms of quality of the produced cups and permanent deformation of the punches due to production process. An experimental campaign was designed, and a statistical approach has been followed to analyse the results.

Content from this work may be used under the terms of the Creative Commons Attribution 3.0 license. Any further distribution of this work must maintain attribution to the author(s) and the title of the work, journal citation and DOI. Published under license by Materials Research Forum LLC.

Materials and Methods

The investigated case study was the deep drawing of a circular blank (*Φ* 86, 1 mm thick) to produce cups with height equal to 18 mm; stainless steel alloy (AISI 304) was used as sheet material. Punches have a hollow cylindrical geometry with a central hole designed as a housing for the screw needed for the assembly on the press (Fig. 1a). Forming die and blankholder have hollow cylindrical geometries too (fig.1b). 45 NiCrMo 16 was selected as material for the forming die and blank holder.

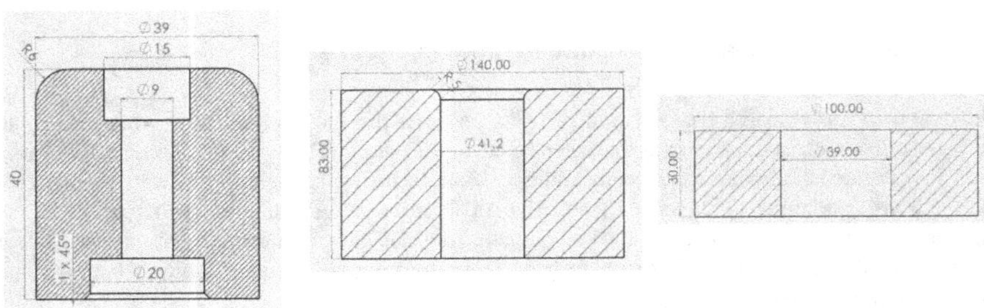

(a) (b) (c)

Fig. 1 Geometry of punch (a), forming die (b) and blank holder (c)

Punches were produced using an FFF printer Mark Two (Markforged, Watertown, MA, USA). A solid filling strategy with two wall layers was adopted, the layer height was set to 0.125 mm. Onyx, which is nylon reinforced with carbon microfibers, was used as material and printed at 245 °C; the other process parameters as nozzle feed or offset distance between layer are blocked and it is not possible to change them from the machine governing software. Under these conditions three different punches have been designed as a function of the orientation on the printing plate and of the wire orientation inside the parts. In particular:

- Geometry H45: punch with axis parallel to the printing plate; wire oriented at 45° and 135° to the drawing direction; fillet radius average roughness equal to 7.2 μm.
- Geometry H90: punch with axis parallel to the printing plate; layer with wire oriented at 0° and 90° to the drawing direction; fillet radius average roughness equal to 8.1 μm.
- Geometry V: punch with axis orthogonal to the print plate; layer orthogonal to the drawing direction; fillet radius average roughness equal to 12.8 μm.

Tab.1 lists the mechanical properties of the involved materials. Fig. 2 shows the wires orientation in the cross section and punches position on the printing plate.

Tab.1 Mechanical properties of involved materials

	Onyx [13]	45 NiCrMo 16	AISI 304
Young Modulus [GPa]	2.4	284	193
Tensile Stress at Yield [MPa]	37	696	190
Tensile Stress at Break [MPa]	40	950	500-700
Tensile Strain at Break [%]	25	11	40
Density [g/cm^3]	1.2	7.84	8

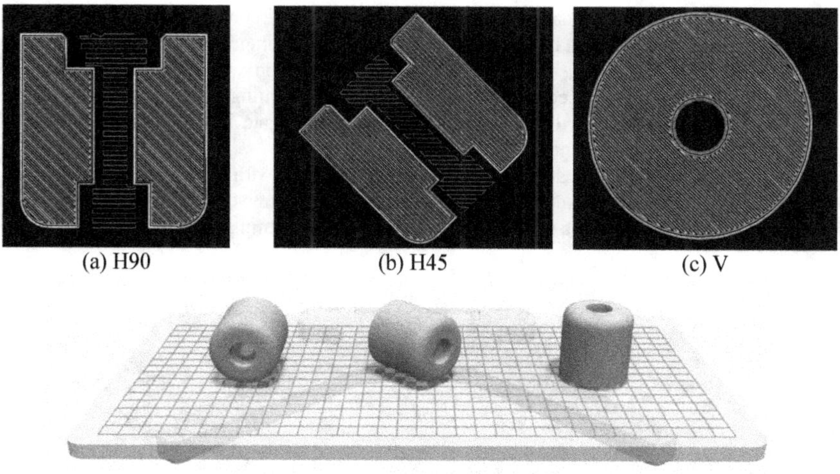

(a) H90 (b) H45 (c) V

(d) left H90, center H45, right V

Fig. 2 Cross section of polymer punches and orientation on the printing plate. (white for part, purple for support, cyan for wall layer)

The press utilized for the experiments was an EVL/400-A (Galdabini, Varese, Italy) with a power of 400 tons, the punch speed was set constant and equal to 10 mm/s and the maximum press load was equal to 180 kN. The blank-holder load was set to 1 kN (recommended value from the industrial practice), the blanks had a thickness of 1 mm and the punch-die clearance distance was set to 1.1 mm. The blanks were manually lubricated with mineral oil. Tab.2 reports the most significand process parameters.

Tab.2 Deep drawing process parameters

Parameter	Value
Punch diameter [mm]	39
Punch fillet radius [mm]	6
Forming die internal diameter	41.2
Forming die fillet radius [mm]	5
Clearance distance [mm]	0.1
Drawing ratio	2.2
Blank diameter [mm]	85
Blank thickness [mm]	1
Blank material	AISI 304
Drawing depth [mm]	18
Replicas	3

After production two different measures have been done both on cups and punches:

- A linear profile analysis (LP) to acquire the internal shape of the cups, the measures of the effective drawing height (h_d) and of the cup fillet radius. On the punches the LP was executed to identify possible plastic deformations of the fillet radius due to the drawing process and evaluate the punch height (h_p) after the cups production.

- A circular profile analysis (CP) to acquire the internal circumference of the cups at 6 mm from the top; these data were used to estimate the average cup radius (R_{med}) and cups roundness tolerances $r_t = R_{max} - R_{min}$. The same method was followed for punches CP analysis but with two acquisitions: a first one in the forming zone (6 mm from the punch top) and a second one 15 mm from the top to investigate if the forming process affect punches radius and roundness.

Data points were acquired using Cyclone Series 2 equipped with the probe SP620 (Renishaw, Wotton-under-Edge, UK). Results have been compared with statistics methods as ANOVA and Tukey range test. Fig.3 shows a scheme of the LP and CP measuring points.

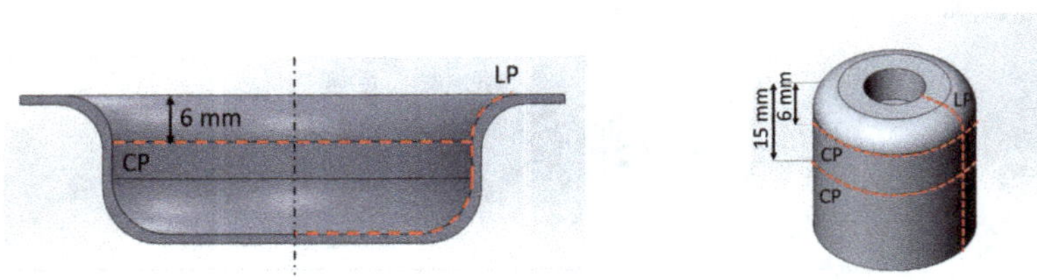

(a) cup cross section (b) punch

Fig.3 Graphical representation of linear and circumferential profiles acquired

Results

In this section the most significant results have been reported; the analyses have been divided in two subsections referring to the cups and punches, respectively.

Cups analysis. The main results of the measures executed on the cups produced with H45, H90 and V punches are reported in Fig. 4. Specifically, fig.4a shows the LP of cups produced with additive punches, as can be observed similar trend was achieved by test H90 and V, on the contrary cups produced by test H45 are characterised by higher fillet radius. Fig. 4b shows a focus on the cup fillet radius. The ANOVA test evaluated on the drawing height shows a significative effect of punch orientation (p-value < 0.001) and Tukey range test highlights that test V produces cups with lower drawing height (Fig. 4c). The ANOVA test of cup radius present a p-value equal to 0.021 so that punch orientation is significative; Tukey range test identified three different cup radius levels with a maximum in test H90 and a minimum in test V (Fig. 4d). Tab.3 presents the average cup radius and the related standard deviation, drawing height and the roundness tolerances; about r_t it is possible to observed that is maximum in test H90 and minimum in V. Moreover, test H90 and H45 have higher cup radius standard deviation (0.150 and 0.148) with respect to test V (0.131).

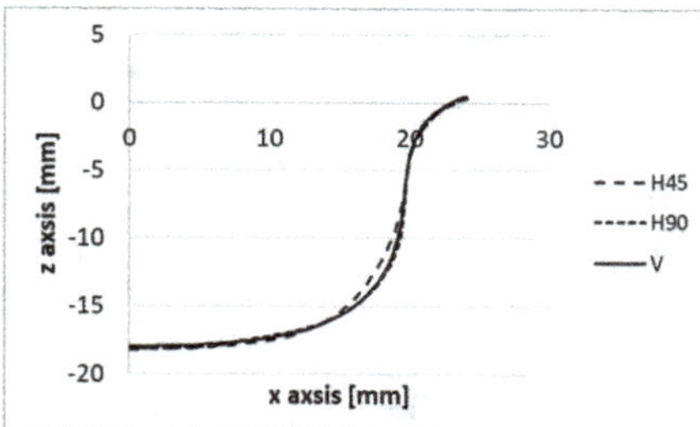

(a) Cups LP

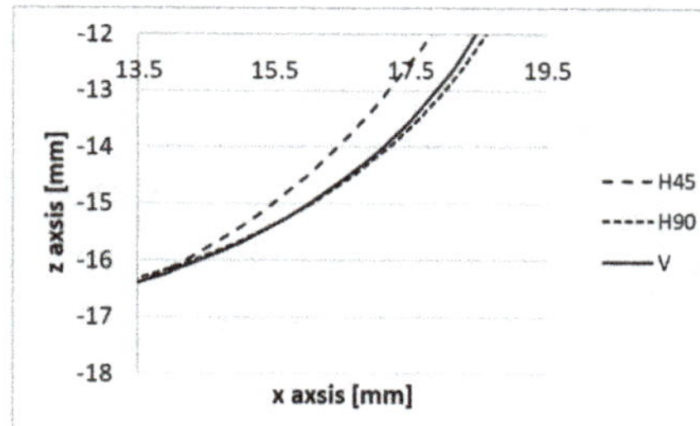

(b) Cups fillet radius LP

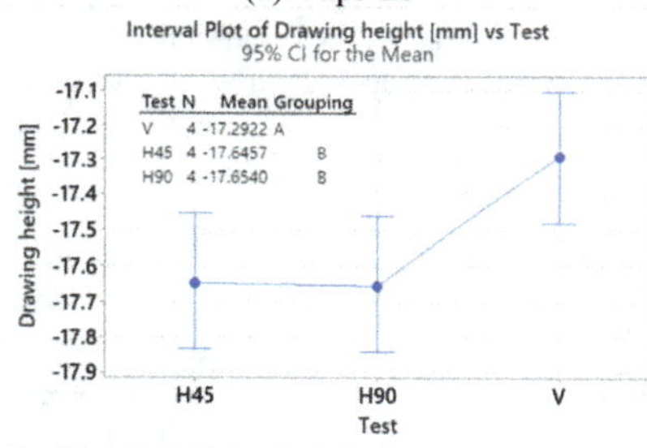

(c) Drawing height Tukey results

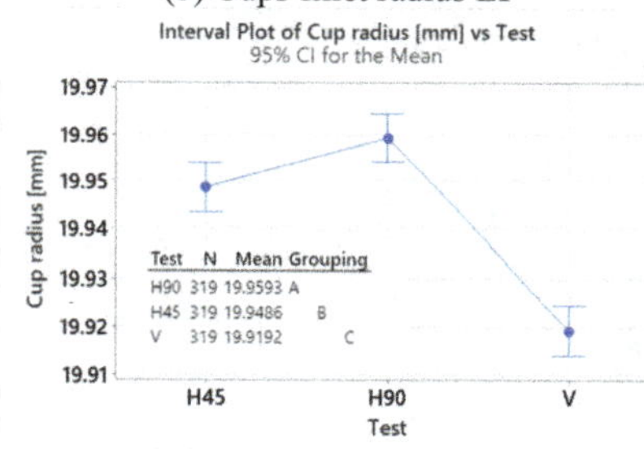

(d) Cup radius Tukey results

Fig. 4 Main results of cups analysis

Tab.3: average cup radius, standard deviation, roundness tolerances, drawing height

Test	R_{med} [mm]	σ [mm]	r_t [mm]	h_d [mm]
H90	19.96	0.053	0.150	17.65
H45	19.95	0.048	0.148	17.65
V	19.92	0.041	0.131	17.29

Punch Analysis after deep drawing process. Results obtained from the punches analysis after drawing highlight that the process affects tools dimensions. Fig. 5a shows how both distance from top and test configuration are significant for the punch radius; as reported in Fig. 5b punch radius far from the forming zone (distance 15 mm) is quite similar for all tested geometries, on the contrary close to the fillet radius (distance 6 mm) the radius reaches lower values for the horizontal tests (H90). The loss of accuracy of horizontal punches is confirmed by Fig. 5c where a magnification of ¼ CP is plotted. These results are also reported in Tab. 4: test H90 has lower punches radius equal to 18.93 ± 0.49 mm at a distance of 6 mm from the top, moreover it has the higher roundness error (1.50 mm at 6 mm for top and 1.14 mm at 15 mm). Punches produced with axis parallel to z (V) are characterised by lower radius reduction on the fillet radius (19.3 ± 0.1) and lower roundness tolerances (0.4 mm for distance equal to 6). The LP analysis presents a better behaviour for vertical punches with respect to the horizontal ones. Fig. 5d shows a magnification of LP with a focus on the punch fillet radius: in this zone a similar trend could be registered for test V and H45 while test H90 has a higher fillet radius and lower punch height. This trend is confirmed by the values reported in Tab.4.

Analysis of Variance

Source	DF	Adj SS	Adj MS	F-Value	P-Value
Distance from top [mm]	1	117.18	117.184	1462.09	0.000
Test	2	14.81	7.405	92.39	0.000
Error	1940	155.49	0.080		
Lack-of-Fit	2	11.49	5.743	77.29	0.000
Pure Error	1938	144.00	0.074		
Total	1943	287.48			

(a) punches radius ANOVA

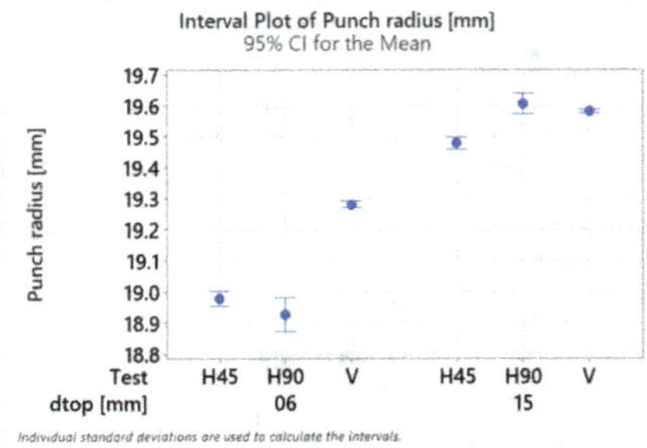

(b) comparison between measured punches radius at different distance from the punch top (d_{top})

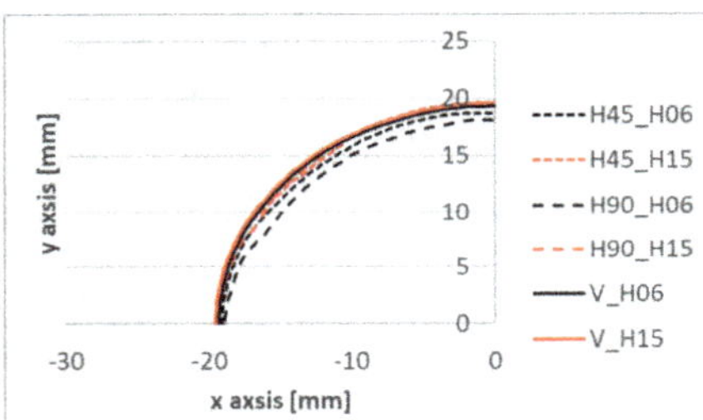

(c) quarter CP of punches

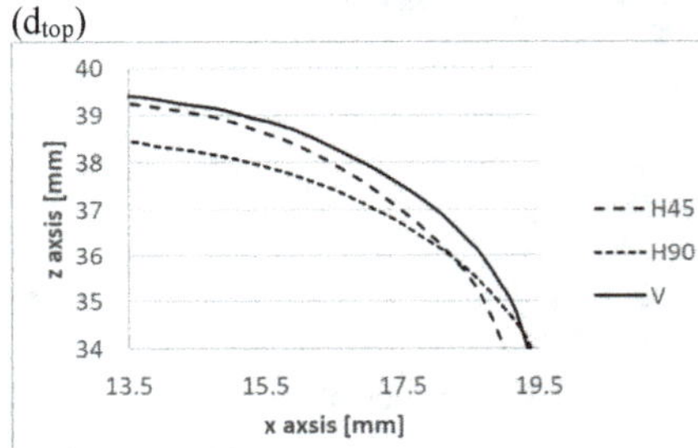

(d) LP of punches fillet radius

Fig. 5 Main results of punches analysis

Tab.4 Average punch radius with standard deviation, roundness tolerances, height

Test	Distance from top [mm]	R_{med} [mm]	σ [mm]	r_t [mm]	h_p [mm]
H45	6.0	18.98	0.230	0.980	39.8
H45	15.0	19.48	0.185	0.762	
H90	6.0	18.93	0.495	1.502	39.3
H90	15.0	19.60	0.294	1.140	
V	6.0	19.28	0.100	0.402	39.7
V	15.0	19.58	0.057	0.256	

Conclusion

In this research the authors investigated the influence of punch orientation on the printing plate produced by FFF technology. The results showed different trends for the tested geometries, in particular:

- Punches produced with axis parallel to z axis (test V) realise cups with lower drawing height, cup radius and cup roundness tolerances; the compression induced by the drawing process reduces punch height and generates a moderate diameter increase (punch radius is 19.5 mm while cup radius is 19.92 mm). This result is confirmed by the punch analysis that identify test V as the best in terms of minimum variation of punch radius acquired at different distances from the top, minimum punch roundness tolerances and higher punch height.
- Punches with horizontal axis and wire orientation at 45° and 135° with respect to the drawing direction (test H45) produced cups having higher heights, linear profile equal to

Materials Research Forum LLC
https://doi.org/10.21741/9781644902417-42

test V but higher cup radius and roundness tolerances. Punch analysis highlights a permanent deformation that increases fillet radius as it can be observed by the linear profile even if the punch height is coherent with test V (fig.5d). A significant difference can be found observing the radius measured at 6 and 15 mm from the top as shown in fig.5b with a roundness tolerance that reaches 0.98 and 0.76 mm, respectively.

- Punches with horizontal axis and wire orientation at 0° and 90° with respect to the drawing direction (test H90) realise higher cup drawing height but larger cup radius, cup roundness tolerances and from their linear profile it is possible to observe an increase of cup fillet radius compared to test H45 and V. Punch analysis shows results similar to test H45 but with higher punch roundness tolerance (1.5 mm).

Summarising all the results it is possible to assert that the punch orientation on the printing plate affects punch performance during deep drawing; horizontal punches guarantee higher drawing height but less accuracy in terms of cup radius and permanent punch deformation. On the contrary vertical punches guarantee closer roundness tolerances, more accurate profiles but lower drawing height. Future researches are ongoing to evaluate the punches behaviour when the number of produced cups increases.

Acknowledge
The authors are grateful to R. Pinti of Pinti Inox S.p.a. – Sarezzo (Brescia) for the experimental campaign and the CMM analysis.

References

[1] D.A. Kai, E.P. de Lima, M.W. Machado Cunico, S.E.G. da Costa, Additive manufacturing: A new paradigm for manufacturing, Proceedings of the 2016 Industrial and Systems Engineering Research Conference, ISERC 2016, (2020) 452-457.

[2] A. Rosochowski, A. Matuszak, Rapid tooling: the state of the art, Journal of Materials Processing Technology, 106 (2000) 191-198. https://doi.org/10.1016/S0924-0136(00)00613-0

[3] C.M. Cheah, C.K. Chua, C.W. Lee, C. Feng, K. Totong, Rapid prototyping and tooling techniques: A review of applications for rapid investment casting, International Journal of Advanced Manufacturing Technology, 25 (2005) 308-320. https://doi.org/10.1007/s00170-003-1840-6

[4] L. Giorleo, M. Bonaventi, Casting of complex structures in aluminum using gypsum molds produced via binder jetting, Rapid Prototyping Journal, 27 (2021) 13-23. https://doi.org/10.1108/RPJ-03-2020-0048

[5] J.R.C. Dizon, A.D. Valino, L.R. Souza, A.H. Espera, Q. Chen, R.C. Advincula, Three-dimensional-printed molds and materials for injection molding and rapid tooling applications, MRS Communications, 9 (2019) 1267-1283. https://doi.org/10.1557/mrc.2019.147

[6] L. Giorleo, B. Stampone, G. Trotta, Micro injection moulding process with high-temperature resistance resin insert produced with material jetting technology: Effect of part orientation, Additive Manufacturing, 56 (2022) 102947. https://doi.org/10.1016/j.addma.2022.102947

[7] D. Tomasoni, S. Colosio, L. Giorleo, E. Ceretti, Design for additive manufacturing: Thermoforming mold optimization via conformal cooling channel technology, Procedia Manufacturing, 47 (2020) 1117-1122. https://doi.org/10.1016/j.promfg.2020.04.128

[8] L. Giorleo, E. Ceretti, Deep drawing punches produced using fused filament fabrication technology: Performance evaluation, Journal of Manufacturing Processes, 84 (2022) 1-9. https://doi.org/10.1016/j.jmapro.2022.09.054

[9] G. Schuh, G. Bergweiler, P. Bickendorf, F. Fiedler, C. Colag, Sheet metal forming using additively manufactured polymer tools, Procedia CIRP, 93 (2020) 20-25. https://doi.org/10.1016/j.procir.2020.04.013

[10] P. Frohn-Sörensen, M. Geueke, T.B. Tuli, C. Kuhnhen, M. Manns, B. Engel, 3D printed prototyping tools for flexible sheet metal drawing, International Journal of Advanced Manufacturing Technology, 115 (2021) 2623-2637. https://doi.org/10.1007/s00170-021-07312-y

[11] M. Geueke, P. Frohn-Sörensen, J. Reuter, N. Padavu, T. Reinicke, B. Engel, Structural optimization of additively manufactured polymer tools for flexible sheet metal forming, Procedia CIRP, 104 (2021) 1345-1350. https://doi.org/10.1016/j.procir.2021.11.226

[12] G. Bergweiler, F. Fiedler, A. Shaukat, B. Löffler, Experimental investigation of dimensional precision of deep drawn cups using direct polymer additive tooling, Journal of Manufacturing and Materials Processing, 5 (2021). https://doi.org/10.3390/jmmp5010003

[13] L. Giorleo, I. Papa, A.T. Silvestri, Pin-bearing mechanical behaviour of continuous reinforced Kevlar fibre composite fabricated via fused filament fabrication, Progress in Additive Manufacturing, 7 (2022) 723-735. https://doi.org/10.1007/s40964-022-00261-2

Sheet Metal 2023 Materials Research Forum LLC
Materials Research Proceedings **25** (2023) 345-352 https://doi.org/10.21741/9781644902417-43

Extrusion as an energy-efficient manufacturing process for thermoplastic organosheets

Bastian Richter[1, a], Benedikt Neitzel[1, b] and Florian Puch[1, 2, c *]

[1]Technische Universität Ilmenau, Plastics Technology Group, 98693 Ilmenau, Germany

[2]Thüringisches Institut für Textil- und Kunststoff-Forschung e.V., 07407 Rudolstadt, Germany

[a]bastian.richter@tu-ilmenau.de, [b]benedikt.neitzel@tu-ilmenau.de, [c]florian.puch@tu-ilmenau.de

Keywords: Fiber Reinforced Plastic, Extrusion, Thermoplastic Composites

Abstract. Organosheets combine the advantages of reinforcement fibers and thermoplastic polymers. By pairing these two materials, composites with outstanding mechanical properties and low densities can be produced. These semi-finished products can be further processed into complex and functionalized components by thermoforming or injection molding. There are a number of different manufacturing processes for continuous fiber reinforced thermoplastics (CFRT), however, most of them require long production times and recurrent melting of the polymer resulting in high energy and manufacturing costs. This study presents a novel extrusion process, that enables a continuous production of reinforced thermoplastic sheets with only one melting step. Due to the high energy efficiency and wide range of processible materials, this process shows a high potential for an economical production of CFRT. To investigate the extrusion process in more detail, the influence of the processing and the flow behavior of the polymer on the impregnation quality and the mechanical properties of the composites were studied. The results showed increasing fiber volume contents with lower polymer viscosities. Furthermore, higher die temperatures and pressures resulted in higher fiber volume contents and thus in higher mechanical properties. The experiments also revealed that a complete impregnation can currently not be achieved without an additional small double belt press due to the line load of the calender, the high viscosity of the melt and the short impregnation time.

Introduction

Organosheets – continuous fiber reinforced thermoplastics (CFRT) – are semi-finished flat sheets that consist of continuous reinforcement fibers and a thermoplastic matrix polymer. The promising material class provides excellent weight specific mechanical properties [1] and can be recycled in contrast to conventional thermoset composites [2]. Due to their suitability for circular economy and the applicability of renewable and biobased fibers and polymers, continuous fiber reinforced thermoplastics enable sustainable lightweight construction [3]. Organosheets can be melted repeatedly [4] and further processed in highly automated mass production processes such as injection molding, thermoforming or other joining technologies [4].

Despite their high lightweight potential [2] and the wide range of well-suited processing options [5], organosheets have not yet been established on the market [6]. The main reason is the high price of the reinforced sheets [6], which is caused by high costs for the textiles [4], limited throughputs of common manufacturing processes [7] and intense energy consumption during the production [8]. To increase the economic attractiveness of the continuous reinforced thermoplastic composites new energy- and cost-efficient manufacturing processes must be developed [9]. In recent years, there have already been increasing efforts to optimize the production speeds of the conventional processes [5]. However, until now, the industrial usage of organosheets remains hampered by their high prices.

Content from this work may be used under the terms of the Creative Commons Attribution 3.0 license. Any further distribution of this work must maintain attribution to the author(s) and the title of the work, journal citation and DOI. Published under license by Materials Research Forum LLC.

One approach to reduce the costs during the manufacturing of CFRT is the reduction of processing and melting steps. State of the art are continuously or semi-continuously working presses that process plastic films or powder pre-impregnated fabrics [4]. These films have to be extruded or granules must be milled into powder prior to the production of the reinforced sheets. In a subsequent process step the semi-finished plastic product is molten again in a press and the impregnation occurs. First, macro-impregnation takes place and the voids between the fiber bundles are filled with melt. Due to the small gaps between the individual fibers, the wetting of the individual filaments – the micro-impregnation – occurs after the completion of the macro-impregnation. Finally, the thermoplastic composite is cooled and solidifies.

In contrast to the state of the art, the extrusion process enables a continuous impregnation of the reinforcement textile with only one melting step. The thermoplastic granules are molten by an extruder and are directly applied onto the dry fabric. Additionally, due to the direct processing of the granules, the process is suited for a wide variety of materials.

Several publications already discussed the influence of the process parameters such as pressure or temperature on the impregnation quality and the mechanical properties of organosheets. However, in these studies, only conventionally used presses were considered. Due to the line load between the calender rolls the extrusion process cannot be directly compared to these manufacturing processes and the existing models cannot be applied to describe the matrix flow into the fiber bundles.

Materials

Three different polypropylenes (PP) were investigated. The three polymers are typically used for injection molding and had comparable densities, melting and crystallization temperatures (Tab. 1). Differential scanning calorimetry (DSC 204 F1 Phoenix, NETZSCH-Gerätebau GmbH, Germany) with a heating rate of 20 K/min was used to determine the melting and crystallization temperatures.

Tab. 1 Overview of the polymers used in this work

	Borealis® BC612WG	Ducor DuPure® SR76	Ineos Rigidex® 450-HP90
Melt flow rate (230 °C / 2.16 kg, ISO 1133)	5 g/10 min	15 g/10 min	90 g/10 min
Melting temperature	175.4 °C	176.4 °C	175.4 °C
Crystallization temperature	106.1 °C	110.3 °C	119.3 °C

As can be seen in Tab. 1, the major difference between the polymers is the melt flow rate. To assess the different flow behavior in more detail, a rotational rheometer (Modular Compact Rheometer MCR 302, Anton Paar GmbH, Germany) with a cone-plate setup was used. The viscosity data for the different temperatures, shear-rates and polymers are shown in Fig. 1. All investigated polymers show a pronounced newtonian plateau and a shear thinning behavior at higher shear rates. Since shear rates below 100 1/s usually occur during calendering, it can be expected that the different viscosities at low shear rates influence the impregnation [9].

A glass fiber textile (style number 92130), supplied by Porcher Industries Germany GmbH, with plain weave was used in all experiments. The warp consisted of five 68 tex rovings and the weft of one 272 tex roving. The rovings consisted of e-glass filaments with a 9 µm diameter and had an areal weight of 390 g/m². The fabric was coated with a volan chromium complex finish (FK144) und had a finish content between 0.08 and 0.28%.

Materials Research Proceedings **25** (2023) 345-352 https://doi.org/10.21741/9781644902417-43

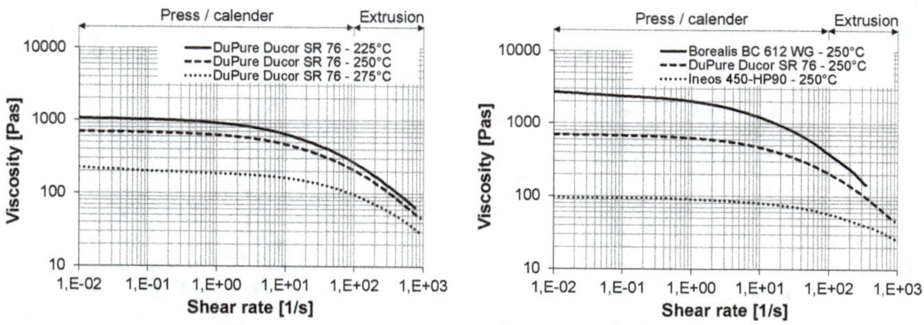

(a) (b)

Fig. 1 Viscosity data for different temperatures (a) and polymers (b)

Experimental

The aim of this work was to investigate the potential of the extrusion process for the manufacturing of CFRT and to determine the influences of the flow behavior of the matrix polymer and the processing on the quality of the reinforced sheets. With the extrusion process, shown in Fig. 2, the polymer is applied directly onto the dry fabric and converted into a reinforced sheet. The granules are molten by a co-rotating twin-screw extruder. A melt pump separates the melt stream into two streams and ensures a consistent output. In the extrusion dies the melt is evenly spread and applied to both sides of the reinforcement textile to minimize the impregnation length. Due to the pressure of the calender rolls, the melt infiltrates the fabric and the macro impregnation between the rovings takes place. Because of the short impregnation time and the line load between the calender rolls, a complete micro-impregnation does not occur at this stage. An additional small double belt press can be used to ensure a complete wetting of all filaments. In contrast to the conventionally used double belt presses, this press only requires a reduced heating power, as the pre-impregnated organosheets enters the machine at an elevated temperature.

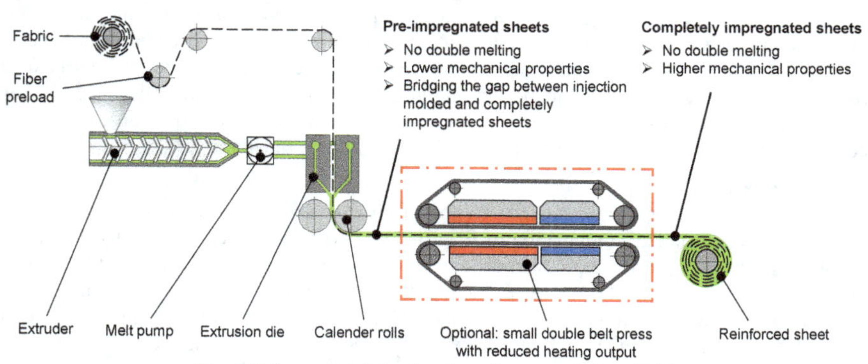

Fig. 2 Schematic of the developed extrusion line [10]

The focus of this study is the influence of the process and material parameters on the quality of the CFRT after the impregnation between the calender rolls. The key components of the extrusion line and the design of the extrusion dies are illustrated in Fig. 3. The dies are positioned directly in front of the calender rolls to minimize the temperature loss until the impregnation of the fabric. Furthermore, the position above the calender prevents an uneven influence of gravity on the melt.

 Materials Research Forum LLC
https://doi.org/10.21741/9781644902417-43

The two dies have an identical, but mirrored design and redirect the melt before the stream exits the die to minimize the distance to the fabric.

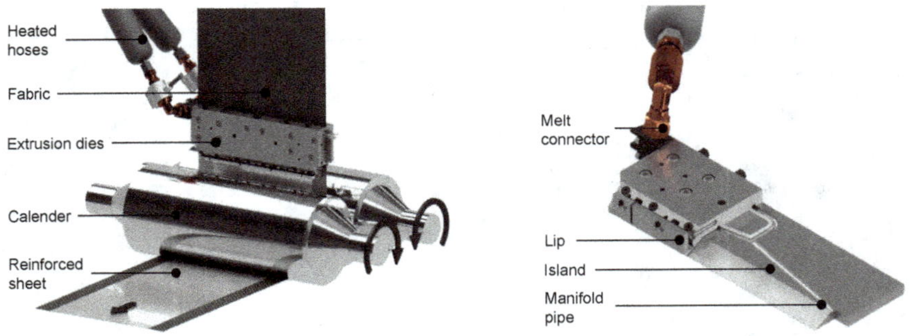

(a) (b)

Fig. 3 Impregnation of the fabric (a) and design of the extrusion dies (b) [10]

In the conducted experiments a full factorial experimental plan was used and the die temperature and the pressure of the hydraulic cylinders driving the calender rolls were varied in two stages. The temperature was varied between 225 and 275 °C, while a hydraulic pressure between 5 and 9 MPa was used during the experiments. Furthermore, the center point of the experimental plan (250 °C, 7 MPa) was investigated to identify non-linear relationships. The material used was Ducor DuPure SR 76.

The influence of the flow behavior on the quality of the CFRT was investigated using three different matrix polymers (Ducor DuPure SR76, Borealis BC612WG and Ineos 450-HP90) at the same operating point. For the comparison of the materials, the center point of the full factorial test plan was selected with a hydraulic pressure of 7 MPa and a die temperature of 250 °C.

The haul-off speed and the extruder throughput were kept constant at 1.2 m/min and 3.1 kg/h in all experiments of this study. Previous studies showed that these two parameters have the greatest influence on the fiber and void volume content, as they directly affect the matrix-fiber-ratio [10]. However, it has been found that the impregnation quality is not influenced by these parameters [10]. The extruder temperature and the temperatures of the heated hoses were kept constant in all experiments, to prevent a thermal degradation of the matrix polymer.

Characterization

The impregnation quality of the obtained composites was evaluated by determining the fiber and void volume content. The combination of fiber and void volume content essentially defines the resulting mechanical properties of fiber reinforced composites for a given fiber orientation [11]. For each experiment ten rectangular specimens with the dimensions of 40 mm x 20 mm were cut out from the center of the composite and evaluated. Samples were taken every 500 mm along the haul-off direction. The weight and the density of the samples were measured with a laboratory balance (Sartorius MSU224S, Sartorius AG Germany, Germany). The burn-out test was carried out in a muffle furnace at 550 °C for a duration of 45 minutes. After the complete thermal decomposition of the resin, the weight of the fibers m_f was determined and the fiber volume content v_f was calculated according to equation 1.

$$v_f = \frac{\frac{w_f}{\rho_f}}{\frac{w_f}{\rho_f}+\frac{1-w_f}{\rho_m}} = \frac{\frac{m_f/m_c}{\rho_f}}{\frac{m_f/m_c}{\rho_f}+\frac{1-m_f/m_c}{\rho_m}}. \tag{1}$$

Where w_f is the fiber mass content, ρ_f and ρ_m are the densities of the fibers and the matrix and m_c is the mass of the composite. With knowledge of the density of the composite ρ_c the void volume content v_v was calculated according to DIN EN 2564.

$$v_v = 1 - \left[w_f \cdot \frac{\rho_c}{\rho_f} + (1-w_f) \cdot \frac{\rho_c}{\rho_m}\right] = 1 - \left[\frac{m_f}{m_c} \cdot \frac{\rho_c}{\rho_f} + \left(1-\frac{m_f}{m_c}\right) \cdot \frac{\rho_c}{\rho_m}\right]. \tag{2}$$

Furthermore, tensile tests were conducted based on DIN EN ISO 527-4 (test specimen geometry type 2b). Ten rectangular specimens with the dimensions 250 mm x 25 mm were again cut out from the center of the composites every 500 mm for each experiment. All mechanical tests were performed in warp direction using a universal testing machine (Shimadzu SFL-50KNAG, Shimadzu Corp., Japan). All tests were performed at room temperature and at a constant testing speed of 5 mm/min.

Results and discussions

Fig. 4 shows the influence of the matrix polymer on the resulting fiber and void volume content. The comparatively high viscosity of the Borealis BC612WG results in a lower fiber volume content of 36%. The two polymers with a lower viscosity and therefore better flow show a higher fiber volume content around 40%, a further improvement with decreasing viscosity could not be determined. When examining the void volume content, no significant influence could be observed due to standard deviations around 1%. Although a higher fiber volume content can be achieved due to the better flow, the lower viscosity does not lead to a significant improvement of the void volume content or a better micro-impregnation. This observation may be attributed to the short impregnation duration while calendering. Due to the cooled calender rolls, the matrix polymer solidifies within a short time.

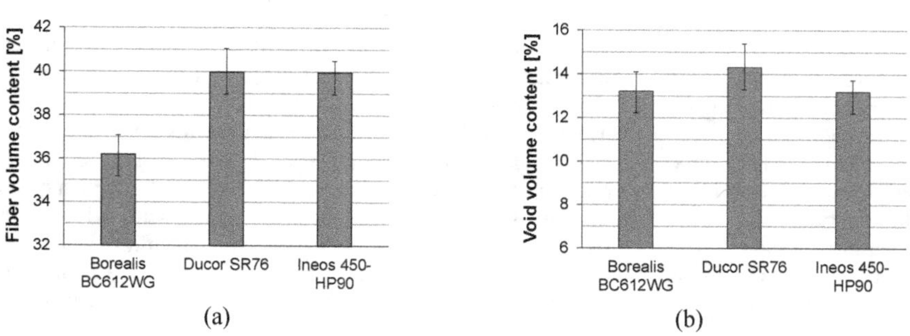

Fig. 4 Fiber (a) and void volume content (b) for different matrix polymers

The influence of the process parameters pressure and die temperature on fiber and void volume content are illustrated in Fig. 5 for Ducor SR76. Both parameters show a significant influence on the fiber volume content, while only the die temperature has a significant influence on the void volume content. The interactions between the two parameters were not significant. The statistical investigation was carried out with an error probability of 1%.

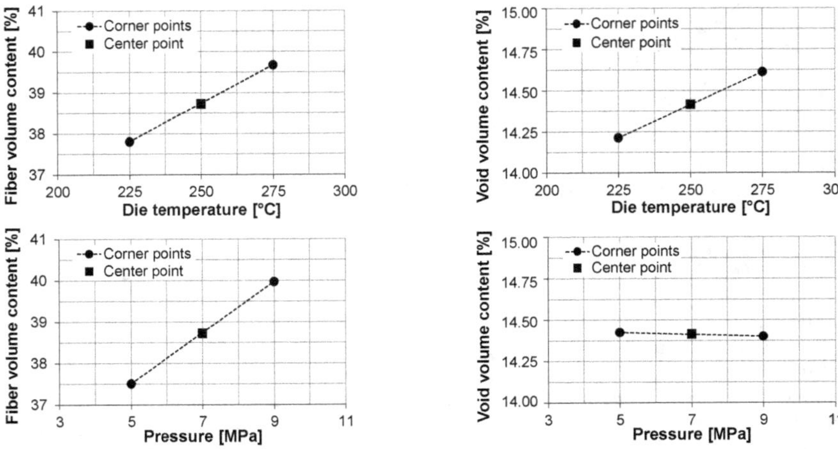

Fig. 5. Effects of die temperature and pressure on fiber and void volume content

A closer look at Fig. 5 shows that the change in pressure has the most distinct effect on the resulting fiber volume content, while a higher pressure does not significantly affect the void volume content. An increase in pressure from 5 to 9 MPa leads to a linear increase of 2.5% in fiber volume content. Contrary to the findings in the literature, a higher die temperature results not in a better impregnation [12], but in both a higher fiber and void volume content. A linear relationship can be observed for fiber and void volume content. At first impression, the higher melt temperature and consequently lower viscosity, as can be seen in Fig. 1, should also lead to a better impregnation and a lower void volume content. However, it turns out that during extrusion, the viscosity also affects the extrusion width. Due to the increased flow and the later solidification of the polymer, a wider melt film is formed in front of the calender rolls, which leads to a higher extrusion width. This reduces the matrix volume in the center of the composite, which results in an increase in the void volume content for identically sized pores. The change in fiber and void volume content is therefore primarily attributed to the changed dimensions and not to better impregnation. Since the width of the extruded sheets also increases at higher pressure, the effects of better impregnation and higher width seem to compensate each other.

The fact that changes in fiber and void volume content primarily results from the changed dimensions and not from changed impregnation can be confirmed when analyzing the relationship between extruded width and fiber volume content. Fig. 6 shows the fiber volume content in context to the normalized width, which corresponds to the ratio between extruded width and die width. Regardless of the material investigated, process and material parameters, which result in a larger extruded width due to better flow behavior and later solidification, also lead to a higher fiber volume content. A linear correlation between the width and fiber volume content can be identified.

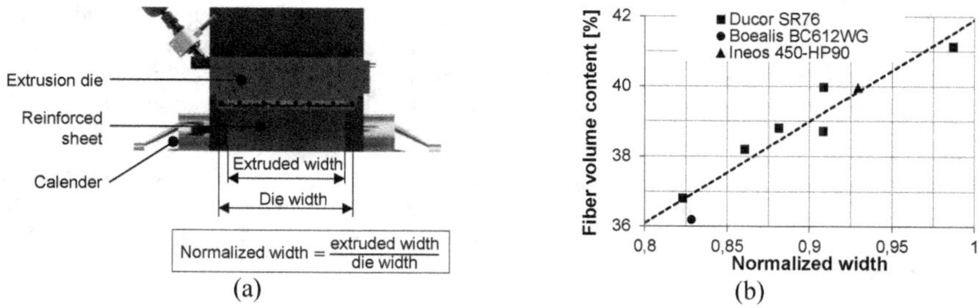

Fig. 6. Definition of the normalized width (a) and correlation between normalized width and fiber volume content (b)

The results of the mechanical characterization show a dependency of the mechanical properties on fiber volume content (fig. 7). Deviations can be attributed to the different mechanical properties of the polymers and slightly varying void contents and fiber waviness of the specimens.

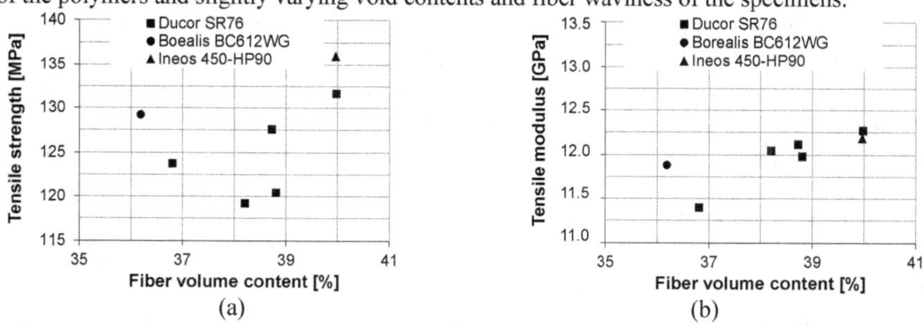

Fig. 7. Tensile strength (a) and tensile modulus (b) as functions of the fiber volume content

The tensile strength ranges from 120 MPa to 137 MPa, while the tensile modulus ranges from 11.4 to 12.3 GPa. A higher fiber volume content leads to both a higher tensile strength and modulus concurrent to the rule of mixtures. Compared to fully impregnated and consolidated organosheets [1], the extruded CFRT show around 33% lower mechanical properties. This can partially be attributed to the respective fiber volume contents. Additionally, the void volume contents of the extruded CFRT are higher and the impregnation is incomplete. Due to entrapped air in the fiber bundles, the force transmission between the fibers is disturbed, leading to premature failure. Further, the individual fibers are not perfectly aligned during the mechanical testing as a slight displacement of the fibers can occur during the extrusion. Previous investigations suggest that sizing and adhesion promoters also have a significant influence on the mechanical properties [2].

Conclusions

This study demonstrates the high potential of the extrusion process for the continuous production of CFRT. The extrusion enables an energy-efficient and economically reasonable manufacturing of pre-impregnated sheets by sparing one melting step. Due to the short line load between the calender rolls, fiber impregnation is still incomplete, even with low viscosity polymers. The investigations of process parameters showed that higher pressures and processing temperatures lead to increased fiber volume contents. Varying flow behavior and pressure, influences the melt film before the calender rolls and the width of the composites. Both tensile modulus and -strength tend to increase with higher fiber volume contents. Compared to fully impregnated organosheets, the extruded and not completely consolidated composites still have lower mechanical properties even at a high fiber volume content of 40%. To achieve complete impregnation, an additional small double belt press can be used, which can be operated with reduced heating power. Due to their lower costs, the pre-impregnated sheets might be able to bridge the price and performance gap between short fiber reinforced injection molded parts and fully impregnated sheets.

Acknowledgments
The results presented were obtained in the context of the research project KK5007902EB0 sponsored by the Federal Ministry for Economic Affairs and Climate Action (BMWK). The authors thank the BMWK for their financial support for the research, authorship, and publication of this article.

References

[1] P. Kiss, W. Stadlbauer, C. Burgstaller, et al., Development of high-performance glass fibre-polypropylene composite laminates, Compos. Part A: Appl. Sci. Manuf. 138 (2020) 106056. https://doi.org/10.1016/j.compositesa.2020.106056

[2] P. Kiss, J. Schoefer, W. Stadlbauer, et al., An experimental study of glass fibre roving sizings and yarn finishes in high-performance GF-PA6 and GF-PPS composite laminates, Compos. Part B Eng. 204 (2021) 108487. https://doi.org/10.1016/j.compositesb.2020.108487

[3] G. Toffe, Life Cycle Evaluation of Manufacturing and Mechanical Properties for Novel Natural Fibre Composites, Ph. D. thesis, University Hertfordshire (2020).

[4] F. Piott, A. Krämer, A. Lück, et al., Increasing the performance of continuous compression moulding by local pressure adaption, Adv. Manuf. Pol. & Compos. Sci. 7 (2021) 1–14. https://doi.org/10.1080/20550340.2021.1888209

[5] M. Christmann, Optimierung der Organoblechherstellung durch 2D-Imprägnierung, Ph. D. thesis, Technische Universität Kaiserslautern (2014).

[6] M. Christmann, L. Medina, P. Mitschang, Effect of inhomogeneous temperature distribution on the impregnation process of the continuous compression molding technology, J. Thermoplast. Compos. Mater. 30 (2017) 1285–1302. https://doi.org/10.1177/0892705716632855

[7] S. Mörl, Kurzfasermodifikation matrixreicher Zonen gewebeverstärkter, thermoplastischer Verbundwerkstoffe mittels In-Situ-Schmelzeimprägnierung, Ph. D. thesis, Universität Bayreuth (2022).

[8] S. Sockol, C. Doerffel, J. Mehnert, et al., Ultrasonic-Impregnation for Fiber-Reinforced Thermoplastic Prepreg Production, KEM 742 (2017) 17–24. https://doi.org/10.4028/www.scientific.net/KEM.742.17

[9] V. A. Bühler, Gradual Impregnation during the Production of Thermoplastic Composites, Ph. D. thesis, Technische Universität München (2017).

[10] F. Puch, B. Richter, Influence of the Processing on the Properties of Continuous Fiber Reinforced Thermoplastic Sheets Prepared by Extrusion, 37th International Conference of the PPS, Fukuoka (2022).

[11] Z. Zhicheng, X. Chunling, L. Ying, et al., Modeling and experimental characterization of power-law fluids impregnation behavior in fabric during compression molding, J. Reinf. Plat. Compos. 36 (2017) 176–185. https://doi.org/10.1177/0731684416677744

[12] J. Studer, C. Dransfeld, J. Jauregui Cano, et al., Effect of fabric architecture, compaction and permeability on through thickness thermoplastic melt impregnation, Compos. Part A: Appl. Sci. Manuf. 122 (2019) 45–53. https://doi.org/10.1016/j.compositesa.2019.04.008

Sheet Metal 2023
Materials Research Forum LLC
Materials Research Proceedings 25 (2023) 353-360
https://doi.org/10.21741/9781644902417-44

An experimental study of bushing formation during friction drilling of titanium grade 2 for medical applications

Hans Vanhove[1, a *], Caroline Mossay[1,b] and Joost R. Duflou[1,c]

[1]Department of Mechanical Engineering, Katholieke Universiteit Leuven / Member of Flanders Make, Celestijnenlaan 300B, B-3001 Leuven, Belgium

[a]hans.vanhove@kuleuven.be, [b]Caroline.Mossay@kuleuven.be, [c]joost.duflou@kuleuven.be

Keywords: Friction, Drilling, Titanium

Abstract. Recent advances towards patient specific, Titanium sheet based medical implants introduce a new challenge for the fixation of these implants to bones. Mainly the use of locking screws requires an implant thickness of approximately 2 mm for screw thread formation. Friction drilling is a hole-making process that displaces material to create a bushing below the sheet rather than extracting material. Screw thread can be formed in this bushing for the medical locking screws. This experimental study explores the influence of axial force, rotational speed and workpiece temperature on the bushing formation during friction drilling of Titanium grade 2 sheets. Finally, the influence of the optimal parameters on the material hardness is characterized.

Introduction

The shift towards patient tailored medical care has been tangible in the advances made in research towards patient specific plate osteosynthesis. Vancleef et al [1] have studied the potential of custom thin-walled implants for clavicle bone reconstruction. Benefits were identified on improved anatomical alignment of bone fragments, due to a better fit, while the reduced implant thickness results in less soft tissue irritation. Although thin-walled implants can reduce irritation and eliminate a potential secondary operation for implant removal, a reduced plate thickness poses a challenge towards plate fixation. While conventional osteosynthesis screws are countersunk in the implant, locking head screws use a threaded head to create angular stability to the bone fragments (Fig. 1).

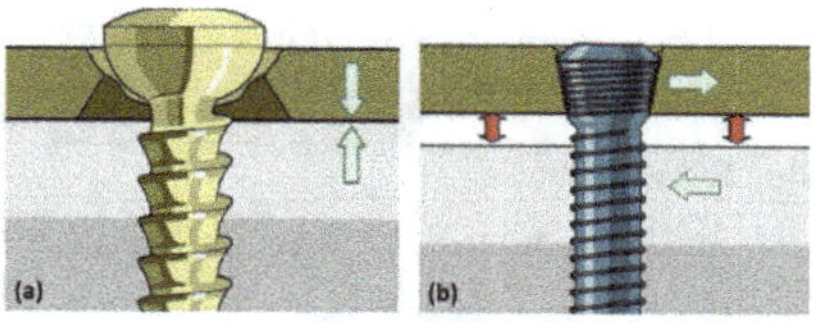

Fig. 1. Medical implant fixation through (a) conventional screws, (b) locking screws [2].

Commercial osteosynthesis locking screws have a head height of approximately 2 mm fitting the current commercial clavicle plates with a thickness varying from 2.3 to 3.4 mm. Vancleef et al [1] have shown, however, that patient specific plates of 1.5mm average thickness suffice to bear the loads acting on the broken bone. Thinner plates with varying thickness throughout the plate, even down to 1mm, are subject of ongoing research.

Friction drilling, a chip-less hole making process, provides an interesting solution to increase plate thickness at the fixation points. The process deforms the workpiece material through friction between the conical rotating tool and the sheet, instead of removing it. The resulting upwards

Content from this work may be used under the terms of the Creative Commons Attribution 3.0 license. Any further distribution of this work must maintain attribution to the author(s) and the title of the work, journal citation and DOI. Published under license by Materials Research Forum LLC.

formed boss and downwards formed bushing can have a combined height up to four times the initial sheet thickness [3].

This process was first proposed/described in 1923 by Jan Clause de Valliere but could only be put into practice 60 years later through the development of Tungsten Carbide (TC) tools.[4]. Since the first experiments, research has been focusing on improving the emerging geometry of bushing and hole through optimizing process parameters. More specifically the bushing quality can be quantified by its height and thickness and occurrence of cracks and petal formation. Ozler et al. [5] concluded that an increased *rotational speed* of the drill (ω) increases the temperature and thus lowers the yield stress, resulting in a lower thrust force. A longer bushing height is accompanied by a thinner bushing and reduced petal formation as a result of the increased forming temperature.

Most research has been performed using a constant *feed rate* (*f*), letting the thrust force vary throughout the different stages of friction drilling a hole [6]. The contact time between drill and sheet reduces with increased feed rate, resulting in a higher thrust force and tool torque. An increased roughness, shortened but thicker bushing and increased petal formation and cracks are the main effects on the final hole. [6]. The presented work uses a *constant thrust force* and thus the forming speed varies as the drill sinks through the sheet.

Friction drilling Titanium is associated with multiple challenges due to its low thermal conductivity and high chemical reactivity. Isolating and accumulating the temperature in the tool/sheet contact zone results in local melting and extreme long and thin bushings [7]. The high workpiece temperature in an oxygen rich environment results in the formation of Titanium oxides which drastically increases drill wear through adhesion and abrasion[3]. This publication will introduce *workpiece pre-heating* for Titanium grade 2, in order to reduce the temperature gradient throughout the sheet and the required heat flux, generated by friction, for plastic bush formation. A more uniform heating eases the superplastic deformation of the material from the sheet into the bushing.

Material and methods

Titanium grade 2 of 1mm thickness is chosen as workpiece material. It is the most used commercially pure Titanium for medical applications due to its excellent strength to weight ratio, biocompatibility, lack of magnetic attraction and wide availability.

Table 1. Properties of Titanium grade 2.

Density [Kg/m³]	4510
Youngs modulus [GPa]	100 - 105
Yield strength [MPa]	276 – 400
Hardness, Vickers [HV]	145 – 165
Thermal Conductivity [W/(mK)]	16 – 18
Melting Point [°C]	1670

A 3.7mm friction drill (Flowdrill M04 – 3.7 Long standard) has been used. It is made of tungsten carbide in cobalt matrix and has a friction angle of 36°. Berulit 935, is selected as lubricant for its high temperature stability, in order to reduce tool wear. It is applied to the tool prior to forming of each hole. Moreover, the drill is also cleaned after each use with a steel wire brush to remove Titanium adhesion to the tool.

Drilling is performed on an Alzmetal Drill press which allows a wide variety of rotation speeds. It is equipped with an actuation disk that allows to attach different weights to provide a constant thrust force on the drill. Fig. 2 depicts the experimental platform.

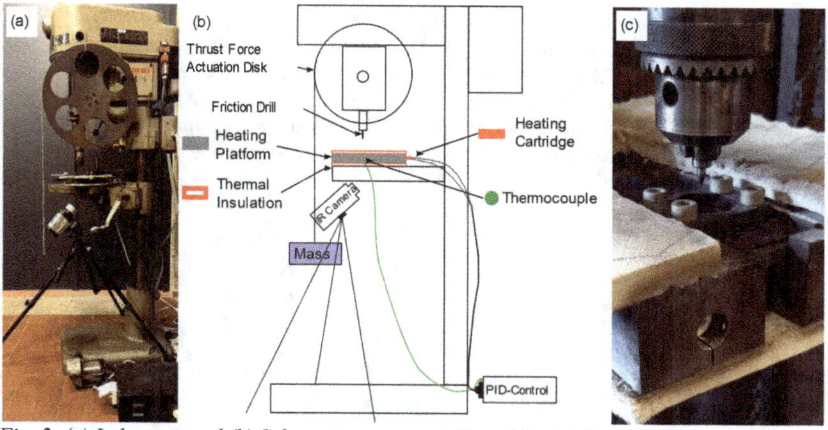

Fig. 2. (a) Lab setup and (b) Schematic representation of the friction drilling platform (c) Closeup of the friction drill and heating platform.

A custom made heating platform has been mounted on the bed of the table drill in order to pre-heat the Titanium sheet. The platform is manufactured from S235 JR and is heated by two Vulstar 1007-26 cartridge heaters. Applying a thermal paste between the platform and heating elements ensures that no air gap is present. Insulation layers are applied at the top and bottom of the platform to prevent the heat from flowing to the machine. Due to the high thermal inertia, the block is kept at constant temperature throughout tests using a PID controller receiving temperature feedback by a K-type thermocouple. Interchangable steel clamping plates, with a low thermal capacity, are used to clamp the Titanium sheet and can be mounted on the steel heating platform. Fig. 3 shows a CAD drawing of the heating platform. An 8mm hole in the clamping plate allows space for the drill to penetrate through the sheet while the plate acts as support of the sheet to localise plastic deformation.

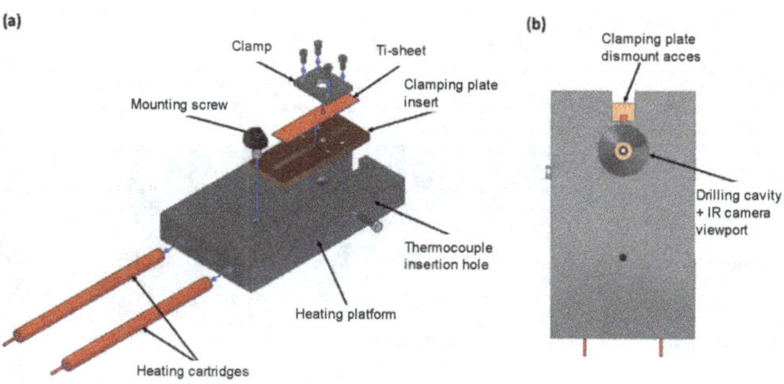

Fig. 3. CAD drawing of the heating platform: (a) Exploded view, (b) Bottom view.

Besides a thermocouple to control the heating block temperature, a FLIR Thermovision A20m IR camera is set up to monitor the backside of the sheet for temperature verification. A secondary thermocouple on the front of the sheet is used to verify the preheating temperature.

The bushings are analyzed using a Nikon LC 60 Dx laser scanning probe, mounted on a Coord3 CMM. The resulting point clouds allow to measure the bushing height and thickness. The total

Materials Research Proceedings **25** (2023) 353-360 https://doi.org/10.21741/9781644902417-44

hole length, which corresponds to the useful length for thread cutting, is defined as the bushing height + the 1mm thickness of the Titanium grade 2 sheet, as described in the materials section. The thickness measurement is performed at 1mm from the back of the sheet, thus at a 1mm bushing height. As possible cracks or petal formation cause variation of height and thickness along the circumference of the bushing, the bushing height is defined as the minimum crack-less height measured. The thickness is defined as an average thickness taken along the circumference of the 1mm mark for the remainder of this paper (Fig. 4).

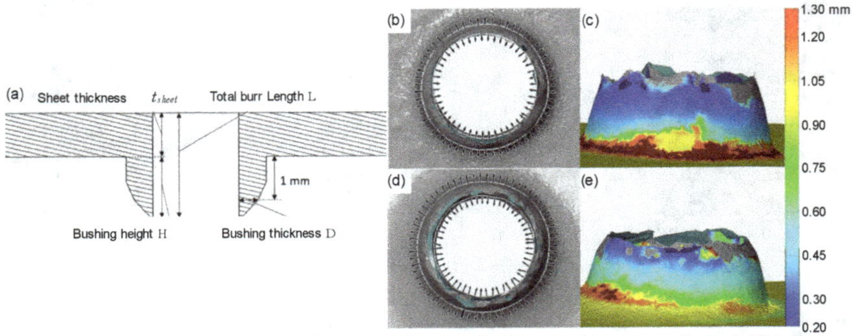

Fig. 4. (a) Dimensions nomenclature for friction drilling in this paper, (b & c) Thickness measurement of cold friction drilled hole, (d & e) Thickness measurement of pre-heated friction drilled hole.

The formed bushing is used to contain and stabilize locking screws used during operations. This requires the bushing to be strong, but still ductile enough to form the thread using roll tapping. Micro hardness measurements have been performed along the cross section of the friction drilled hole. First the workpiece is cut in half by means of wire electrical discharge machining. After embedding and polishing, a Shimadzu HMV-2000 micro hardness measuring platform is used with 500gram of weight for indentation. Fig. 5 shows the indent locations along the bushing and the sheet surrounding the hole. A total of five indents have been made for each sample respecting the proper distance from the side as well as from each other. The indents are measured using a Hirox KH-8700 digital microscope.

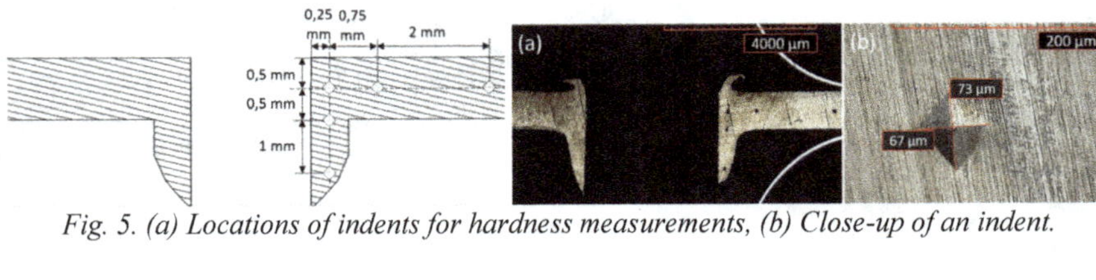

Fig. 5. (a) Locations of indents for hardness measurements, (b) Close-up of an indent.

The thrust force, rotational speed and workpiece pre-heating temperature are considered in a factorial analysis varying between two levels. The thrust force is limited on the lower side by a lack of penetration while the upper boundary is defined by premature breakage of the bushing. Both cases are most stringent limited when no pre-heating is used. Analogously, a lack of penetration through lack of heat defines the bottom limit of the rotational speed. The upper limit here however is constrained by machine capacity. The table drill used for the experiments has a maximum rotation speed of 2800rpm. Finally, the lower- and upper limit for the variable workpiece pre-heating are respectively the room temperature (20°C) and the highest stable temperature the described setup can achieve (525°C). Table 2 summarizes the input parameters for the factorial analysis.

Table 2 variable levels for factorial analysis

	Low	High
Thrust force [N]	361	722
Rotational speed [rpm]	1400	2800
Temperature [°C]	20	525

Results and discussion

The factorial analysis focusses on the total hole length, cracks and bushing thickness as responses. A minimal total crack-less hole length of 2mm is envisioned in order to contain the total locking screw head. Longer bushings are acceptable as they can be ground down to the correct length. The thickness should however be as large as possible for structural purposes.

All experiments were performed 5 times and had a minimal bushing height (Hmin) of 1mm, resulting in a total hole length exceeding 2mm. The interval plot in Fig. 6 depicts a 95% confidence interval for the minimal bushing height. A trend of increasing bushing height can be observed with increasing rotational speed and increasing workpiece pre-heating temperature. Both will increase the final total workpiece temperature, improving superplasticity. The effect of increased thrust force is however less obvious as rising the thrust force increases the occurrence of cracks. A higher thrust force also increases the forming speed and lowers the superplastic effect.

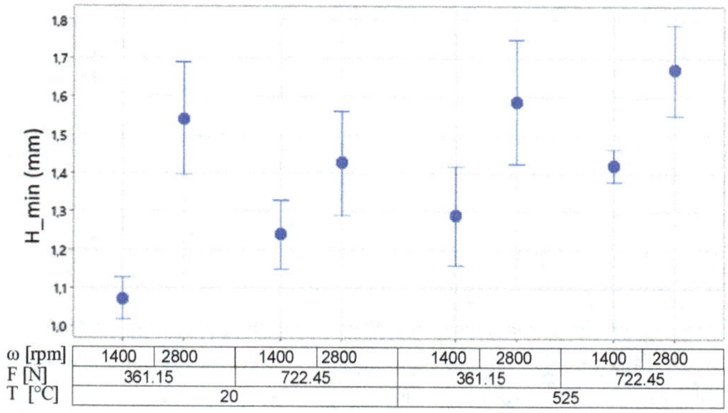

Fig. 6. Interval plot of the minimal bushing height.

As all experiments fit the requirement of minimal 1mm bushing height, the thickness can be analysed using Minitab. The Pareto graph shown in Fig. 7 shows the thrust force (F) as most significant variable for determining the bushing thickness. It is followed by the workpiece pre-heating temperature (T), the temperature interaction-thrust force (T·F) and rotational speed (ω). The thrust force-rotational speed interaction barely crosses the significance line, while other interactions are deemed not significant.

Materials Research Proceedings **25** (2023) 353-360 https://doi.org/10.21741/9781644902417-44

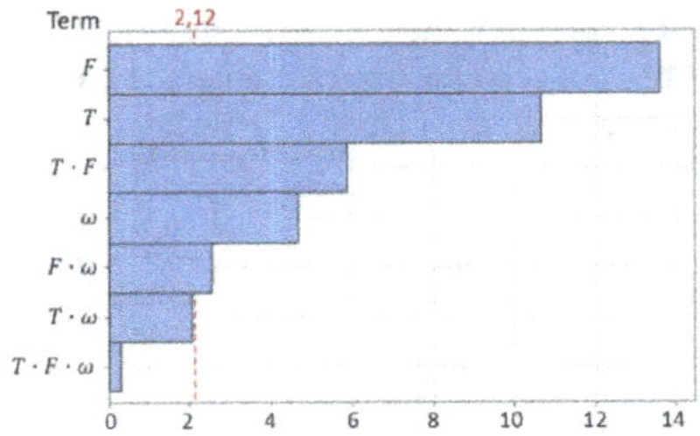

Fig. 7. Pareto chart of the standardized effects for the bushing thickness.

An analysis of the significant effects shows a considerable increase in bushing thickness when increasing the workpiece pre-heating temperature. This was expected as pre-heating the sheet reduces the temperature gradient and aids in increasing the volume of material at elevated temperature, which is thus subjected to superplastic forming. An increase of thrust force also has a beneficial effect on the bushing thickness and follows earlier observations described in literature [5], where an increased feed rate (and accompanying thrust force) resulted in thicker bushings due to the increased radial force applied on the material. This, however, also causes shorter bushings and an increased probability of crack formation as negative side effects. The rotational speed has a reducing effect on the bushing thickness due to increased temperature, which results in an increased material flow into a longer bushing, at the cost of bushing thickness.

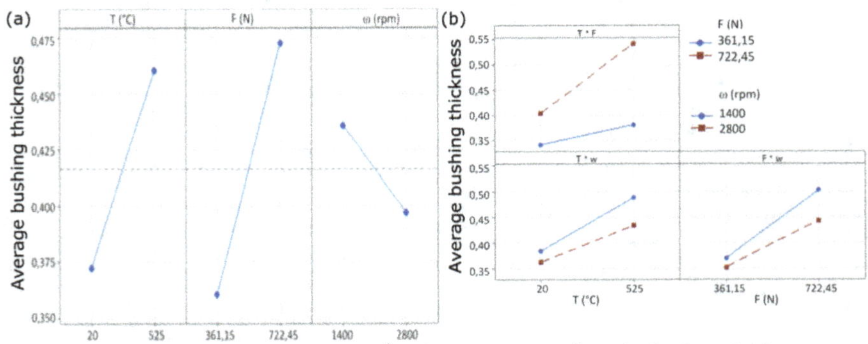

Fig. 8. (a) Main effects and (b) 2nd order interaction effects for bushing thickness.

Fig. 8b shows a clear interaction between workpiece pre-heating and thrust force. The effect of temperature increase is much more pronounced at high thrust force than at low thrust force. This again confirms the effect on bushing thickness of involving more material in the plastic deformation. Fig. 4 shows the difference in bushing thickness between room temperature and pre-heated workpiece material.

A reverse and reduced interaction can be seen for T·ω and F·ω. In these cases the effect of workpiece pre-heating and thrust force respectively are reduced when the rotational speed is increased.

Hardness measurements for the bushings and surrounding of the friction drilled holes were performed on bushings made with and without pre-heating of the workpiece. The parameters used are summarized in Table 2.

Materials Research Forum LLC
https://doi.org/10.21741/9781644902417-44

Table 2. Parameters for pre-heated and room temperature friction drilling targeting maximal bushing thickness.

	Pre-heated	Room temperature
Temperature [°C]	535	20
Thrust force [N]	1017	722
Rotational speed [rpm]	1400	1400

Fig. 5 shows the five different measurement locations along the cross section of the sheet. Three along the horizontal axis and three along the bushing, sharing the corner location. Each test was performed at least 5 times of which the average is depicted in Fig. 9. The intervals represent the 95 confidence interval for these values. As discussed in the material section, the reference values for the hardness of Titanium grade 2 is 166-165HV. This matches the values obtained 3mm away from the drilled hole. The measured hardness increases when moving closer to the hole and the heat affected zone. A more linear increase is observed in the pre-heated samples as compared to the exponential profile seen in the room temperature samples.

The hardness along the bushing shows a similar evolution between pre-heated and non-pre-heated samples. In both cases the measured hardness was higher at the 1mm location than in the 0.5mm corner point and the 2mm point, further down the bushing. The cold formed bushings have a higher hardness over the entire bushing compared to the pre-heated ones.

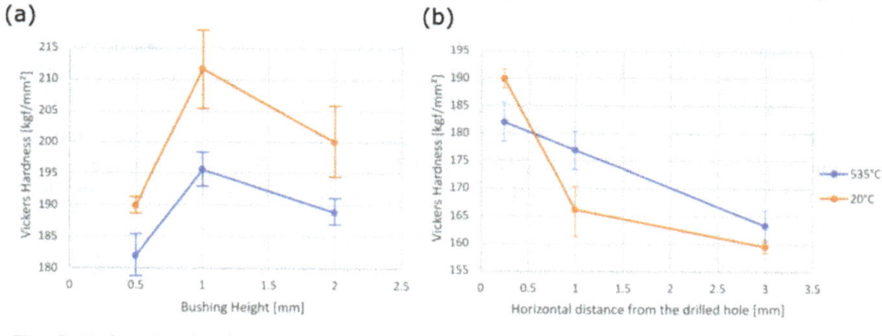

Fig. 9. Vickers hardness profile (a) along the created bushing, (b) along the undeformed sheet.

Conclusion and future work

This paper summarizes an explorative study on the influence of rotational speed, thrust force and pre-heating on the bushing formation in friction drilling of grade 2 Titanium. It has been shown that the use of pre-heating has a positive influence on both bushing height as well as bushing thickness. An increased thrust force mainly has a varying effect on the crack free bushing height due to increased occurrence of cracks. It is however, beneficial for the bushing thickness. Increasing the rotation speed induces more heat into the sheet tool contact zone, resulting in higher but thinner bushings for Titanium grade 2. Finally, a comparison is made between the hardness of a workpiece formed at room temperature and pre-heated conditions, where forming at room temperature results in a higher hardness. Future work will focus on the mechanical characterisation of the bushings, thread cut into them and strength and stiffness of the connection with the medical locking screws.

Aknowledgements
The study performed in the framework of the STIFF project, facilitated by the KU Leuven C2 research fund.

References

[1] S. Vancleef, M. Wesseling, J.R. Duflou, S. Nijs, I. Jonkers, J. Vander Sloten, Thin patient-specific clavicle fracture fixation plates can mechanically outperform commercial plates: An in silico approach, Journal of Orthopeadic Research, vol. 40, no.7, (2021) 1695-17006. https://doi.org/10.1002/jor.25178

[2] P. Szypryt, D. Forward, The use and abuse of locking plates, Orthopaedics and trauma, vol. 23, no. 4, (2009) 281–290. https://doi.org/10.1016/j.mporth.2009.07.002

[3] S. Dehgan, M. I. S. Ismail, M. K. A. Ariffin., and B. T. H. T. Baharudin, Experimental investigation on friction drilling of Titanium alloy, Engineering Solid Mechanics, vol. 46, no. 12, (2018) 135–142. https://doi.org/10.5267/j.esm.2018.2.002

[4] J. E. France, J. Buick Davison, and P. A. Kirby, Strength and rotational stiffness of simple connections to tubular columns using flowdrill connectors, Journal of constructional steel research, vol. 50, no. 1, (1999) 15–34. https://doi.org/10.1016/S0143-974X(98)00236-3

[5] L. Ozler and N. Dogru, An experimental investigation of hole geometry in friction drilling, Materials and Manufacturing Processes, vol. 28, no. 4, (2013) 470– 475. https://doi.org/10.1080/10426914.2012.746699

[6] M. T. Kaya, A. Aktas, B. Beylergil, and H. K. Akyildiz, "An experimental study on friction drilling of st12 steel," Transactions of the Canadian Society for Mechanical Engineering, vol. 38, no. 3, (2014) 319–329. https://doi.org/10.1139/tcsme-2014-0023

[7] S. Dehghan, I. M., A. M., and B. B., "Friction drilling of dicult-to-machine materials: Workpiece microstructural alterations and toolwear," Fundamental Study on Friction Drilling of Difficult-to-Machine Materials, vol. 9, no. 9, (2019) 1– 14. https://doi.org/10.3390/met9090945

Quality and reliability

Sheet Metal 2023 Materials Research Forum LLC
Materials Research Proceedings **25** (2023) 363-370 https://doi.org/10.21741/9781644902417-45

Revisiting image-based quality evaluation of laser cut edges

Masoud Kardan[1,a]*, Nikita Levichev[1,b], Alberto Tomás García[1,c] and Joost R. Duflou[1,d]

[1]Department of Mechanical Engineering, KU Leuven, Celestijnenlaan 300, 3001 Heverlee, Belgium, Member of Flanders Make

[a]masoud.kardan@kuleuven.be, [b]nikita.levichev@kuleuven.com, [c]alberto.tomasgarcia@kuleuven.be, [d]joost.duflou@kuleuven.be

Keywords: Laser, Cutting, Quality Evaluation

Abstract. The optimization of laser cutting process parameters relies on productivity and obtained edge quality. Whereas maximizing cutting speed is a default procedure to increase the process performance, quality assessment of a cut edge is a non-trivial task. Both contact-based and image-based approaches can be used to quantify the quality of a cut surface. Since contact-based techniques are time-consuming and typically require expert knowledge, the development of simple and fast image-based approaches could improve the performance of sheet metal workshops. Due to the numerous quality characteristics that have to be considered, a significant challenge remains to establish a versatile approach for image-based quality evaluation. Within this paper, the quality assessment of laser cut edges by means of image processing techniques is analyzed. Additionally, the potential for employing visual evaluation to assess all quality indicators in a comprehensive measuring strategy is explored. Finally, the role of the presented approaches in shifting toward intelligent manufacturing is briefly discussed.

Introduction

In recent years, laser cutting productivity and thickness limitations have been shifted to higher values as a result of the increase in the available laser power, the development of novel optical components and the improvement of machine dynamics. When cutting thick plates, the quality assessment of a cut edge becomes more important since a large cut surface increases the variation of quality characteristics along the thickness. At the same time, whereas roughness and dross attachment are typically regarded in the literature as the main quality aspects [1, 2], the evaluation of cut edge quality in an industrial setting goes beyond measuring those characteristics. According to ISO 9013 [3], the main characteristics of an edge created by thermal cutting are perpendicularity (or angularity in bevel cutting), surface smoothness and dross attached to the lower edge of the cut surface. Measurements of roughness and perpendicularity are commonly performed by contact-based profilometry, while there is no standardized method for dross quantification. However, using such contact-based methods usually consumes considerable time and requires specific equipment and skilled operators to carry out, thus limiting extensive quality assessment within sheet metal workshops and preventing the collection of relevant production data. Such status restricts taking most of the benefits envisaged in the frame of Industry 4.0 that relies on the extensive use of data for continuous process improvement [4]. Therefore, the first step toward Industry 4.0 for quality evaluation of the cut edges is developing approaches with fewer possible human interactions with the process operations.

A few investigations have addressed the quality quantification of laser cutting parts using image-based strategies. Obtaining surface topography of laser cut edges using a focus-variation microscope and subsequent evaluation of the cut quality with areal surface roughness according to ISO 25178 [5] has been illustrated in [6]. An image-based technique to estimate the roughness of

Content from this work may be used under the terms of the Creative Commons Attribution 3.0 license. Any further distribution of this work must maintain attribution to the author(s) and the title of the work, journal citation and DOI. Published under license by Materials Research Forum LLC.

laser cutting samples has been described in [7]. In addition, an algorithm to predict the roughness of cutting edges by using an artificial neural network and applying image processing has been developed in [8]. A technique for quantitative identification of cut edge quality factors using an image analysis algorithm has been presented in [2]. An image processing algorithm to automatically evaluate the dross area in laser fusion cutting has been proposed in [9]. Despite numerous studies dedicated to measuring cut edge quality factors separately, there is no systematic approach to evaluate and quantify laser cutting quality considering all characteristics together.

In this work, quality evaluation of the laser cut edges employing an optical microscope with focus variation is investigated and compared with contact-based approaches. Moreover, a comprehensive method to measure all quality characteristics within a single measurement approach is proposed. Finally, a quality index is recommended to consider all quality factors with different weight parameters, which can be used further in the framework of Industry 4.0.

Roughness

One of the key factors determining the quality of a laser cut edge is roughness. According to ISO 9013, roughness of the cut edge should be evaluated as the arithmetic mean of the height between the peak and the valley over five sampling lengths (Rz5).

New techniques combining image processing and machine learning have been recently developed to estimate roughness [7, 8, 10]. But, as discussed in [11], these methods are extremely sensitive to lighting conditions and require a large database. Moreover, since roughness varies significantly along the plate thickness, especially for thick plates [9], the complex 3D geometry of the cut surface can hardly be represented by single line measurements. At the same time, contact-based profilometry does not provide enough information regarding dissimilarities in cut edges obtained by different processes. On the contrary, optical devices with focus variation can be used to extract full information about the surface topography of cut edges. This information may be further processed to quantify areal surface roughness and emphasize the difference between various laser cutting technologies.

Fig. 1 shows the surface condition and measured roughness using a profilometer and optical microscope with focus variation for 15 mm cut edges of stainless steel and mild steel created by the fusion and flame cutting processes, respectively. Contact-based (Fig. 1a, b) and optical (Fig. 1c, d) measurements have been performed for the same measuring area using a Mitutoyo Formtracer CS-3200 profilometer and a Keyence VHX-6000 microscope, respectively. To measure the roughness over an area using a profilometer, the edge of the sample can be divided into several sections, for which roughness is evaluated independently and then averaged (Fig. 1a, b). As a result, the average roughness of the cut edge, as well as the worst roughness, can be extracted. The profile roughness is calculated employing a Gaussian filter with λc and λs equal to 2.5 mm and 8 μm respectively. Computed average roughness can be considered as a value to represent the overall surface conditions of the cut edge. In this case, ignoring some areas between the measuring lines would be inevitable. However, increasing the number of measuring lines to cover the area as much as possible would significantly increase the measurement time.

In contrast to line measurements, the whole edge surface is considered by means of surface topography, thus eliminating the possibility of disregarding potentially important edge areas. Different roughness parameters can be used to quantify surface roughness using optical measurements. The maximum height of the surface (Sz) is sensitive to individual peaks and valleys which may not be significant for the whole surface characterization [12]. To evaluate the surface roughness, a Gaussian filter with λs and λc equal to 20 μm and 2.5 mm is applied. According to Fig. 1, both surfaces have comparable roughness conditions in terms of average Ra and Rz5, whereas it is not evident for Sz. As a result, Sz cannot be considered a reliable value implying areal surface roughness. Another areal roughness parameter is the arithmetic mean height (Sa), defined as the arithmetic mean of the absolute value of the height within an evaluation area. It provides a more representable value for a surface and a better criterium for comparison between different surfaces since it is less sensitive to individual peaks and valleys.

As seen in Fig. 1, a comparison between contact-based profilometry and surface topography using an optical microscope with focus variation illustrates that optical evaluation provides more absolute information for different surfaces and a roughness value (Sa) for the whole area, considering all peaks and valleys.

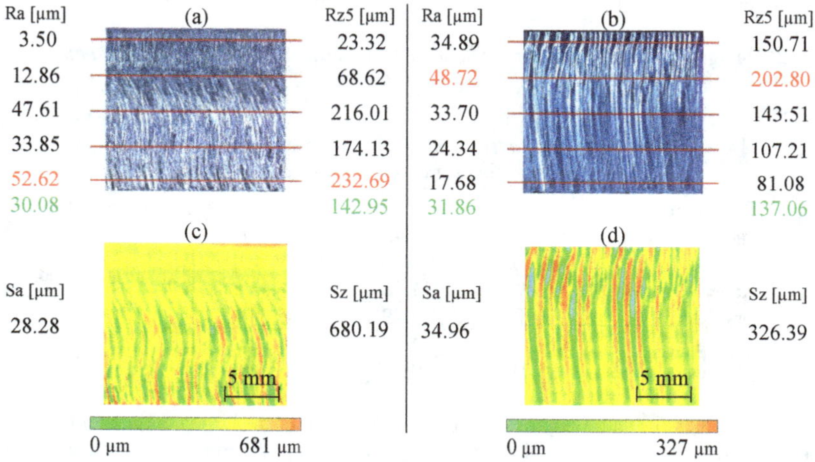

Fig. 1. An example of roughness measurements: (a) and (b) contact-based measurements for fusion and flame cut edge, respectively; (c) and (d) optical measurements for fusion and flame cut edge, respectively. Red and green color values present the worst and average roughness of the cut edge, respectively.

Fig. 2 shows a comparison of different surface conditions created by laser fusion cutting. Table 1 summarizes the measured surface parameters employing both contact-based and optical techniques for the samples presented in Fig. 2. The roughness distribution is more uniform on the surface in Fig. 2a, while the rougher region is near the bottom edge of the surface in Fig. 2b. Despite a higher value for the worst roughness on the surface in Fig. 2b using profilometry, the average roughness is lower. Considering areal roughness, Sz is affected by isolated peaks and valleys which can be seen by comparing the surfaces of Fig. 2b and Fig. 2c. Based on visual observation, the surface in Fig. 2c is rougher than the surface in Fig. 2b, but both have comparable values for Sz. Among all roughness parameters in Table 1, Sa considers the whole cut edge area and can be a more robust indicator of surface conditions.

Although contact-based profilometers provide more accurate measurements in a specific location, the obtained areal roughness values using optical measurements are more reliable due to considering all details of a region of interest. However, when roughness measurement aims to define the post-process requirement, other feature parameters (for instance, S10z) can be used to shift the importance towards extreme peaks and valleys.

Materials Research Proceedings **25** (2023) 363-370 | https://doi.org/10.21741/9781644902417-45

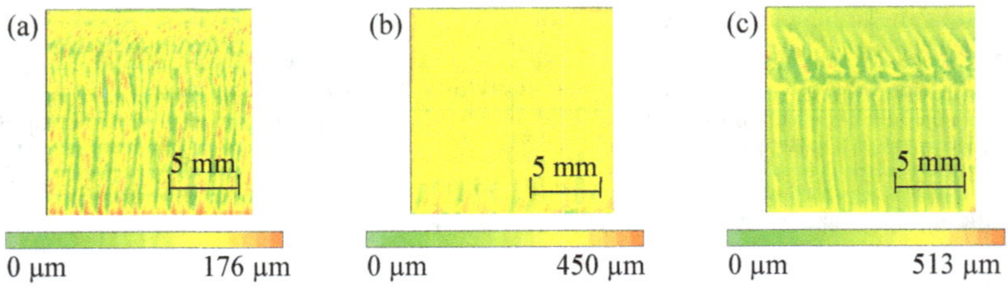

Fig. 2. Topography measurements: (a) homogeneous surface roughness; (b) non-homogeneous surface roughness; (c) rough surface.

Table 1. Measured roughness parameters using both contact-based and optical approaches for surfaces presented in Fig. 2.

Parameter	Fig. 2a	Fig. 2b	Fig. 2c	Measurement type
Average Rz5 [µm]	87.25	62.45	174.59	Contact-based
Worst Rz5 [µm]	105.29	114.91	295.93	
Average Ra [µm]	15.49	11.48	36.33	
Worst Ra [µm]	18.87	23.79	66.64	
Sa [µm]	13.83	14.38	39.07	Optical
Sz [µm]	175.01	449.65	512.86	

By correlating the cutting process parameters with the areal roughness of the cut edge, roughness prediction could be realized with machine learning algorithms, for instance, by using one of the already proposed techniques [7, 8, 13]. This approach could be further extended to be used for modeling the surface topography of the cut edge depending on the process parameters [14].

Dross attachment

Melting drops that are not able to escape during the laser cutting process, lead to the formation of dross on the bottom side of the cutting edge. Although the perfect cut edge is a dross-free surface, the attachment of dross might become inevitable, especially in the case of fusion cutting of thick plates [9]. Moreover, there is no clear definition for the acceptable amount of dross. Therefore, it is necessary to measure dross as a quality factor and accordingly determine the post-processing requirements based on industrial demands. Different techniques based on image processing have been proposed to measure the average dross height [15, 16]. Employing these methods, the cut edge is divided into a few sections and the maximum dross height is determined in each area. Then, the average dross height along the entire cut edge can be calculated. However, dross area can also be measured instead of dross height [9].

Fig. 3 shows two cut surfaces with different amounts of dross and compares both dross height and dross area measurements. Whereas the measured dross height is only increased by 34 % changing from the surface in Fig. 3a to Fig. 3b, the measured dross area is increased by 115 %. Therefore, dross area measurements provide a more reliable value of the dross amount compared to the height measurement approach. The principle of the algorithm for measuring the dross area has been thoroughly explained in [9]. Although the measuring length can be divided into different areas to provide an average value when considering dross height, the algorithm for measuring dross area is more robust. The latter method measures the amount of dross considering two dimensions while the other approach is based on a one-dimensional measurement. All in all, when

Materials Research Proceedings **25** (2023) 363-370 https://doi.org/10.21741/9781644902417-45

the total amount of dross is the goal of measuring, the dross area assessment is more accurate. However, both approaches can be considered when defining post-processing requirements.

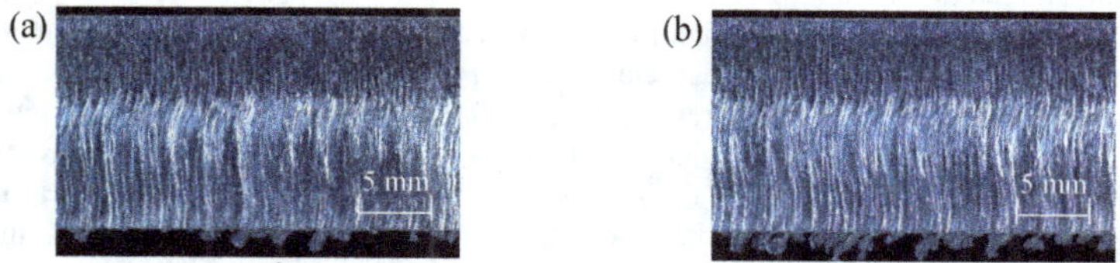

(a) Dross area [mm²/mm]: 0.307
Dross height [mm]: 1.56

(b) Dross area [mm²/mm]: 0.665
Dross height [mm]: 2.09

Fig. 3. Dross measurements using two different approaches for two surfaces with (a) non-homogeneous dross and (b) homogeneous dross.

Perpendicularity

According to ISO 9013, the perpendicularity of the cut surface is the distance between two parallel straight lines with a right angle to the reference surface in which the cut edge profile is enclosed [3]. The above-mentioned definition is shown in Fig. 4b, where Δa is the reduced cut face to allow for melting of the top and lower cut edge, a is the area for evaluating the perpendicularity, and u is the perpendicularity tolerance.

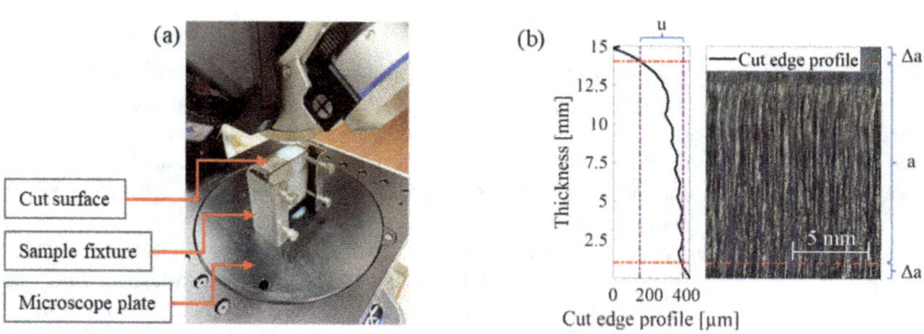

Fig. 4. Perpendicularity evaluation using the proposed image-based technique: (a) the setup used to avoid slope correction; (b) the average cut edge profile of the entire measuring area.

To evaluate perpendicularity, an image-based approach is proposed rather than complex contact-based measurements. The principle of this technique is similar to the roughness measurement using an optical microscope with focus variation. However, the reference surface should be located perpendicular to the microscope plate (Fig. 4a) to avoid slope correction after scanning the cut surface. A sample fixture can be used for the latter purpose as seen in Fig. 4a. After scanning the cut edge using a microscope, the 3D surface geometry can be captured in a JPEG image. Furthermore, this data can also be exported in a matrix representing the height of each point on the cut surface. The rows of the matrix are in the cutting direction and the columns are in the thickness direction. Afterward, instead of evaluating the perpendicularity in different vertical locations along the cut length, the average of the cut edge profile is calculated. Considering laser cutting as a stable process, the striation patterns are repeated periodically along the cutting length [16, 17]. Therefore, the average cut edge profile can be measured to ignore the possible extreme areas which have resulted from irregularities in the dynamic flow of the molten material in the cutting process. However, another statistical analysis is possible on the matrix to achieve the minimum and maximum height, standard deviation, etc.

To extract the average cut edge profile, the matrix is transformed into a column vector where each component is the average of the associated row in the matrix. Then, the average cut edge profile is extracted based on the previously calculated column vector. Furthermore, the perpendicularity tolerance (u) can be evaluated within the a area. Fig. 4b shows the average cut edge profile of the evaluated surface. By means of this approach, Δa can be determined for different thicknesses using [3]. Within the a area, the distance of the two parallel lines, in which the cut edge profile is inscribed, can be calculated and defined as the perpendicularity of the cut edge.

Quality index

Since the quality evaluation is affected by different factors, a multicriteria quality index is proposed instead of assessing surface characteristics separately. Although merging cutting performance parameters, including cutting speed, has been explored in the literature [16, 18], there is no recommendation for the quality index for laser cutting. As a result, a quality index (Eq. 1) has been developed considering roughness, dross area and perpendicularity with changeable weight parameters ($a + b + c = 1$).

$$\mathrm{QI} = a\frac{R_m}{R_t} + b\frac{DA_m}{DA_t} + c\frac{U_m}{U_t} \quad (1)$$

where R is roughness, DA is dross area and U is perpendicularity. Indices "m" and "t" refer to the measured and tolerance (and or reference) values respectively for the particular parameter. Tolerances for roughness and perpendicularity can be adjusted in accordance with the levels specified in [3] when QI is used as a rejection criterion. However, QI might also be taken as a target value. In this situation, the optimum results from laser cutting with the recommended process settings for each individual machine can be used as a reference.

Table 2 lists the selected references, measured values and weight parameters for different quality factors for a 15 mm thick stainless steel surface processed with an industrial fiber laser cutting machine. Since cutting results are not comparable with each other when employing different machines due to different process parameters and conditions, the references in Table 2 have been defined based on the optimal values obtained during laser cutting with nominal process parameters on a particular machine.

In this work, due to the priorities in industrial applications, weight parameters were selected using the following condition (Eq. 2):

$$\begin{array}{l} a = 0.35, b = 0.35, \text{and } c = 0.30 \ \text{ if } DA_m \le \text{Threshold} \\ a = 0.55, b = 0.05, \text{and } c = 0.40 \ \text{ if } DA_m > \text{Threshold} \end{array} \quad (2)$$

Considering the threshold, the post-process requirements can be defined to remove the dross. If dross is less than the threshold, post-processing is not required and the importance of dross is much higher. The opposite is expected when the dross area is higher than the threshold. However, this statement does not lead to increasing the quality index value for a cut edge with an amount of dross lower than the threshold. Since there is no standard value for this threshold, a value of 0.05 mm^2/mm could be considered based on the recommendation of industrial users and experienced operators [18].

Table 2. Measured data, tolerances and weight parameters for one sample.

	R [µm]	DA [mm²/mm]	U [mm]
Measured (m)	28.28	0.347	0.356
Reference (t)	13.83	0.147	0.235
Weight parameters (a,b,c)	0.55	0.05	0.40

By replacing the parameters from Eq. 1 with variables from Table 2, the calculated quality index is 1.85. The smaller the quality index, the higher the quality of the cut surface.

All quality parameters required in the proposed quality index can be evaluated using one optical microscope with focus variation, where a sample with a sufficient length of the cut edge is positioned in a fixture to ensure that the reference surface of the sample is perpendicular to the microscope plate. After scanning the cut surface using the microscope with focus variation, the extracted surface topography can be used to quantify areal roughness (with slope correction), dross and perpendicularity.

The time for scanning the surface is dependent on the scanning area, magnification, and the number of focus variations at each point. However, this time can be equal to the roughness measurement using the contact-based approach illustrated in Fig. 1. Finally, the calculated quality index can be used as the objective function for an optimal selection of process parameters.

Conclusion

The measurement of different quality characteristics of laser cut edges through image-based methods has been discussed in this contribution. The proposed techniques can provide a more robust and faster quality assessment compared to conventional contact-based approaches. For instance, measuring areal roughness gives more detailed information about surface topography and can be used to clarify cut surfaces produced through different technologies. Perpendicularity evaluation using an optical microscope with focus variation provides more reliable values compared to contact-based approaches since it considers the whole cut edge surface.

Image-based approaches can also be used to evaluate other quality factors, such as cutting kerf and heat-affected zone. This study, however, took into account the three main characteristics based on current standards.

The proposed quality index can be used as the objective function for training machine learning algorithms for process parameter optimization based on overall quality requirements. Compared to contact-based measurements, the approach proposed in this paper allows faster and more automatic collection of quality information, enabling the use of larger data sets for the training of the required machine learning algorithms for process parameter optimization. Whereas the proposed techniques are applied for offline measurements, further studies can be dedicated to the development of online methods which can be used for real-time optimization.

References

[1] N. Levichev, J. R. Duflou, On multi-sensor monitoring of fiber laser fusion cutting, IOP Conf. Ser. Mater. Sci. Eng. 1135 (2021) 012014. https://doi.org/10.1088/1757-899X/1135/1/012014

[2] M. Pacher, L. Monguzzi, L. Bortolotti, M. Sbetti, B. Previtali, Quantitative identification of laser cutting quality relying on visual information, Lasers in Manufacturing Conference (2017)

[3] DIN ISO 9013. Thermal cutting - Classification of thermal cuts - Geometrical product specification and quality tolerances. (2017).

[4] A. R. Torna, T. H. J. Vanekera, Mass Personalization with Industry 4.0 by SMEs: a concept for collaborative networks, Procedia Manuf. 28 (2019) 135-141. https://doi.org/10.1016/j.promfg.2018.12.022

[5] DIN ISO 25178. Geometrical product specifications (GPS) - Surface texture: areal. (2021).

[6] E. Librera, G. Riva, H. Safarzadeh, B. Previtali, On the use of areal roughness parameters to assess surface quality in laser cutting of stainless steel with CO_2 and fiber sources, Procedia CIRP. 33 (2015) 532-537. https://doi.org/10.1016/j.procir.2015.06.069

[7] L. Tatzel, F. Puente León, Image-based roughness estimation of laser cut edges with a convolutional neural network, Procedia CIRP. 94 (2020) 469-473. https://doi.org/10.1016/j.procir.2020.09.166

[8] A. Tomás García, N. Levichev, V. Vorkov, et al. Roughness prediction of laser cut edges by image processing and artificial neural networks, Procedia Manuf. 54 (2021) 257-262. https://doi.org/10.1016/j.promfg.2021.07.040

[9] N. Levichev, G. Costa Rodrigues, V. Vorkov, J. R. Duflou, Coaxial camera-based monitoring of fiber laser cutting of thick plates, Opt. Laser Technol. 136 (2021) 106743. https://doi.org/10.1016/j.optlastec.2020.106743

[10] J. Stahl, C. Jauch, M. Tuncel, M. Huber, Investigation of different illumination scenarios for the evaluation of thermally cut edges with convolutional neural networks using a mobile device. In: Proceedings of IS&T International Symposium on Electronic Imaging 2021. https://doi.org/10.2352/ISSN.2470-1173.2021.6.IRIACV-314

[11] N. Levichev, A. Tomás García, M. Kardan, D. Cattrysse, J. R. Duflou, Experimental validation of a machine learning algorithm for roughness quantification in laser cutting, Procedia CIRP. 113 (2022) 575-580. https://doi.org/10.1016/j.procir.2022.09.175

[12] F. Blateyron, The Areal Field Parameters. In: R. Leach (Eds), Characterization of Areal Surface Texture, Springer, Berlin, Heidelberg, 2013, pp. 15-43. https://doi.org/10.1007/978-3-642-36458-7_2

[13] Y. Zhang, J. Lei, Prediction of laser cutting roughness in intelligent manufacturing mode based on ANFIS, Procedia Eng. 174 (2017) 82-89. https://doi.org/10.1016/j.proeng.2017.01.152

[14] A. F. Courtier, M. McDonnell, M. Praeger, et al. Modelling of fibre laser cutting via deep learning, Opt. Express. 29 (2021) 36487-36502. https://doi.org/10.1364/OE.432741

[15]M. Pacher, L. Franceschetti, S. C. Strada, et al. Real-time continuous estimation of dross attachment in the laser cutting process based on process emission images, J. Laser Appl. 32 (2020) 042016. https://doi.org/10.2351/7.0000145

[16] C. Goppold, T. Pinder, P. Herwig, Transient beam oscillation with a highly dynamic scanner for laser beam fusion cutting, Adv. Opt. Technol. 5 (2016) 61-70. https://doi.org/10.1515/aot-2015-0059

[17] N. Levichev, T. Staudt, M. Schmidt, J. R. Duflou, Hyperspectral imaging and trim-cut visualization of laser cutting, CIRP Ann. Manuf. Technol 70 (2021) 207-210. https://doi.org/10.1016/j.cirp.2021.04.015

[18] M. Kardan, N. Levichev, J. R. Duflou, Experimental and numerical investigation of thick plate laser cutting using dynamic beam shaping, Procedia CIRP. 111 (2022) 740-745. https://doi.org/10.1016/j.procir.2022.08.115

Sheet Metal 2023 Materials Research Forum LLC
Materials Research Proceedings **25** (2023) 371-378 https://doi.org/10.21741/9781644902417-46

Computer vision for industrial defect detection

Johannes Landgraf[1,a *], Moritz Kompenhans[1,b *], Tassilo Christ[1,c *], Theresa Roland[2,d] and Adrian Heinig[3,e]

[1]d-fine GmbH, An der Hauptwache 7, 60313 Frankfurt, Germany

[2]voestalpine Metal Forming GmbH, voestalpine Straße 1, 4020 Linz, Austria

[3]voestalpine Automotive Components Dettingen GmbH & Co. KG, Daimlerstraße 29, 72581 Dettingen/Erms, Germany

[a]Johannes.Landgraf@d-fine.de, [b]Moritz.Kompenhans@d-fine.de, [c]Tassilo.Christ@d-fine.de [d]Theresa.Roland@voestalpine.com, [e]Adrian.Heinig@voestalpine.com

Keywords: Artificial Intelligence, Visual Inspection, Defect Detection

Abstract. Despite the continuous progress in computer vision, its application to many industrial tasks like the detection and size measurement of non-trivial defects is still a demanding problem. In this paper, typical challenges, workflows, and key performance indicators are discussed and the application of AI-based Semantic Image Segmentation methods is demonstrated to the detection of minute damages on metal surfaces using the d-fine vision toolbox. A performance improvement for a public data set above prior results is reported and the successful transfer of the approach to real-world sheet metal parts produced by voestalpine Automotive Components Schmölln GmbH is shown.

Introduction

Computer vision is an essential enabler of the end-to-end digitalization of production processes. Use cases like object tracking, pattern recognition, event detection or visual servoing (vision-based robot control) are among the fundamentals of intelligent, autonomous, and self-optimizing production processes [1]. Industrial computer vision can detect problems in an autonomous, continuous, and reliable manner and supply feedback into production processes and machines directly, e.g., in case of high failure rates due to incorrect machine calibration.

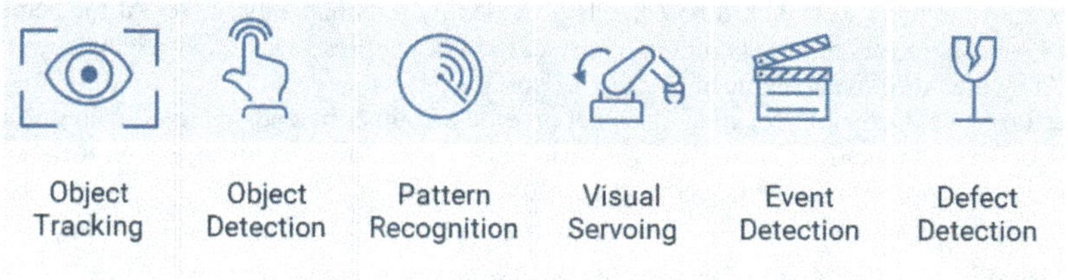

Fig. 1: Industrial computer vision applications

Virtually all modern applications of computer vision are based on neural networks with so called convolutional architectures [2] playing the most prominent role. When applied properly, this technology allows to automate analyses and processes that hitherto required a human's ability to interpret data. The following is an overview of the most important steps and the challenges involved.

Content from this work may be used under the terms of the Creative Commons Attribution 3.0 license. Any further distribution of this work must maintain attribution to the author(s) and the title of the work, journal citation and DOI. Published under license by Materials Research Forum LLC.

Data quality. Stable imaging conditions and good image quality are crucial for the success of computer vision. Deficiencies can only be partially compensated for by smart preprocessing methods and image resolution and lighting conditions must be carefully controlled.

Preprocessing. In most computer vision use-cases, the available data is imbalanced. The training data contain only a small fraction of NOK-parts and do not cover possible damage manifestations. To generate a more representative image base and to at least partially remedy these deficiencies algorithmic preprocessing can generate additional synthetic images (see e.g. [3] for an overview) with altered lighting and noise conditions, aspect ratios and sub-crops.

Deep Learning architecture. (Convolutional) Neural networks (CNNs) are inspired by the connectivity structure of neurons in the mammalian brain but should not be taken to represent a cognitive process. Their performance depends on a suitably chosen architecture, the quality of the input data and the careful tuning of the training strategy. [4] provides an excellent introduction into the architectural principles and the way CNNs build a hierarchical representation of image content.

Postprocessing. Algorithmic segmentation of the image into regions of different characteristics is not sufficient in productive applications. The results must be transformed into a suitable format for (quality-) experts and decision makers. This includes both a visual representation of the identified damages and performance dashboards for error rates or product quality assessments.

Benchmarking computer vision systems

Four practical requirements have proven crucial for a successful defect detection scheme, namely: robustness, data efficiency, accuracy and performance.

Robustness. Computer vision systems are affected by lighting situation and perspective and are negatively impacted by contamination and reflections. A good system should be robust with respect to such real-life disturbances and deviations from ideal conditions.

Data efficiency. Defect types, part shapes as well as location and orientation of the components to be inspected may vary between images. An ideal vision tool can be trained on a limited set of components with few images taken from only a subset of defect types and can still apply its "cognitive" abilities to new and hitherto unseen parts, defect types and perspectives within a given production process.

Accuracy. To ensure that the system can take over tedious and repetitive QA tasks, detected errors must coincide well with defect labels provided by a human supervisor. At the same time False Positives (errors reported without an actual error being present) and False Negatives (errors missed by the AI) must be reduced as much as possible.

Performance. Detection accuracy does not provide a competitive advantage if data throughput is low due to computational complexity. In particular for real-time applications on the shopfloor the achievable processing rate is a critical quantity.

Related work

Deep learning and computer vision provide robust solutions to the tasks shown in Fig. 1, including object tracking [5], object recognition [6] and visual servoing [7].

Vision-based quality inspection offers the possibility to perform non-destructive inspection processes in industry. Huang et al. [8] provide an overview of inspection tasks in the semi-conductor industry that can potentially be automated. Defect detection in an industrial context using CNNs on X-ray images has been investigated in [9]. The detection of cracks is part of many research projects. Detecting cracks in concrete using deep learning has been studied by [10]. Surface inspection is an important part of industrial quality assurance. The authors of [11] investigate the use of deep learning for this application and present essential design requirements.

In [12], U-Nets were presented to segment anomalies in biomedical data. U-Nets comprise a contracting and an expanding data path (see left side of Fig. 2 for an example) and are also used in this work. Their output is a mask with probabilities for each pixel indicating whether the pixel is or is not part of a labelled region (e.g., "DEFECT" / "OK"). The horizontal and vertical dimensions of the output are identical to the input dimensions of the image such that the results can be visually overlaid for ease of interpretation later. [13] has achieved state of the art results using U-Nets for magnetic tile inspection.

Deep learning enables very good detection rates but requires very large amounts of data. Especially in the industrial environment it can be difficult to generate and annotate these correctly for training.

In [14] the authors address the data efficiency by applying networks trained in other domains to new data sets (transfer learning). In [15] a special loss function ("focal loss") for training is introduced that emphasizes hard to learn image areas to increase network performance. Finally, in [16] a special sampling method is proposed that increases the weight of entire images that contain hard to learn content.

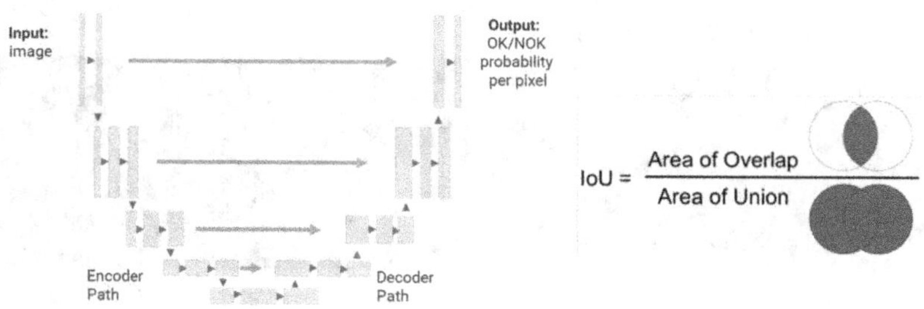

Fig. 2: Left: A simplified U-shaped neural network architecture. Right: Intersection over Union of a training mask and a network output mask as metric for the evaluation of image segmentation methods

Approach

The objective of this work is to design a defect detection system that provides for robustness, data efficiency and computation performance while achieving state-of-the-art results for accuracy and to demonstrate its applicability not only to laboratory data but to real-world images from an industry-grade production process. As described in [15], the training of semantic segmentation algorithms suffers twofold from data imbalance: First, "OK"-parts are overrepresented. Second, the defects to be detected are very small compared to the size of both the component itself and the image on which they are to be found. For this purpose, this paper investigates the visual inspection capabilities of a U-Net trimmed for data efficiency in combination with a modified focal loss function, emphasizing the hard to train image regions and hiding out the component background.

To cope with the small number of input images, the training data set is enriched using various geometric and photometric data augmentations and innovative sampling strategies that enhance the "NOK"-regions' fraction in the data set by transferring damaged regions between images.

The used U-Net architecture includes five skip connections that copy input data from early layers to later layers. This allows the model to keep local information together and prevents information loss. Careful tuning of the network details ensures that the system is already highly data efficient at the architectural level.

Finally, a post-processing routine incorporating domain-specific knowledge about typical defect sizes enhances the performance of the semantic segmentation approach and suppresses false positives. This comprises i.a. a threshold set upon the probability level with which the network considers a pixel to be part of a defective area. To finally measure the recognition accuracy, it is common practice to compare the manually provided defect mask and the output mask of the network. Their agreement is usually quantified as the ratio between the intersection of the two areas and their union ("IOU-metric") [17] – see right side of Fig. 2.

Example use case: ball screw drives

To demonstrate the performance of the image segmentation and defect detection pipeline and to determine quantitative KPIs, a publicly available data set of ball screw drive images is used and the workflows and the results obtained are discussed.

Data set. The data set "Industrial Machine Tool Element Surface Defect Dataset" [18] by the Karlsruhe Institute of Technology (KIT) contains images of ball screw drives (BSDs). BSDs are roller bearings used for translating rotary motion into linear motion and "one of the most wear-prone machine tools" [19]. The data set consists of 394 images each with at least one defect, which was labelled by a human expert using a polygonal mask along the boundary of the damage region (see Fig. 3). The defect regions are of different sizes and the images show soiling as it is typical for industrial manufacturing environments.

Fig. 3: Image taken from the KIT ball screw drive data set showing a defect (left) and the corresponding defect mask (right).

Data split and preprocessing. The data is split into subsets as proposed by the authors [18]. The training proceeds in so called "epochs" during with the system tries to optimize recognition results on the training data set. The performance is measured independently after each training epoch using the validation data set. The final performance KPIs reported below however are determined on the test data set which has never been presented to the network before.

Training details. The system has been trained on Google Cloud Platform for 62 hours using a Nvidia V100 GPU with 16 GB RAM.

Results. Fig. 4 shows the results achieved when applying the network to previously unseen error samples. Defects on the test data set are detected with a mean IoU of 45.3%, which is a noticeable improvement compared to the previously published value of 31.6% (see [19]). For all practical applications it is necessary to decide at some point whether a region with a "defect"-label is to be classified as an actual defect or to be rejected as a false positive candidate (as may be done with very small defect regions in practice). Fig. 5, shows the classification results of the approach for a set of 60 selected test-images. Each image is represented by a dot on a two-dimensional map, with the horizontal and vertical extension of the damaged region (measured relative to the total image size) used as coordinates. Orange dots indicate correctly detected damages, blue ones indicate missed ones. The toolchain presented in this paper detects even small defects more reliably when compared to results shown in [20]. The undetected defects have a small relative size compared to the image's dimension. A higher resolution or an inspection using image sub-crops might help to further improve the performance.

Materials Research Forum LLC
https://doi.org/10.21741/9781644902417-46

Fig. 4: Left: Training images with damaged regions - defect masks provided during the training phase are shown in red; Right: Separate validation with unseen data - damage masks identified

by the trained network are shown in amber.

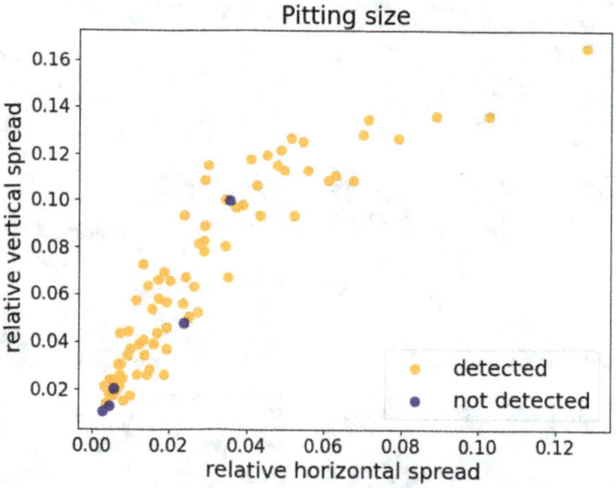

Fig. 5: Detected and missed defects and their relative horizontal and vertical spread in the image. The solution almost perfectly detects all but the most minute damages.

Application to sheet metal components

To demonstrate the feasibility of defect detection outside the lab with real-world shop-floor data with comparable challenges, the method is applied to several sheet metal parts that belong to the automotive chassis. The parts were produced and photographed in the press plant of voestalpine in Schmölln, Germany.

Data set and enrichment. The images for this analysis were taken on a separate inspection island from a single camera perspective under reproducible and calibrated lighting conditions. The dataset contains 179 images taken from 3 different sheet metal component shapes. 107 images were used for training, 36 for validation and 36 for testing. The high resolution of the images and the need to detect also small defects mandate that the defect recognition is applied to image sub-crops to ensure reliable detection. Also, the use of image cropping allows to increase the proportion of damaged regions during training and thus to equalize the ratio between "OK" and "NOK" images in the data set.

Foreground segmentation and damage detection. To find the damaged sections, the AI-based segmentation method has been applied twofold: First it masks the component part ("foreground") and separates it from the uninformative image "background". The images on the left of Fig. 6 show typical masks resulting from this step with reddish pixels indicating the image sections that the AI ascribes to the part. This step also works reliably when new shapes are presented to the background segmentation step. Damage detection itself proceeds subsequently. Typical results for images from the test data set are shown in the right side of Fig. 6. Here pixels to which the AI ascribes a "damaged" label are marked in green. Pixels in blue are "missed pixels": They have been labelled manually as "damage region" before the training phase but have not been recognized by the AI.

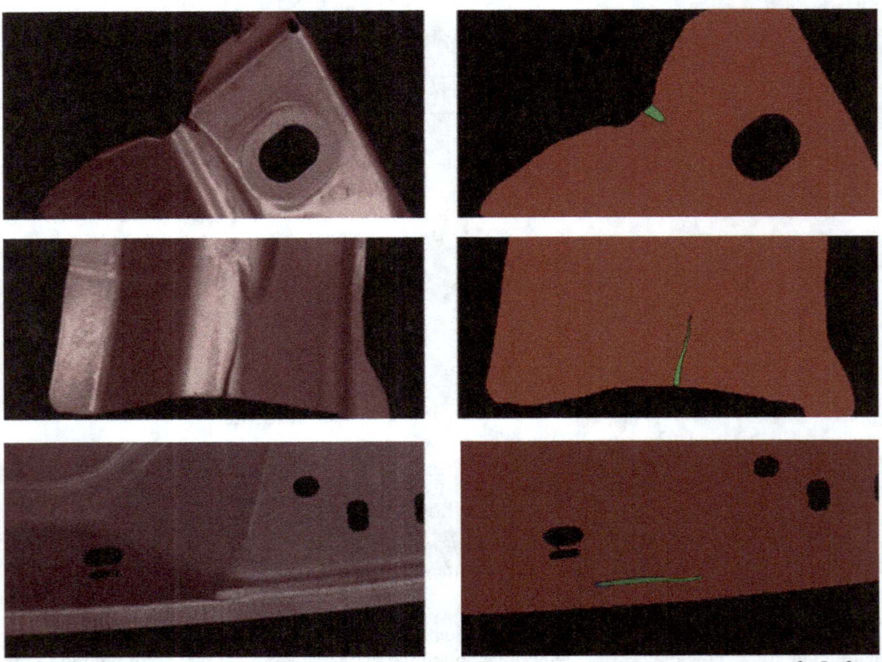

Fig. 6: A sheet metal components inspected with the two-stage segmentation approach. Left: foreground separation; right: damage detection

Results. While applying the methodology allows to train a neural network with the very small data set, a statistical analysis of the results would not provide meaningful and reliable performance numbers. As described in the review paper [21], crack detection algorithms are usually trained on data sets that are about 100 times larger in scale. To provide quantitative, statistically sound results a data set of at least five times the one available at present would be needed. A qualitative analysis shows however, that the trained network detects damages also in the test data set if they are optically similar to defects presented during training. Using sufficiently well resolved image sub-crops, also very small defects can be detected that are hardly visible to the human eye on a full-scale image. As the training of the damage detection step is performed on image sub crops it should be agnostic with respect to the specific component shape under inspection. This is an important feature for real life applications because it substantially reduces the need to retrain the AI for new

Materials Research Forum LLC
https://doi.org/10.21741/9781644902417-46

shapes and different camera perspectives. However, a larger data set is required to verify this unambiguously.

Summary
Using convolutional neural networks, an AI-based image segmentation method is developed which detects even small defects on sizable metallic surfaces. The method is data efficient in that it requires only a very limited amount of training data and thus greatly reduces the labelling effort required before training. It generalizes well to unseen data and outperforms prior results on a publicly available data set. The developed preprocessing, training and postprocessing cascade is applied to real world images from a sheet metal plant. While statistical performance KPIs can not be given due to the limited amount of available data promising qualitative results are achieved. A fully quantitative evaluation of the performance KPIs on a larger data set is required next to develop the system into a flexible, scalable AI system for end-of-line quality control for sheet metal parts and other components with metallic surfaces.

Acknowledgement
The authors would like to thank the staff of voestalpine Automotive Components Schmölln GmbH for explanation of their image acquisition setup and the provision of the data set and the voestalpine data science team for fruitful technical exchange.

References

[1] B. Vogel-Heuser, T. Bauernhansl, M. ten Hompel, eds., Handbuch Industrie 4.0 Bd.4, Allgemeine Grundlagen., Springer, 2017. https://doi.org/10.1007/978-3-662-53254-6

[2] Y. Lecun, Y. Bengio, Convolutional networks for images, speech, and time-series, in: M.A. Arbib (Ed.), The Handbook of Brain Theory and Neural Networks, MIT Press, 1995.

[3] C. Shorten, T.M. Khoshgoftaar, A survey on Image Data Augmentation for Deep Learning, J. Big Data. 6 (2019) 60. https://doi.org/10.1186/s40537-019-0197-0

[4] I. Goodfellow, Y. Bengio, A. Courville, Deep learning, MIT press, Cambridge, MA, USA, 2016.

[5] S. Mojtaba Marvasti-Zadeh, L. Cheng, H. Ghanei-Yakhdan, S. Kasaei, Deep Learning for Visual Tracking: A Comprehensive Survey, ArXiv E-Prints. (2019) earXiv:1912.00535.

[6] J. Redmon, S. Divvala, R. Girshick, A. Farhadi, You Only Look Once: Unified, Real-Time Object Detection, ArXiv E-Prints. (2015) earXiv:1506.02640. https://doi.org/10.1109/CVPR.2016.91

[7] Q. Bateux, E. Marchand, J. Leitner, F. Chaumette, P. Corke, Visual Servoing from Deep Neural Networks, ArXiv E-Prints. (2017) earXiv:1705.08940.

[8] S.-H. Huang, Y.-C. Pan, Automated visual inspection in the semiconductor industry: A survey, Computers in Industry. 66 (2015) 1–10. https://doi.org/10.1016/j.compind.2014.10.006

[9] M. Ferguson, R. Ak, Y.-T.T. Lee, K.H. Law, Detection and Segmentation of Manufacturing Defects with Convolutional Neural Networks and Transfer Learning, ArXiv E-Prints. (2018) earXiv:1808.02518. https://doi.org/10.1520/SSMS20180033

[10] D. Einarson, D. Mengistu, Deep Learning Approaches for Crack Detection in Bridge Concrete Structures, in: 2022 International Conference on Electronic Systems and Intelligent Computing (ICESIC), 2022: pp. 7–12. https://doi.org/10.1109/ICESIC53714.2022.9783576

[11] D. Martin, S. Heinzel, J. Kunze von Bischhoffshausen, N. Kühl, Deep Learning Strategies for Industrial Surface Defect Detection Systems, ArXiv E-Prints. (2021) earXiv:2109.11304. https://doi.org/10.24251/HICSS.2022.146

[12] O. Ronneberger, P. Fischer, T. Brox, U-Net: Convolutional Networks for Biomedical Image Segmentation, ArXiv E-Prints. (2015) earXiv:1505.04597. https://doi.org/10.1007/978-3-319-24574-4_28

[13] Y. Huang, C. Qiu, Y. Guo, X. Wang, K. Yuan, Surface Defect Saliency of Magnetic Tile, in: 2018 IEEE 14th International Conference on Automation Science and Engineering (CASE), 2018: pp. 612–617. https://doi.org/10.1109/COASE.2018.8560423

[14] J. Yosinski, J. Clune, Y. Bengio, H. Lipson, How transferable are features in deep neural networks?, ArXiv E-Prints. (2014) earXiv:1411.1792.

[15] T.-Y. Lin, P. Goyal, R. Girshick, K. He, P. Dollár, Focal Loss for Dense Object Detection, ArXiv E-Prints. (2017) earXiv:1708.02002. https://doi.org/10.1109/ICCV.2017.324

[16] A. Shrivastava, A. Gupta, R. Girshick, Training Region-based Object Detectors with Online Hard Example Mining, ArXiv E-Prints. (2016) earXiv:1604.03540. https://doi.org/10.1109/CVPR.2016.89

[17] T.T. Tanimoto, An Elementary Mathematical Theory of Classification and Prediction, International Business Machines Corporation, 1958.

[18] T. Schlagenhauf, M. Landwehr, J. Fleischer, Industrial Machine Tool Element Surface Defect Dataset, 2021. https://doi.org/10.1016/j.dib.2021.107643

[19] A. Haberkern, Leistungsfähige Kugelgewindetriebe durch Beschichtung, Universität Karlsruhe, Institut für Werkzeugmaschinen und Betriebstechnik, 1998

[20] T. Schlagenhauf, M. Landwehr, Industrial machine tool component surface defect dataset, Data in Brief. 39 (2021) 107643. https://doi.org/10.1016/j.dib.2021.107643

[21] S.D. Nguyen, T. Son, T. Van Phuc, H. Lee, M. Piran, V.P. Le, Deep Learning-Based Crack Detection: A Survey, International Journal of Pavement Research and Technology. (2022). https://doi.org/10.1007/s42947-022-00172-z

Materials Research Proceedings 25 (2023) 379-386 https://doi.org/10.21741/9781644902417-47

Evaluating material failure of AHSS using acoustic emission analysis

Eugen Stockburger[1,a,*], Hendrik Vogt[1,b], Hendrik Wester[1,c], Sven Hübner[1,d], and Bernd-Arno Behrens[1,e]

[1]Institut für Umformtechnik und Umformmaschinen, Leibniz Universität Hannover, An der Universität 2, 30823 Garbsen, Germany

[a]stockburger@ifum.uni-hannover.de, [b]vogt@ifum.uni-hannover.de, [c]wester@ifum.uni-hannover.de, [d]huebner@ifum.uni-hannover.de, [e]behrens@ifum.uni-hannover.de

*Corresponding author

Keywords: Fracture Analysis, High Strength Steel, Acoustic Emission

Abstract. Driven by high energy prices and strict legal requirements on CO_2 emissions, high-strength sheet steel materials are increasingly gaining importance in the automotive industry regarding electric vehicles and their battery range. Simulation-based design of forming processes can contribute to exploiting their high potential for lightweight design. However, previous studies show that numerical simulation with conventional forming limit curves does not always provide adequate prediction quality. Failure models that take the stress state into account represent an alternative prediction method for the shear-dominated failure, that frequently occur in high-strength steels during forming. The failure behaviour of the sheet materials can be determined by different specimen geometries for a wide range of stress states and by using an optical measurement system to record the local strain on the surface of the specimen at the location of failure. However, for many high-strength steels, critical damage or failure initiation already occurs inside the specimen. Therefore, a method is needed that allows detection of failure initiation at an early stage before the crack becomes visible on the surface of the specimen. One possible method is the use of acoustic emission analysis. By coupling it with an imaging technique, the critical strains leading to failure initiation inside the specimen can be determined. In the presented paper, butterfly tests are performed for a wide range of stress states and measured with an optical as well as an acoustical measurement system. The tests are analysed regarding the failure initiation using a mechanical, optical as well as acoustical evaluation method and compared with each other.

Introduction

Strict laws on vehicle emissions and high fuel prices have been leading to changes in vehicle concepts for years. Throughout the automotive industry, there is a clear trend towards e-mobility. In order to compensate for the weight of the battery and increase the driving range, manufacturers are increasingly relying on modern lightweight materials, including advanced high-strength steels (AHSS) such as complex phase (CP) or dual phase (DP) steels to reduce the weight of their vehicles [1]. The mechanical properties of AHSS are controlled by a specific mixture of different hard microstructural components such as retained austenite, ferrite, bainite, martensite and non-metallic precipitates. Thus, it is possible to cover yield strengths and tensile strengths far above those of conventional deep-drawing and low-alloy steels while maintaining a good formability.

In order to optimally utilise the high potential of AHSS for reducing vehicle weight at reasonable costs, a reliable design of forming and crash behaviour of the sheet metal components is required by use of numerical simulation. Therefore, precise modelling of the material behaviour under process conditions is needed. With regard to sheet metal forming, the yield condition, the hardening behaviour and the forming capacity must be included in the modelling. For the description of the forming capacity of sheet metals, the use of forming limit curves (FLC) is state

Content from this work may be used under the terms of the Creative Commons Attribution 3.0 license. Any further distribution of this work must maintain attribution to the author(s) and the title of the work, journal citation and DOI. Published under license by Materials Research Forum LLC.

of the art. A FLC determines strain states that can be withstood by the sheet metal without necking or fracture. However, it is only valid for linear strain paths and a spectrum of strain states between uniaxial and biaxial tension [2]. Due to the shear-dominated material failure of AHSS, the prediction quality of a FLC is not adequate for that. An alternative to the FLC in sheet metal forming simulation are failure models which take the stress state into account. In most failure models, fracture is described by the failure strain as a function of the stress state [3]. Many models already exist for the mathematical description of these failure surfaces for AHSS, such as the Modified Mohr-Coulomb (MMC) failure model [4]. In order to characterise the material failure, material tests such as uniaxial tensile tests with different notches or holes, shear tests and Nakajima tests are performed to investigate the failure under different stress states from compression to shear to tension. Multiple tests are necessary to describe the material failure realistically in a wide range of different loadings and to parametrise the parameters of the failure model [4]. Those failure characterisation tests are often monitored with an optical measurement system, which is used to determine the fracture initiation on the specimen surface and to calculate the plastic strain. To improve the accuracy of the failure characterisation, a method capable of detecting a critical fracture initiation inside the specimen that remains invisible to the optical measurement system can be used [5]. Furthermore, sometimes the fracture initiation on the specimen surface is hardly to determine optically due to the stress state or because of the stochastic pattern applied on the specimen surface. In this context, acoustic emission (AE) analysis has a high potential for an improvement in failure characterisation due to enhanced fracture detection [6].

Therefore, in this paper an existing testing rig for butterfly specimen with optical measurement system is extended with an acoustical measurement system. Seven loading angles of the butterfly tests, generating seven different stress states, are investigated for two AHSS. The tests are monitored with an optical as well as an acoustical measurement system. Finally, the failure initiation is determined and compared by the three different evaluation methods mechanical, optical as well as acoustical.

Material Characterisation with Acoustic Emission

AE analyses were used prior in the literature to characterise different material properties of metals. Shen et al. investigated the behaviour of titanium during a tensile test with a broadband and a resonant sensor [7]. The yielding and fracture of the material was characterised by a larger amplitude and a higher number of counts exceeding a threshold value. In contrast to the broadband sensor, the resonant sensor was also able to identify necking due to a higher amplitude. AE signals during plastic deformation of a low-carbon steel were analysed by Muravev et al. [8]. The localisation of acoustic emissions in combination with speckle interferometry enabled tracking of the movement of Lüders bands at yielding. Furthermore, it was possible to predict necking and fracture. In [9], Chuluunbat et al. used AE to analyse the fracture behaviour of the steel X70 during tensile tests with notched specimens. The fracture development was recorded simultaneously with a high-speed camera. At the time of fracture development inside the specimen, the AE activity increased rapidly. Changes of the process parameters temperature and strain rate also influenced the AE activity. Panasiuk et al. used the acoustic emission method and Kolmogorov-Sinai entropy to successfully determine the yield point of the aluminium alloy EN AW-7020 during tensile testing [10]. A new acoustic emission activity detection method based on short-term waveform features was invented by Piñal-Moctezuma et al. and applied to tensile tests of sheet metal [11]. In [12], Petit et al. observed the strain localisation features and mechanisms during tensile tests of C35 by coupling AE analysis and electronic speckle pattern interferometry. The AE behaviour of DP steels with different martensite contents was investigated with tensile tests by Khamedi et al. [13]. It was shown that for DP steels with a low martensite content, the damage mechanism was dominated by decohesion of ferrite and martensite. In specimens with a high martensite content, fractures also occurred in the martensite. The different damage mechanisms had a characteristic

Materials Research Proceedings **25** (2023) 379-386 https://doi.org/10.21741/9781644902417-47

frequency range that could be identified by wavelet analysis. Pathak et al. investigated the influence of the stress state on micro-void nucleation for a CP800 and a DP780 using 3D micro-tomography for interrupted tests and in situ digital image correlation during testing [14]. For both steels, little void nucleation was observed under shear deformation but extensive void damage was observed under biaxial tension.

Materials and Methods

The investigated AHSS HCT780C (CP800) with 1.6 mm sheet thickness and HCT980X (DP1000) with 1.5 mm sheet thickness are provided from voestalpine Stahl GmbH. The CP800 is a complex phase steel consisting of martensite, bainite as well as ferrite and the DP1000 is a dual phase steel consisting of martensite as well as ferrite. To investigate the failure behaviour of the steels CP800 and DP1000, butterfly tests are conducted, which are detailed described by Behrens et al. in [15]. The experimental set-up is shown in Fig. 1 and consists of a rotatable testing rig, which is installed in the tensile testing machine S100/ZD from DYNA-MESS Prüfsysteme GmbH.

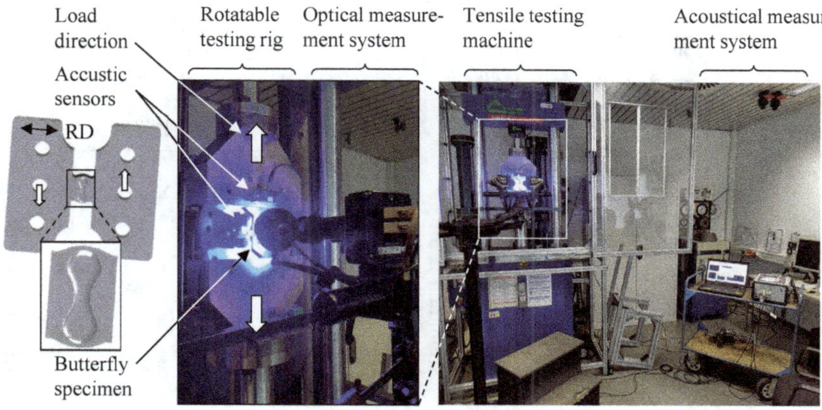

Fig 1: Experimental set-up for butterfly tests with optical and acoustical measurement systems

It further contains the optical measurement system Aramis from Carl Zeiss GOM Metrology GmbH and the acoustical measurement system AMSY-6 from Vallen Systeme GmbH. Piezoelectric sensors of the same company as the measuring system are used to record the acoustic emissions. To receive the signals off the specimen, a resonance sensor (VS900-RIC) and a broadband sensor (AE2045) are used, which are both applied on the rotatable testing rig. Thus, a frequency band of 100-2200 kHz can be investigated. Furthermore, a guard sensor (VS900-RIC) is attached to the tensile test machine as well to record and filter the machine-induced noise. In the following, the resonance sensor will be referred to as AE1 and the broadband sensor as AE2. When applying the AE sensors, the coupling agent Echotrace from Karl Deutsch Prüf- und Messgerätebau GmbH is used. With the rotatable testing rig, the butterfly specimen can be aligned at -3°, 12.5°, 28°, 43.5°, 59°, 74.5° and 90° to the loading direction, which results in different stress states in the centre of the specimen ranging from shear loading to uniaxial tension. During the experiments, the testing rig is fixed against rotation. The seven angles are tested until failure with a velocity of 0.02 mm/s (quasi-static) for both AHSS at room temperature and each test is repeated five times for statistical validity. While testing, the forming force is measured by the tensile testing machine, the displacement of the specimen by the optical measurement system and the acoustic emission by the acoustical measurement system. The recording frequency of the optical measurement system was 20 Hz and the sound emissions are recorded event-triggered with a sampling rate of 5 MHz.

Each specimen failure is evaluated based on the three methods mechanical, optical and acoustical, as shown in Fig. 2. The mechanical method assumed the failure for each specimen, when a 30 % drop of the forming force appeared in about 0.5 s. The threshold value was chosen based on the inertia of the tensile testing machine. Using the optical method, the fracture initiation on the specimen surface is determined by the images from the optical measurement system. The acoustical signals are characterised on the basis of the form of the signal and their parameters. At the failure initiation, only burst signals (clearly recognisable beginning and end) are obtained. These are first classified based on the amplitude reached. Other certain features such as time and frequency are extracted from the signals and those suitable for differentiation are determined. These included rise-time and the centre of gravity of the frequency spectrum (FCOG). The three methods are compared regarding the failure initiation by evaluating the displacement and plastic strain before fracture determined by each method.

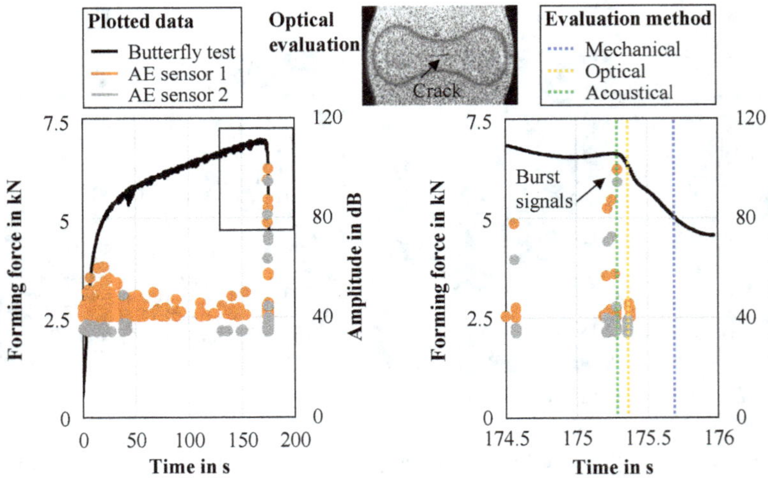

Fig 2: Mechanical, optical and acoustical evaluation methods of material failure

Experimental Results and Comparison of the Evaluation Methods

Fig. 3 displays the forming force-time curve of the performed butterfly tests for CP800 on the left and DP1000 on the right under variation of the loading angle from -3° to 90°. For both materials, the maximum forming force is the highest at a loading angle of 90° and reduces continuously until a loading angle of 12.5°. The reason for this effect is that at a loading angle of 90° the stress state in the specimen is uniaxial tension, which is more and more superimposed with shear loading reducing the loading angle until mostly shear loading is present at -3°. The forming force at -3° is at the same level as 12.5°, but the material failure occurs later. Comparing the CP800 and the DP1000, the maximum forming force for the DP1000 is higher than for the CP800 at every loading angle due to the higher strength of the DP1000.

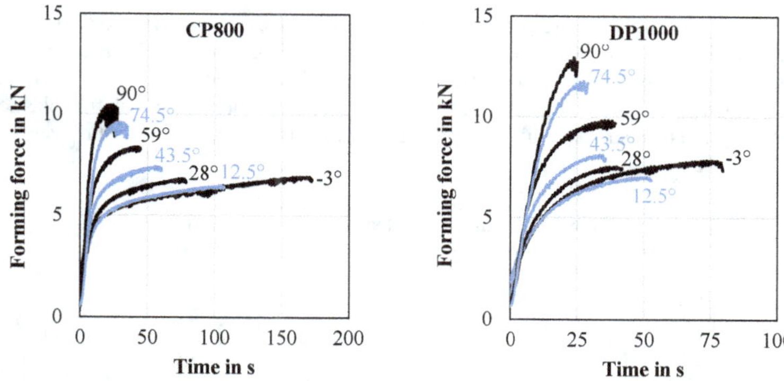

Fig 3: Forming force-time curves of the butterfly specimen for CP800 (left) and DP1000 (right)

The optical measured and calculated plastic strain of the butterfly tests before failure is shown in Fig. 4. At the top, the plastic strain distribution according to von Mises is plotted for CP800 and on the bottom for DP1000 for the seven loading angles determined by the optical evaluation method. As clearly visible at every loading angle for both materials, the plastic strain concentrates on the middle area of the specimen due to the reduced thickness in the specimen centre. Hence, no fractures appeared along the edges of the specimen. In contrast to the maximum forming force, the maximum plastic strain decreases for higher loading angles for both materials. Due to the higher formability of CP800 compared to DP1000, the achieved plastic strain is higher for CP800 at every loading angle.

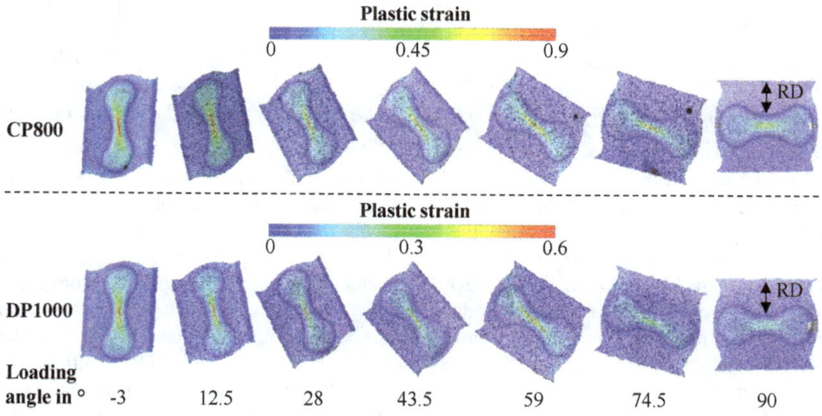

Fig 4: Plastic strain distribution before failure from the optical measurements for CP800 (top) and DP1000 (bottom)

To investigate the influence of the evaluation methods on the results of the butterfly tests, the displacement in load direction and the plastic strain in the centre of the specimen before failure averaged from the five experiments are shown with standard deviation at the top of Fig. 5 for CP800. As to be expected, the displacement before failure is the highest using the mechanical evaluation method and is reduced with the optical evaluation method. The acoustical evaluation method determines slightly lower or almost the same displacements compared to the mechanical evaluation method, except for the loading angle -3°. Analysing the plastic strain, the effect of the different displacements can be seen well. Although the displacement from the mechanical evaluation method is only slightly higher than from the other methods, the plastic strains determined by the mechanical evaluation method significantly exceed the plastic strains from the other methods for all loading angles.

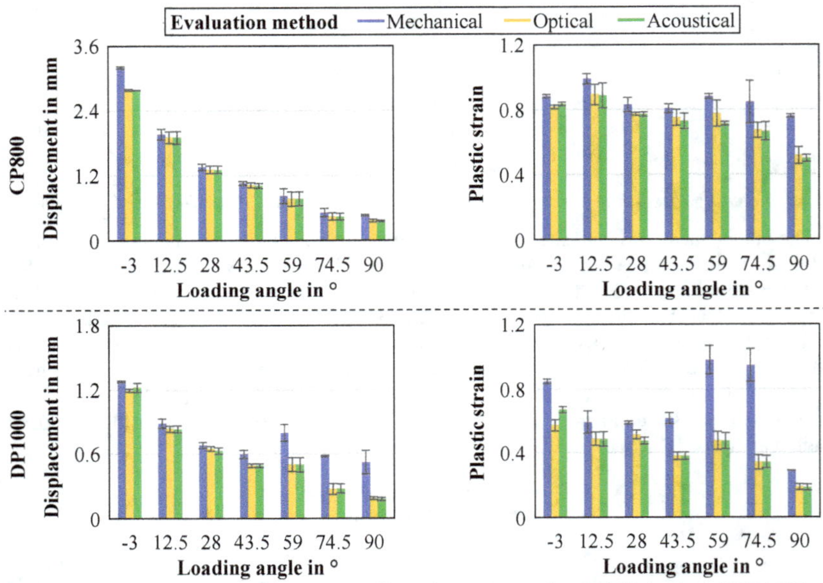

Fig 5: Displacement before failure as well as plastic strain for CP800 (top) and DP1000 (bottom)

Comparing the optical and the acoustical evaluation method, the acoustical evaluation method shows slightly lower values than the optical method likewise to the displacements. Only for a loading angle of -3° the plastic strain of the acoustical evaluation method is higher than the plastic strain of the optical evaluation method. Apparently, the begin of failure under shear loading does not produce an AE signal, which could be measured as well as distinguished from machine-induced noise with the used sensors and amplifiers. The measured AE signals appear after the fracture can already be seen on the specimen surface and propagates in the material. This effect is due to the failure mode of this specific stress state: at the loading angle -3°, a shear stress state and shear failure are present. For the other loading angles, the tension stress state is dominant, which leads to normal failure initiated by the formation and growth of voids. A similar behaviour can be seen for DP1000 at the bottom of Fig. 5. Here the plastic strain before failure using the mechanical evaluation method is often significantly higher than the values determined by the other two methods. Especially for the loading angles 59° and 74.5° of DP1000 the plastic strain before failure of the mechanical evaluation method is much higher than for the other two methods. This is attributed to the fact that the high strength of the DP1000 causes an abrupt failure of the material, especially for the tension stress states at the loading angles 59° and 74.5°.

Summary and Outlook

To increase the determination accuracy of the failure initiation and hence the prediction accuracy of failure models, a test rig for butterfly specimen with optical and an acoustical measurement system was developed. The measurement setup was used to determine the failure initiation of two AHSS under different stress states induced in butterfly specimens by different loading angles using the rotatable butterfly test rig. The failure of the respective specimens was evaluated using optical and acoustical measurements. As a reference, the mechanical failure was determined based on a sudden decrease in the forming force. First, the influence of the loading angle on the required forming force and strain was determined. For low angles, a low load capacity and a high maximum plastic strain according to von Mises were recorded. As the loading angle increases, the load increases and the maximum plastic strain decreases. Subsequently, the time of failure initiation determined by the different measurement methods was compared. The optical and acoustical signals lead to earlier detection of the failure initiation compared to the mechanical signal. However, the acoustical signal can only achieve a small improvement in the determined failure time compared to the optical signal. In the case of a shear stress at -3°, the time of failure determined by means of the acoustical signal was even later than the optical measurement.

In further investigations, an improved recording of the failure initiation in the specimens is to be achieved by replacing the AE measurement amplifier. With a different amplification it is possible to detect signals with higher amplitudes. In addition, a coupling of the measuring systems and the test rig is planned so that the test is interrupted at an early stage with the aid of an automatic failure criterion. Furthermore, the experiments will be numerically modelled and the stress state will be determined for each evaluation method. Failure models like the MMC model will be calibrated based on the results. It is assumed that the failure models for the different methods vary significantly due to a difference in the stress state at the specimen surface and in the specimen centre. Thus, failure models with higher accuracy can be created to represent processes more precisely in the numerical simulation.

Acknowledgment

The authors gratefully acknowledge the support of the German Research Foundation (Deutsche Forschungsgemeinschaft, DFG) within the Project 385276585 ''Improving the failure characterisation of advanced high-strength steel sheets by coupling measurement systems for optical forming analysis with acoustic emission technology''.

References

[1] N. Fonstein, Dual-phase steels, in: Automotive steels - Design, Metallurgy, Processing and Applications, Elsevier, 2017, pp. 169–216.

[2] B.-A. Behrens, D. Rosenbusch, H. Wester, E. Stockburger, Material Characterization and Modeling for Finite Element Simulation of Press Hardening with AISI 420C, J. of Materi Eng and Perform 31 (2022) 825–832. https://doi.org/10.1007/s11665-021-06216-y

[3] B.-A. Behrens, K. Dröder, A. Hürkamp, M. Droß, H. Wester, E. Stockburger, Finite Element and Finite Volume Modelling of Friction Drilling HSLA Steel under Experimental Comparison, Materials 14 (2021). https://doi.org/10.3390/ma14205997

[4] Y. Bai, T. Wierzbicki, Application of extended Mohr–Coulomb criterion to ductile fracture, Int J Fract 161 (2010) 1–20. https://doi.org/10.1007/s10704-009-9422-8

[5] E. Martinez-Gonzalez, I. Picas, D. Casellas, J. Romeu, Detection of crack nucleation and growth in tool steels using fracture tests and acoustic emission, Meccanica 50 (2015) 1155–1166. https://doi.org/10.1007/s11012-013-9858-9

[6] B.-A. Behrens, S. Hübner, K. Wölki, Acoustic Emission – A promising and challenging technique for process monitoring in sheet metal forming, in: Journal of Manufacturin Processes, Volume 29, 2017, pp. 281–288. https://doi.org/10.1016/j.jmapro.2017.08.002

[7] G. Shen, L. Li, Z. Zhang, Z. Wu, Acoustic Emission Behavior of Titanium During Tensile Deformation, in: Advances in Acoustic Emission Technology, Springer, 2015, pp. 235–243. https://doi.org/10.1007/978-1-4939-1239-1_22

[8] T.V. Murav'ev, L.B. Zuev, Acoustic emission during the development of a Lüders band in a low-carbon steel, Tech. Phys. 53 (2008) 1094–1098. https://doi.org/10.1134/S1063784208080203

[9] T. Chuluunbat, C. Lu, A. Kostryzhev, K. Tieu, Investigation of X70 line pipe steel fracture during single edge-notched tensile testing using acoustic emission monitoring, Mater. Sci. Eng. A 640 (2015) 471–479. https://doi.org/10.1016/j.msea.2015.06.030

[10] K. Panasiuk, L. Kyziol, K. Dudzik, G. Hajdukiewicz, Application of the Acoustic Emission Method and Kolmogorov-Sinai Metric Entropy in Determining the Yield Point in Aluminium Alloy, Materials 13 (2020). https://doi.org/10.3390/ma13061386

[11] F. Piñal-Moctezuma, M. Delgado-Prieto, L. Romeral-Martínez, An acoustic emission activity detection method based on short-term waveform features: Application to metallic components under uniaxial tensile test, Mechanical Systems and Signal Processing 142 (2020) 106753. https://doi.org/10.1016/j.ymssp.2020.106753

[12] J. Petit, G. Montay, M. François, Strain Localization in Mild (Low Carbon) Steel Observed by Acoustic Emission - ESPI Coupling during Tensile Test, Exp Mech 58 (2018) 743–758. https://doi.org/10.1007/s11340-018-0379-2

[13] R. Khamedi, A. Fallahi, A. Refahi Oskouei, Effect of martensite phase volume fraction on acoustic emission signals using wavelet packet analysis during tensile loading of dual phase steels, Materials & Design 31 (2010) 2752–2759. https://doi.org/10.1016/j.matdes.2010.01.019

[14] N. Pathak, J. Adrien, C. Butcher, E. Maire, M. Worswick, Experimental stress state-dependent void nucleation behavior for advanced high strength steels, Int. J. Mech. Sci. 179 (2020) 105661. https://doi.org/10.1016/j.ijmecsci.2020.105661

[15] B.-A. Behrens, D. Rosenbusch, H. Wester, P. Althaus, Comparison of three different ductile damage models for deep drawing simulation of high-strength steels, IOP Conf. Ser.: Mater. Sci. Eng. 1238 (2022) 12021. https://doi.org/10.1088/1757-899X/1238/1/012021

Dynamic conformity assessment for joining force monitoring using Bayes filters

Lorenz Butzhammer[1,a *], Fabian Kappe[2,b], Gerson Meschut[2,c] and Tino Hausotte[1,d]

[1]Friedrich-Alexander-Universität Erlangen-Nürnberg, Chair of Manufacturing Metrology (FMT), Nägelsbachstr. 25, 91052 Erlangen, Germany

[2]Laboratory for Material and Joining Technology (LWF), Paderborn University, Pohlweg 47-49, 33098 Paderborn, Germany

[a]Lorenz.Butzhammer@fmt.fau.de, [b]fabian.kappe@lwf.upb.de, [c]meschut@lwf.upb.de, [d]Tino.Hausotte@fmt.fau.de

Keywords: Metrology, Joining, Force

Abstract. Monitoring force-displacement or force-time curves is a widely used quality control technique in the field of mechanical joining. For online monitoring of self-piercing riveting, envelope curves are often used to define a tolerance zone for the measured setting force. However, the measurement uncertainty is typically not considered and the force curve of a joint can be wrongly rated as non-conform due to measurement errors and noise. In this article, we present a method for dynamical online filtering and uncertainty determination for noisy force curves using two types of Bayesian filters. The methodology is based on a Bayesian probability framework using a priori information for the process curve and sensor noise. To investigate the general feasibility of the method, force measurements with different noise levels are simulated and processed. The conformity is further assessed taking the uncertainty of the filtered signal into account. The results show that the Bayes filter technique is principally able to reduce noise for well-known characteristics of the process curve and sensor noise. Advantages over common filtering techniques, especially for experimental conditions with less known characteristics, are still to be verified. The methodology could be used in future for closed-loop controls to adapt process parameters dynamically.

Introduction

Mechanical joining is established in a variety of production processes in industry, i.e. in the automotive sector. Different joining processes such as self-piercing riveting (SPR) or clinching are increasingly adapted to further material combinations and process conditions. These processes are often used to join multi-material structures as thermal joining processes reach their process limits due to the different material properties. In case of SPR, a joint of two or more sheets is created using an auxiliary joining part (the rivet). The process generally consists of a punch, a blank holder, the rivet, the parts to be joined and the die (see Fig. 1). The process itself can be divided into four stages. First, the sheets are placed between the punch and the die and the blank holder fixes them. Subsequently, the rivet receives a feed from the punch and cuts through the punch-sided joining part. The joint is created by upsetting of the rivet in the die-sided sheet, which leads to a force and form fit connection. The process sequence and a typical force-displacement curve are shown schematically in Fig. 1.

To ensure a high load-bearing capacity of mechanical joints in series production, quality monitoring is important. On the one hand, quality-relevant parameters such as the interlock or the minimum die-side material thickness are measured. These can only be detected in macro sections,

Content from this work may be used under the terms of the Creative Commons Attribution 3.0 license. Any further distribution of this work must maintain attribution to the author(s) and the title of the work, journal citation and DOI. Published under license by Materials Research Forum LLC.

which leads to the destruction of the joint. On the other hand, a widespread non-destructive method to detect or prevent from failed joints is to record force-displacement or force-time curves.

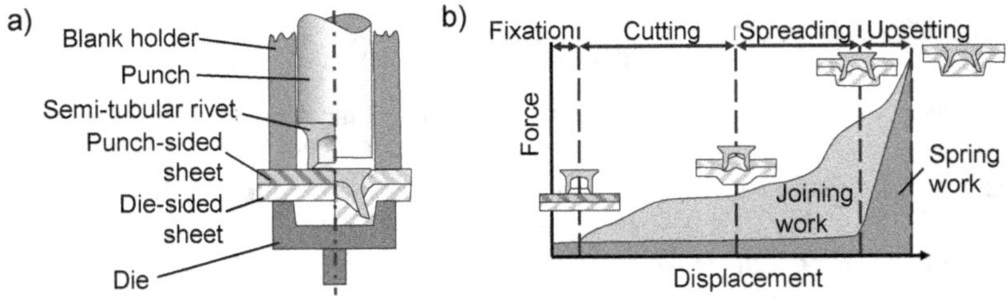

Fig. 1: Tooling for SPR and process sequence for SPR (a) as well as the exemplary joining force-displacement curve (b) according to [1].

In the case of self-piercing riveting, the guideline DVS/EFB 3410 [1] proposes two ways to rate if the force curve is in tolerance, window monitoring and tolerance monitoring. For the former method, 2D windows are defined in the force-displacement diagram that have to be crossed by the curve. This method is also used e.g. for clinch joining [2]. In case of tolerance monitoring, which considers the full process curve, envelope curves are defined as tolerance zone in which the measured curve must be located. In an on-line application, the process is stopped if the curve is out of tolerance. Several advantages such as reduced rejection rates can be achieved in this way.

In this article, we put a focus on tolerance monitoring. For this method, it is not fully clear how to define a proper tolerance band and how to consider the uncertainty of the measurement during on-line monitoring. Increased noise of the sensor can lead to a non-conformity decision although the process actually is in tolerance. Here, we investigate the use of recursive Bayesian filtering, taking a priori information about the process curve (assessed by experiments and statistical means) and about the measurement uncertainty into account. Bayesian filters can principally be used to dynamically correct (filter) incoming sensor readouts and estimate the uncertainty of the resulting filtered signal [3]. There are numerous application examples throughout different disciplines, such as GPS tracking [4], mobile robot localization [5] and dynamic coordinate measurements [6]. We use the Extended Kalman Filter (EKF) and a particle filter (PF) (see e.g. [9]), which are widely used and are able to handle nonlinear systems.

We investigate the performance of the filters for simulated measurements with different known noise levels, which represents an idealised case, but allows testing the general feasibility of the method. The conformity of the filtered measurements is finally rated using methods based on the internationally accepted Guide to the Expression of Uncertainty in Measurement (GUM).

Methodology

A rough summary of the methodology is given here before details are explained in the following subsections. The basic idea was to experimentally obtain a reference force-time curve (nominal curve) from repeated error-free joining events that can be used as a priori information for process measurements of future joining events. Experimental data is obtained by a self-piercing riveting process for two different material combinations. The a priori information is then fed to a Bayes filter for simulated sensor readouts, as well as the information of the known noise level. The experimentally obtained force-time curves are also used to define a conformity band around the reference curve. The filtered signal and its uncertainty is then used to judge if non-conformity can be stated, given a specified conformance probability.

Reference Curve and Tolerance Interval. The reference curves were generated by calculating mean curves from ten experimental joining events. In order to create the experimental data, self-piercing riveted joints were created using a specially developed joining system [8]. It was selected due to the open interfaces and the access to all data without internal prior filtering. The system is equipped with an inner and outer punch as well as a blank holder. In this investigation, only the inner punch as well as the blank holder were used to create the joints. The motion of the inner punch, with which the data required for the investigation are generated, is realised using a servo-electric drive. Joining forces of up to 80 kN and setting speeds of up to 80 mm/s can be realised in this way.

The maximum permissible error of the readout setting force values are given as ±0.5 % of the nominal force. The accuracy was verified with a load cell (HBM C6 R, F 04063), which was calibrated with a universal testing machine (Zwick Z100, accuracy class 0.5). However, the uncertainty of the load cell due to the calibration is in a similar range in comparison to the tested device, so the significance of the verification is limited. This is not seen as a problem as possible measurement errors are significantly lower than process variations and the simulated noise levels.

Two different sheet material combinations were used, aluminium for both sheets (Al-Al) and steel in combination with aluminium (St-Al). Consequently, two times ten curves were obtained and used for two different reference curves for the different material combinations. Due to the targeted application case of Bayesian online filtering with discrete time steps, force-time curves were used for this article. However, force-displacement curves were obtained as well and could also be used. The curves for the single measurements are shown in the left column of Fig. 2.

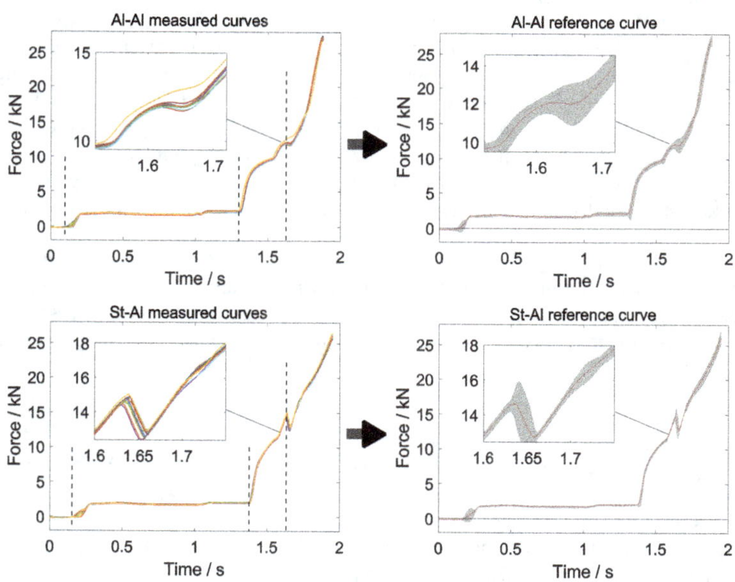

Fig. 2: Measured curves (left column) and calculated mean as reference curve (right column) with 99 % prediction interval (grey area) determined from the point-by-point empirical standard deviations.

The force monitoring shows a characteristic curve for the two investigated material combinations respectively. The first increase in force is caused by the application of the blank holder. The pneumatic drive keeps this force constant throughout the entire joining process. The second increase in force is attributed to the penetration of the rivet into the punch-sided joining part. Since the aluminium material has a greater overall thickness, the increase in force occurs a little earlier considering this joint. The force increases constantly until the punch-sided joining part is cut. This can be detected in the slight drop of the curves (see Fig. 2, detail). In the case of the multi-material joint, the cutting phase can be detected clearly. Due to the softer aluminium material, this is less pronounced in the Al-Al joint. Furthermore, a higher force is required for punching through the steel joint part than the aluminium joint part. The cutting phase is followed by a constant increase in the joining force. In this process phase, the interlock is formed. After the maximum is reached and the joint is embossed, the punch and the blank holder return to their initial position and the joint can be removed from the die.

For data post-processing, the ten single curves respectively were synchronised and resampled with a time step of 10 ms. In principle, this leads to an additional uncertainty contribution. As the observed influence was low compared to the process variation, this contribution was however not considered for the calculation of the tolerance interval. For each point in time, the arithmetic mean was used to extract a reference curve for the process (right column of Fig. 2). Approximating an underlying normal distribution, we additionally calculated the 99 % prediction interval (significance level $\alpha = 0.01$) for single force values using the empirical standard deviation at each point in time, multiplied with the value of the Student's t-distribution $t = 3.25$ for $\nu = n - 1 = 9$ degrees of freedom and a probability of $1 - \alpha/2$. This prediction interval was used as tolerance interval for our later investigations.

Simulation of Noisy Force Curves. For the investigations in this article, we simulated noisy sensor readouts for each material combination starting with one of the ten curves obtained by experiment (see section Reference Curve and Tolerance Interval). A random number following a normal distribution $\mathcal{N}(0, \sigma_w^2)$ is added to each force value of this starting curve, simulating white noise. We used standard deviations of $\sigma_w = 0.1$ kN, $\sigma_w = 0.5$ kN and $\sigma_w = 1.0$ kN. In sum, the resulting simulated curves have an error component with regard to the reference curve due to the deviation of the underlying single curve from the mean curve (simulated process variation, which will be unknown to the filter) and a random error component according to the simulated readout noise.

Bayes Filter. Recursive Bayesian state estimation builds the theoretical background for the used filters. We give a short outline for the theory, which can be found in more detail e.g. in [9] and with respect to Bayesian tracking by means of PF in [10]. In case of a Markov process, the current state of a process only depends on its previous state, which can be expressed for discrete time steps by a process equation

$$\boldsymbol{x}_n = \boldsymbol{f}_{n-1}\,(\boldsymbol{x}_{n-1}, \boldsymbol{v}_{n-1}), \tag{1}$$

where $\boldsymbol{x}_n$ is the state vector at time step n and $\boldsymbol{f}_n$ the state transition function, which also depends on the process noise $\boldsymbol{v}$ [10]. In our case, the state space is one-dimensional, consisting of the setting force as a single parameter and the state transition is simply modelled by the force increment ΔF_{n-1} obtained from the reference curve. Therefore, we have $\boldsymbol{x}_n = F_n = F_{n-1} + \Delta F_{n-1 \to n} + v_{n-1}$. The measurements at each discrete point in time are accordingly given by

$$\boldsymbol{y}_n = \boldsymbol{h}_n(\boldsymbol{x}_n, \boldsymbol{w}_n), \tag{2}$$

where $\boldsymbol{w}_n$ is the measurement noise [10]. For our investigations, the most simple model of a direct measurement with additive white noise is used, i.e. $\boldsymbol{y}_n = F_n + w_n$.

The goal is now to estimate the current state in form of a posterior probability density function $p(\boldsymbol{x}_n \,|\boldsymbol{y}_{1:n})$, depending on all previous measurements, but in a recursive way. For this purpose, the following equation can be derived using Bayes law [10]:

$$p(\boldsymbol{x}_n \,|\boldsymbol{y}_{1:n}) \;=\; \frac{p(\boldsymbol{y}_n|\boldsymbol{x}_n)\,p(\boldsymbol{x}_n|\boldsymbol{y}_{1:n-1})}{p(\boldsymbol{y}_n|\boldsymbol{y}_{1:n-1})}. \tag{3}$$

Eq. 3 represents the correction step for a Bayesian filter. The estimation of the current state based on all former measurements $p(\boldsymbol{x}_n|\boldsymbol{y}_{1:n-1})$ can be obtained from the Chapman-Kolmogorov equation [10]

$$p(\boldsymbol{x}_n \,|\boldsymbol{y}_{1:n-1}) \;= \int p(\boldsymbol{x}_n|\boldsymbol{x}_{n-1})\; p(\boldsymbol{x}_{n-1}|\boldsymbol{y}_{1:n-1})\; \mathrm{d}\boldsymbol{x}_{n-1}. \tag{4}$$

It can be seen that Eq. 4 and 3 give the link between the posterior density function of the last step $p(\boldsymbol{x}_{n-1}|\boldsymbol{y}_{1:n-1})$ and the posterior density function of the current step $p(\boldsymbol{x}_n|\boldsymbol{y}_{1:n})$. First, the prediction (eq. 4) is done using the process model and then a correction according to eq. 3 is done using the current measurement.

The given equations represent a conceptual solution, but can generally not be solved analytically. For Gaussian distributions, the Kalman filter can be used in case of linear system equations and the extended Kalman filter (EKF) for non-linear systems by means of linearization [9]. Particle filters (PF) on the other hand are a class of sequential Monte Carlo methods, which can handle non-Gaussian distributions [10]. Particles representing possible process states are propagated and used to estimate a posteriori density function, which can in turn be used to get a state estimate and variance.

While the system equations of our model could generally be solved with a basic Kalman filter, we used the EKF and a PF to be able to implement models that are more complex in future. The implementation was done using the EKF and PF classes of the Control System Toolbox of Matlab R2020b (The MathWorks, Inc.). The process noise was roughly estimated with $\sigma_v = 0.01$ kN and given as input for the filters. In case of the measurement noise, the values used for the simulation were also passed to the filters. Both filter types were initialised with the first value of the reference curve. For the particle filter, 10^6 particles were used and initialised with a variance of $4 \cdot \sigma_w$. A Gaussian measurement likelihood function was used. The state estimate was obtained by the sample mean after the correction step. The standard uncertainty of the filtered curve was obtained from the corrected state estimation error variance.

Conformity Assessment. Regarding conformity assessment, we used the methodology given in DIN EN ISO 14253-1 [11], which was originally developed with regard to geometric product specification. The method is based on the Guide to the expression of uncertainty in measurement (GUM) [12] and one of its supplements [13]. With regard to the discussed application of on-line monitoring, the task is to identify a curve progression that runs out of tolerance. In such a case, the joining process would be stopped. This should only be done if there is a sufficient probability that the force is not conform. Consequently, the measurement is tested for non-conformity. Considering the measurement uncertainty, the tolerance interval should be extended with guard bands to reach a certain probability of non-conformance [11]. For an assumed normal distribution, an extension factor (to be multiplied with the standard uncertainty) of 1.65 must be used to calculate the acceptance limits for a probability of non-conformance of 95 %.

To sum up, the Bayes filter is applied to reduce the measurement noise considering a priori knowledge. The estimated force values as an output of the filter are dynamically tested for non-conformity by checking if the filtered force value is out of the acceptance interval, which is formed by the tolerance interval plus guard intervals that depend on the estimated standard uncertainty of the filtered signal. As the filters need a certain time to settle with a reduced uncertainty, the time span from $t = 0$ to $t = 0.1$ s was not considered for conformity assessment.

Results and Discussion

The simulated noisy measurements are shown in the left column of Fig. 3 for the material combination St-Al. The insets show zoomed-in views around the point where the cutting through the punch-sided sheet starts and the point where the punch-sided joining part is cut.

It can be seen that a noise level of $\sigma_w = 0.5$ kN and $\sigma_w = 1$ kN leads to a measured signal that distinctly exceeds the tolerance interval. It is even hard to identify the different stages of the piercing riveting process.

In the right column of Fig. 3, the filtered version of the measurement signal is presented. The extended Kalman filter was used for the figure, but throughout our investigations, the particle filter led to nearly identical results. This can be expected as Gaussian distributed noise was simulated and modelled. Differences would be expected if the probability distributions had major deviations from the Gaussian shape. Significant differences only exist in the starting phase up to around $t = 0.1$ s until the filters show a stable behaviour with small variance. As already mentioned in the previous chapter, the non-stable time span was excluded from conformity assessment. For the use in real applications, it should be ensured that the filter is stable at the onset of the joining process.

The zoomed-in views show that the filtered signal is close to the true values. It has to be emphasised that these true values deviate from the reference curve and were therefore not directly used as a priori information. In Fig. 3, the true curve is not in the centre of the tolerance interval, which was constructed symmetrically around the reference curve.

When the non-conformity evaluation is applied, it shows that all filtered curves are within the acceptance interval (also for the filtered Al-Al curves, which are not shown here). Non-conformity can therefore not be stated with a probability of 95 % or higher. This is the desired result as the noisy curves were simulated from a curve that was used to construct the tolerance interval. It should however be noted that using a different curve from the experimental repetitions as starting point for simulation could result in a different decision. For real applications, a greater tolerance interval might be used to cover acceptable variations from different process conditions. Long-term data and a comparison with other quality inspection techniques might be a sound basis for this, although the concrete implementation is an open question.

While we performed our simulations and data processing off-line, the method in principle can be used for real-time filtering. However, we did not evaluate run times and the real-time capability.

The investigated noise levels are higher than for typical sensor outputs from closed measurement systems used for self-piercing riveting. If the acceptable process variation is multiple times greater than the sensor noise, the achievable improvements due to the filtering might not be significant. Unknown systematic measurement errors might further deteriorate the results. A Bayesian framework could however be interesting in special applications, e.g. if information from multiple sensors has to be fused.

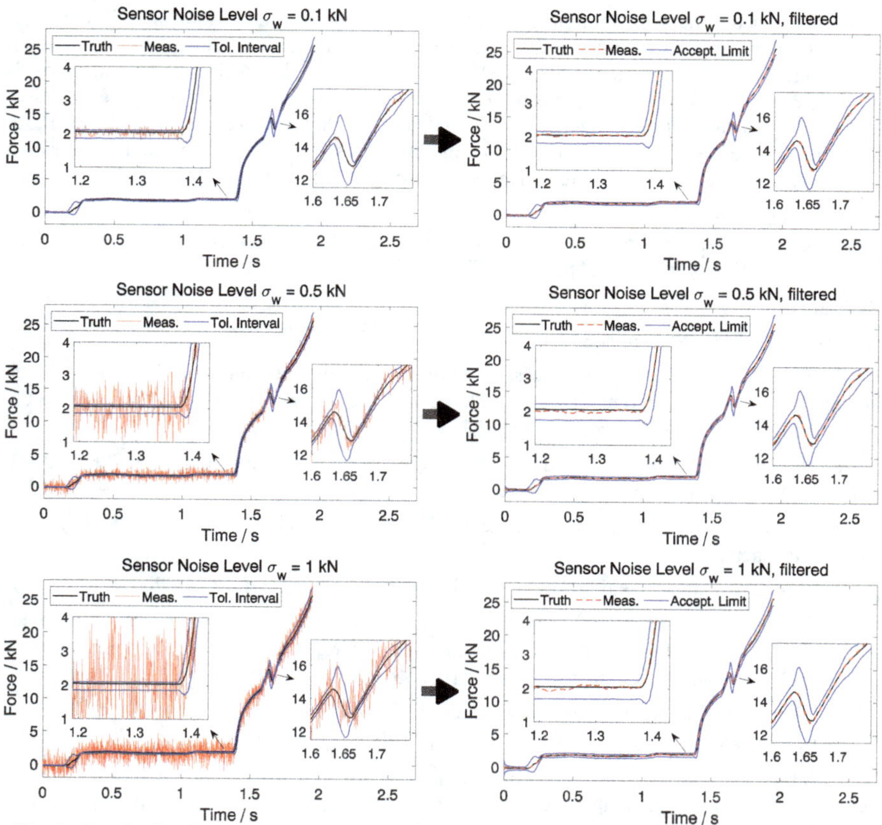

Fig. 3: Results for the material combination St-Al. Left: Simulated process curves with different noise level (red curve) and indicated tolerance interval (blue curves). Right: Filtered signal (extended Kalman filter, red curve) and indicated acceptance limits (Tolerance interval extended with guard bands that depend on the uncertainty output from the Kalman filter, blue curve) for a probability of non-conformance of 95 %.

Summary

In this article, a methodology to increase the accuracy of dynamic force measurements for mechanical joining by Bayesian filtering and to assess the conformity in the sense of quality monitoring, taking the measurement uncertainty into account is presented. A reference curve was statistically obtained from acceptable joints and used as a priori information for an extended Kalman filter and a particle filter. The filter performance was demonstrated by means of simulated measurements with different noise levels. It was shown that the measurement noise can effectively be reduced and the conformity properly assessed. The presented investigations should only be seen as a proof of principle that Bayes filters can theoretically be used to reduce noise and estimate the uncertainty for force-time or force-displacement curves obtained by mechanical joining. The noise levels investigated here do not represent typical conditions for commercial devices. For real applications, the behaviour and dependencies of sensors are additionally more complex and the noise characteristics as well as specific systematic errors are not necessarily known. The filter performance deteriorates the more the assumed characteristics of the sensors and of the process deviate from reality. This can even lead to results that are inferior to filters that rely on observation-only inference [14]. Regarding future work, it remains to be shown that the methodology is advantageous for real noisy data and is outperforming non-Bayesian filters. For joining processes and sensors that cannot be accurately characterised, there might be no improvement. However, for well-characterised systems, Bayesian filtering might be a useful technique for future application cases, where process parameters are dynamically adapted in a closed-loop control.

Acknowledgements
Funded by the Deutsche Forschungsgemeinschaft (DFG, German Research Foundation) – TRR 285 – Project-ID 418701707 - subprojects C02 and C05.

References

[1] DVS/EFB 3410, Stanznieten - Überblick, 2005.

[2] Y. Tan, O. Hahn, and F. Du, Process monitoring method with window technique for clinch joining, ISIJ International, vol. 45, no. 5, pp. 723–729, 2005. https://doi.org/10.2355/isijinternational.45.723

[3] S. Särkkä, Bayesian Filtering and Smoothing. Cambridge: Cambridge University Press, 2013. https://doi.org/10.1017/CBO9781139344203

[4] E. D. Kaplan and C. J. Hegarty, Understanding GPS: principles and applications. Artech House Publishers, 2005.

[5] S. Thrun, W. Burgard, and D. Fox, Probabilistic Robotics. The MIT Press, 2005.

[6] E. Garcia, T. Hausotte, and A. Amthor, Bayes filter for dynamic coordinate measurements - accuracy improvment, data fusion and measurement uncertainty evaluation, Measurement, vol. 46, no. 9, pp. 3737–3744, 2013. https://doi.org/10.1016/j.measurement.2013.04.001

[7] D. Simon, Optimal state estimation: Kalman, H [infinity] and nonlinear approaches. Hoboken, N.J: Wiley-Interscience, 2006. https://doi.org/10.1002/0470045345

[8] F. Kappe, S. Wituschek, M. Bobbert, and G. Meschut, Determining the properties of multi-range semi-tubular self-piercing riveted joints, Production Engineering, vol. 16, no. 2-3, pp. 363–378, 2022. https://doi.org/10.1007/s11740-022-01105-2

[9] A. J. Haug, Bayesian Estimation and Tracking. John Wiley & Sons, Inc., 2012. https://doi.org/10.1002/9781118287798

[10] M. S. Arulampalam, S. Maskell, N. Gordon, and T. Clapp, A tutorial on particle filters for online nonlinear/non-Gaussian Bayesian tracking, IEEE Transactions on Signal Processing, vol. 50, no. 2, pp. 174–188, 2002. https://doi.org/10.1109/78.978374

[11] DIN EN ISO 14253-1, Geometrische Produktspezifikationen (GPS) - Prüfung von Werkstücken und Messgeräten durch Messen - Teil 1: Entscheidungsregeln für den Nachweis von Konformität oder Nichtkonformität mit Spezifikationen (ISO/DIS 14253-1:2016), 2018.

[12] JCGM 100:2008, Evaluation of measurement data - guide to the expression of uncertainty in measurement, 2008.

[13] JCGM 106:2012, Evaluation of measurement data - the role of measurement uncertainty in conformity assessment, 2012.

[14] T. Li, J. M. Corchado, J. Bajo, S. Sun, and J. F. De Paz, Effectiveness of Bayesian filters: An information fusion perspective, Information Sciences, vol. 329, pp. 670–689, 2016. https://doi.org/10.1016/j.ins.2015.09.041

Sustainability

Sustainability

SheMet 2023

Sheet Metal 2023 Materials Research Forum LLC
Materials Research Proceedings **25** (2023) 397-404 https://doi.org/10.21741/9781644902417-49

Approach for a sustainable process chain in manufacturing of fasteners for mechanical joining

Benedikt Uhe[1,a *], Clara-Maria Kuball[2,b], Marion Merklein[2,c] and Gerson Meschut[1,d]

[1]Laboratory for Material and Joining Technology (LWF), Paderborn University, Pohlweg 47-49, 33098 Paderborn, Germany

[2]Institute of Manufacturing Technology (LFT), Friedrich-Alexander-Universität Erlangen-Nürnberg, Egerlandstraße 11-13, 91058 Erlangen, Germany

[a]benedikt.uhe@lwf.upb.de, [b]clara-maria.kuball@fau.de, [c]marion.merklein@fau.de, [d]gerson.meschut@lwf.upb.de

Keywords: Forming, Joining, Sustainability

Abstract. With regard to environmental protection, the sustainability of production processes is decisive. Mechanical joining technologies like self-piercing riveting are of special importance with regard to realising lightweight constructions in the automotive industry. However, the production of self-piercing rivets is costly, time-consuming and energy-intensive, as the rivets conventionally must be heat treated and coated in order to ensure an adequate strength, ductility and corrosion resistance. Within this paper, it is shown by the example of a newly established rivet manufacturing process how the sustainability of fastener production can be increased. The general approach in this context is the use of alternative, high strain hardening stainless steels as rivet material, which allows the omission of the post treatment of the rivets after forming. The shortening of the process chain enables a more sustainable rivet production. Thus, not only the energy consumption is reduced, but also costs, which is why the novel manufacturing process is also of interest from an economic point of view.

Introduction

The protection of the environment and the natural resources as well as the tackling of climate change are challenging issues for the manufacturing industry. Hence, new industrial solutions that enable a reduction of CO_2 emissions and energy consumption have to be found [1]. To reach a reduction of emissions in car traffic, the lightweight design of car bodies is an established method. The lightweight design causes an increasing use of multi-material structures made of high-strength steel and aluminium. To join these materials the use of mechanical joining techniques like the self-piercing riveting (SPR) is emerging [2]. The process stages of SPR and the characteristic joint parameters are shown in Fig. 1.

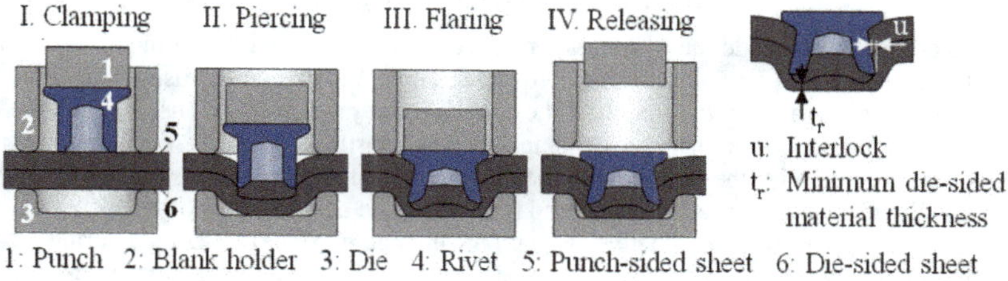

Fig. 1. Characteristic of the SPR process according to [3].

Content from this work may be used under the terms of the Creative Commons Attribution 3.0 license. Any further distribution of this work must maintain attribution to the author(s) and the title of the work, journal citation and DOI. Published under license by Materials Research Forum LLC.

 Materials Research Forum LLC
https://doi.org/10.21741/9781644902417-49

The rivet is pressed into the clamped sheets and pierces the punch-sided material. After that, the rivet flares and an interlock is created within the die-sided sheet. The interlock is the most important property of the joint, because it is an indicator for the expectable joint strength [4]. The joining result depends on the rivet and the die, which are selected in accordance to the strength, the ductility and the thickness of the sheets to be joined [5]. Rivets for SPR differ in geometry, material and the surface condition. As a function of the materials used for car bodies, several different rivet geometries were developed in the past. Due to the challenges when joining high-strength steel, especially on the die side, the authors have designed a new rivet geometry for the joining of two challenging material combinations consisting of high-strength steel HCT780X and aluminium EN AW-5083 in former studies [6]. Based on this rivet geometry an innovative rivet with special properties and a completely new approach of the rivet manufacturing process are developed, as explained within this paper.

Approach for a sustainable process chain

A holistic view of the process chain related to SPR refers not only to the joining process itself, but includes the entire life cycle from the material production process to the use of the finally joined components, as illustrated in Fig. 2. There is a great potential to improve the process efficiency and sustainability of the rivet manufacturing process in particular.

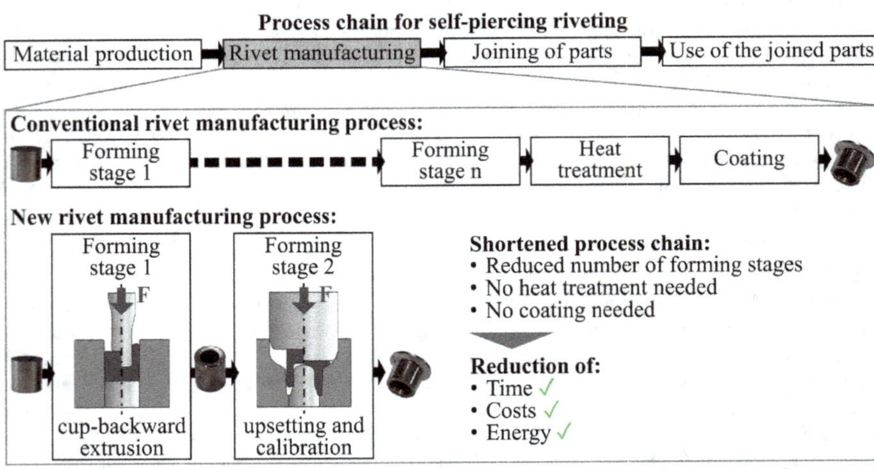

Fig. 2. Entire SPR process chain and comparison between new and conventional rivet manufacturing process according to [7].

Rivets are generally made out of high-strength boron steels [5] and are formed in multistage cold forming processes [2]. To ensure an adequate strength and a good corrosion resistance, the rivets must be heat treated and coated after the rivet forming process [8], which makes the entire rivet production time-consuming, energy-intensive and costly. Especially the heat treatment process, more precisely the hardening and tempering of the rivets, is energy-intensive. Unfortunately, the authors have no insight into the energy and cost structure of the manufacturing companies. This is why only a qualitative evaluation with regard to the rivet manufacturing process is possible. However, it is known from [9] that about 40 % of the energy that is used in the industrial environment in Germany is attributed to heat treatment processes. This is underpinned by the results of a scientific analysis carried out by Mendioka et al. [10] concerning the energy consumption of a steel foundry, which reveals that more than a third of the total gas of the company

Materials Research Proceedings **25** (2023) 397-404 https://doi.org/10.21741/9781644902417-49

is consumed due to the heat treatment. As a logical consequence, there is a need to avoid the heat treatment of cold formed components. Considering in addition the wide use of mechanical joining operations and the rising energy prices, a more sustainable and economical rivet manufacturing process is needed.

The approach presented in this paper relates to the shortening of the rivet manufacturing process chain by eliminating the need of the post treatment of the rivets in order to increase the process efficiency and the sustainability. The key aspect for this is the use of high strain hardening stainless steels as rivet material. An overview of a selection of potential rivet materials is given in Fig. 3.

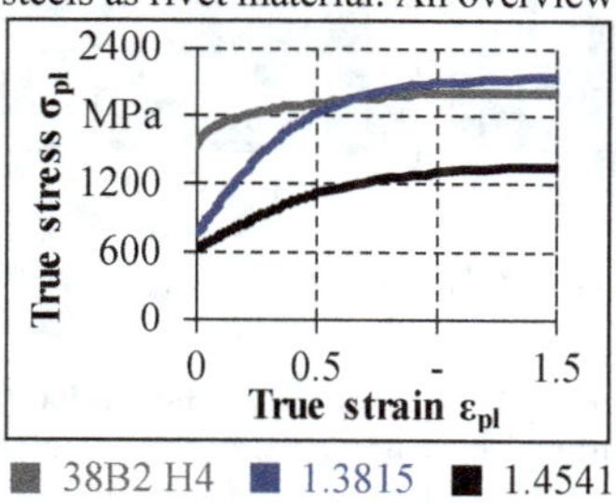

Material	Mass fraction of chemical components [m%]						
	C	Mn	Cr	Ni	Ti	B	N
38B2	0.38	0.76	0.27	X	X	0.004	X
1.3815	0.08	18.86	17.07	0.58	X	X	0.76
1.4541	0.02	1.55	17.14	9.19	0.27	X	X

Fig. 3. True stress - true strain curves and chemical composition of new rivet materials in comparison with conventional rivet material 38B2 according to [3].

In the context of the material choice, high nitrogen steels such as the featured steel 1.3815 represent a promising option. The chosen material 1.3815 is an established high nitrogen steel, which was already used for highly stressed retaining rings for turbo-generators due to its excellent material properties [11]. Steels of this type are characterised by an exceptionally high nitrogen content and provide a high strength, ductility and corrosion resistance [12]. Due to the high strain hardening, a sufficient rivet strength for joining can already be achieved through the forming process. Therefore, the usual heat treatment and coating of the rivets becomes obsolete leading to cost, time and energy savings. Thus, the benefits of forming as an efficient and sustainable technology are fully exploited. Furthermore, the comparison between the conventional and the new process in Fig. 1 also reveals that the new process consists of two forming stages in contrast to the usual up to six stages in industry. Consequently, energy and costs required for the manufacture of additional forming tools including the tool materials can be saved. It must be pointed out that the production of materials such as high nitrogen steels involves more expenditure compared to conventional steels, which is opposed to the cost and energy savings by omitting the heat treatment and the coating. There is a need for future research and development in relation to the material production processes in order to exploit the potential for energy and cost savings in this field as well. Nevertheless, the new approach contributes to an increase in economic efficiency and sustainability of the rivet manufacture. As SPR is used in the automotive sector in particular, this is also an important step on the way to fully sustainable mobility in the future, because the production history of each individual vehicle component counts in this case.

Forming of Rivets with Graded Mechanical Properties

The new rivet manufacturing process, as presented by the authors in [7], consists of two forming stages, cup-backward extrusion in stage 1 in combination with the forming of the rivet head and foot in stage 2. The mechanical properties of rivets, besides the rivet geometry, are decisive for the joining process and for achieving an adequate joint formation. This applies in particular to challenging joining tasks such as the joining of high-strength steels on the die side. The hardness analysis of the new rivets made of 1.3815 (Fig. 4a) reveals that over the whole rivet cross section a mean hardness of about 600 HV 0.02 is reached. Thus, the hardness range of conventional rivets made of boron steel, which is 480 HV ± 30 HV in the heat treated condition [6], is not only completely covered, but even extended by the new rivets manufactured without heat treatment.

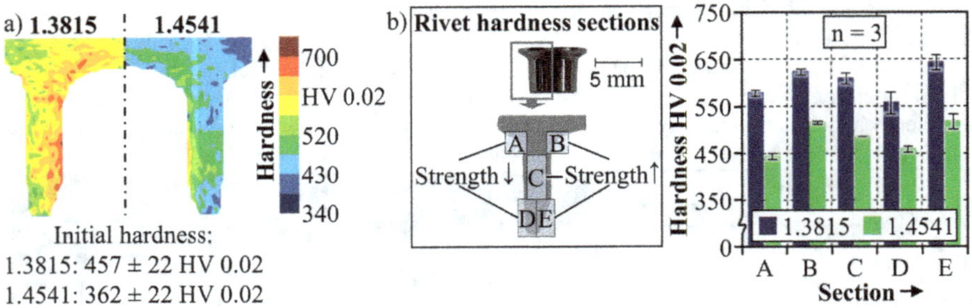

Fig. 4. a) Rivet hardness after forming according to [7] and b) Local hardness in dependence of the selected rivet sections.

A particular potential of the new rivet manufacturing process lies in the fact that the hardness gradation after forming remains intact in contrast to conventional heat treated rivets exhibiting homogenous mechanical properties. This provides the chance to specifically adapt the hardness according to the requirements during the joining process, either by selecting different forming sequences or by replacing the rivet material. In this case, the presented two-stage forming process is chosen in order to create a beneficial hardness distribution with regard to the special requirements arising in particular when joining high-strength steels. This results in varying hardness values in different rivet sections, as depicted in Fig. 4b. Inside high-strength materials the flaring process is problematic [5]. The interlock formation that is essential for the joint formation and the resulting joint strength is influenced by the flaring. A too low flaring can be the reason for an insufficient interlock. Against this backdrop, the achieved hardness gradation of the new rivets is advantageous. For joining with conventional rivets, a risk for the occurrence of cracks in the transition area between the outer rivet shank and the rivet head in particular was found in [6]. In addition to the high ductility of high nitrogen steel, which is even maintained for high strains, the comparatively lower hardness in section A is beneficial before the background of the risk of cracks. In accordance with this assumption, no cracks after joining using the new rivets made of 1.3815 could be found. To avoid the excessive deformation of the rivet shank, what is known to be a high risk when joining high-strength steel [2], the sections B and C are characterised by a relatively high hardness. The mean hardness in the outer area of the rivet foot, section D, is lower compared to the hardness in section E and thus facilitates the rivet flaring and the interlock formation. Moreover, the comparatively high hardness in section E ensures that the foot chamfer is not deformed while cutting the upper sheet, especially in case of using high-strength steel on the punch side, because the chamfer is important for the subsequent beginning of the flaring. Thus, the gradation of the rivet hardness can be used to support the joining.

The results of the hardness analysis for another potential rivet material, the stainless steel 1.4541, are also shown in Fig. 4. As the initial hardness and the strain hardening of 1.4541 differs from the high nitrogen steel 1.3815, different hardness levels are achieved in dependency of the material. The transferability of the approach demonstrates the potential for manufacturing a wide range of rivets with graded mechanical properties, which can be set depending on the requirements of the joining process. This enables the production of customised rivets to realise joining processes, which are not possible using conventional rivets.

Joining of High-Strength Materials

The use of a new rivet material provokes some issues for the joining process. One benefit of the corrosion resistance of the new rivet material is that the coating can be omitted. However, it is known from [13] and [14] that the surface conditions of the rivet and the sheets have an influence on the friction between the rivet and the sheet materials during the joining process. Because of the changed friction properties, the rivet deformation and the generation of the interlock are affected. However, a study concerning the influence of the rivet surface condition came to the finding that the friction influence is limited and especially the friction properties of uncoated rivets and rivets plated with the common coating Almac® are very similar [3].

As stated at the beginning, the new rivet is designed for the joining of material combinations consisting of high-strength steel and aluminium to cover a wide and relevant range of joining parts. The validation of the approach is conducted for the material combinations specified in Fig. 5. For the comparison to a conventional rivet, the rivet type from [6] is used. This rivet has a geometry, which fulfils the requirements of both material combinations. It is made of boron steel 38B2 and quenched and tempered to a hardness level of 480 ± 30 HV. The surface is coated with Almac®.

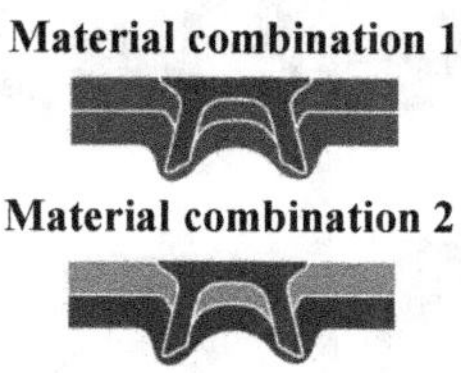

	Material	Sheet thickness (mm)	Yield strength YS (MPa)	Tensile strength TS (MPa)
■	HCT780X	1.5	564.1 ± 2.2	860.8 ± 1.3
■	EN AW-5083	1.5	157.3 ± 1.0	291.5 ± 0.6

Tensile test; n = 5; rolling direction = 0°

Fig. 5. Material combinations and mechanical properties of the sheet materials to be joined.

The comparison between the conventional rivets and the new type of rivet made of 1.3815 and 1.4541 is initially made by evaluating the rivet deformation and the joining results based on multi-step joining tests. In the cross-section of the joints, the deformation of the rivets can be analysed within the different process stages (Fig. 6). Due to geometrical deviations of the rivets made of 1.3815 and 1.4541 compared to the target geometry that are caused by the forming process, the rivet head diameter is much lower than with conventional rivets. Further deviations are existing within the rivet foot. However, the geometrical deviations do not impede the joining process. Instead, it can be demonstrated that the joining result with rivets made of 1.3815 fulfils the common quality criteria of [4] and that the new rivet has a competitive joining performance compared to conventional rivets. In addition to the examination of the geometrical deformation of the rivets, the hardness within the rivets is measured for each step so that the change in the material condition can be analysed, too. As the material condition is almost homogenous within the conventional rivet, due to the heat treatment, the material hardness within the new rivet type is locally different. In the first process step, the initial hardness distribution is still existent. Compared to the conventional rivet, the hardness within the outer rivet foot is lower than with the rivet made of 1.3815. Nevertheless, as the hardness within the conventional rivet increases by approx. 15 % during the joining process, the increase of the hardness within the rivet foot of the rivet made of 1.3815 is much higher and at the end of the process a comparable hardness is reached. The hardness distribution within the rivet made of 1.4541 is similar to the hardness distribution within the rivet made of 1.3815, but the hardness level is much lower. The lower material strength of 1.4541 causes the compression of the rivet and no joint can be created. Therefore, it can be seen, that the rivet material strength has a significant influence on the deformation behaviour of the rivets and the joining result.

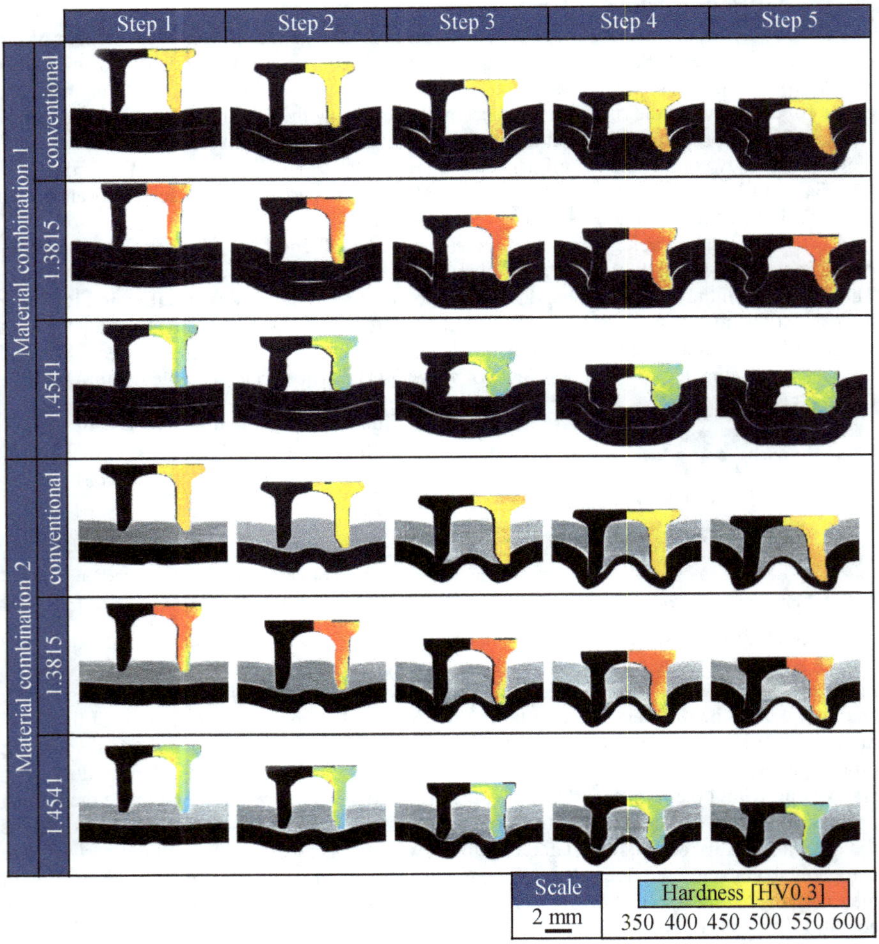

Fig. 6. Joining result with conventional rivets and rivets made of 1.3815 and 1.4541 in accordance to [3].

However, as the competitive joining of the two material combinations with rivets made of high nitrogen steel 1.3815 could be proved by the illustrated joining tests, the joint strength has to be determined, too. Therefore, in [15] the strength of joints riveted with conventional rivets and rivets made of 1.3815 was analysed by experimental tests. Tests under quasistatic load were conducted. The reached maximum force until failure of the joint and the kind of failure were evaluated. By doing so, it could be found that the smaller rivet head diameter of the rivets made of 1.3815 provokes a lower joint strength of approx. 25 % when testing material combination 2. The punch-sided aluminium sheet is teared over the rivet head at a reduced force compared to a rivet with a bigger head diameter. Because of that, the elimination of the geometrical deviations of the new rivets has to be a goal of further research.

Summary and Outlook

As an approach for sustainable manufacturing, the potential of a novel rivet manufacturing process to save costs, time and energy is pointed out. The usually necessary treatment of the rivets after forming, including hardening and tempering as well as coating, can be omitted due to the use of high strain hardening stainless steel as rivet material. Thus, the entire process chain is shortened. Additionally, the new approach makes it possible to produce rivets with graded mechanical properties. This is a special feature compared to conventional rivets, which provide homogeneous properties due to the heat treatment. This opens up the opportunity of producing customised rivets with a varying local hardness. In this paper, the achieved hardness distribution of rivets formed from the high nitrogen steel 1.3815 and the stainless steel 1.4541 using the new rivet manufacturing process is discussed with regard to the requirements of the subsequent joining. Furthermore, the impact of the changed rivet properties on the joining process is assessed. In this way, a holistic evaluation of the process route is ensured. Material combinations of high-strength steel and aluminium alloy can be successfully joined with the new rivets. Nevertheless, it must be stated that the geometrical deviations of the rivets must be reduced in order to ensure an adequate joint strength. Therefore, future work will focus on the further development of the tool concept and on improved forming strategies to extend the existing process limits due to the high tool loads. Further research regarding tailored mechanical rivet properties will be conducted as well to improve the performance of riveted joints. In the future, the approach can be transferred to various fastener production processes to reduce energy consumption, costs and time. This contributes to a reduction of emissions and the protection of natural resources, which is an important factor for the increase in economic efficiency ensuring the competitiveness of manufacturers.

Acknowledgement

The authors would like to thank the German Research Foundation (DFG) for their support of the research project "Forming and joining of semi-tubular self-piercing rivets made of high-strength steel with adapted mechanical properties and numerical analysis of the process chain" (ME 1840/8-4, Project-ID: 328853593), on which this paper is based.

References

[1] Information on https://www.europarl.europa.eu/news/en/headlines/society/20180703STO07129/eu-responses-to-climate-change

[2] D. Li, A. Chrysanthou, I. Patel, G. Williams, Self-piercing riveting - a review, The International Journal of Advanced Manufacturing Technology 92 (2017) 1777-1824. https://doi.org/10.1007/s00170-017-0156-x

[3] B. Uhe, C.-M. Kuball, M. Merklein, G. Meschut, Self-piercing riveting using rivets made of stainless steel with high strain hardening, in: G. Daehn, J. Cao, B. Kinsey, E. Tekkaya, A. Vivek, Y. Yoshida (Eds.), Forming the Future, The Minerals, Metals & Materials Series, Springer, Cham, 2021, pp. 1495-1506. https://doi.org/10.1007/978-3-030-75381-8_124

[4] Technical Bulletin DVS 3410, Self-pierce Riveting - Overview, DVS Media, Düsseldorf, 2019.

[5] K.-i. Mori, Y. Abe, A review on mechanical joining of aluminium and high strength steel sheets by plastic deformation, International Journal of Lightweight Materials and Manufacture 1 (2018) 1-11. https://doi.org/10.1016/j.ijlmm.2018.02.002

[6] B. Uhe, C.-M. Kuball, M. Merklein, G. Meschut, Improvement of a rivet geometry for the self-piercing riveting of high-strength steel and multi-material joints, Production Engineering 14 (2020) 417-423. https://doi.org/10.1007/s11740-020-00973-w

[7] C.-M. Kuball, B. Uhe, G. Meschut, M. Merklein, Process-adapted temperature application within a two-stage rivet forming process for high nitrogen steel, Proceedings of the Institution of Mechanical Engineers Part L - Journal of Materials-Design and Applications 236 (2022) 1285-1301. https://doi.org/10.1177/14644207211068693

[8] K.-i. Mori, N. Bay, L. Fratini, F. Micari, A.E. Tekkaya, Joining by plastic deformation, CIRP Annals - Manufacturing Technology 62 (2013) 673-694. https://doi.org/10.1016/j.cirp.2013.05.004

[9] Information on https://www.ema-indutec.com/en/service/additional-services/environmental-protection/umweltschutz

[10] I. Mendioka, M. Sorli, A. Armijo, L. Garcia, L. Erausquin, M. Insunza, J. Bilbao, H. Friden, A. Björk, L. Bergfors, R. Skema, R. Alzbutas, T. Iesmantas, Energy efficiency optimisation in heat treatment process design, in: C. Emmanouilidis, M. Taisch, D. Kiritsis (Eds.), Advances in Production Management Systems. Competitive Manufacturing for Innovative Products and Services, APMS 2012, IFIP Advances in Information and Communication Technology, vol. 397, Springer, Berlin, Heidelberg, 2013, pp. 127-134. https://doi.org/10.1007/978-3-642-40352-1_17

[11] R. Ritzenhoff, A. Hahn, Corrosion resistance of high nitrogen steels, in: H. Shih(Ed.), Corrosion Resistance, IntechOpen, London, 2012, pp.55-80. https://doi.org/10.5772/33037

[12] V.G. Gavriljuk, H. Berns, High nitrogen steels: structure, properties, manufacture, applications, Springer, Berlin, 1999. https://doi.org/10.1007/978-3-662-03760-7

[13] L. Han, A. Chrysanthou, Evaluation of quality and behaviour of self-piercing riveted aluminium to high strength low alloy sheets with different surface coatings, Materials and Design 29 (2008) 458-468. https://doi.org/10.1016/j.matdes.2006.12.020

[14] D. Li, Influence of aluminium sheet surface modification on the self-piercing riveting process and the joint static lap shear strength, The International Journal of Advanced Manufacturing Technology 93 (2017) 2685-2695. https://doi.org/10.1007/s00170-017-0710-6

[15] B. Uhe, C.-M. Kuball, M. Merklein, G. Meschut, Strength of Self-Piercing Riveted Joints with Conventional Rivets and Rivets Made of High Nitrogen Steel, in: Proceedings of the ESAFORM 2021 - 24th International Conference on Material Forming. Liège, BE. https://doi.org/10.25518/esaform21.1911

Sheet Metal 2023
Materials Research Forum LLC
Materials Research Proceedings **25** (2023) 405-412
https://doi.org/10.21741/9781644902417-50

A study on economic tooling concepts for dry deep-drawing using environmentally benign volatile lubricants

Gerd Reichardt[1, a *], Mathias Liewald[1, b] and Kim Riedmüller[1, c]

[1]Institut for Metal Forming Technology, University of Stuttgart, Holzgartenstrasse 17, D-70174 Stuttgart, Germany

[a]gerd.reichardt@ifu.uni-stuttgart.de, [b]mathias.liewald@ifu.uni-stuttgart.de, [c]kim.riedmueller@ifu.uni-stuttgart.de

Keywords: Deep Drawing, Tribology, Economic Efficiency

Abstract. Dry processes represent promising approaches in forming technology to improve environmental aspects and human health by avoiding harmful substances and additives of conventional lubricants. Among many different approaches for dry forming investigated within the priority program SPP 1676, this paper addresses the use of volatile media as lubricants such as CO_2 or N_2. These volatile media are introduced directly into the interstice between sheet metal and tool surface via microinjectors. Indeed, this does not contradict the principle dry forming approach, as dry forming is defined as a process in which no residues are left on the surfaces [1]. This is ensured by the complete evaporation of the volatile media. The general feasibility of this novel tribological system has already been demonstrated in previous research work. However, despite good tribological results, the manufacturing costs of the required tool inserts and the media consumption per component are not yet economically competitive with conventional tribological systems. Therefore, this paper focuses on new designs of dry deep-drawing tools utilising volatile lubricants, considering the economics of different manufacturing processes.

Introduction

The use of lubricants has always been mandatory in sheet metal forming. The purpose of lubricants includes the reduction of wear phenomena such as abrasion and adhesion, both on the tool and on the components, as well as the reduction or modification of friction forces. Mineral oils, water/oil emulsions and wax-type dry lubricants are usually used for this purpose. These lubricants often contain ingredients that are harmful to health and environment such as chlorinated paraffins, which are intended to improve the properties of the lubricants. On the one hand, these lubricants have to be applied onto the surface of the sheet metal, on the other hand, they have to be removed from the component´s surface after forming in order to enable subsequent processes such as painting, welding or bonding. These efforts can be avoided by using volatile, additive-free lubricants. Fig. 1 shows the approach for applying such volatile lubricants in deep drawing processes. Here, CO_2 or N_2 are introduced directly into the friction zone under high pressure via microinjectors integrated into the tool surface. After opening the tools, the lubricants evaporate without leaving any residues.

In a basic research project lasting several years, various parameters of dry forming with volatile media were investigated experimentally, starting with simple strip drawing tests. In this way, the influences of different microinjector designs and arrangements, structured tool surfaces, surface coatings, sheet metal materials, varying contact normal stresses and media pressures were investigated [1].

Content from this work may be used under the terms of the Creative Commons Attribution 3.0 license. Any further distribution of this work must maintain attribution to the author(s) and the title of the work, journal citation and DOI. Published under license by Materials Research Forum LLC.

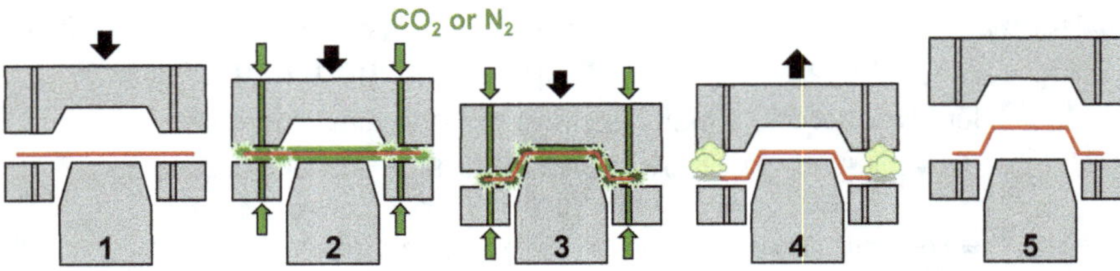

Fig. 1: Process scheme of dry deep drawing lubricated by volatile media. According to [2]

Subsequently, the tribological conditions at tool radii were also investigated experimentally using specific stretch bending tests. Thus, the influence of different radii, injection angles of the volatile lubricants, locally deviating contact stresses and media pressures were analysed. As a result, the critical tribological area of tool radii could be optimised.

The findings gained were used to realise a deep-drawing tool of a rectangular cup for single-stroke investigations. Here, a two-part shell design (cf. Fig. 2b) was used to simplify the integration of feeding and distribution channels into the die and blank holder. Afterwards, these tool inserts were utilized to experimentally determine the process window for CO_2 (liquid) and N_2 (gaseous, 60 bar), N_2 (gaseous, 100 bar) as well as for a conventional mineral oil-based drawing oil (ZO3368) as lubricants. The result of this investigations demonstrate the clear increase of the process range and therefore the process reliability for all volatile lubricants [3].

At the end of the fundamental investigations carried out, wear tests were performed with a new, series-oriented laboratory endurance run of a rectangular cup tool and the most promising volatile lubricating medium $CO_{2,\ liquid}$. The results were compared to experimental series with the conventional mineral oil ZO3368. During the endurance test series with CO_2 and the mineral oil ZO3368, 1000 rectangular cups each were drawn from the sheet metal material DC05+ZE. Neither adhesions on the tool surface nor abrasions on the surface of the parts could be detected for all 1000 deep-drawn rectangular cups using liquid CO_2 as volatile lubrication [4].

Despite the promising findings regarding the tribological performance, the tool insert design used in the basic investigations (two-part shell) reveal some economic disadvantages. For example, the valves controlling the media flow could not be mounted directly onto the tool inserts, which resulted in an increase in media consumption per part due to longer supply tubes to the tool. The dismountable sealing concept between the two shell parts of blank holder and die was also associated with high efforts in cost and time regarding manufacturing and assembly. Overall, the manufacturing costs of the tooling inserts used for the investigations on volatile lubrication described above were more than four times higher than those of the conventional tool inserts for the reference test series with ZO3368.

Tooling concepts

Hence, the aim of the study described as follows was, on the one hand, to reduce high tool costs and, on the other hand, to improve economic aspects of the new lubrication system by developing new tool concepts for blank holders and dies in dry deep drawing processes using volatile lubricants. For this purpose, based on design catalogues and literature research a morphological list of alternative features for tools with media supply was elaborated. Based on this, the most valuable concepts were selected and specified in detail.

When elaborating the morphological list, five essential tool features were identified by abstracting the deep-drawing tools for dry forming using volatile lubricants. These tool features are in detail the basic tool material, the design, the joining method of tool components, the sealing method and the manufacturing process of the microinjectors. A brief excerpt of alternatives of identified features are listed in Table 1. Based on this list, six solution concepts were extracted, which are colour-coded in Table 1. Each concept was named on the basis of the individually selected design, as this ensures a clear identification.

Table 1: Morphological listing of feature alternatives and solution concepts

Feature	Alternatives/ characteristics					
	1	2	3	4	5	6
Material	Steel	Powder	Cast iron			
Design	Laser deposite welding	Shell design	Layer laminate (LOM)	3D - SLM	Casting	Monobloc
Joining method	No joining required	Screwing	Bonding	Soldering		
Sealing method	No sealing required	Static sealing	Joining pressure-tight			
Manu.microinj.	EDM	Laser drilling	Convent. drilling	Integr. with 3D-SLM		

Solution concepts:	3D-SLM	Shell design	Laser deposite welding	Layer laminate	Monobloc	Casting

Rating of the solution methods

The evaluation of the six solution concepts was carried out on the basis of selected criteria corresponding to the general requirements for forming tools with integrated media channels and microinjectors defined and developed during the former basic research investigations. For each criterion, a score of zero points (not suitable) to ten points (fully suitable) were assigned. For scoring, numerous literature sources, product examples, expert opinions and the experience gained from the basic investigations were used in order to ensure the most objective and balanced rating of an appropriate tool design. The relevant results of the evaluations, including all criteria, are provided in Table 2.

Table 2: Evaluation of all proposed solution concepts

Solution concepts \ Evaluation criteria	Stiffness	Tightness	Producibility	Fatigue strength	Strength	Production time	Geometric accuracy	Flexibility of design	Repair capability	Total score	Rating
3D-SLM	10	9	10	8	9	8	9	10	3	**76**	**1**
Shell Design	8	10	10	10	9	5	10	7	5	**74**	**2**
Laser Deposite Welding	10	10	10	8	8	7	8	7	3	**71**	**3**
Laminated Object	6	8	10	8	7	8	6	8	4	**65**	**4**
Casting	6	8	8	9	7	3	7	9	3	**60**	**5**
Monobloc	10	10	0	10	10	0	8	2	3	**53**	**6**

In general, the four best-ranked solution concepts show a beneficial manufacturability from a technical point of view. In the case of casting (5th rank), there are limitations in terms of manufacturability with regard to channels with small cross-sections and cross-sectional changes in the form of distortion and cracking as well as shrinkage. Due to the small channel structures, only a complex and expensive fine casting process can therefore be considered for this solution method. The solution as a monobloc with complex deep-hole drilling and the small wall thicknesses is extremely questionable and can only be implemented to a very limited extent. Therefore, the manufacturability of this method is rated with a score of zero points. In the following, only the four best-ranked solution concepts are considered and specified in more detail.

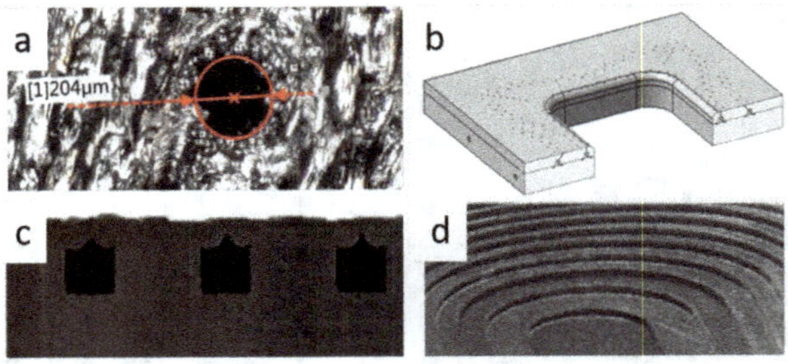

Fig. 2: a: 3D-LMF [5], b: Shell design [6], c: LDW cover layer [7], d: LOM [8]

1st rank solution concept "3D-SLM". The metal 3D printing process received the highest score in the evaluation of the different solution concepts. Due to the high geometric accuracy, laser metal fusion (LMF) is intended for this application [9]. With this process, the microinjectors presumably can be introduced directly into the tool components during printing, thus avoiding subsequent laser drilling or EDM in the three-dimensional tool body. According to current preliminary tests, bore-shaped structures with a minimum diameter of 0.1 to 0.2 mm can be produced with the LMF process using appropriate metal powders and equipment (see Fig. 2a). Due to the surface roughness of printed surfaces (approx. 30-55 µm with LMF), however, a machining of the tool surfaces will be required. A limitation of this concept could be the availability of high-strength materials in powder form. Available tool steels (e.g. 1.2709) today only achieve a hardness of approx. 43-52 HRC [10]–[12]. Hence, the characterisation of the variety of shapes of the tool inserts in relation to the economic efficiency by printing with regard to the achievable strength and surface topologies is in the focus of this solution concept. Furthermore, the relatively high nickel content of 1.2709 leads to a higher adhesion tendency. Here, the use of relatively new, qualified tool steels in powder form taken from the group of 8% chromium steels (e.g. W360) promises a remarkable improvement. The achievable working hardness for this group of steels is approx. 55-57 HRC [10], [13], so that almost the same results could be achieved with printed tool inserts as with conventional tool steels.

2nd rank solution concept "Shell Design". The shell design is frequently used for deep-drawing tools in press hardening technology and is therefore established and well known as a design principle in industrial application. Press hardening tools have a segmented design into separate shells to enable the manufacturability of complex internal cooling devices. The cooling channels, necessary for quenching the blanks, are usually located in the basic tool structure of the punch and die. This is closed with a cover layer (upper shell) often made of hot work steel having a high wear resistance. The inlet and outlet holes to the cooling channels often require deep-hole drilling [14]–[16]. The design principle of segmentation in shells already was successfully used for the basic investigations in the above-mentioned dry forming project (c.f. Fig. 2b). Here, however, cost-intensive deep-hole drilling was necessary for the media supply channels in the two-part shell design. A further optimisation of the shell design is therefore a three-part segmentation to simplify the insertion of supply channels.

The advantages of the shell design are a relatively high rigidity and a relatively simple and cost-efficient production of the individual segments. The sealing of the shells to each other has so far been carried out in a complex way by dismountable sealing systems. In order to improve an economic manufacturing process, non-demountable joining processes such as soldering or gluing will be considered in future.

3rd rank solution concept "Laser Deposit Welding". In this solution, the intended tool inserts are not to be produced completely by laser deposition welding (LDW). Rather, LDW aims to completely cover the media channels introduced by machining in a conventional steel plate as depicted in Fig. 2c. In this way, a relatively high degree of design freedom can be achieved for the feed and distribution channels introduced into the tool inserts. Properties such as hardness and strength can be positively influenced by suitable additive materials applied to the surface of the tool insert by means of LDW.

Due to the application of individual welding paths, however, a geometric notch effect (c.f. Fig. 2c) occurs in the freely spanned areas of the channels, which is supplemented by a metallurgical notch effect due to the successive microstructure solidification during welding. This reduction in component-strength can be countered by a high tensile strength and residual ductility of the additional material of the cover layer. Even with the LDW concept, the microinjectors still have to be introduced subsequently by electrical discharching, laser drilling or conventional micro-machining. Sealing of the individual layers is not necessary due to the metallurgically-bonded microstructure [7].

4th rank solution concept "Laminated Object Manufacturing". The laminated object manufacturing (LOM) process is a well-known concept for prototype and small-series forming tools. Here, the tool inserts are constructed from individual, stacked sheet metal laminates (see Fig. 2d). The idea is to divide the complex tool geometry into individual, easy-to-make laminates. These laminates can be produced inexpensively using common processes such as laser cutting, water jet cutting or punch nibbling [8], [17]–[19]. The advantage is that with the horizontal arrangement of the laminates the feed and distribution channels can be freely designed avoiding expensive drilling of deep boreholes. The disadvantages of this technology are given by the additional machining of the stair-step effect occurring between the individual sheet metal layer and the reduced structural rigidity of the tool body compared to solid tools. The compressive strength due to the media pressure in the tool inserts can be ensured by a sufficiently dimensioned cover plate (possibly made of tool steel).

Furthermore, all laminates must be joined together in a pressure-tight manner by surface bonding, welding or soldering, as otherwise the volatile media would escape. Especially bonding with epoxy resin and furnace brazing will be investigated because of their simple and cost-effective application. The two-dimensional joining of the individual layers also increases the structural rigidity of the tool inserts, as shear stresses can be transferred between the individual layers. The internal pressures in the channels of the media are a maximum of 10 MPa (= 100 bar), and the areas exposed to the internal pressure are very small. According to Volkersen's calculation method [20] and the geometric boundary conditions, the shear stresses occurring in the joint layer amount to $\tau max \approx 5$ MPa. The equivalent stress (Tresca) including the normal stress reaches a calculated maximum of approx. 17 MPa. This strength can be easily achieved with commercially available epoxy adhesives and solders [21].

Economic considerations

From an economic point of view, the equipment investments for oil application, component cleaning and the disposal costs for the cleaning residues may be found as the main cost drivers of conventional tribo-systems in sheet metal forming. These costs are completely eliminated by using dry forming with volatile media. Especially when using CO_2 as volatile lubricant, a cleaning effect of anti-corrosion oils attached to sheet metal materials could be observed. This cleaning effect has to be quantified in future investigations by recording the basic oiling and the residual oil quantity after forming. This could also enable application in the fields of electrical systems, food technology as well as medical and pharmaceutical engineering.

However, according to current purchasing costs, the use of conventional lubricants is significantly cheaper than the volatile media to be used. In the previous experimental investigations on volatile lubricants, approx. $9 \cdot 10^{-2}$ kg CO_2 was used per component. This is not only due to the outflow under high pressure during forming, but also to the blind volume of the supply line from the valves and the distribution channels in the tool. Above all, the blind volume due to the distance of the valves from the tool surface need to be significantly reduced in a new tool design in order to reduce the amount of volatile lubricant used.

Regarding the different tool designs, no final evaluation can be made yet without actual production and endurance testing. However, some expected trends and correlations already can be outlined. In the 3D-LMF process, utilise material volume will correlate directly with costs. For an economical application, it is essential to keep the mass of the tool inserts as low as possible while maintaining high strength and stiffness by optimising the structure of the tool inserts. If the microinjectors can be integrated into the tool inserts during the printing process, the economic efficiency of the 3D printing process will increase significantly. The costs for the shell design and the laser deposition welding design will be close to each other. For the shell design the joining and sealing technology is more complex and for the laser deposition welding the process is more cost-intensive. The sheet metal laminates will be moderately cheaper in terms of the sheet metal material used. Here, the expenditure for joining will be decisive for the total costs.

During future endurance tests planned, all four best ranked solution concepts will be investigated in the same progressive die in order to avoid an influence of the overall tool design. The aim of these tests is to quantify tool life, wear and friction force for the different tooling concepts.

Outlook

At present, a progressive die is being designed for the planned experimental investigation of the tooling concepts, in which the tool inserts considered for the dry deep-drawing operation will be installed (c.f. Fig. 3). A force measurement for the friction-sensitive punch force of the drawing stage will be integrated into this tool structure in order to measure the friction forces. The wear will be measured by means of the surface changes of the active tool surfaces using a confocal microscope. For this purpose, the tool inserts will be removed from the tool at defined intervals and measured.

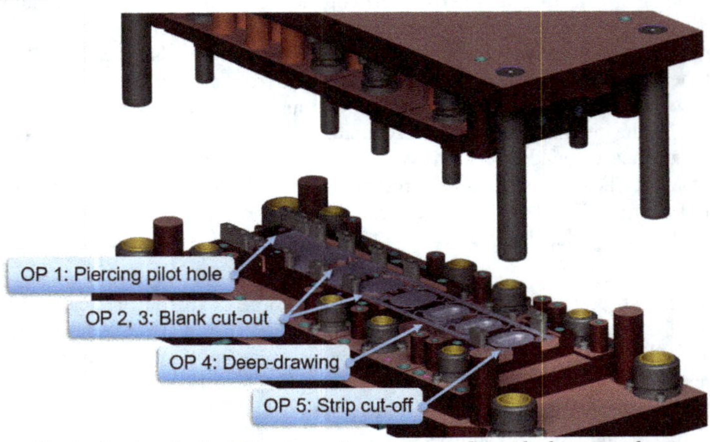

Fig. 3: Design draft of the planned progressive die with sheet metal strip

In order to cover a broader field of application of the novel lubrication system, three sheet metal materials (DC05+ZE, CR280BH and DP500) will be investigated. For each combination of lubricant, sheet metal material and tool design, 5000 components are scheduled. Thus, a total of 120.000 cups will be deep-drawn during the scheduled tests. To keep the material costs as low as possible, a relatively small oval cup geometry has been selected. The geometry of the cup and the design of the progressive die are shown in Fig. 3.

As shown in Fig. 3, the first operation in the progressive die is the piercing of the pilot holes. Subsequently, the blank geometry having four retaining sheet stripes is cut out on the progressive strip in operation two and three. In the fourth operation, the parts are deep-drawn and then cut off at the strip. The drawing and shearing stages are completely separated from each other by their own tool frames in order to adjust the part holder forces for shearing and the blank holder force for deep-drawing independently of each other and to remove the individual operation stages individually. This design ensures an easy access for surface measurements on the blank holder and die.

Summary

In this study, methods for the economic manufacturing of tool inserts for dry deep-drawing with volatile lubricants were presented based on a methodical approach. By means of a criteria-based evaluation, four suitable tooling concepts were selected, which will be examined and evaluated in detail in future investigations. The results of these investigations will show to which extent the preliminary evaluation of the concepts presented here needs to be adjusted. With the selected four solution methods for the deep-drawing tool inserts, the economic efficiency and thus also the acceptance in industrial application of volatile lubrication will improve significantly. These new approaches will be one step further to more economic, sustainable and ecologic manufacturing processes.

Acknowledgement

The research project „Werkzeugbauweisen für das Tiefziehen mit flüchtigen Schmiermedien" of the “Europäische Forschungsvereinigung EFB e.V.” is carried out in the framework of the industrial collective research programme (IGF no. 22067N). It is supported by the Federal Ministry for Economic Affairs and Climate Action (BMWK) through the AiF (German Federation of Industrial Research Associations e.V.) based on a decision taken by the German Parliament.

References

[1] F. Vollertsen and F. Schmidt, “Dry metal forming: Definition, chances and challenges,” Int. J. Precis. Eng. Manuf. - Green Technol., vol. 1, no. 1, pp. 59–62, 2014. https://doi.org/10.1007/s40684-014-0009-0

[2] M. Liewald, C. Wörz, and K. R. Riedmüller, “Characterization of a novel aerostatic lubrication system for deep drawing processes,” CIRP Ann., vol. 70, no. 1, pp. 239–242, 2021. https://doi.org/10.1016/j.cirp.2021.03.013

[3] C. Wörz, G. Reichardt, M. Liewald, E. Zahedi, and R. Weber, “Dry deep drawing of a rectangular cup assisted by volatile media injected from laser-drilled microholes,” Dry Met. Form. OAJ FMT 4, vol. 4, pp. 1–8, 2018.

[4] G. Reichardt et al., “Friction and Wear Behavior of Deep Drawing Tools Using Volatile Lubricants Injected Through Laser-Drilled Micro-Holes,” JOM, vol. 74, no. 3, pp. 826–836, 2022. https://doi.org/10.1007/s11837-021-05028-8

[5] C. Maucher, “Test Specimens in 3D-LMF Additive Manufacturing of Special Features.” Institute for Machine Tools, University of Stuttgart, Stuttgart, 2020.

[6] G. Reichardt et al., “Tribological system for cold sheet metal forming based on volatile lubricants and laser structured surfaces,” Dry Met. Form. OAJ, vol. 6, no. Final Project Report, p. 38, 2020.

[7] R. Vollmer, “Optimierung mittels Laserauftragsschweißen hergestellter Beschichtungen für die Blechumformung,” Technischen Universität Graz, 2016.

[8] M. Kleiner and R. Krux, Entwicklung eines Verfahrens zur schnellen Herstellung von Tief- und Streckzieh Werkzeugen aus Blechlamellen für die Prototypen- und Kleinserienfertigung (Rapid Tooling), vol. P 384. Düsseldorf: Studiengesellschaft Stahlanwendungen e.V., 2001.

[9] A. Huskic, U. Giedenbacher, Jochen Pschebezin, and N. Wild, “Rapid Tooling für Umform-werkzeuge,” RTejournal, 2012.

[10] Böhler-Edelstahl, “Böhler-Edelstahl Additive Manufacturing,” Böhler W360, 2020. [Online]. Available: https://www.bohler-edelstahl.com/de/additive-manufacturing/. [Accessed: 13-May-2020].

[11] 3D-Laserdruck GbR, “Materialdatenblatt 1.2709.” 3D-Laserdruck GbR, Reutlingen, pp. 1–4, 2020.

[12] voestalpine Böhler Edelstahl GmbH, “Materialdatenblatt Böhler E185.” voestalpine Böhler Edelstahl GmbH, p. 2, 2020.

[13] Trumpf, Laser- und Systemtechnik GmbH, “Das beste Pulver für Ihre TruPrint,” TruServices Materials, 2019. [Online]. Available: www.trumpf.com/s/additivemanufacturing. [Accessed: 28-Apr-2020].

[14] H. Hoffmann, R. Neugebauer, and G. Spur, Handbuch Umformen. München: Carl Hanser Verlag, 2012. https://doi.org/10.3139/9783446430044

[15] R. Kolleck and R. Veit, “Technologien und Fertigungsmittel für die temperierte Blech-umformung,” in 13. Dresdner Werkzeugmaschinen-Fachseminar Thermisch - Thermisch unter-stützte Formprozesse, 2008.

[16] R. Kolleck, S. Pfanner, and E. . Warnke, “Development of Cooled Tools for Press Hardening of Boron Steel Sheets,” Key Eng. Mater., vol. 344, no. figure 1, pp. 225–232, 2007. https://doi.org/10.4028/www.scientific.net/KEM.344.225

[17] T. Nakagawa, “Advances in prototype and low volume sheet forming and tooling,” J. Mater. Process. Technol., vol. 98, no. 2, pp. 244–250, 2000. https://doi.org/10.1016/S0924-0136(99)00205-8

[18] A. Techel, “Lamellierte Werkzeuge im Formen- und Werkzeugbau Metal Laminated Tooling – Prozesskette zur schnellen Werkzeugfertigung (MELATO),” wt Werkstattstech. online, vol. 94, no. 11/12, pp. 663–667, 2004. https://doi.org/10.37544/1436-4980-2004-11-12-663

[19] S. Yoo and D. F. Walczyk, “Advanced design and development of profiled edge laminae tools,” J. Manuf. Process., vol. 7, no. 2, pp. 162–173, 2005. https://doi.org/10.1016/S1526-6125(05)70093-1

[20] O. Volkersen, “Die Schubkraftverteilung auf die Verbindungselemente langer Laschenver-bindungen,” Technische Hochschule Wien, 1943.

[21] M. Rasche, Handbuch Klebtechnik. Carl Hanser Verlag GmbH & Co. K, 2012. https://doi.org/10.3139/9783446431980

Sheet Metal 2023 Materials Research Forum LLC
Materials Research Proceedings **25** (2023) 413-420 https://doi.org/10.21741/9781644902417-51

Optimization of the sheet hydroforming process parameters to improve the quality of reshaped EoL components

Antonio Piccininni[1, a *], Angela Cusanno[1,b], Giuseppe Ingarao[2,c], Gianfranco Palumbo[1,d] and Livan Fratini[2,e]

[1]Dept. of Mechanics, Mathematics and Management, Politecnico di Bari, Via Orabona 4, 70125 Bari

[2] Department of Engineering, University of Palermo, Viale delle Scienze, Palermo, 90128, Italy

[a]antonio.piccininni@poliba.it, [b]angela.cusanno@poliba.it, [c]giuseppe.ingarao@unipa.it, [d]gianfranco.palumbo@poliba.it, [e]livan.fratini@unipa.it

Keywords: Aluminium, Hydroforming, Optimization

Abstract. The reshaping of End-of-Life (EoL) components by means of sheet metal forming process has been considered largely attractive, even from the social and economic point of view. At the same time, EoL parts can be often characterized by non-uniform thicknesses or alternation of work-hardened/undeformed zones as the results of the manufacturing process. Such heterogeneity can hinder a proper reshaping of the EoL part and residual marks on the re-formed blanks can be still present at the end of the reshaping step. In a previous analysis, the authors evaluated the effectiveness of reshaping a blank with a deep drawn feature by means of the Sheet Hydroforming (SHF) process: it was demonstrated that residual marks were still present if the deep drawn feature was located in a region not enough strained during the reshaping step. Starting from this condition and adopting a numerical approach, additional investigations were carried out changing the profile of the load applied by the blankholder and the maximum oil pressure. Numerical results were collected in terms of overall strain severity and residual height of the residual marks from the deep drawn feature at the end of the reshaping step. Data were then fitted by accurate Response Surfaces trained by means of interpolant Radial Basis Functions, subsequently used to carry out a virtual optimization managed by a multi-objective genetic algorithm. Optimization results suggested the optimal value of the output variables to reduce the marks from the deep drawn feature without the occurrence of rupture.

Introduction

Production of just five key materials (steel, cement, paper, aluminum, and plastic) accounts for over half of all the greenhouse gas (GHG) emissions released by industry worldwide each year [1]. Primary aluminum production process is the most energy and emissions intensive among the aforementioned materials [2]. Aluminum production industry is responsible for about 3% of the world's 9.4 Gt of direct industrial CO_2 emissions in 2021 [3]. Since 1971, the global demand for aluminum has increased by nearly six times; although during the Covid-19 pandemic the aluminum production fell flat, it has since started growing quickly once again and global demand is likely to continue growing in response to increasing global population and GDP 2021 [3]. The main approach for reducing the primary production and decoupling the resource depletion from the economic growth is the implementation of Circular Economy Strategies [4].

The basic idea is to turn material scraps and End-of-Life (EoL) products/components directly into reusable materials or, better yet, into new products/components. Although recycling is the most commonly applied strategy when metals are concerned, it allows only the material out of a EoL components to be recovered. Actually, to further reduce the environmental impact of Industry sector it is necessary moving towards more virtuous Circular Economy strategies able to recover both material and function from EoL components. Specifically for metals, Cooper and Allwood

Content from this work may be used under the terms of the Creative Commons Attribution 3.0 license. Any further distribution of this work must maintain attribution to the author(s) and the title of the work, journal citation and DOI. Published under license by Materials Research Forum LLC.

proposed a framework for metal reuse, in which four main strategies – relocation, cascade, remanufacturing, and reform/reshaping – are identified [5]. Remanufacturing and Reshaping envisages manufacturing processes to be applied to reuse EoL components. In this respect manufacturing scientists have to find to processes or rethink conventional processes for processing/reprocessing EoL components. In this respect sheet metal based components might represent a significant share of EoL sector. It is worth remarking that sheet metal based components account for a quite large share of the global semi-finished production. In the case of aluminum alloys, sheets (including strips, foil production) account for about 45% of the yearly global demand [6]. Therefore, finding out reuse options for sheet metal based EoL components would enable significant environmental impact reduction.

Overall the Reshaping/Reforming approach has been overlooked by the scientific community so far. Forming processes applied specifically as reuse strategy for sheet metal based components are covered in very few scientific papers so far. Brosius et al. [7] described in a review paper how a demounted automotive engine-hood can be reshaped into a rectangular sheet metal component by sheet hydroforming process. Takano et al. [8] applied Single Point Incremental Forming (SPIF) on a flattened sheet. The Reshaping they propose includes the flattening of a previously bent sheet and a subsequent incremental forming step. Some of the authors of the present paper have recently proposed a novel approach to reshape sheet metal based EoL components [9,10]. Specifically, SPIF was directly applied on a deep-drawn square box to change its shape. However, the SPIF approach presented by the authors allows to add a feature on an EoL components rather than providing a brand new shape. To tackle this challenge, the authors have recently proposed a new Reshaping idea based on Sheet Hydroforming (SHF) process. In this study the authors replicated a reshaping route: a deep drawing process was adopted to impart a square feature; subsequently, SHF was performed to remove the feature and obtain a brand new shape. Results proved that hydroforming can be successfully applied for reshaping proposes [11]. Nevertheless, the developed analyses revealed that reshaping process design is more challenging than conventional forming one as new design objectives and constraints need to be considered making the process engineering more complex. The present paper, propose a multi objective design methodology addressed to the new design challenges characterizing Reshaping by hydroforming processes.

Methodology

A deep drawn aluminum blank (AA5754 H22, initial thickness equal to 0.5 mm) was considered as the EoL component to be completely reshaped by means of the hydroforming process, as shown in Fig. 1. The hydroforming already proved to be a suitable technological solution to give a new shape to the mentioned EoL component [11]. Nevertheless, it was also reported that when the deep drawn feature is located in the BOTTOM position – i.e. in the blank portion that comes in contact with the shallow step of the die cavity – its removal at the end of the reshaping step was much more complicated. As a consequence, using the same numerical scheme of previous analyses by the same authors, the main process parameters of the reshaping step – i.e., the maximum oil pressure and the blankholder force (BHF) – were optimized. In this way, the final geometry of the hydroformed part was improved and the residuals of the deep drawn feature removed as much as possible.

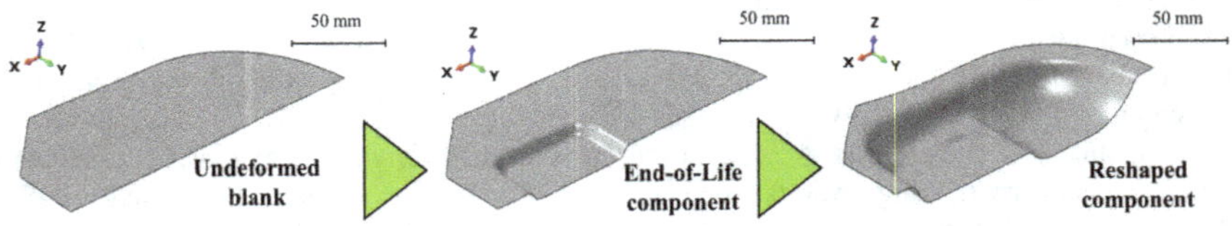

Fig. 1 The sequence of the simulated manufacturing steps

Materials Research Proceedings **25** (2023) 413-420
https://doi.org/10.21741/9781644902417-51

The numerical approach is briefly recalled in the following. The deep drawing step and the reshaping by hydroforming were simulated using the commercial code Abaqus (explicit solver) along with the springback analysis, modelling half of the system and using the mass scaling approach to speed up the numerical runs. The reliability of the numerical results was ensured by checking that the system's kinetic energy resulted always negligible (less than 20%) with respect to the system's internal energy. Tools were modelled as discrete three-dimensional (3D) shell rigid bodies, whereas the blank as a 3D shell deformable body (5 through-thickness integration points) and meshed using 4-node and 3-node triangular shell elements with reduced integration and hourglass control (S4R and S3R, respectively) with an average mesh size of 1 mm. Friction was modelled using the Coulomb's formulation and setting the coefficient to 0.125 [12]. Flow stress curve and Forming Limit Curve (FLC) [11] were implemented in the FE model; the latter allowed to activate a damage criterion - i.e., FLC damage initiation criterion (FLDCRT) - according to which when output variable FLDCRT overcame the threshold value of 1, the model predicted the occurrence of rupture.

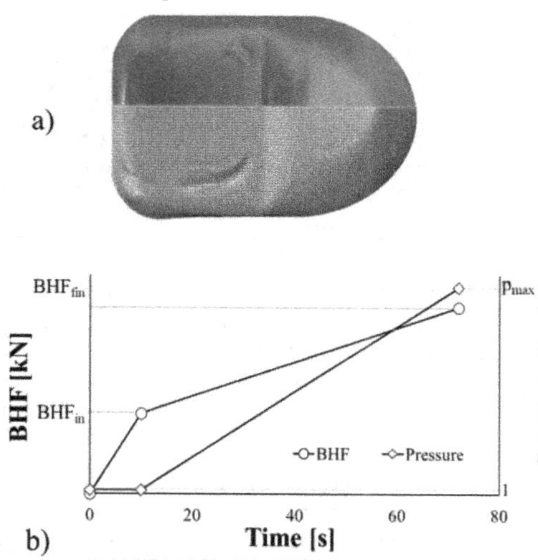

ID	BHF_{in}	BHF_{fin}	BHF_{avg}	P_{max}
1	23	87	55	52
2	23	87	55	104
3	23	87	55	156
4	23	87	55	208
5	33	97	65	52
6	33	97	65	104
7	33	97	65	156
8	33	97	65	208
9	43	107	75	52
10	43	107	75	104
11	43	107	75	156
12	43	107	75	208
13	53	117	85	52
14	53	117	85	104
15	53	117	85	156
16	53	117	85	208

c)

Fig. 2 Additional FE runs: a) the reference ID06 working condition [11]; b) BHF and oil pressure profiles and c) list of the investigated reshaping conditions.

In the present work, 16 conditions were numerically investigated starting from the reference one, labelled as ID06 in [11] and shown in Fig. 2a, which was characterized by the deep drawn feature located in the bottom position, the BHF linearly increasing from 13 to 77 kN and the oil pressure up to 52 bar (with a pressure rate of 1 bar/s). The investigated conditions were arranged according to a full factorial plan changing only the process parameters related to the reshaping by hydroforming. In particular, the forming time was kept unchanged and the initial and final values of the BHF increased over 4 levels; on the other hand, the maximum oil pressure was increased up to 208 bar. Fig. 2b shows the simulated linear profiles of the mentioned process parameters, whereas Fig. 2c reports the list of the simulated conditions. Numerical results were analyzed in terms of the maximum value of the FLDCRT variable over the whole blank $FLDCRT_{max}$ and the height of the residual from the residual deep drawn feature (DelH), graphically explained in Fig. 3. The so-collected data were then used to train accurate Response Surfaces (RS) using the Radial

Basis Functions (RBF) interpolant algorithms available in the modeFRONTIER integration platform. More in details, the data set was initially subdivided into the training set (containing 90% of the 16 designs) and the validation set (based on the remaining 10%). RS were trained on the training set and their accuracy checked on the validation set; subsequently, they were used as the base for the virtual optimization aimed at simultaneously minimizing $FLDCRT_{max}$ and the DelH variables. The virtual optimization, managed by a multi-objective genetic algorithm (MOGA-II) evolved throughout 1000 successive generations.

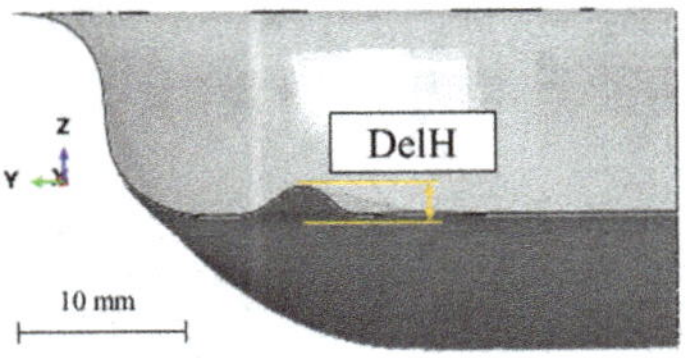

Fig. 3 Definition of the DelH output variable

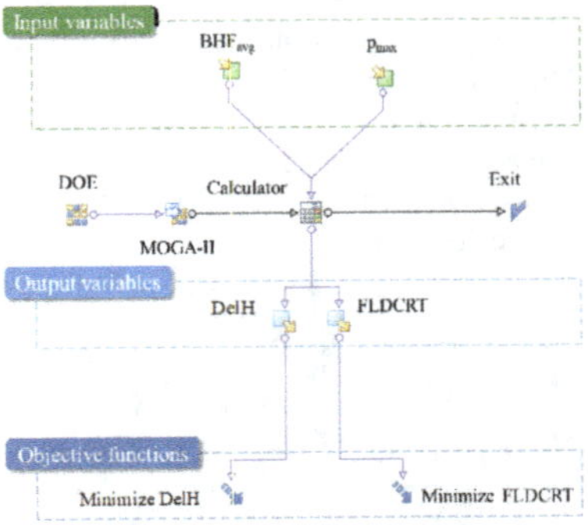

Fig. 4 Workflow of the optimization procedure

The workflow describing the virtual optimization procedure based on the trained RS is schematically shown in Fig. 4. At the end of the optimization round, optimal designs could be evaluated, represented by the average values of BHF and p_{max} which allowed to balance a significant reduction of the deep drawn feature without leading to a premature fracture of the component (i.e., a FLDCRT value lower than 1).

Results

Response Surfaces (RS) were created on the so-called "training set" containing approximately 90% of the DoE data set.. Among the five Radial Basis Functions available for the training of the RS, the Duchon's Polyharmonic Splines algorithm was chosen as the most precise on the DelH output variable, as shown by the errors and the R-squared values calculated on the "validation set" (containing the remaining 10% of the data set) and listed in Tab. 1.

Table 1 Evaluation of the RBF algorithms' fitting capabilities (DelH output variable): Duchon's Polyharmonic Spline (PS), Wendland's Compactly Supported (CS), Hardy's MultiQuadrics (MQ), Inverse MultiQuadrics (IMQ), Gaussians (G).

RBF Algorithm	Mean Absolute Error	Mean Relative Error	Mean Normalized Error	R-squared
PS	2.32E-2	1.90E-2	3.62E-2	0.993
CS	4.81E-2	3.18E-2	7.52E-2	0.973
MQ	5.41E-2	3.64E-2	8.46E-1	0.967
IMQ	8.92E-2	5.88E-2	1.39E-1	0.905
G	1.49E-1	1.01E-1	2.33E-1	0.756

As for the second output variable ($FLDCRT_{max}$), the Wendland's Compactly Supported resulted to be the most accurate RBF algorithm (as proved by the data listed in Tab. 2).

Materials Research Forum LLC
https://doi.org/10.21741/9781644902417-51

Table 2 Evaluation of the RBF algorithms' fitting capabilities ($FLDCRT_{max}$ output variable): Duchon's Polyharmonic Spline (PS), Wendland's Compactly Supported (CS), Hardy's MultiQuadrics (MQ), Inverse MultiQuadrics (IMQ), Gaussians (G).

RBF Algorithm	Mean Absolute Error	Mean Relative Error	Mean Normalized Error	R-squared
CS	1.51E-2	2.02E-2	4.54E-2	0.991
MQ	1.89E-2	2.18E-2	5.69E-2	0.977
IMQ	3.75E-2	4.52E-2	1.13E-1	0.922
PS	3.95E-2	6.75E-2	1.19E-1	0.909
G	7.31E-2	9.10E-2	2.20E-1	0.734

As Fig. 5a suggests, increasing the load applied by the blankholder and the maximum oil pressure leads to a more effective annihilation of the feature from the previous manufacturing step. Nevertheless, higher values of the input parameters may result detrimental for the outcome of the process since those operative conditions may lead to the blank rupture (FLDCRT higher than one) as confirmed by the RS in Fig. 5b.

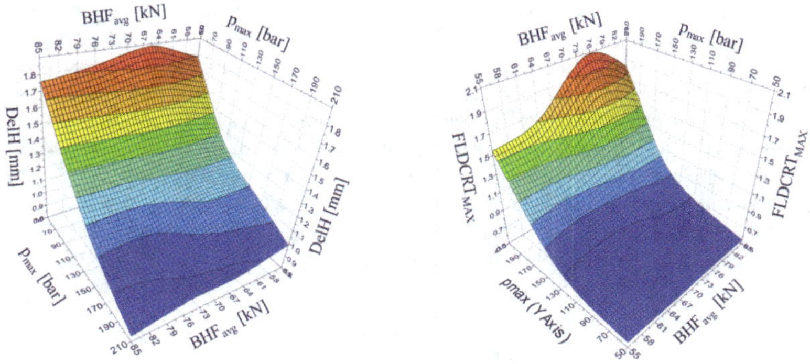

a) b)

Fig. 5 Trained response surfaces (RS) on the output variables: a) DelH and b) FLDCRTmax

The indications coming from the trained RS were absolutely in line with the general guidelines formulated by the authors in a previous work [11]: any evidence from the previous manufacturing step – if located in a blank region that is not going to be excessively deformed during the reshaping step – can be eliminated if the blank is more severely stretched during the second manufacturing step, which means to increase the load exerted by the blankholder or, alternatively, the maximum oil pressure. Nevertheless, such conditions may not be favorable for the overall soundness of the final component: in fact, an increase in the restraining forces may more easily lead to the occurrence of rupture over the blank region in contact with the deeper step of the die cavity. The RS were subsequently used as the base of a virtual optimization, managed by a multi-objective genetic algorithm (MOGA-II) [13]. On each of the defined output variable, a single objective function was defined with the final aim of minimizing simultaneously the FLDCRTmax and the DelH. In such a way, the optimization procedure would have led to the definition of the working window in which the residuals from the previous manufacturing step could be minimized without the occurrence of rupture in the reshaped blank.

Materials Research Proceedings **25** (2023) 413-420 https://doi.org/10.21741/9781644902417-51

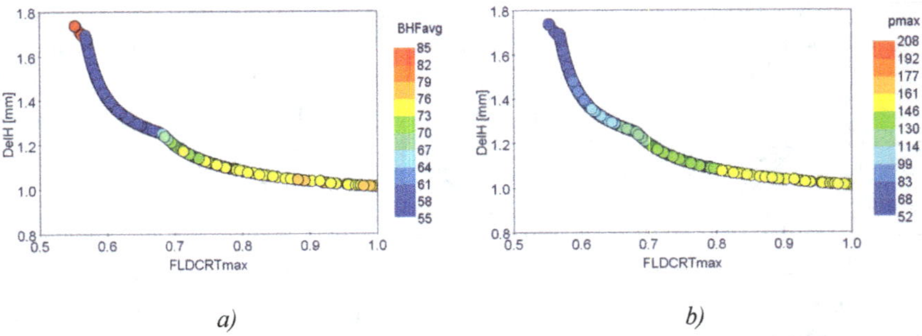

Fig. 6 Pareto designs showing the optimal conditions in terms of a) BHFavg and b) pmax

The Pareto front is shown in Fig. 6 (a and b) collecting all the design considered optimal, i.e. capable of satisfying the two mentioned objective functions. The distribution of the Pareto designs is limited to conditions leading to maximum values of the FLDCRT variable lower than 1 (i.e., all those conditions leading to a final sound component). The result of the optimization suggests that if the final goal is to reduce as possible the appearance of the feature from the previous step without the occurrence of rupture, an average value of BHF_{avg} between 75 and 80 kN and a maximum pressure around 150 bar has to be set. The height of the feature is reduced down to 1 mm and the formed component is globally sound. To evaluate the accurateness of the optimization prediction, one of the Pareto designs (BHF_{avg} equal to 75 kN and maximum oil pressure equal to 160 bar) was simulated. The comparison between the numerical results (in terms of maximum FLDCRT and DelH) and the predicted values, shown in Fig. 7, clearly confirms the accuracy of the predicted optimized values (and, in turns, of the trained RS).

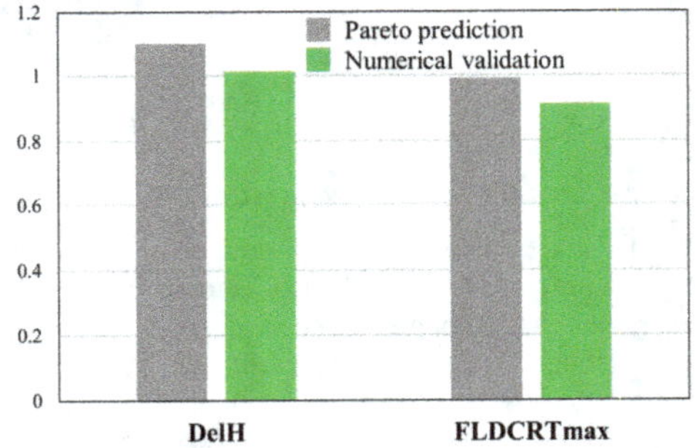

Fig. 7 Accuracy of the optimization prediction

Moreover, the optimized process parameters further confirmed their effectiveness when the appearance of the deep drawn feature at the end of the reshaping step was compared to the one from the ID06 (see Fig. 8, a and b): the increased oil pressure combined with higher restraining force led to the reduction of the DelH (from 1.8 mm to 1.1 mm) without affecting the soundness of the reshaped component (maximum FLDCRT lower than 1).

Materials Research Forum LLC
https://doi.org/10.21741/9781644902417-51

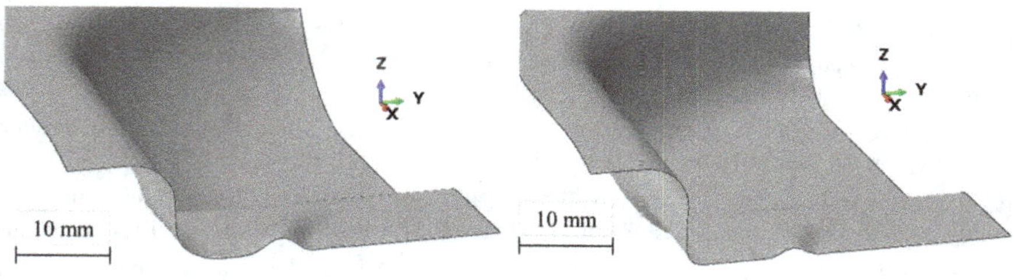

a) b)

Fig. 8 Appearance of the deep drawn feature at the end of the reshaping step: a) refernce ID06 condition and b) optimal Pareto condition

Conclusions

In the present paper, a follow-up of the reshaping methodology has been presented. In particular, starting from a condition characterized by the deep drawn feature located in the bottom positions, the hydroforming process parameters were varied to investigate the possibility of optimizing the blankholder force and the maximum oil pressure to completely remove the residuals from the deep drawn feature. Results from the optimization round, computationally attractive since based on the virtual generation of designs based on accurate Response Surfaces, confirmed the guidelines the authors already formulated in the previous analyses. More in details, the removal of the existing feature is much more effective if the blankholder force and the maximum oil pressure is increased, mainly related to the more severe blank stretching. At the same time, the increase of the restraining force becomes detrimental since inducing a more severe strain conditions in the rest of the blank (where the stretching is more pronounced since being in contact with the deep step of the die cavity). It was then concluded that the overall quality of the reshaped component can be undoubtfully improved and the residual marks from the deep drawn feature can be reduced but not eliminated until the feature is located in the bottom position. In fact, if an increase of the blankholder force may reduce more evidently the deep drawn feature but the integrity of the reshaped component cannot be ensured. Overall it is worth remarking that reshaping processes rise new challenges at the design stages. Along with conventional design objectives (homogenous thinning distribution with absence of fracture, absence of wrinkling and springback distortions minimization), the quality in terms of existing feature removal need to be added as output indicator. Also, the feature might be placed just in a narrow zone of the components to be reshaped. For actual industrial parts, restraining forces differentiation coupled with optimization techniques might be mandatory for a correct implementation of Reshaping processes. Along with the technological challenges to be addressed, the cost effectiveness of such an approach might represent a further issue to be faced. As a matter of fact, it might happen that the costs of the needed operations prior to reshaping (safe disassembly, inspection and decocting) might be too high with respect the value of the recovered sheet metal based EoL component. In this respect, a thorough comparative cost analysis with respect conventional recycling route would provide more clarity about the industrial applicability of the reshaping approach.

References

[1] G. Oberhausen, Y. Zhu, D.R. Cooper, Reducing the environmental impacts of aluminum extrusion, Resour. Conserv. Recycl. 179 (2022) 106120. https://doi.org/10.1016/j.resconrec.2021.106120

[2] M.F. Ashby, Materials and Environment, third ed., Butterworth-Heinemann, Oxford, 2021.

[3] Information on https://www.iea.org/reports/aluminium

[4] T. Tolio, A. Bernard, M. Colledani, S. Kara, G. Seliger, J. Duflou, Design, management and control of demanufacturing and remanufacturing systems, CIRP Ann. 66(2017) 585–609. https://doi.org/10.1016/j.cirp.2017.05.001

[5] D.R. Cooper, J.M. Allwood, Reusing Steel and Aluminum Components at End of Product Life, Environ. Sci. Technol., 46 (2012) 10334–40. https://doi.org/10.1021/es301093a

[6] J.M. Cullen, J.M. Allwood, Mapping the Global Flow of Aluminum: From Liquid Aluminum to End-Use Goods, Environ. Sci. Technol., 47 (2013) 3057–64. https://doi.org/10.1021/es304256s

[7] A. Brosius, M. Hermes, N.B. Khalifa, M. Trompeter, A.E. Tekkaya. Innovation by forming technology: motivation for research, Int. J. Mater. Form., 2 (2009) 29–38. https://doi.org/10.1007/s12289-009-0656-9

[8] H. Takano, K. Kitazawa, T. Goto, Incremental forming of nonuniform sheet metal: Possibility of cold recycling process of sheet metal waste, Int. J. Mach. Tools Manuf., 48 (2008) 477–82. https://doi.org/10.1016/j.ijmachtools.2007.10.009

[9] G. Ingarao, O. Zaheer, D. Campanella, L. Fratini, Re-forming end-of-life components through single point incremental forming, Manuf. Lett., 24 (2020) 132–5. https://doi.org/10.1016/j.mfglet.2020.05.001

[10] G. Ingarao, O. Zaheer, L. Fratini, Manufacturing processes as material and energy efficiency strategies enablers: The case of Single Point Incremental Forming to reshape end-of-life metal components, CIRP J. Manuf. Sci. Technol., 32 (2021) 145–53. https://doi.org/10.1016/j.cirpj.2020.12.003

[11] A. Piccininni, A. Cusanno, G. Palumbo, O. Zaheer, G. Ingarao, L. Fratini, Reshaping End-of-Life components by sheet hydroforming: An experimental and numerical analysis, J. Mater. Process. Technol. 306 (2022) 117650. https://doi.org/10.1016/j.jmatprotec.2022.117650

[12] J. Domitner, Z. Silvayeh, A. Shafiee Sabet, K.I. Öksüz, L. Pelcastre, J. Hardell, Characterization of wear and friction between tool steel and aluminum alloys in sheet forming at room temperature, J. Manuf. Process 64 (2021) 774–84. https://doi.org/10.1016/j.jmapro.2021.02.007

[13] S. Poles, E. Rigoni, T. Robič, MOGA-II Performance on Noisy Optimization Problems. Proc. Int. Conf. Bioinspired. Optim. Methods Their Appl. (2004) 51–62.

Warm and hot formingy

Effect of process variables on interface friction characteristics in strip drawing of AA5182 alloy in warm forming temperature range

Archit Shrivastava[1, a *] and D. Ravi Kumar[1, b]

[1]Department of Mechanical Engineering, Indian Institute of Technology Delhi, New Delhi, India

[a]mez178604@iitd.ac.in, [b]dravi@mech.iitd.ac.in

* Corresponding author

Keywords: Metal Forming, Friction, Strip Drawing Test

Abstract. Warm forming is widely used to enhance formability of aluminum alloy sheets in order to manufacture components with complex shapes. However, forming of aluminum alloy sheets by various sheet metal forming processes such as deep drawing and stretch forming involves sliding, drawing or stretching of sheet materials over the tool surfaces. Warm forming results in change of frictional characteristics at the tool-blank interface during forming. Higher friction leads to poor formability, non-uniform strain distribution, higher forming load, and poor surface finish of the component. So it is important to investigate the effect of process variables on friction at the interface in warm forming of aluminum alloy sheets. In this work, the tribological behavior of an Al-Mg-Mn alloy (AA5182) has been studied by performing strip drawing experiments in the warm forming temperature range (100-250 °C) in lubricated condition. Experiments were conducted to investigate the effect of temperature, normal force, and drawing speed on the coefficient of friction. A significant impact on the friction coefficient is observed by the change in boundary conditions as a result of variation in process variables with temperature being the most influential. The results have been compared with frictional characteristics in strip drawing at ambient temperature.

Introduction

Aluminum alloys have proven to be the ideal choice for weight reduction in the automotive and aerospace industries due to their higher strength-to-weight ratio than steels, good mechanical properties, and excellent corrosion resistance [1]. Complex sheet metal parts are required for automotive applications [2], and these are manufactured by processes such as deep drawing, stretch forming and bending [3]. Deep drawing is one of the most important sheet metal forming processes, and it involves radial drawing of a blank into the die cavity with a punch, as shown in Fig. 1 (a). Deep drawing is primarily used to create parts with a large depth, such as fuel tanks, oil sumps, gas cylinders, automotive panels etc. As the blank material in the flange region slides over the dies during deep drawing, friction between the blank and the tools affects drawability, uniformity of strain distribution, and drawing load [4]. However, the aluminum alloys have low formability at ambient temperature compared to the elevated temperature [5]. Jang et al. [6] studied the tensile deformation behavior of AA5182 aluminum alloy in the warm working temperature range of 150 – 350 °C. Warm forming of AA5182 aluminum alloy at temperatures below the recrystallization temperature combines advantages of both cold working and hot forming to a certain extent. D. Raja satish et al. [7] observed a similar behavior for AA5182 in the temperature range of 200 – 300 °C. The drawability of AA5182 has also been observed to increase in warm deep drawing. However, the frictional characteristics in deep drawing are very critical in warm forming because temperature, sliding speed, lubrication and binder force affect the interface friction in the flange region.

The friction and galling in forming processes are known to be affected by various parameters in the sliding contact region including contact pressure, sliding velocity, temperature, tool

Content from this work may be used under the terms of the Creative Commons Attribution 3.0 license. Any further distribution of this work must maintain attribution to the author(s) and the title of the work, journal citation and DOI. Published under license by Materials Research Forum LLC.

geometry, the sliding surface of the contact pair [8], and lubricant [9]. The adhesion of aluminum alloys to tool surfaces is enhanced during warm forming. This causes increased friction, degrades the surface quality of the formed part, and damages the tool [10]. As a result, lubrication and application of coatings has been a standard practice to reduce metal adhesion, tool wear, and friction [11]. Januszkiewicz et al. [12] studied the friction and wear behavior of AA5182 aluminum alloy ring rubbing against SAE (AISI) 52100 type bearing steel ball at various temperatures (up to 300 °C) and different applied forces (4 N and 24 N) in ball on ring frictional tester. The author noticed a transfer layer growing on the tool surface and severe scratches on AA5182 at a critical temperature of 230 °C. The findings demonstrated that temperature significantly impacts adhesion and wear.

Some test methods for assessing friction characteristics that have been developed over the years include pin-on-disk [13], ball-on-disk [14], block-on–disk [15], ball–on–plate [16], and strip drawing [17]. The parameters used in friction testing must be similar to the actual process parameters for reliable estimation of friction coefficient. The contact condition in strip drawing is similar to that encountered in the flange region during deep drawing. Assumption of a constant friction coefficient cannot accurately reflect the friction and galling behavior at the interface and predict the forming behavior in the warm forming process. Therefore, in the present work, strip drawing tests were carried out to investigate effect of important process variables on friction coefficient value under lubricated condition in the warm deep drawing temperature range of AA5182 aluminum alloy sheets. Tribological tests were carried out by varying temperature, normal pressure, and drawing speed. The friction coefficient is determined for different contact conditions and the results are compared with those at ambient temperature (25 °C).

Methodology

Material and specimens. Specimens of 1.2 mm thick AA5182 (Al–Mg–Mn) alloy sheets were used in the present work. This grade of aluminum is widely used in the automobile industry for sheet metal stamping due to its work hardening ability [2]. Samples of 500 mm x 65 mm were sheared from the metal strips for carrying out strip draw tests. The initial average surface roughness of the specimens was 0.95 ± 0.05 µm.

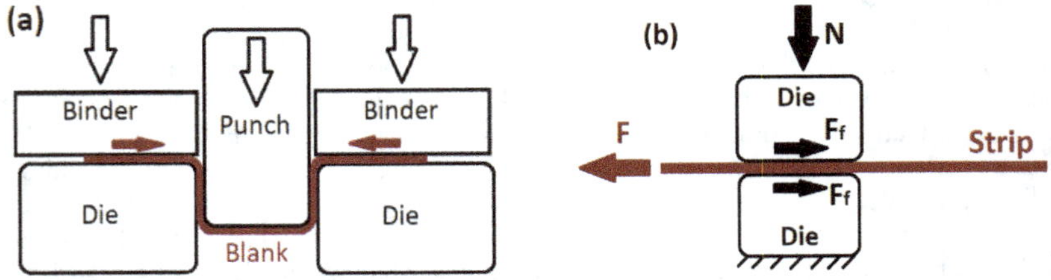

Fig. 1. Schematic of (a) deep drawing process (b) strip drawing test

Strip drawing tests. Fig. 1(b) depicts the schematic of the strip drawing test used in the present work. One end of the strip was clamped with wedge grips and a predefined normal force was applied on the strip placed between the dies. The strip was pulled using the wedge grips over a distance at a specified speed, and the draw force was measured. The friction coefficient was calculated using half of the draw force due to the frictional force acting on both sides of the strip. Flat dies of cross-section area 100 mm x 55 mm made up of high carbon, high chromium tool steel were used in the tests. The die surfaces had an average surface roughness (Ra) of 0.30 ± 0.05 µm after being ground and polished. The dies and the samples were cleaned with acetone before performing the experiments to remove any dirt that might affect the friction coefficient measurement. Oil-based lubricants KTL N16 and SHF 430 were used for testing at ambient and elevated temperatures respectively.

The amount of lubricant used was equivalent to 2.0 g/m2, which is the amount typically used in sheet metal stamping operations. The dies were preheated by cartridge heaters. Insulating isolation plates and cooling plates were installed to keep the surrounding electrical and mechanical components from overheating. When the desired temperature of the dies was reached, the strip was heated to a sufficiently high temperature taking into account the loss of temperature while placing it between the dies. The strip was pulled over a distance of 200 mm. From the force-displacement data obtained over the sliding distance range of 50 mm to 150 mm, the average friction coefficient was determined. The experiments were repeated three times for each combination to ensure the reproducibility of the results.

Process parameters. The strip drawing experiments were conducted to determine friction coefficient at different temperatures, speeds and forces. Although the recrystallization temperature of aluminum alloy is generally in the range of 300 °C to 400 °C, the material's strength decreases rapidly beyond 300 °C. Thus, to balance both strength and ductility, the recommended warm forming temperature of aluminum alloys is in the range of 100 °C to 250 °C. Thus, the strip draw tests in this work were carried out in this range. The blank holding pressure applied in the flange area during warm deep drawing of AA5182 aluminum alloys usually varies from 1.5 MPa to 2.5 MPa. Therefore, the normal force in strip drawing experiments is determined based on this pressure range and the die contact area (5500 mm2). Thus, the strips were drawn with a normal force in the range of 7.5 kN to 12.5 kN. The strips were drawn by varying the speed in the range of 2.5 mm/s to 10 mm/s to capture the effect of speed also on the tribological behavior.

Results and Discussion

Effect of temperature. The variation of friction coefficient with temperature in strip drawing experiments under lubricated condition is shown in Fig. 2(a) at two normal forces with a constant drawing speed of 10 mm/s. At ambient temperature, the friction coefficient is found to be 0.12 and 0.15 at 7.5 kN and 10 kN normal forces, respectively. The tests at elevated temperatures indicated the friction coefficient in the temperature 100-125 °C is lower than at ambient temperature. This could be due to material softening and lower shear stress required to draw the strip. At 10 kN normal force, with further increase in temperature, friction coefficient is found to increase before becoming nearly constant. The susceptibility of adhesion increases with increase in temperature leading to a gradual change in the interface condition and rise in friction coefficient. The friction coefficient sharply increased to 0.22 when the temperature is raised to 150 °C and it further increased to 0.23 at 250 °C. The variation of the friction coefficient at 7.5 kN exhibits a similar behavior, with a minor shift in the overall shape of the curve. The friction coefficient is minimum (0.078) at 125 °C at a normal force of 7.5 kN.

The variation of friction coefficient with the sliding distance for different temperatures is depicted in Fig. 2(b). These experiments were carried out at a normal force of 7.5 kN and a drawing speed of 10 mm/s. The friction coefficient is nearly constant along the sliding distance at ambient temperature but slightly increased at higher temperatures. An increase in temperature causes the lubricant's viscosity to decrease at the contact interface, which reduces the lubricant film thickness and its lifespan along the sliding distance. A sharp increase in friction coefficient has been observed towards end of the sliding at 250 °C possibly, due to inadequate lubricant at the interface at later stages of strip drawing leading to dry sliding condition and galling. Similar behavior has been observed by Januszkiewicz et al. [12] on AA5182 aluminum alloys at temperatures over the critical temperature, where there is a significant metal transfer to the tool and severe scratches on the workpiece has been observed.

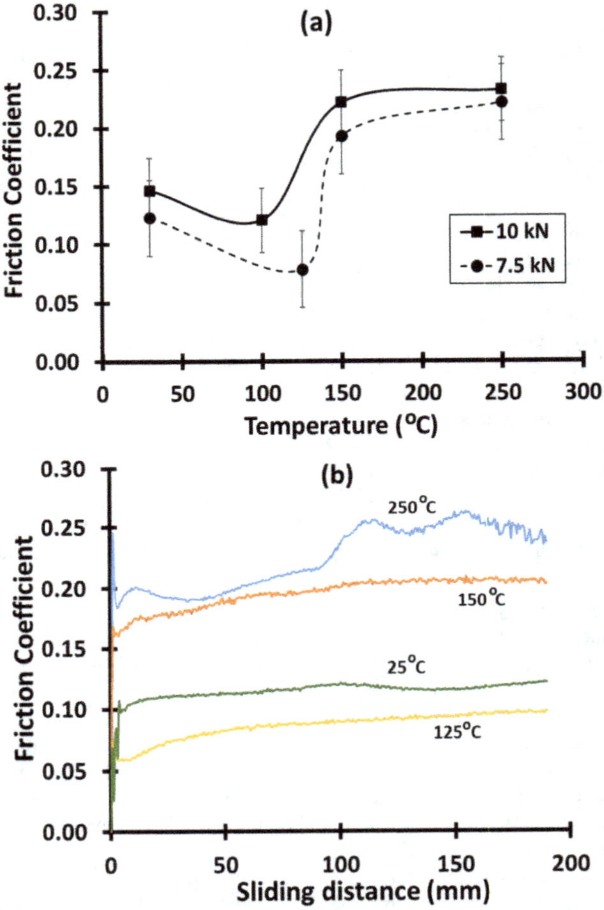

Fig. 2. (a) Variation of average friction coefficient with temperature at two different normal forces at 10mm/s and (b) Variation of friction coefficient with the sliding distance during strip drawing experiments at different temperatures

Effect of normal force. The variation of average friction coefficient value with normal force at elevated temperatures is shown in Fig. 3. Initially when the force is increased from 7.5 kN to 10 kN. The friction coefficient is raised by localized sticking due to high contact stresses produced under higher normal force. A similar trend is observed at the ambient temperature also. But at elevated temperatures, the friction coefficient is observed to decrease with further increase in the force to 12.5 kN. At high temperatures, the affinity of adhesion increases as the normal force increases. Asperities will undergo excessive plastic deformation, so higher shear forces are required due to stronger workpiece-tool interface bonding. But the decrease in friction coefficient could be attributed to subsurface shear deformation which requires a lower shear force [18].

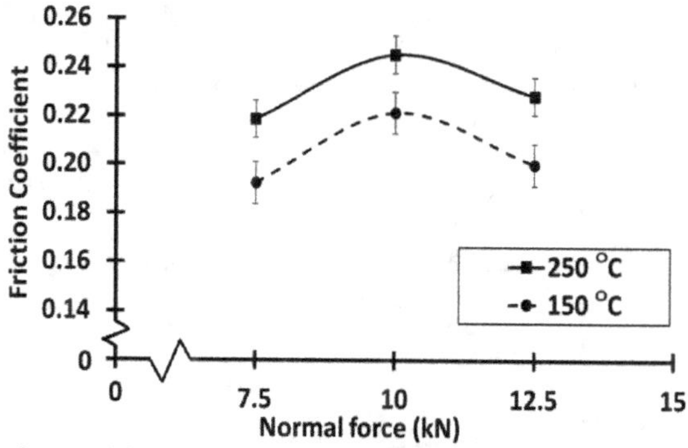

Fig. 3. Effect of normal force on friction coefficient in strip drawing in warm forming temperature range of Aluminum

Effect of speed. The effect of drawing speed on the friction coefficient at different temperatures is shown in Fig. 4. As drawing of the strip begins, the lubricant is drawn to the surface from the valleys, which act as a lubricant reservoir. As less amount of lubrication being distributed in the contact area at low speeds, higher friction coefficient has been observed due to stick/slip phenomenon of the contacting asperities. The flow rate of the lubricant from valleys to the surface increases with speed, thus reducing the friction coefficient. However, the decrease in friction coefficient with increasing speed at higher temperatures is lower than at ambient temperature. The lubrication mechanism's effectiveness at the contact interface plays an important role. Increasing tendency of oxidation and sticking friction that cause adhesion with increase in temperature might offset the decrease in friction coefficient with increase in drawing speed.

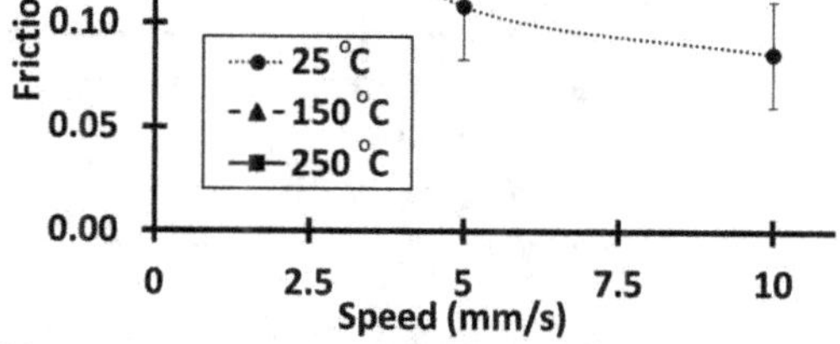

Fig. 4. Effect of drawing speed on friction coefficient in strip drawing at three different temperatures

The variation of drawing speed on friction coefficient with sliding distance at 150 °C and 250 °C, is shown in Fig. 5 (a) and (b) respectively. The peak static friction coefficient at 150 °C is 0.17 ± 0.01 while it is 0.2 ± 0.01 at 250 °C. This increase in static friction coefficient is due to higher adhesion with increase in temperature. The force required to draw the strip should overcome the static frictional force. The pulling force required to draw the strip decreases once the sliding

Materials Research Proceedings 25 (2023) 423-430
https://doi.org/10.21741/9781644902417-52

begins. The kinetic frictional force is found to depend on the sliding speed. Fig. 5 shows that the kinetic friction coefficient dropped to 0.15 at 10 mm/s. As the sliding continues, increase in adhesion affinity leads to increase in kinetic frictional forces. This leads to a rise in the friction coefficient again with sliding distance, as shown in Fig. 5.

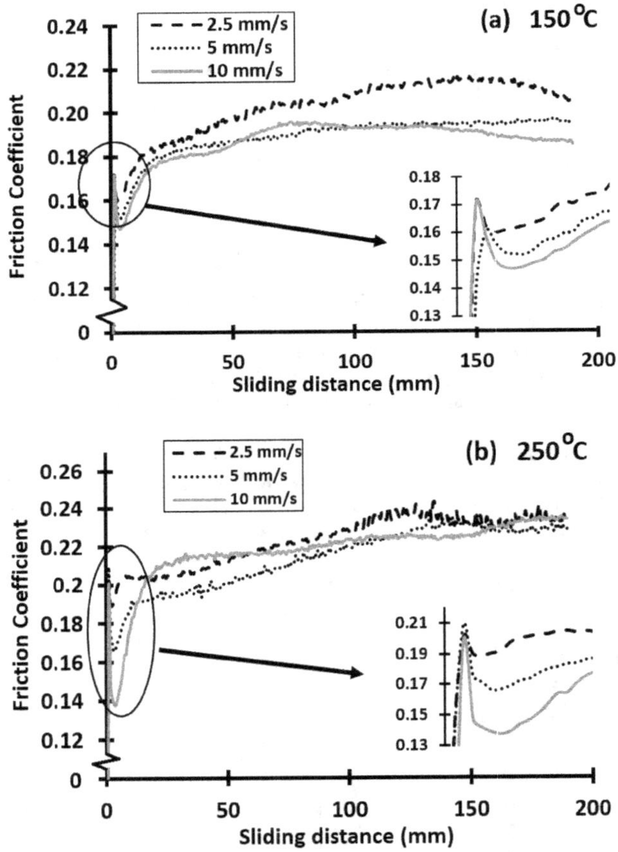

Fig. 5. Variation of friction coefficient with sliding distance for different speeds in strip drawing experiments at (a) 150 °C and (b) 250 °C

Conclusions

The effect of process variables on friction coefficient at the tool-strip interface in the warm forming temperature range of AA5182 aluminum alloy has been examined in the conditions similar to that exist in the flange region in warm deep drawing. The following are the main findings from the results and discussion presented in the previous section:

The friction coefficient is significantly influenced by the temperature. It decreased initially when compared to the ambient temperature. As the temperature rises further, adhesion becomes more susceptible increasing the friction coefficient. The peak static friction coefficient is also observed to increase with the increase in temperature. The friction coefficient is found to increase when the normal force is increased due to increased area of contact and localized sticking. But at elevated temperatures friction coefficient decreased with further increase in normal force due to subsurface shear deformation. The friction coefficient decreases as the drawing speed increases but the extent of decrease in warm forming temperature range is lower when compared to that at ambient temperature.

The outcome of this work can be used to improve the accuracy of predictions from numerical simulations of warm deep drawing of this alloy by incorporating the effect of process variables on friction instead of considering a constant friction coefficient value for a given blank temperature.

Acknowledgement

The authors would like to acknowledge the support received from Prof. Marion Merklein and her research group at the Lehrstuhl für Fertigungstechnologie, Friedrich-Alexander-Universität, Erlangen, Germany, for the research carried out in this article. Archit Shrivastava would like to acknowledge the assistantship received from the Ministry of Human Resource Development (MHRD), Government of India for this research work.

References

[1] P. Rambabu, N. Eswara Prasad, V.V. Kutumbarao, R.J.H. Wanhill, Aluminium Alloys for Aerospace Applications, in: N. Prasad, R. Wanhill (Eds.), Aerospace Materials and Material Technologies, Indian Institute of Metals Series. Springer, Singapore, 2017, 29–52. https://doi.org/10.1007/978-981-10-2134-3_2

[2] J.C. Benedyk, Aluminum alloys for lightweight automotive structures, in: P.K. Mallick (Eds.), Materials, Design and Manufacturing for Lightweight Vehicles, Woodhead Publishing Series in Composites Science and Engineering, 2010, 79-113. https://doi.org/10.1533/9781845697822.1.79

[3] Z. Marciniak, J.L. Duncan and S.J. Hu, Mechanics of Sheet Metal Forming, Second ed., Butterworth-Heinemann, 2002.

[4] T. Altan, A.E. Tekkaya (Eds.), Sheet metal forming: processes and applications, ASM international, 2012. https://doi.org/10.31399/asm.tb.smfpa.9781627083171

[5] M. Tisza and Z.S. Lukacs, High strength aluminium alloys in car manufacturing, *IOP Conf. Ser.: Mater. Sci. Eng.*, 418 (2018) 012033. https://doi.org/10.1088/1757-899X/418/1/012033

[6] D.H., Jang, W.J. Kim, Warm Temperature Deformation Behavior and Processing Maps of 5182 and 7075 Aluminum Alloy Sheets with Fine Grains, Met. Mater. Int., 24 (2018) 455–463. https://doi.org/10.1007/s12540-018-0061-3

[7] D. Raja Satish, D. Ravi Kumar and M. Merklein, Effect of temperature and punch speed on forming limit strains of AA5182 alloy in warm forming and improvement in failure prediction in finite element analysis: A case study, The Journal of Strain Analysis for Engineering Design, 52:4 (2017) 258-273. https://doi.org/10.1177/0309324717704995

[8] J. Noder, R. George, C. Butcher, and M. J. Worswick, Friction characterization and application to warm forming of a high strength 7000-series aluminum sheet, Journal of Materials Processing Technology, 293 (2021) 117066. https://doi.org/10.1016/j.jmatprotec.2021.117066

[9] V. Prakash, D. Ravi Kumar, Performance evaluation of bio-lubricants in strip drawing and deep drawing of an aluminium alloy, Advances in Materials and Processing Technologies, 8:1 (2020) 1044-1057. https://doi.org/10.1080/2374068X.2020.1838134

[10] R. Lachmayer, B. Behrens, T. Ehlers, P. Müller, P. Althaus, M. Oel, E. Farahmand, P.C. Gembarski, H. Wester, S. Hubner, Process-Integrated Lubrication in Sheet Metal Forming, *J. Manuf. Mater. Process*, *6* (2022) 121. https://doi.org/10.3390/jmmp6050121

[11] J. Tenner, K. Andreas, A. Radius, M. Merklein, Numerical and experimental investigation of dry deep drawing of aluminum alloys with conventional and coated tool surfaces, Procedia engineering, 207 (2017) 2245-2250. https://doi.org/10.1016/j.proeng.2017.10.989

[12] K. R. Januszkiewicz, A. R. Riahi, and S. Barakat, High temperature tribological behaviour of lubricating emulsions, Wear, 256 (2004) 1050-1061. https://doi.org/10.1016/j.wear.2003.06.001

[13] G. Maculotti, E. Goti, G. Genta, L. Mazza, M. Galetto, Uncertainty-based comparison of conventional and surface topography-based methods for wear volume evaluation in pin-on-disc tribological test, Tribology International, 165 (2022) 107260. https://doi.org/10.1016/j.triboint.2021.107260

[14] S. Swirad, P. Pawlus, The Effect of Ball Burnishing on Tribological Performance of 42CrMo4 Steel under Dry Sliding Conditions. *Materials*, 13:9 (2020) 2127. https://doi.org/10.3390/ma13092127

[15] T. Bjork, J. Bergström, S. Hogmark, Tribological simulation of aluminium hot extrusion, Wear, 224:2 (1999) 216-225. https://doi.org/10.1016/S0043-1648(98)00332-9

[16] J. Lu, Y. Song, L. Hua, P. Zhou, G. Xie, Effect of temperature on friction and galling behavior of 7075 aluminum alloy sheet based on ball-on-plate sliding test, Tribology International, 140 (2019) 105872. https://doi.org/10.1016/j.triboint.2019.105872

[17] P. Schwingenschlogl, M. Weldi and M. Merklein, Investigation of the influence of process parameters on adhesive wear under hot stamping conditions, *J. Phys.: Conf. Ser.* 896 (2017) 012048. https://doi.org/10.1088/1742-6596/896/1/012048

[18] G.E. Dieter, D. Bacon, Mechanical metallurgy, SI Metric ed., McGraw-hill, New York, 1988, 544-545.

Sheet Metal 2023
Materials Research Forum LLC
Materials Research Proceedings **25** (2023) 431-438
https://doi.org/10.21741/9781644902417-53

Investigation of the influence of workpiece-side parameters on the layer formation of zinc-coated boron-manganese steel

Franz He[1, a *] and Marion Merklein[1,b]

[1]Lehrstuhl für Fertigungstechnologie, Friedrich-Alexander Universität Erlangen-Nürnberg, Egerlandstr. 13, 91058 Erlangen, Germany

[a]franz.he@fau.de, [b]marion.merklein@fau.de

Keywords: Hot Stamping, Coating, 20MnB8

Abstract. The hot stamping process has been established as a technology for the production of ultrahigh-strength steel parts for safety-relevant components in lightweight construction for the automotive sector. Thanks to the reduced overall thickness combined with high tensile strength, it is possible to realize lightweight design concepts with improved crash behavior. Boron-manganese steels are usually used for this purpose. Due to initial process temperatures above 800 °C, hot stamping is considered a lubricant-free process. In addition to high friction and wear in the process, surface scaling and the need for tool repairs are the result. In light of these phenomena, hot stamping materials are coated to protect them from corrosion. Until now, aluminum-silicon-based coatings have been primarily used for the direct hot stamping route. While zinc-based coatings have so far mainly found usage in the indirect process route, they are now also become a valid alternative for the direct process route. In previous investigations, a significant influence of workpiece-side parameters on the formation of the coating during the austenitization process was found for aluminum-silicon (AlSi) coatings. In light of this, a similarly significant influence is suspected for zinc-based coatings. The parameters heating rate, furnace temperature and dwell time in the furnace and the effect on the coating formation of zinc-coated 20MnB8 during austenitization will be investigated. The resulting findings will form the basis for further experiments to investigate the influence of the parameters on friction and wear in the industry near strip drawing tests.

Introduction

Ever since the oil crisis in 1970 and the resulting shortage of raw materials, the demand for lighter and more energy-efficient automotive vehicles has increased, which has resulted in the increased use of ultra-high-strength steels [1]. In this regard, cold forming faces challenges due to springback, tool wear, and high forces in the process during the production of geometrically complicated components [1]. Forming at high temperatures allows an improvement of the formability with simultaneously reduced forces in the process as well as a higher dimensional stability [2]. For this purpose, the hot stamping process has been developed, which is based on a process combination of hot forming of fully austenitized blanks and hardening of the semi-finished product in the die within one process step [3]. Due to the martensitic microstructural transformation in the quenching process, the manufactured components achieve tensile strengths of up to 1500 MPa, which allows for improved crash properties with simultaneously reduced component weight [4]. Nowadays, hot-stamped components are used in impact beams, bumpers and side members [5]. Due to process temperatures above 850 °C necessary to achieve full austenitization in the steel material, lubricants are not stable during hot stamping. High friction and wear are the result which are significantly influenced by the process parameters [6]. Two process routes are available for hot stamping, the direct and indirect variants [7]. For both variants, the established boron-manganese steel 22MnB5 is widely in use [8]. In its initial state, this steel has a ferritic-pearlitic microstructure with a tensile strength of 500 to 700 MPa and an elongation

Content from this work may be used under the terms of the Creative Commons Attribution 3.0 license. Any further distribution of this work must maintain attribution to the author(s) and the title of the work, journal citation and DOI. Published under license by Materials Research Forum LLC.

at break of over 10% [9]. When heated above the material-specific Ac3 temperature of 850 °C, a complete transformation of the initial microstructure to austenite takes place [10]. The resulting reduction in strength and increase in elongation at break require low forces in the forming process and provide good conditions for the production of components with complicated geometries. Quenching above the critical cooling rate of 27 K/s results in martensite transformation, allowing improved mechanical properties to be obtained [4]. The boron-manganese steel can be used without an additional coating in the process, but a scale layer is formed during the austenitizing process due to the reaction with oxygen in the furnace atmosphere. The brittle and rough oxide layer impairs downstream process steps and results in a reduced heat transfer rate between the contact partners and thus reduced cooling rates [11]. Furthermore, the die surface is damaged by abrasive oxide particles, which is why the scale layer has to be removed by sand or shot blasting [12]. For this reason, semi-finished coatings are used for hot stamping to prevent the formation of an oxide layer. For direct hot stamping, aluminum-silicon (AlSi) based coatings are commonly used [10]. An unwanted material transfer of AlSi particles in the process leads to a deterioration of the tool surface, which results in a deviation from the dimensional accuracy of the component on the one hand, and different local heat transfer coefficients on the other hand, resulting in a reduction of the component quality [13]. Furthermore, the AlSi coating only provides a barrier effect and no cathodic protection [14], which limits its use in areas where electrolytic corrosion is of importance [1].

Zinc-based coatings have been mainly used for indirect hot stamping [3], where they exhibit a combination of high strengths in the base material while also providing a cathodic protection effect [5]. Due to the high risk of evaporation, run-off and oxidation of zinc coatings, due to the low melting temperature of 420 °C and evaporation temperature of 907 °C [3], the material was considered unsuitable for direct hot stamping [15]. This led to liquid metal-induced stress corrosion, where zinc diffuses into the material structure at high temperatures leading to material separation [2]. Due to recent developments the use of zinc-coated boron-manganese steel has also been made viable for direct hot stamping. By adding oxygen affine elements such as aluminum, a protective oxide skin ($Al2O3$) is formed on the component surface, which prevents oxidation of the zinc coating and inhibits the formation of a liquid phase by progressive formation of a higher-melting Zn-Fe layer [15]. In previous studies, the influence of process parameters during austenitizing on the layer formation of AlSi coatings was investigated. A significant influence of the furnace dwell time was found [16]. In light of these findings, the influence and interdependencies of process parameters on the coating formation are also investigated for the austenitization of zinc-coated boron manganese steels for direct hot stamping in order to determine significant parameters. The results of this work provide a data basis for further experiments to investigate the friction and wear behavior in the direct hot stamping process.

Methodology and experimental setup

Material and test procedure. Specimens with dimensions of 70 mm x 50 mm were used for the experiments. A 20MnB8 boron-manganese steel with a thickness of 1.5 mm coated with the ZF180 zinc coating was used as the test material. The coating was specially developed for the usage in direct hot stamping and prevents the formation of a liquid phase due to an intermediate precooling step [17]. This increases the time window required for cooling when using 22MnB5, consequently the required critical cooling rate of 27 K/s to reach martensite transformation cannot be achieved. The redistribution of alloying elements for the 20MnB8 base material shifts the ferrite, pearlite, and bainite regions, reducing the required cooling rate to 20 K/s [12] and ensuring a stable cooling corridor for martensite formation over a longer period [17]. The Gleeble 3500 GTC thermomechanical simulator from Dynamic Systems Inc. was used to study the influence of the heating rate. Here, the ends of the specimen are clamped in jaws through which an electric current flow is introduced. The resulting resistance causes local heating of the specimen [18]. The temperature is monitored during the test run in the center of the specimen by welded-on type K thermocouples. Based on the findings during the investigation of the heating rate, further experiments on the furnace dwell time and temperature parameters were carried out in an ME87/13 annealing and hardening furnace from RHODE. The furnace can reach a maximum temperature of up to 1300 °C. A homogeneous furnace atmosphere and temperature distribution is ensured by three-sided chamber heating on the sides and bottom. A two-layer insulation structure reduces the power loss at the walls outside the furnace. The dimensions of the usable area in the furnace are 350 mm x 250 mm x 1000 mm [19].

Parameters and methodology. The parameters selected for the present study are based on the findings of past investigations done on AlSi coated 22MnB5. The heating rate, furnace temperature and dwell time for austenitization in the furnace were considered as critical influencing parameters [15]. The parameters were varied in three steps and repeated three times (n=3) to ensure reproducibility. The experiments were arranged based on the design of experiments and randomized for execution to prevent systematic errors. The exact parameters and variation levels can be found below in Table 1.

Table 1: Parameters and levels for the heating tests

Parameter	Levels
Furnace temperature [°C]	880; 910
Furnace duration [s]	50; 125; 200
Heating rate [K/s]	5; 25; 50

After the test was carried out, the surface roughness Rz was recorded on the perthometer by means of tactile roughness measurement in accordance with DIN EN ISO 3274 and then cold-embedded, ground and polished for metallographic examination. The micrographs were analyzed for layer formation and layer height using a confocal light microscope, the Keyence VK-X200. In order to analyze the detected phase fractions, the elemental composition was also recorded using energy dispersive X-ray spectroscopy (EDS).

Results and discussion

Heat rate. The heating rate was investigated on a Gleeble 3500 GTC thermomechanical simulator. The heating rates were set at 5, 25 and 50 K/s. After reaching the target temperature of 910 °C, the specimen was held for 60 seconds and then cooled in air. The micrographs, shown in Fig. 1, show similar layer formation for all three heating rates. The base material of 20MnB8 can be found in the lower part of the image. The coating appears above the base material and separates into lighter and darker gray phases 1 and 2, followed by the embedding material.

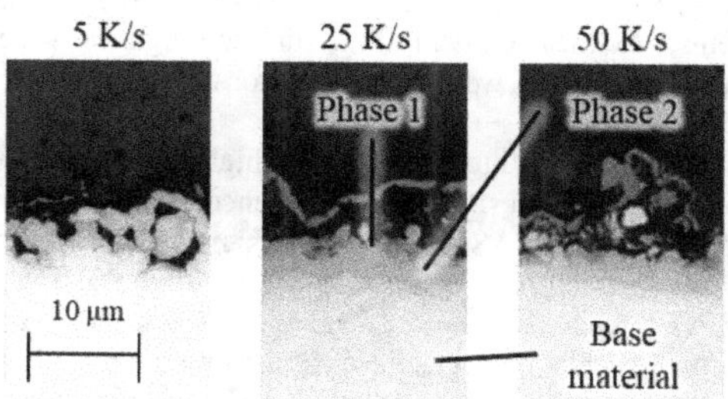

Material	
Base material	20MnB8
Coating	ZF180
Process parameters	
Furnace temperature	910 °C
Soaking time	60 s
n_{Strips}	3

Fig. 1: Layer formation of the ZF180 coating at different heating rates

Based on the micrographs, no significant change in the coating formation due to the different heating rates can be detected. At heating rates >5 K/s, the test showed an increased susceptibility to reaction with impurities on the surfaces and a tendency for the coating to detach during the heating process, as shown in Fig. 2. The coating detachment can be attributed to an emergence of liquid zinc phases.

Fig. 2: Exemplary coating delamination due to liquid zinc phases at heating rates >5 K/s

The liquid zinc phases at heating rates >5 K/s occur due to the reduced time required to reach the target temperature. Melting of the zinc coating is inhibited by continuous diffusion of iron from the base material into the coating during the heating process, since the melting temperature of the Ze-Fe composition in the coating at 5 K/s is higher than the current furnace temperature. Since diffusion is a time-dependent parameter in addition to temperature, the phenomenon that not enough iron diffuses into the coating can occur at higher heating rates and the zinc phase exceeds its melting point. Due to the poor coating properties at higher heating rates and the non-significant influence on the coating formation, the influence of the heating rate will not be investigated further. Since the heating rate of the Gleeble shows a linear increase but conventional heating in roller furnaces, as used in the hot stamping industry, show a nonlinear temperature profile, further tests were carried out on an annealing and curing furnace. The heating rate at the furnace is 5 K/s and corresponds with the industrial standard.

Furnace temperature and dwell time. To investigate the influence of the furnace temperature, samples were heated to 880 °C and 910 °C. Furthermore, the duration in the furnace was varied between three dwell times (50 s, 125 s and 200 s) after reaching the furnace temperature. The specimens were placed on brackets in the furnace for the experimental procedure. This avoids direct contact with the furnace surface and altered heat conduction conditions, and ensures homogeneous heating by convection and thermal radiation through the furnace atmosphere. After the heat treatment, the samples were manually removed from the furnace and cooled down to ambient temperature in air. The surface roughness of the specimens was recorded after the heat treatment using perthometer measurements. The measurement of the surface roughness Rz can be found in Fig. 3. Additional tests with the same settings were carried out for the material 22MnB5 with the AlSi coating AS150 for comparison.

Compared to conventional 22MnB5 with AlSi coating, significantly higher roughness values are present for 20MnB8 with ZF180 coating. While no significant difference in roughness values can be seen for the parameter combinations with the AS150 coating, a significant difference in roughness Rz can be observed when the furnace temperature increases from 880 to 910 °C when using the ZF180 coating. The reason for this can be traced to the added alloyed elements aluminum and silicon which support the reaction with ambient oxygen in the furnace atmosphere. In the austenitizing process, these form a protective oxide skin (Al_2O_3) that prevents oxidation and melting of the zinc coating [15].

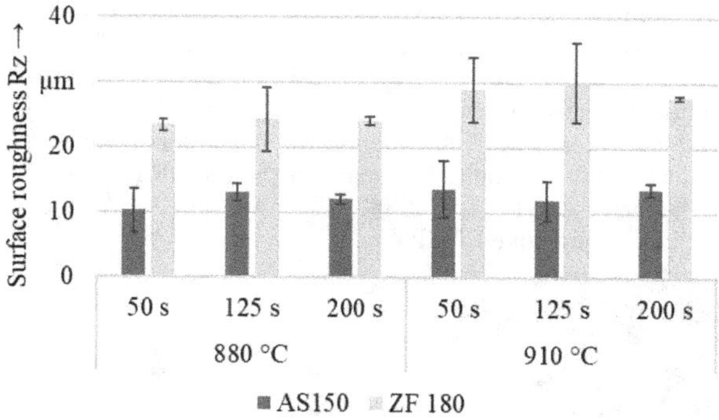

Fig. 3: Influence of the furnace duration and dwell time on surface roughness Rz

To further investigate the influence of the furnace parameters, micrographs of the samples were taken. With the help of EDS measurements, the occurring phase fractions were identified, these are shown for a dwell time of 50 s at a furnace temperature of 880 and 910 °C in Fig. 4. As already shown in Fig. 1, the same composition is visible with a ferritic base material followed by the coating and the embedding material. The coating splits into two phases, this occurs due to iron diffusion from the base material into the coating. This causes the formation of a zinc-rich gamma phase (Phase 1) with small amounts of iron, which occurs due to diffusion during galvanization of the material [5]. Phase 2 consists of an iron-rich zinc ferrite phase with a reduced zinc content compared to Phase 1 [12]. No traces of zinc elements can be found in the base material.

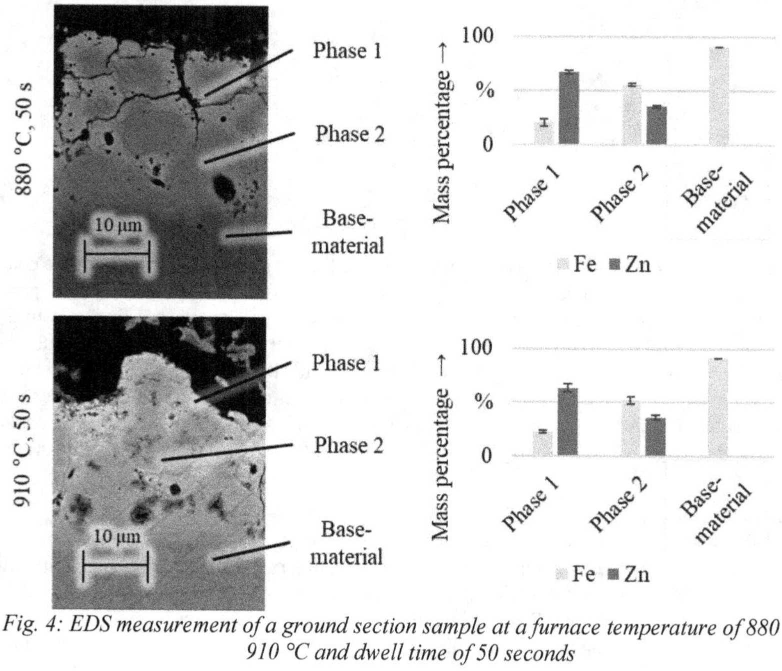

Fig. 4: EDS measurement of a ground section sample at a furnace temperature of 880 and 910 °C and dwell time of 50 seconds

With the further analysis of micrographs, it becomes clear that both an increase in the furnace temperature and the dwell time have a visible influence on the layer formation. An increase in furnace temperature from 880 to 910 °C results in a slight increase in the area fraction of Phase 2 and reduction of Phase 1. The same effect is more pronounced and better visible at a higher dwell time in the furnace for the sample. The composition of the phases however does not change significantly. The micrograph of samples held in the furnace at 880 and 910 °C for 300 seconds is shown in Fig. 5. The increase in Phase 2 at 880 °C is pronounced enough to almost completely displace Phase 1 due to the prolonged dwell time.

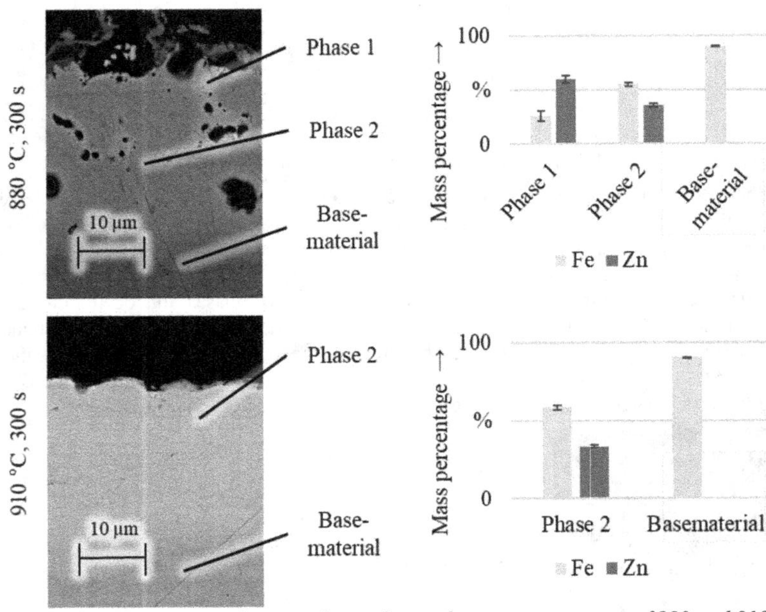

Fig. 5: EDX measurement of a ground sample at a furnace temperature of 880 and 910 °C and dwell time of 300 seconds

The lack of zinc elements in the base material suggests that no diffusion of zinc from the coating into the base material takes place within the parameter steps. Accordingly, the diffusion process occurs only one-way from the base material towards the coating. The process finds a saturation point at 910 °C and 300 seconds, where Phase 1 has completely turned into Phase 2 due to the ongoing diffusion process. The micrographs of samples with these parameter settings show no further zinc-rich gamma phase (Phase 1). These findings are in agreement with the results of Kondratiuk et al. [14] and Knezar et al. [15], who also conducted studies in the same subject area. A change in the height of the coating (Phase 1 and 2) could not be detected for any parameter combination due to the lack of zinc diffusion into the base material.

Summary and outlook

In this work, the layer formation of zinc-coated boron-manganese steel was analyzed under hot stamping conditions. The heating treatment parameters heating rate, furnace temperature and dwell time were varied and investigated with regard to their influence on the surface roughness Rz and layer formation in cross-sections. While a heating rate above 5 K/s does not cause any significant change in the coating components, it can still lead to flaking of the coating due to the formation of a liquid zinc phase and thus cancel the corrosion protection. The furnace temperature has shown a significant effect on the surface roughness, while both furnace temperature and dwell time resulted in an increase of the iron-rich zinc phase and decrease of the zinc-rich gamma phase up to disappearance in the coating formation. The present work forms a basis for further investigations on the influence of heat treatment parameters on friction and wear in hot strip drawing tests for a realistic representation of the industrial hot stamping process. Furthermore, interactions with other relevant parameters with regard to friction and wear behavior should also be analyzed.

Acknowledgment
This study is part of the project P1291 "Workpiece-side Impact on the Tribological Conditions within Hot Stamping" funded by the Research Association for Steel Application (Forschungsvereinigung Stahlanwendung e.V. FOSTA). The authors would like to thank the Research Association for Steel Application for the financial support. Further thanks go to all cooperating industrial partners for their support and cooperation in this project.

References

[1] Hardell J, Kassfeldt E and Prakash B, 2008 Friction and wear behaviour of high strength boron steel at elevated temperatures of up to 800°C Wear 9-10 788–99. https://doi.org/10.1016/j.wear.2006.12.077

[2] Geiger M; Merklein M, 2007 Sheet Metal Forming - A New Kind of Forge for the Future KEM 9–20. https://doi.org/10.4028/www.scientific.net/KEM.344.9

[3] Lechler J, Merklein M and Geiger M, 2006 Beschreibung des mechanischen Werkstoffverhaltens beim Warumumformen höchstfester Vergütungstähle 1. Erlanger Workshop Warmblechumformung 1 13–30

[4] Karbasian H; Tekkaya A E, 2010 A review on hot stamping Journal of Materials Processing Technology 15 2103–18. https://doi.org/10.1016/j.jmatprotec.2010.07.019

[5] Autengruber R, Luckeneder G and Hassel A, W 2012 Corrosion of press-hardened galvanized steel Corrosion Science 12–19. https://doi.org/10.1016/j.corsci.2012.04.048

[6] Venema J, Hazrati J, Matthews DTA, Stegeman R A and van den Boogaard A H, 2018 The effects of temperature on friction and wear mechanisms during direct press hardening of Al-Si coated ultra-high strength steel *Wear* 149–55. https://doi.org/10.1016/j.wear.2018.04.006

[7] Bruschi S; Ghiotti A, 2014 Hot Stamping Comprehensive materials processing ed S. Hashmi(Amsterdam: Elsevier) pp 27–54. https://doi.org/10.1016/B978-0-08-096532-1.00303-4

[8] Jüttner S, Metallschutzgasschweißen von pressgehärteten höchstfesten Stählen mit unterschiedlichen Beschichtungskonzepten : Schlussbericht zu dem IGF-Vorhaben der Forschungsstelle(n) Otto-von-Guericke Universität Magdeburg/Institut für Werkstoff- und Fügetechnik (IWF): Otto von Guericke University Library, Magdeburg, Germany)

[9] thyssenkrupp Issue: June 2020, version 0 MBW® Product information for manganese-boron steels for hot forming, Information on: https://www.thyssenkrupp-steel.com/media/content_1/publikationen/produktinformationen/mbw/thyssenkrupp_mbw_product_information_steel_en_10-2016.pdf [26.01.2022]

[10] Dick P, Scheiker T and Mattes U, 2008 Warmumformung bei Daimler - Ein heiß diskutiertes Verfahren 3. Erlanger Workshop Warmblechumformung 41–55

[11] Olah Neto A, Verran G O, Pissolatto G C and Dutra R R, 2016 Effect of Necking Behavior and of Gap Size on the Cooling Rate in Hot Stamping Mat. Res. 5 1138–43. https://doi.org/10.1590/1980-5373-MR-2015-0654

[12] Kelsch R, Sommer A, Radlmayr K, Schwinghammer H, Faderl J, Kurz T, Luckeneder G, 2017 Hot Forming of Zinc Coated Press Hardening Steel. Characterization of Forming Behaviour and New Process Routes for Mass Production. *HOT SHEET METAL FORMING of HIGH-PERFORMANCE STEEL CHS²* ed M. Oldenburg, et al.) pp 337–344

[13] Wang K, Gui Z, Liu P, Wang Y and Zhang Y, 2014 Cracking Behavior of Al-Si Coating on Hot Stamping Boron Steel Sheet Procedia Engineering 1713–18. https://doi.org/10.1016/j.proeng.2014.10.218

[14] Kondratiuk J, Kuhn P, Labrenz E and Bischoff C, 2011 Zinc coatings for hot sheet metal forming: Comparison of phase evolution and microstructure during heat treatment Surface and Coatings Technology 17-18 4141–53. https://doi.org/10.1016/j.surfcoat.2011.03.002

[15] K. Knezar, T. Manzenreiter, J. Faderl, K. M. Radlmayr K, 2007 Formhärten von feuerverzinktem 22MnB5: ein stabiler und reproduzierbarer Prozess 2. Erlanger Workshop Warmblechumformung ed M. Geiger, et al.(Bamberg: Meisenbach) pp 131–148

[16] He F; Merklein M, 2021 Investigation of the impact of heat treatment on the layer formation of AlSi-coated boron-manganese steel IOP Conf. Ser.: Mater. Sci. Eng. 1 12009. https://doi.org/10.1088/1757-899X/1157/1/012009

[17] Radlmayr K.; Kelsch R.; Sommer A.; Rouet C.; Kurz T.; Faderl J, *THE HOT FORMING OF GALVANIZED STEELS,* Information on: https://www.voestalpine.com/division_stahl/content/download/51623/646391/file/Whitepaper-The-hot-forming-of%20galvanized-steels-voestalpine.pdf [10.10.2022]

[18] Gleeble® 3500-GTC System, Information on: https://www.gleeble.com/products/gleeble-systems/gleeble-3500.html [06.10.2021]

[19] Helmut RHODE GmbH: Technical Data Chamber Furnace ME 87/13, Information on: https://www.rohde.eu/en/industry/products/chamber-furnaces/annealing-and-hardening-furnaces/me-87/13?c=146https://www.rohde.eu/de/Pixelpdfdata/Articlepdf/id/81/onumber/IND-ME-87-13 [10.10.2022]

Sheet Metal 2023 Materials Research Forum LLC
Materials Research Proceedings **25** (2023) 439-446 https://doi.org/10.21741/9781644902417-54

Warm forming of thermoplastic fibre metal laminates

Mohamed Harhash[1,2,a*], Wei Hua[1,b], Gerhard Ziegmann[2,3,c] and Heinz Palkowski[1,d]

[1]TU Clausthal, Institute of Metallurgy, 38678 Clausthal-Zellerfeld, Germany

[2]TU Clausthal, Clausthal Centre for Material Technology, 38678 Clausthal-Zellerfeld, Germany

[3]TU Clausthal, Institute of Polymer Materials and Plastics Engineering, 38678 Clausthal-Zellerfeld, Germany

[a]mohamed.harhash@tu-clausthal.de, [b]wei.hua@tu-clausthal.de, [c]ziegmann@puk.tu-clausthal.de, [d]heinz.palkowski@tu-clausthal.de

Keywords: Forming, Deep Drawing, Thermoplastic Fibre Metal laminates

Abstract. In this paper, the forming behaviour of sheet-like metal/polymer/metal (MPM) and thermoplastic fibre metal laminates (TFMLs) is introduced. TFMLs are based on thin metallic cover sheets and fibre-reinforced thermoplastic matrix (here polyamide 6). With this material combination, the specific mechanical, structural, thermal and acoustic properties can be improved and designed compared to the monomaterials and laminates without fibre reinforcements. However, the restricted formability of TFMLs at room temperature is a strong limitation. Therefore, the approach of this study is concerned with the fundamental description of the influence of warm forming on the degree of the forming improvement of pre-bonded TFMLs sheets experimentally via the investigation of the deep drawability and determining the forming limit curves compared to the formability of steel and MPM sheets. Two elevated test temperatures (200 and 235 °C) besides the room temperature are considered. The results of this approach revealed that warm forming could lead to over 300 % improvement of the forming limit curve (FLC) level of TFML; however, an ignorable difference between 200 and 235 °C is found. For steel and MPM sheets, increasing the temperature showed a slight improvement. Regarding deep drawing, similar results like for FLC were found, where the drawing depth of TFML could be increased from approx. 15 mm up to at least 40 mm before cracking. However, other failure types arose like wrinkling and core squeezing-out. Therefore, a one-step deep drawing approach for TFMLs is foreseen, where the adhesion and forming processes take place simultaneously.

Introduction

Thermoplastic fibre-metal-laminates (TFMLs) are composed of metallic cover sheets (or even interlayers) and layer(s) of fibre reinforced thermoplastic polymers [1]. With such a set-up several advantages can be gained, e.g. improved mechanical properties at lower density [2]. This materials group is developed to overcome the formability drawbacks of the thermoset-based FML such as glass-reinforced aluminium laminate (GLARE) or aramid-reinforced aluminium laminates (ARALL) [3]. Moreover, the potential TFMLs application is increasing due to their recyclability in relation to the thermoset-based FMLs. TFMLs were considered in some projects, e.g. the LEIKA project, where several steel/fibre reinforced polyamide (PA) components can be applied as crash absorbers or as battery carriers in e-vehicles [4]. Besides, lab-scale TFMLs like carbon fibre-reinforced polyamide/aluminium laminate (CAPAAL) and carbon-fibre-reinforced polyetheretherketone/titanium foil laminate (CAPET) showed improved behaviour under bending conditions [5]. Based on this, TFMLs proved their suitability for automated series production in semi-finished products or even directly to finished components starting from single layers. For lightweight applications, metal/polymer/metal (MPM) sandwich laminates, with nonreinforced

Content from this work may be used under the terms of the Creative Commons Attribution 3.0 license. Any further distribution of this work must maintain attribution to the author(s) and the title of the work, journal citation and DOI. Published under license by Materials Research Forum LLC.

polymer core, were also developed such as Litecor [6] and Hylite [7].The deep-drawability of MPM materials is reduced directly with increasing the core layer thickness due to the tensile stresses exerted by the soft polymer core on the outer metal cover near the edge of the punch and the corresponding increased void volume fraction within the metal [8]. Regarding TFMLs, the mobility of the brittle fibre-reinforcement in the core layer is restricted together with the limited plastic deformation of the core layer can lead directly to a sharp increase in tensile stress on the outer steel sheet and premature cracking. Therefore, it is necessary to study the deep-drawability of TFMLs under controlled thermal conditions for identifying and defining the processing conditions of TFMLs with a polymer matrix close to or even above its melting point. In a previous study, the deep drawing of 190 mm diameter TFML (steel/GF-PA6)/steel) blanks utilizing a 100 mm diameter punch was studied, where cracking-free samples could only be obtained utilizing high holding forces of 200-300 kN (i.e. 10-15 MPa holding pressure) for drawing depths of less than 17 mm, where severe wrinkling and thickness irregularities took place [9]. In another study on Al/GF-PA6/Al laminates, forming at a temperature of 270 °C (significant higher than the melting point of PA6, i.e. 220 °C) leading to a significant polymer squeeze-out during deep drawing and thermal degradation of the PA6 matrix [10]. On this basis, there is still a need to understand and identify the forming behaviour of such TFMLs aiming to maximise their limits at reduced processing conditions (temperatures and holding pressures). This can be a further source of energy saving during processing TFMLs.

The current paper is concerned with the warm forming of MPMs and TFMLs. For this purpose, the forming behaviour is characterized via standard methods (tensile testing, deep drawing and FLC determination) at room temperature and two elevated temperatures, which are selected in relation to the melting temperature (T_m) of the polyamide matrix, which is approx. 220 °C. In this regard, the warm forming behaviour of MPMs and TFMLS was characterized at 200 °C (below T_m) and 235 °C (slightly higher than T_m) in order to define the forming temperature at which less defects can be reached. The investigations of this paper focus on the different testing methods on one combination of MPM and TFML, however other combinations are studied previously [11].

Materials and experimental work

The used TFMLs are three-layered ones containing steel cover sheets together with non-reinforced or fibre-reinforced PA6 core layers. The steel sheet grade TS290 is electrolytically-galvanized and has a thickness of 0.3 mm. The fibre-reinforced core used is a roving glass fibre-reinforced PA6 consolidated composite organosheets (Tepex® dynalite 102-RG600(2)/47%, LANXESS Germany GmbH) with a fibre content of 47 vol-% in thicknesses of 0.5 and 1.0 mm, where its weaving style is twill 2/2 [12]. With this, two laminates are investigated: a) metal/PA 6/metal with a core thickness of 1.0 mm, abbreviated as MPM-1.0 and b) metal/Tepex®/metal with the same core thickness of 1.0 mm, abbreviated as TFML-1.0. The steel/core adhesion for the semi-finished laminates is achieved via hot-pressing at 245 °C, 0.3 N/mm² and 5 min holding time. Prior to hot-pressing, the materials were prepared following the scenario described in [13], where the steel sheets are grinded, tempered (at 440 °C for 1 min), cleaned with acetone, applying a co-polyamide-based adhesive agent, and finally activated at 250 °C for 3 min. With this, an adhesion strength in terms of the single lap shear strength for MPM and TFML of about 23 and 15 MPa could be obtained, respectively [13].

To achieve good warm forming conditions, the tools being in contact with the blank were designed. A schematic representation of the FLC determination and deep drawing tools is shown in Fig. 1. The forming die and punch parts are warmed-up before the test in a separate furnace, however the blank holder is in-situ heated via implemented heating cartridges. Insulating ceramic discs are utilized to minimize the heat loss and protect the forming machine from overheating. The testing temperatures (200 and 235 °C) could be successfully controlled with a tolerance of ±5 K [14]. The testing conditions are summarized in Table 1.

Materials Research Proceedings **25** (2023) 439-446
https://doi.org/10.21741/9781644902417-54

Lubrication is a dominant factor affecting the forming limits. For FLC determination, a combination of Teflon and additional thermoplastic films with inlayers of grease were utilized at RT with a total thickness of 2 mm, however only Teflon foils were used at the elevated forming temperatures with a total thickness of 2 mm. For deep drawing, a 0.05 mm thermoplastic film was used at RT, however a combination of Teflon (0.05 mm) and Molycote® spray are utilized for the elevated temperature testing condition. For determining the limiting strains, a digital image correlation system was used, where heat-resistant sprays for the stochastic pattern were applied for a successful strain evaluation.

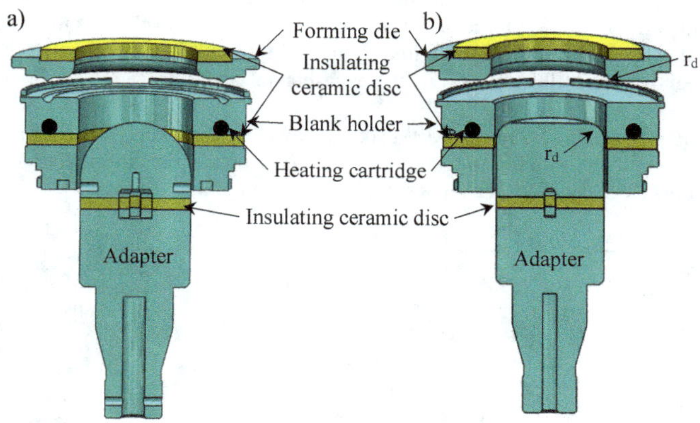

Fig. 1: Schematic representation of: a) the 100 mm$^{\varnothing}$ semi-spherical punch for FLC determination, and b) the 100 mm$^{\varnothing}$ flat cylindrical deep drawing punch.

The FLC curves were determined based on the line-fit method, where the thinning rate is considered for determining the localisation and therewith the major and minor strains for the FLC can be obtained [15]. Four test samples geometries with different widths are utilized for different stress states, namely 200, 150, 100, and 50 mm for equi-biaxial, in-between, plane strain and uniaxial strain conditions, respectively.

Table 1: Test conditions for the FLC and deep drawing experiments.

Test	D_0 [mm]	d_0 [mm]	r_p [mm]	r_d [mm]	v [mm/s]	F_h [kN]	T [°C]
FLC	200	100	-	10	1.5	100	RT, 200, 235
Deep drawing	180	100	15	15	0.5	15	RT, 200, 235

D_0: blank diameter, d_0: punch diameter, r_p: punch corner radius, r_d: die radius, v: test speed, F_h: holding force, T: test temperature

Results and discussion

Firstly, the influence of the temperature on the tensile properties of the steel sheet TS290 is determined and the corresponding results are summarized in Table 2. It can be clearly observed that the strength-related properties (yield strength (YS), ultimate tensile strength (UTS) and elastic modulus (E)) are reduced with increasing the testing temperature up to 235 °C with a slight increase in the strain at failure (A_{80}). This result is expected to have an impact on reducing the forming forces at the elevated test temperatures and improving the fracture displacement, too.

Table 2: Temperature-dependent mechanical properties of the TS290 steel sheet.

Steel grade	T [°C]	E [GPa]	UTS [MPa]	YS [MPa]	A_{80} [%]
TS290	RT	178 ± 2	410 ± 1	283 ± 2	19.5 ± 0.5
	235	168 ± 5	379 ± 1	257 ± 1	20.8 ± 0.9

Forming limits curves (FLC): FLCs for the monolithic steel sheet and the different laminates (MPM and TFML) at different temperatures are presented in Fig. 2. To indicate the forming forces and displacements at failure, Fig. 2a) depicts the force-displacement curves for the samples for the equi-biaxial strain path, i.e. the 200 mm diameter samples. Similar to the behaviour of the tensile properties, the maximum forming force of the steel is reduced from 30 kN down to 24 kN with an increase of the displacement from 33 to 35 mm. As a result, the FLC level of the steel at 235 °C (Fig. 2b)) is slightly higher than at RT in the equi-biaxial straining range (right-side of the diagram).

The same tendency was found for the MPM sheets; however, the level of the forming forces is unsurprisingly higher than those for only one steel sheet. The punch displacement of MPM is less than the ones of the steel sheets indicating a possible reduction of the FLC level.

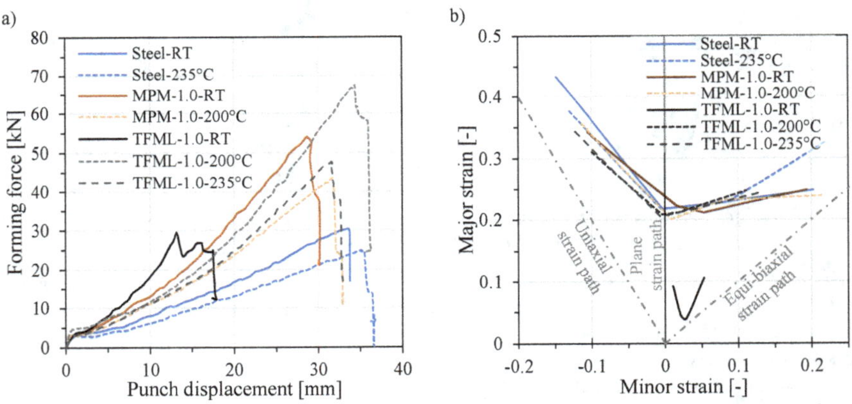

Fig. 2: a) Force-displacement curves for the samples with the 200 mm diameter (i.e. equi-biaxial strain) and b) the obtained FLCs for different materials and testing temperatures.

This is in accordance with the literature, where MPMs have slightly lower forming limits in relation to their monolithic metallic cover sheets [16], where the non-reinforced polymer core can be plastically-uniformly matching the formability of the cover steel sheets. Only for thicker cores (at core volume fraction > 60 %), the FLC level of such thick-cored MPM can be reduced in the right-side of the FLC, as found recently for MPMs with polypropylene core [17], too. Like the steel sheets, there is no significant difference between the FLC of MPM at RT and 200 °C (Fig. 2b)).

Regarding the FLC levels for the TFMLs, as can be seen from Fig. 2b), the FLC level at RT is extremely low and the main failure strains are located in the plane strain regions. This can be attributed to the restricted and localised formability of the organosheets (2/2 twill fabric) at RT and its influence on the plastic deformation of the outer steel sheet, instead of the uniform straining for the steel and MPM sheets, leading to earlier failure; this is in accordance with the results stated in [18]. This can be clearly observed through the displacement value in Fig. 2a), where only a depth of 13 mm before cracking can be reached compared to approx. 35 mm at the elevated forming temperature. The FLC level of TFMLs increases remarkably with increasing the

Materials Research Forum LLC
https://doi.org/10.21741/9781644902417-54

temperature to 200 °C or 235 °C, reaching the level of the MPMs or even the monolithic steel sheet, which is the maximum FLC level achievable. The difference between the FLC levels determined at 200 and 235 °C is negligible. Therefore, forming at 200 °C is sufficient extending the forming limits of TFML together with less squeezing-out of the polymer matrix, which took place in the melted matrix at 235 °C. It can be stated that at RT the forming/draping limit of the organosheet dominates the FLC level of its TFMLs; however, at the elevated temperatures, the cover sheet dominates the FLC level.

Deep drawing results: the deep drawing results in terms of force-displacement curves and the forming strains are depicted in Fig. 3. A criterion was defined indicating a satisfactory forming which is a forming depth of 40 mm. This limit was reached by the steel sheets at 235 °C as well as at RT (see Fig. 3a). This criterion was fulfilled by the MPM sheet at RT, too. This is in accordance with the FLC results, where MPM with the non-reinforced core show a FLC level next to that one of the steel sheet. Specific deep-drawn cups are presented in Fig. 4.

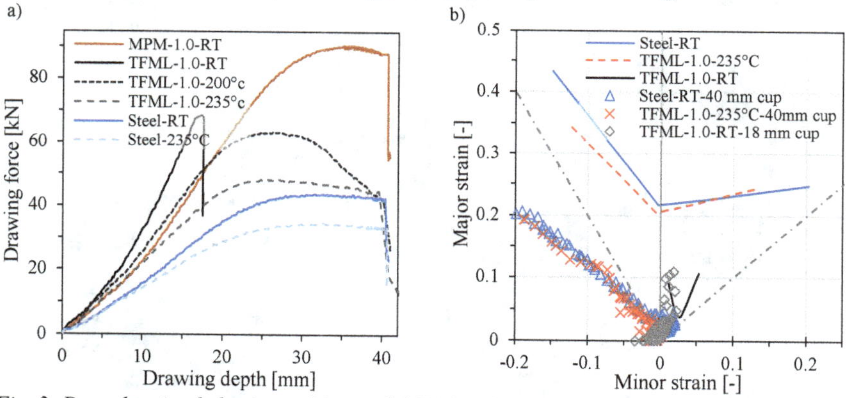

Fig. 3: Deep drawing behaviour of the steel, MPM and TFML sheets at different temperatures in terms of: a) deep drawing force-displacement and b) forming strains compared to the FLC.

With deep drawing of TFML at temperatures above the melting point of PA6, i.e. at 235 °C as shown in Fig. 4e), severe wrinkling on the inner cover sheet and waviness on the outer side took place. The reason is that the molten polymer matrix does not provide the stability of the cover against the circumferential compressive stresses of the blank in the flange and cup wall regions. A geometrical factor could play a role in wrinkling which is the larger gap size (3.3 mm) and the die radius (r_d = 15 mm), where a large gap to sheet thickness ratio often leads to waviness during drawing of the sheet [19]. As shown in Fig. 3a), the drawing force of TFML is reduced at 200 °C and 235 °C compared to RT, which is due to the softening of the core material and, consequently, the increase in fibre mobility/drapability in the PA6 matrix. The maximum drawing depth of the TFML at RT was about 15 mm. It should be mentioned that the drawing depth of the monolithic organosheet (Tepex) core is the same like in the TFMLs at room temperature [14]. This limit was increased to > 40 mm drawing depth at 235 °C, which could be attributed to fibre fracture in the flange area [14]. This highlights the dominancy of the organosheet core determining the forming limit.

A comparison of the strain distribution on the drawn samples with the previously obtained FLCs is shown in Fig. 3b) in order to examine the crack probability as well as to examine the validity of the FLC itself. The minor-major strain points of the failed TFML at RT are overlapped in the diagram. The results show that these strain values are beyond the FLC level indicating cracking and confirming the applicability of the obtained FLC. For the MPM and TFML at elevated temperature, no failure by cracking is expected, where most of the strain values are below the uniaxial tensile strains in the direction of the compressive and shear regions confirming the wrinkles formed.

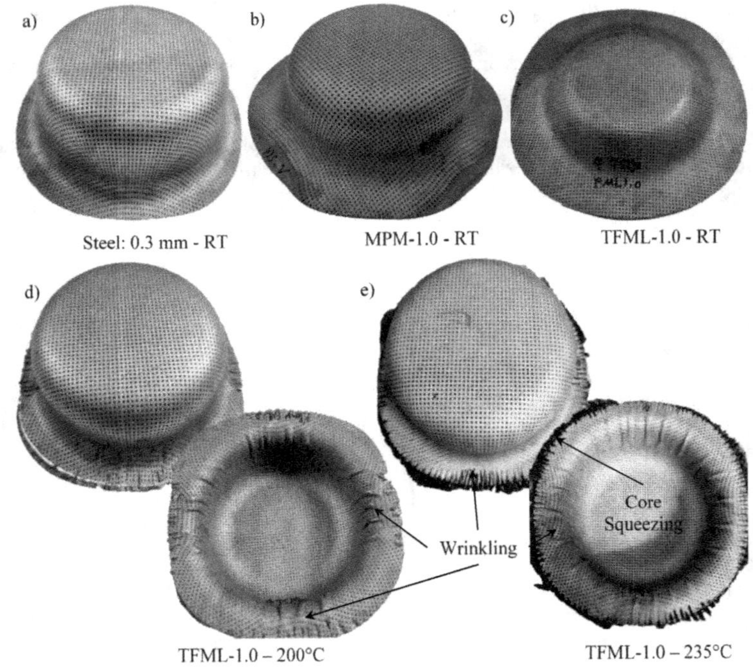

Fig. 4: Drawn cups at RT of: a) the steel sheet, b) MPM-1.0 and c) TFML-1.0, in addition to the TFML drawn cups at: d) 200 °C and e) 235 °C showing wrinkling and squeezing-out.

In order to examine the critical drawing depth, at which wrinkling starts forming, two-step drawing operation was carried out up: 1st drawing: up to 25 mm, and 2nd step: up to 40 mm [14]; it was found that no wrinkling was observed at the 25-mm step at the elevated temperatures.

Summary and conclusions

The influence of the forming temperature on the forming limits of monolithic steel, MPM and TFML sheets was determined in terms of forming limit curves and deep drawing. The following conclusions can be drawn:

- The FLC of the MPMs is comparable to the one of the monolithic steel cover sheets, regardless the testing temperature.
- The FLC of TFML is remarkably improved compared to forming at RT at elevated temperatures near the melting point of the polymer matrix. Forming close below and above the melting point (200 °C and 235 °C) doesn't show significant differences. Therefore, forming at 200°C can be favoured, where less defects like core squeezing-out and thickness irregularities are expected. The organosheets dominate the FLC-level at RT and the steel cover sheets are dominant at the elevated temperatures.

- Under deep drawing conditions, TFML at RT is subjected to localized tensile stresses exerted by the fibre-reinforced core on the outer cover of TFML, leading to early failure. The strain distribution of the drawn parts at RT are dominated by the fibre-reinforced core. At the elevated temperatures of TFML and MPM, the cover sheets dominate the failure conditions. Drawing at 235 °C leads to unavoidable thickness irregularities. Therefore, deep drawing of pre-bonded TFML is favoured at 200 °C. Wrinkling is more susceptible by forming above the melting temperature of the polyamide matrix (i.e. at 235 °C).

Acknowledgments
The authors would like to thank the German Research Foundation (Deutsche Forschungsgemeinschaft, DFG) for the financial support of the project (no 330043166). Moreover, the authors are thankful to thyssenkrupp Steel Europe GmbH for supplying the steel materials and SI Coatings GmbH for the adhesion promoter.

References

[1] F. Schubert, N.K Minar, M. Sause, A. Monden, Thermoplastic Fiber Metal Laminates for Automated Production, Lightweight des worldw. 12 (2019) 12–7. https://doi.org/10.1007/s41777-019-0031-6

[2] T. Sinmazçelik, E. Avcu, M.Ö Bora, O. Çoban, A review: Fibre metal laminates, background, bonding types and applied test methods, Mater. Des. 32 (2011) 3671–85. https://doi.org/10.1016/j.matdes.2011.03.011

[3] A. Vlot, Impact loading on fibre metal laminates, Int. J. Impact Eng. 18 (1996) 291–307. https://doi.org/10.1016/0734-743X(96)89050-6

[4] Gude M, Lieberwith H, Meschut G, Tekkaya AE, Zäh M (eds.), FOREL-Studie 2018: Ressourceneffizienter Leichtbau für die Mobilität: Wandel - Prognose - Transfer (German). Plattform FOREL; 2018.

[5] D. Nestler, H. Jung, S. Arnold, B. Wielage, S. Nendel, L. Kroll, Thermoplastische Hybridlaminate mit variabler Metallkomponente, Mat.-wiss. u. Werkstofftech. 45 (2014) 531–6. https://doi.org/10.1002/mawe.201400259

[6] L. Keßler, R. Dams, A. Marx, M. Theiss, Erweiterung der Simulation für die virtuelle Fertigungskette von Sandwich-Werkstoffen am Beispiel LITECOR®. In: EFB, editor. 35. EFB-Kolloquium - Tagungsband, 1st ed. Hannover: EFB; 2015, p. 33–42.

[7] W. Hufenbach, J. Jaschniski, T. Weber, D. Weck, Numerical and experimental investigations on HYLITE sandwich sheets as an alternative sheet metal, Arch. Civ. Mech. Eng. 8 (2008) 67–80. https://doi.org/10.1016/S1644-9665(12)60194-0

[8] M.H Parsa, M. Ettehad, P.H Matin, Al Ahkami, S. Nasher, Experimental and Numerical Determination of Limiting Drawing Ratio of Al3105-Polypropylene-Al3105 Sandwich Sheets, J. Eng. Mater. Technol. 132 (2010) 31004. https://doi.org/10.1115/1.4001264

[9] A.E. Tekkaya, M. Hahn, L. Hiegemann, C. Weddeling, N. Ben Khalifa, Umformen faserverstärkter thermoplastischer Kunststoff-Halbzeuge mit metallischen Deckblechen für den Leichtbau. In: EFB, editor. 35. EFB-Kolloquium - Tagungsband, 1st ed. Hannover: EFB; 2015, p. 185–199.

[10] B.-A. Behrens, K. Dilger, K. Lippky, L. Kaempf, S. Hübner, S. Hartwig et al., Combined deep drawing and fusion bonding of structural FRP-metal hybride parts, Procedia Manuf. 29 (2019) 296–304. https://doi.org/10.1016/j.promfg.2019.02.141

[11] M. Harhash, T. Fischer, M. Grubenmann, W. Hua, J. Heingärtner, M. Kuhtz et al., Top-hat crashboxes of thermoplastic fibre-metal-laminates processed in one-step thermoforming: Experimental and numerical study, Compos. B Eng. 226 (2021) 109367. https://doi.org/10.1016/j.compositesb.2021.109367

[12] LANXESS Germany GmbH, Tepex® Material Data Sheet - Tepex® dynalite 102-RG600(x)/47% - Roving Glass - PA6 Consolidated Composite Laminate. [September 15, 2021]; Available from: https://lanxess.com/-/media/Project/Lanxess/Corporate-Internet/Products-and-Solutions/Brands/Tepex/Data-sheets/MDS102RG600x47.pdf.

[13] W. Hua, T. Fischer, M. Harhash, G. Ziegmann, H. Palkowski, Study on processing galvanized steel/PA6 sandwich composites by hot-pressing: Influence of interface properties and surface treatment on the adhesion strength, Compos. Struct. 236 (2020) 111779. https://doi.org/10.1016/j.compstruct.2019.111779

[14] W. Hua, Forming behaviour of sandwich materials made of steel covers and polyamide cores with or without glass fibre reinforcements. Clausthal-Zellerfeld, Germany: TU Clausthal; 2022.

[15] M. Merklein, A. Kuppert, M. Geiger, Time dependent determination of forming limit diagrams, CIRP Annals - Manufacturing Technology. 59 (2010) 295–8. https://doi.org/10.1016/j.cirp.2010.03.001

[16] M. Harhash, Forming behaviour of multilayer metal/polymer/metal systems [PhD Thesis]. Clausthal-Zellerfeld, Germany: TU Clausthal; 2017.

[17] C.K Kella, Formability and Springback Analysis of Aluminum/Polypropylene/Aluminum Sandwich Laminates: My University; 2022. https://doi.org/10.3390/jmmp6060152

[18] T. Mennecart, S. Gies, N. Ben Khalifa, A. Tekkaya, Analysis of the Influence of Fibers on the Formability of Metal Blanks in Manufacturing Processes for Fiber Metal Laminates, J. Manuf. Mater. Process. 3 (2019) 2. https://doi.org/10.3390/jmmp3010002

[19] C.P. Singh, G. Agnihotri, Study of Deep Drawing Process Parameters: A Review, Int J Sci Res Publ. 5 (2015) 1–15.

Sheet Metal 2023
Materials Research Proceedings 25 (2023) 447-454
Materials Research Forum LLC
https://doi.org/10.21741/9781644902417-55

Numerical-experimental study of a tailored press-hardening technology with intermediate pre-cooling to manufacture an automotive component in advanced high strength steel

Maria Emanuela Palmieri[1, a *] and Luigi Tricarico[1,b]

[1] Department of Mechanics, Mathematics and Management, Polytechnic University of Bari, Via Orabona 4, 70125, Bari, Italy

[a]mariaemanuela.palmieri@poliba.it, [b]luigi.tricarico@poliba.it

* corresponding author: Palmieri Maria Emanuela

Keywords: Hot Stamping, Heat Treatment, High Strength Steel

Abstract. The development of new vehicles generations lighter and safer is still a challenge. Therefore, several companies are looking for innovative and efficient solutions able to combine hard and soft zones in the same structural part. In this work, attention is paid on the press hardening process with the integration of an intermediate pre-cooling, which involves a secondary furnace where the temperature for hard zones is maintained at the complete austenitization one and soft zones are cooled. A numerical-experimental analysis was carried-out to understand how process parameters, i.e., the pre-cooling temperature and the time taken for the pre-cooling phase, influenced the mechanical properties of the component. As a case study, an automotive B-Pillar in USIBOR®1500 was chosen. For this study, a Finite Element (FE) model was developed for the numerical simulation of the press-hardening technology, varying process parameters. The FE thermal cycles were experimentally reproduced on samples using a Gleeble®3180 thermo-mechanical physical simulator. After physical simulation, tensile tests, hardness tests and metallographic analysis were carried out on treated samples, with the aim of measuring their mechanical properties.

Introduction

In automotive field, some components require higher stiffness, strength, or energy absorbing capacity in a specific area [1]. In press hardened components, to achieve tailored properties (soft and hard zones on a single component) several technologies have been developed. In the following, examples of tailored process technologies, which exploit heat treatment to produce different microstructures and mechanical properties, are listed. First, the Tailored Blank Heating (THB) technology allows to obtain soft zones by controlling the blank temperature in the furnace before the process. Specifically, these soft areas are not fully austenitized [2]. Second, soft zones can be realized after the press hardening process by partially annealing the fully hardened part by means of laser or induction [3, 4]. Third, to obtain tailored properties while the press hardening process is ongoing, Tailored Tool Tempering (TTT) or Intermediate Pre-Cooling (IPC) techniques should be adopted. In the TTT approach, the fully austenitized blank is placed on the tool, but the part is cooled at different cooling rates through heating cartridges and cooling channels in tools in correspondence of the hard and soft zone, respectively [5, 6]. IPC approach, instead, is a relatively new tailored technology which employ a secondary furnace (tempering station) where the temperature of hard zones is maintained equal to that set in the first furnace (complete austenitization temperature), but soft zones are cooled [7]. This technology, compared to the conventional ones, is well suited to realize freely distribute soft zones on the component, e.g., soft flanges around the part and soft spots in the middle of the part. To this end, several intermediate cooling systems have been developed by different furnace makers. As an example, AP&T patented

Content from this work may be used under the terms of the Creative Commons Attribution 3.0 license. Any further distribution of this work must maintain attribution to the author(s) and the title of the work, journal citation and DOI. Published under license by Materials Research Forum LLC.

the application of multi-layer furnaces combined with the TemperBox® one [8]. The blanks are austenitized in the multi-layer furnace. Before the press hardening process, the blanks are moved into another furnace (the TemperBox®) where re-heating is done with masking. Masked areas are cooled below approximately 700 °C, meanwhile the unmasked areas are re-heated to 930 °C (complete austenitization temperature) [9]. In this work, the IPC technology was investigated to manufacture an automotive B-Pillar in USIBOR®1500 steel (22MnB5) which require two soft zones and one hard. Specifically, the influence of pre-cooling temperature in the second furnace ($T_{precooling}$) and the time taken during this cooling on the mechanical properties of the component ($t_{precooling}$) were evaluated.

Methodology

The effect of the tempering station process parameters, i.e. $T_{precooling}$ and $t_{precooling}$, on mechanical properties of the component was investigated. The scheme of the process is reported in Fig. 1.

For this study the following numerical-experimental methodology was adopted:

(i) First, a Finite Element (FE) model for the numerical simulation of the press hardening process with IPC technology was developed in AutoForm R10. As a case study, an automotive B-Pillar was considered (Fig. 2). For this component, one hard zone and two soft zones are required, as highlighted in Fig. 2. The model has been implemented by considering a blank in Al-Si coated 22MnB5 steel of 1.3 mm thick with a ferritic-pearlitic initial microstructure. Twelve FE simulations were run by varying $T_{precooling}$ between 600 °C - 700 °C and $t_{precooling}$ between 30 s - 210 s. The process parameters set in the FE model were reported in Table 1. FE results in terms of the hardness on the final component were collected and then elaborated with the kriging technique [10] to obtain response surfaces of the hardness as a function of the pre-cooling temperature and time taken during the pre-cooling phase. Finally, FE thermal cycles in the hard zone and in one of the two soft zones (Soft region 2) were derived.

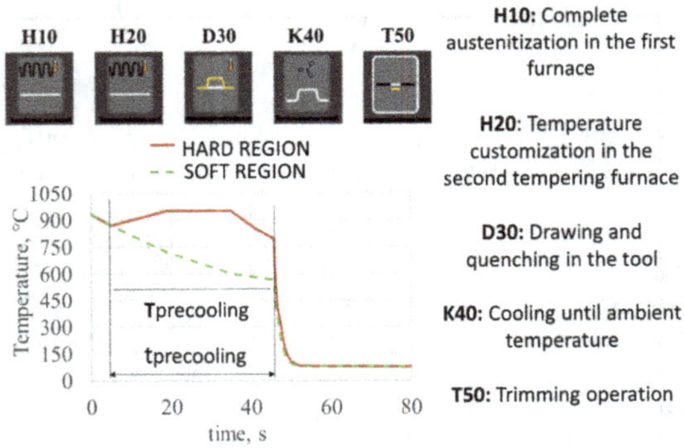

Fig. 1: Scheme of the process investigated

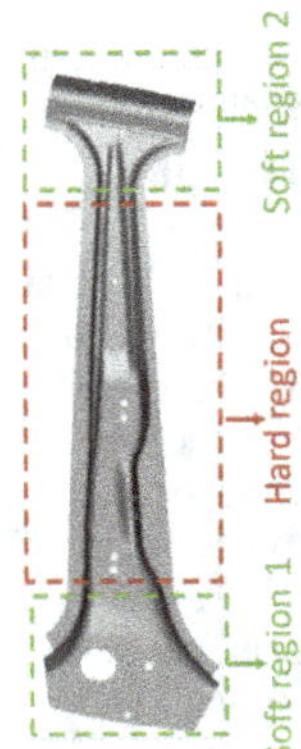

Fig. 2: B-Pillar investigated in this study

(ii) Second, by setting the pre-cooling temperature at 600 °C, the thermal cycles, at different pre-cooling times, acquired for the soft and hard regions were physically simulated by means of the Gleeble®3180 system. For such tests, the Usibor®1500 specimens 1.3 mm thick, shown in Fig. 3 were adopted. During physical simulation tests, the specimens were instrumented with three thermocouples: one in the center of the specimen (control thermocouple) and the other two at a distance from the center of 5 mm and 15 mm, respectively. The four thermal cycles of the soft region, one for each pre-cooling time, were

Materials Research Forum LLC
https://doi.org/10.21741/9781644902417-55

physically simulated on eight specimens. Four of these eight specimens were subjected to the metallographic and hardness analyses, the other four were adopted for the tensile tests. These last specimens, however, were previously machined to obtain the specimen in Fig. 4. The geometry of the specimen shown in Fig. 4 guarantees the location of the break in the central point (point in which control thermocouple was welded). The same procedure was adopted for the termo-mechanical cycles of the hard region.

Table 1. Process parameters in FE model

Process parameter	**Value**
Temperature of tools (T_{tool}) [°C]	80
Initial temperature of the blank (T_{blank}) [°C]	950
Pre-cooling Temperature ($T_{precooling}$) [°C]	600 °C – 650 °C – 700
Pre-cooling time ($t_{precooling}$) [s]	30 – 90 – 150 – 210
Transport time from tempering furnace to press ($t_{transport2}$) [s]	5
Transport time from austenitization furnace to tempering furnace ($t_{transport1}$) [s]	5
Friction coefficient	0.4
Forming velocity (v) [mm/s]	200
HTC to Tool [mW/(mm^{2}K)]	3.5
Quenching Force (F_{quench}) [kN]	1.2E+04

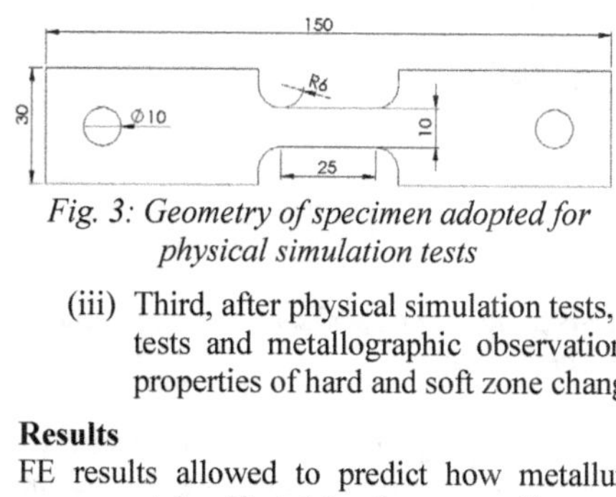

Fig. 3: Geometry of specimen adopted for physical simulation tests

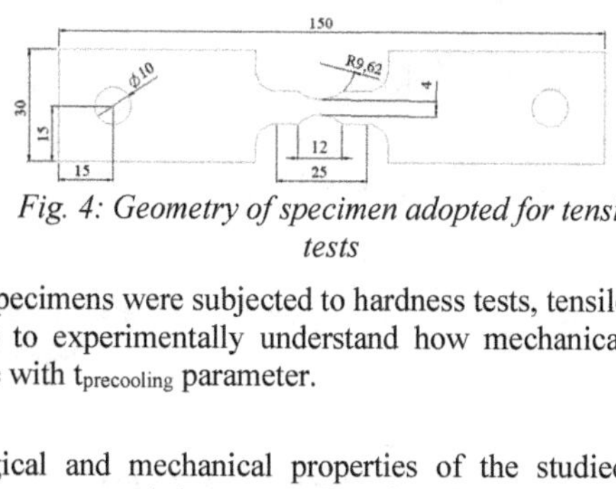

Fig. 4: Geometry of specimen adopted for tensile tests

(iii) Third, after physical simulation tests, specimens were subjected to hardness tests, tensile tests and metallographic observations to experimentally understand how mechanical properties of hard and soft zone change with $t_{precooling}$ parameter.

Results

FE results allowed to predict how metallurgical and mechanical properties of the studied component is affected by the pre-cooling temperature and the pre-cooling time in the second tempering furnace, by means of AutoForm-PhaseChange Plug-In and AutoForm-Thermo Plug-In. For example, for $T_{preccoling}$ of 600 °C, FE simulations predict, in the soft region (Soft region 2), a microstructural composition equal to 56.45 % of martensite, 3.9 % of ferrite and 39.65 % of bainite in correspondence of $t_{precooling}$ equal to 30 s. On the other hand, for $t_{precooling}$ equal to 210 s, the estimated microstructural composition is equal to 57.27 % of ferrite and 42.73 % of pearlite. In the hard region, instead, for each value of the pre-cooling time and pre-cooling temperature, a completely martensitic microstructure was observed with a mean value of hardness equal to 490 HV10. Hardness results obtained in the soft region 2 with the FE simulation were collected and processed using the kriging technique to obtain the metamodel shown in Fig. 5. The increase in the pre-cooling time in the second tempering furnace leads to a reduction in hardness. In the explored range, the pre-cooling temperature has smaller effects on hardness. For this reason, $T_{preccoling}$ was set equal to 600 °C for the subsequent experimental analysis. Specifically, for $T_{preccoling}$ equal to 600 °C, thermal cycles of the process for each investigated pre-cooling time were

Materials Research Proceedings **25** (2023) 447-454
https://doi.org/10.21741/9781644902417-55

derived from FE model. These thermal cycles are shown in Fig. 6a and Fig. 6b for the hard region and soft region 2, respectively.

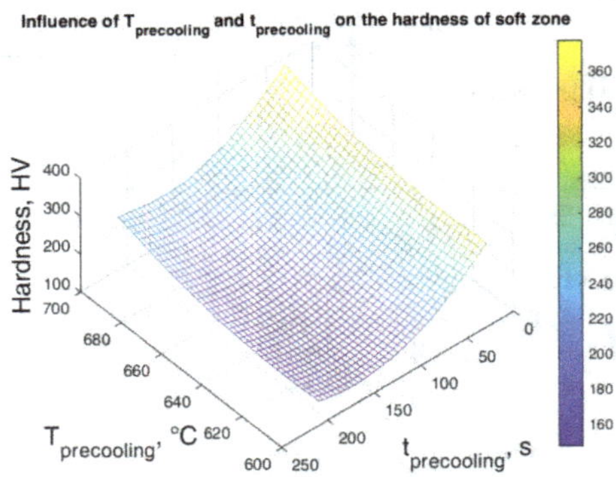

Fig. 5: Metamodel of the hardness in the soft region 2 as a function of $T_{preccoling}$ and $t_{preccoling}$

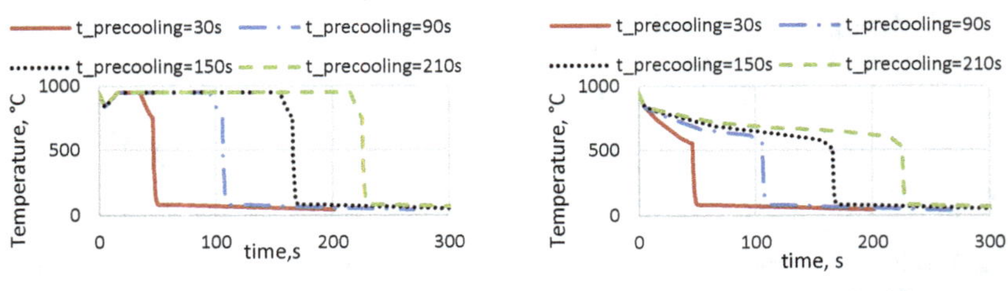

(a) (b)

Fig. 6: FE thermal cycles at different values of pre-cooling time for: (a) the hard region and (b) the soft region 2

FE thermal cycles in Fig. 6 were physically simulated using the Gleeble®3180 system on Usibor®1500 specimens, as described in the methodology section. To experimentally evaluate the hardness and the microstructure in the specimen center after physically simulating thermal cycles, the specimens were cut in the central section, mounted in an epoxy resin and then grinded and polished. Vickers hardness tests (load: 2 kg, dwell time: 10 s) were carried out, instead, for metallographic observations, the specimens were etched with a 2 % nital solution and observed by the Eclipse MA200 Nikon light microscope. The hardness mean values near the point in which control thermocouple was welded and the relative error bars are shown as a function of the pre-cooling time in Fig. 7a and Fig. 7b for the soft region and the hard region, respectively. From Fig. 7a, it can observe that an increase in the time taken in the pre-cooling phase in the second tempering furnace leads to a reduction in hardness in the soft region. Starting from a pre-cooling time equal to 150 s, the hardness remains quite constant and equal to the material hardness in the supply state. The hardness trend as a function of the pre-cooling time is in agreement with the FE one. However, the difference between numerical and experimental results of hardness is about 20 %. The results collected in Fig. 7b show a slight reduction of hardness with increasing the pre-cooling time also for the hard region. This issue is more evident if the hardness value related to $t_{precooling} = 30$ s and the one obtained with $t_{precooling}$ = 210 s are compared. In the hard region, the hardness trend as a function of the pre-cooling time does not match with the FE one. In fact, FE model estimated a constant hardness value for each value of the pre-cooling time.

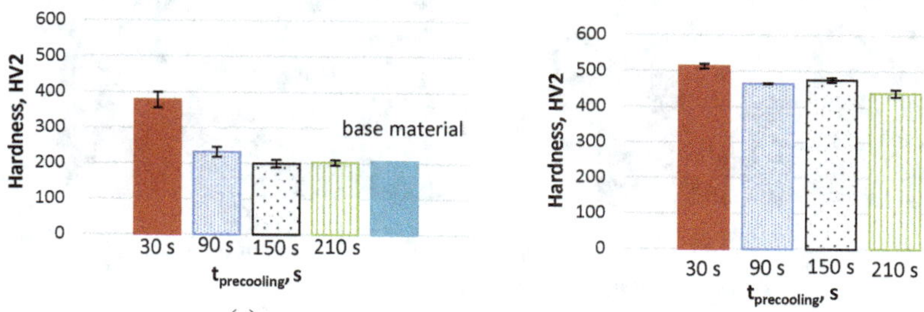

Fig. 7: Experimental hardness values for each investigated pre-cooling time in correspondence of (a) the soft region and (b) the hard region

The phenomena determining the reduction of hardness are different for the soft and hard parts. In the soft region, the longer the time taken to cool this region, the lower the cooling rate. Low values of cooling rate lead to intersect the continuous cooling transformation (CCT) curves in correspondence of most ductile microstructural phases such as ferrite, perlite and bainite. In this way, the percentage of martensite is reduced, obtaining lower hardness values. In the hard region, instead, long $t_{precooling}$ leads to an increase in the total time the blank remains at high temperature (950 °C). Geiger et al. [11], in fact, verified that the permanence at high temperature can cause the phenomenon of the austenite grain growth. Furthermore, it was reported that with increasing austenite grain size, the size of the lath martensite increases [12]. These aspects were deeper investigated and confirmed by means of metallographic analyses. In Fig. 8 the micrographs corresponding to the thermal cycles of the soft region for $t_{precooling}$ equal to 30 s (Fig. 8a) and $t_{precooling}$ equal to 210 s (Fig. 8b) are observed. Confirming FE results, from Fig. 8a a mixed ferritic-pearlitic-martenitic-bainitic microstructure is observed for $t_{precooling}$ equal to 30 s, while from Fig. 8b, a ferritic-pearlitic microstructure was recorded for a $t_{precooling}$ of 210 s. Fig. 9 shows the micrographs in the hard region for the minimum cooling time (Fig. 9a) and for the maximum pre-cooling time (Fig. 9b) investigated. By comparing these two figures qualitatively, it is possible to observe larger martensitic laths in Fig. 9b respect to those observed in Fig. 9a. Results of metallographic analyses confirm that the reduction in hardness is due to an increase in the grain size, caused by a great value of dwell time at the complete austenitization temperature.
By calculating the softening level in the soft region by means of equation 1, the histogram in Fig. 10 is obtained.

$$softening\ level\ (\%) = \frac{Hardness_{hard\ zone} - Hardness_{soft\ zone}}{Hardness_{hard\ zone}} \quad (1)$$

Results in Fig. 10 show that the new tailored press-hardening technology with intermediate pre-cooling allows to differentiate the mechanical properties of the component. For $t_{precooling}$ = 30 s a softening of 26% is obtained. The maximum softening equal to 58% is recorded for $t_{precooling}$ = 150 s.

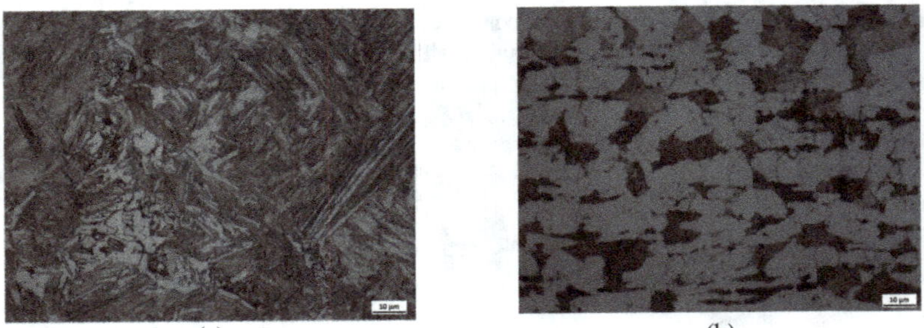

(a) (b)

Fig. 8: Micrographs (1000X) corresponding to the thermal cycles of the soft region for $t_{precooling}$ equal to 30 s (a) and $t_{precooling}$ equal to 210 s

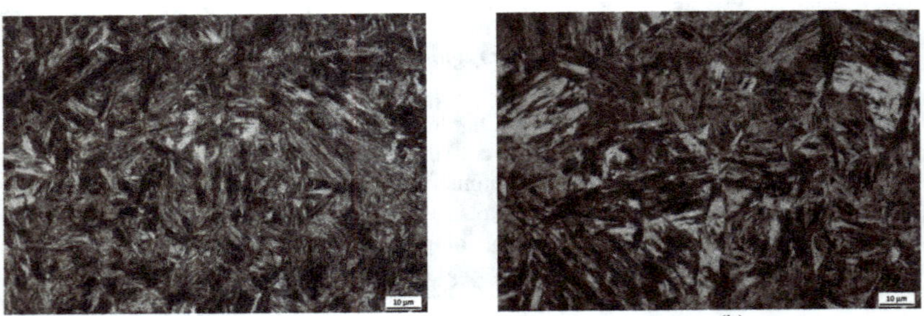

(a) (b)

Fig. 9: Micrographs (1000X) corresponding to the thermal cycles of the hard region for $t_{precooling}$ equal to 30 s (a) and $t_{precooling}$ equal to 210 s

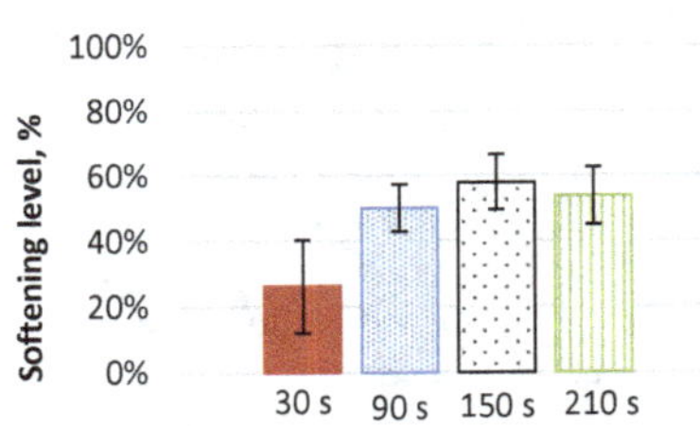

Fig.10: Softening level achieved with the investigated tailored press-hardening technology as a function of the pre-cooling time

To verify the influence of the pre-cooling time in terms of Ultimate Tensile Strength (UTS) and elongation at break (EL) in the soft region, tensile tests were performed as described in methodology section. These tests were assisted by the Digital Image Correlation (DIC) system ARAMIS 3D (provided by GOM) to evaluate the local elongation of the specimen. Fig. 11a shows the obtained stress-strain curves evaluated in the rupture point. As an example, Fig. 11b shows the deformation distribution on the specimen subjected to the thermal cycle of the process with a $t_{precooling}$ = 30 s and the relative rupture point (R). From Fig. 11a it is possible to observe that as the pre-cooling time increases, the UTS is reduced, in agreement with the hardness tests results. The high ultimate tensile strength at $t_{precooling}$=30 s is justified by the presence of a martensitic microstructure fraction. The EL, instead, tends to increase with the increase in the pre-cooling time.

The elongation at break is about 10 %, for $t_{precooling}$ = 30 s and approximately 40 % for a pre-cooling time equal to or higher than 90 s. In the maximum softening condition, the elongation at break is comparable with the one achieved with the tailored tool tempering approach [10].

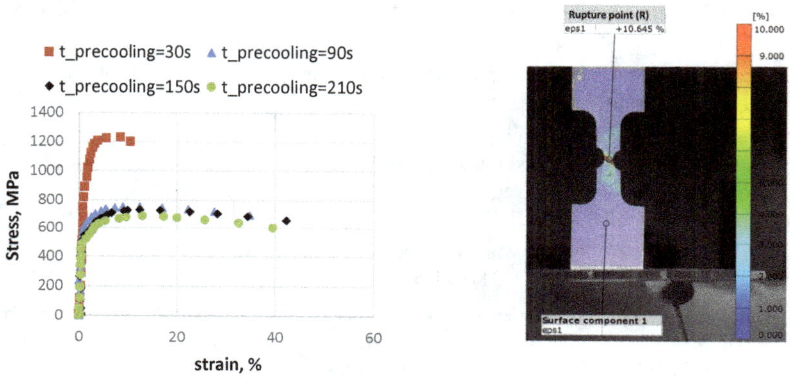

(a) (b)

Fig. 11: (a) stress-strain curves measured in the rupture point and (b) the deformation distribution on the specimen subjected to the thermal cycle with a $t_{precooling}$ = 30 s

Conclusions

The following conclusions can be drawn:

- The combination between physical and numerical simulation allowed to study the new press-hardening process with the intermediate pre-cooling phase for manufacturing automotive component with tailored properties;
- FE analysis had revealed that the most influent parameter on the component properties at the end of the process is the time taken during the taken for the pre-cooling phase in the second furnace;
- Experimental analysis in the soft region showed, in agreement with FE results, that as the pre-cooling time increases, there is a reduction in hardness, ultimate tensile strength and an increase in the elongation at break. By means of the metallographic analysis, this behaviour was justified by the microstructural changes. Specifically, as the pre-cooling time increases the austenite is transformed in ductile microstructural phases such as ferrite, perlite and bainite. Instead, experimental analysis in the hard region showed that the pre-cooling time influenced also its mechanical properties. In fact, the greater the pre-cooling time, the greater the high temperature dwell time in the hard region. The increase in the dwell time at high temperature caused a slight increase in the size of the martensitic lath, so a slight decrease in the hardness;
- The comparison between tailored tool tempering approach and the one investigated in this work showed that the softening level and deformability reached in the soft region in the condition of maximum softening are comparable. However, the new explored approach can be of greater interest to automotive companies. In fact, the manufacturing cycle time to reach greater softening is lower in the tailored press-hardening technology with intermediate pre-cooling, since the dwell time in the second tempering furnace is a downtime in which the press machine is not involved.

Acknowledgments
The authors are grateful to PNRR - Spoke 11 (Innovative Materials and Lightweighting) project for funding this research, to AutoForm and the engineer De Bellis Sergio Andrea for his support in laboratory activities.

References

[1] E. Billur. Hot Stamping of Ultra High-Strength Steels. From a technological and business perspective. Cham, 2019. https://doi.org/10.1007/978-3-319-98870-2

[2] J. Zhou, et al. Numerical simulation and experimental investigation of tailored hot stamping of boron steel by partial heating. J Mater Res Tech 14 (2021) 1347-1365. https://doi.org/10.1016/j.jmrt.2021.07.025

[3] S.A. Gunnarsdòttir, et al. Towards simulation of geometrical effects of laser tempering of boron steel before self-pierce riveting. Proc CIRP 44 (2016) 304-309. https://doi.org/10.1016/j.procir.2016.01.025

[4] D.K. Kim, et al. Induction System with Deformation Control Unit for Local Softening of Hot-Stamped Parts. Int J Auto Tech 22.3 (2021) 621-629. https://doi.org/10.1007/s12239-021-0058-7

[5] M.E. Palmieri, L. Tricarico. Influence of Conformal Cooling Channel Parameters on Hot Stamping Tool and Press-Hardening Process. In: Key Engineering Materials. Trans Tech Publications Ltd (2022) 645-654. https://doi.org/10.4028/p-5kczh8

[6] M.E. Palmieri, F.R. Galetta, and L. Tricarico. Study of Tailored Hot Stamping Process on Advanced High-Strength Steels. J Manuf Mat Process 6.1 (2022) 11. https://doi.org/10.3390/jmmp6010011

[7] Information on https://formingworld.com/hotforming-tailored-temporing-methodologies/, accessed October 2022.

[8] Information on https://www.aptgroup.com/company/news/apt-launching-temperbox%C2%AE-new-cycle-time-neutral-production-solution-enables-tailored, accessed October 2022.

[9] Information on https://ahssinsights.org/forming/press-hardened-steels/phs-tailored-products/, accessed October 2022.

[10] S.N. Lophaven, H.B. Nielsen, J, Søndergaard. Aspects of the matlab toolbox DACE. IMM, Informatics and Mathematical Modelling, The Technical University of Denmark, 2002.

[11] M. Geiger, M. Merklein, C. Hoff. Basic investigations on the hot stamping steel 22MnB5. Advanced Materials Research. Vol. 6. Trans Tech Publications Ltd, (2005) 795-804. https://doi.org/10.4028/www.scientific.net/AMR.6-8.795

[12] Y. Liu, et al. Effect of heat treatment on microstructure and property of Cr13 super martensitic stainless steel. Journal of iron and steel research international 18.11 (2011) 60-66. https://doi.org/10.1016/S1006-706X(11)60118-0

Examination of controlled thermal radiation exchange for the production of tailored properties on press-hardened components

Alborz Reihani[1, a *], Sebastian Heibel[1,b], Thomas Schweiker[1,c] and Marion Merklein[2,d]

[1] Mercedes-Benz AG, Bela-Baenyi Str., 71059 Sindelfingen, Germany

[2] Lehrstuhl für Fertigungstechnologie, Friedrich-Alexander Universität Erlangen-Nürnberg, Egerlandstr. 13, 91058 Erlangen, Germany

[a]alborz.reihani_masouleh@mercedes-benz.com, [b]sebastian.heibel@mercedes-benz.com, [c]thomas.schweiker@mercedes-benz.com, [d]marion.merklein@lft.uni-erlangen.de

Keywords: Hot Stamping, Heat Treatment, Tailored Properties

Abstract. Through the development of furnace technologies and tool concepts, partial press hardening has become an established process in the automotive industry for setting tailored properties on press-hardened components. Tailored properties on press-hardened components offer the potential to improve crash performance as well as to facilitate downstream joining processes and mechanical treatments. One approach to adjust these properties through local pre-cooling is based on a special oven chamber with an integrated masking technology, called TemperBox®. The integrated mask protects the blank locally from incident heat radiation and simultaneously absorbs the blank's own radiation. An important parameter during this process is the pre-cooling time in the TemperBox®. The pre-cooling time can influence the phase transformation and the component properties. Therefore, blanks made of the conventional press hardening steel 22MnB5 AS150 are partially precooled in the TemperBox® for different periods of time and the component properties are investigated on the basis of formed hat profiles. In addition, a simulation approach for calculating the heat radiation exchange with the finite element software LS-Dyna® is shown. Based on the simulative and experimental investigations, it is demonstrated that there is a dependency between pre-cooling time and the components properties.

Introduction

In modern automotive engineering, press-hardened steels are increasingly used in highly stressed structural components. Through a combination of forming in the austenitized state and rapid cooling, conventional press hardening steels such as 22MnB5 achieve tensile strengths up to 1600 MPa [1], which are used in all Mercedes-Benz models. In addition to the intention of achieving higher strengths and thus fully exploiting the potential for lightweight construction, the demand for steels with higher ductility is increasing. These enable targeted energy absorption as a result of plastic crash deformation. In particular, the connection areas of the B-pillar to the sill or to the roof frame as well as axially loaded side member structures have to be made more ductile. In order to integrate ductile areas on the press-hardened component, four different semi-finished product concepts and component concepts are available according to the current state of the art: Tailored welded blanks, tailored rolled blanks, patchwork blanks and tailored tempering parts [2]. For the production of tailored tempering parts, a variety of partial heat treatment processes can be distinguished due to technological developments in furnace and tooling technology. The three basic process routes in partial heat treatment are partial austenitization before forming, locally varying cooling during and after forming, and partial tempering after press hardening [2]. Through controlled heat treatment in the respective process routes, component areas with lower strength and simultaneously higher ductility can be produced. The high ductility on the component enables

Content from this work may be used under the terms of the Creative Commons Attribution 3.0 license. Any further distribution of this work must maintain attribution to the author(s) and the title of the work, journal citation and DOI. Published under license by Materials Research Forum LLC.

improved energy absorption and deformation capacity as a result of crash-like deformation and at the same time facilitates the downstream joining operations.

In this work, the influence of partial austenitization before the forming phase is investigated. A chamber furnace from AP&T AB is used for the experiments. This furnace has two chambers for heating and homogeneous austenitization of the blank and the TemperBox® for partial austenitization. The time- and temperature-dependent influence of thermal radiation plays a key role in achieving tailored properties on press-hardened components. In order to calculate this influence in advance, an approach for simulating the heat radiation exchange in LS-Dyna® is demonstrated and validated by experimental investigations.

Process for controlled phase transformation before hot stamping

In industry, controlling the temperature of semi-finished products during the heating phase is an established method for adjusting tailored properties [3]. The use of new furnace technologies enables a controlled phase transformation on the blank after the heating phase. A furnace technology for partial austenitization can be found in the TemperBox®, in which a cooled mask made of aluminium protects the blank from thermal radiation and at the same time absorbs the thermal energy of the blank. The process steps of partial press hardening with the TemperBox® are shown in Fig. 1a. First, the blank is trimmed and, depending on the blank geometry, material and sheet thickness, heated for several minutes and thus homogeneously austenitized. Afterwards, the austenitized blank is automatically positioned in the TemperBox® and partially cooled.

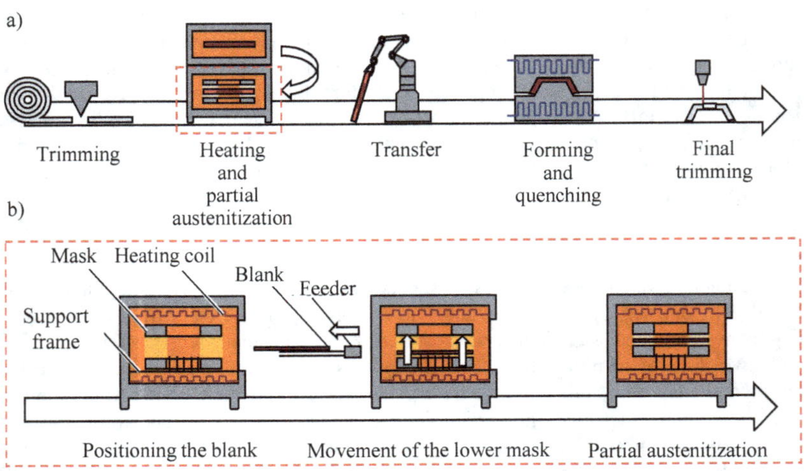

Fig. 1: Process steps in partial press hardening with TemperBox®

The process steps within the TemperBox® are shown in Fig. 1b. In the first step, the homogeneously austenitized blank is positioned on the support frame within about 4 to 6 seconds. Then the lower movable mask moves in the direction of the upper fixed mask and picks up the blank via centring cones. Depending on the geometry, material and sheet thickness, the blank is then partially precooled for about 45 to 140 s by the cooled aluminium mask. While the side of the mask facing the heating coils has a smooth surface to prevent the mask from heating up, the heat radiation is absorbed on the side of the mask facing the blank due to the higher roughness. After partial austenitization, the blank is positioned in the tool, for instance with a robot or a furnace feeder, and then formed and partially quenched. During the forming and quenching process, blank cooling begins as a result of thermal energy extraction due to tool contact. This thermal energy extraction or thermal conduction leads to the diffusionless transformation of the austenitized area of the blank into a martensitic microstructure. The effect of heat conductivity on the phase transformation of the blank depends on pre-cooling time and the phase transformation from austenite to ferrite, pearlite and bainite that has taken place beforehand. After the quenching process is completed, the components are cooled in air and trimmed to size. In the following, the basics of calculating the heat radiation exchange are described and application-related aspects of partial austenitization with the TemperBox® are presented.

Calculation of radiation exchange

In hot stamping, different heat transfer mechanisms occur depending on the process step. During the transfer phase, the semi-finished product temperature decreases due to heat radiation and convection with the environment, and in the forming and quenching phase, additional heat conduction occurs due to tool contact. Technological developments in plant and tooling technology extend the complexity of the thermal problem. For instance, in the TemperBox®, the exchange of heat radiation between surfaces in a closed space takes on a separate role. In comparison to the heat transfer mechanisms convection and heat conduction, in which the heat energy is transported from warm to cold areas, heat radiation also involves an exchange of heat energy from cold to warm areas [4]. This heat radiation exchange depends in particular on the orientation and position of the radiating surfaces, their temperatures and radiation properties [5]. A geometric quantity that is used to simplify the calculation of the heat radiation exchange is the view factor. The view factor describes the proportion of heat radiation that is transferred from one surface to another, depending on the position and orientation [4]. The starting point for the derivation of the view factor is the photometric fundamental law. For two surfaces in empty space shown in Fig. 2, the photometric fundamental law gives the radiant flow $d^2\varphi_{12}$ from the surface increment dA_1 to the surface increment dA_2 as a function of the intensity L_1, the distance S and the angles θ_1 and θ_2 [4]:

$$d^2\varphi_{12} = L_1 \frac{\cos\theta_1 \cos\theta_2}{S^2}\, dA_1 dA_2. \tag{1}$$

The integration of the photometric fundamental law (Eq. 1) over two finite surfaces and the assumption that the radiation intensity L_1 remains constant over the surface, the radiation flow can be described with the following equation:

$$\varphi_{12} = L_1 \int_{A_1}\int_{A_2} \frac{\cos\theta_1 \cos\theta_2}{S^2}\, dA_1 dA_2. \tag{2}$$

With the relation $\varphi_1 = \pi L_1 A_1$ as the radiant flux emitted from the surface A_1, the view factor is obtained as:

$$F_{12} := \frac{\varphi_{12}}{\varphi_1} = \frac{1}{A_1}\int_{A_1}\int_{A_2} \frac{\cos\theta_1 \cos\theta_2}{\pi S^2}\, dA_1 dA_2. \tag{3}$$

From the equation Eq. 3 it can be seen that the view factor is a geometric quantity that determines the proportion of radiation that is emitted from surface A_1 to surface A_2.

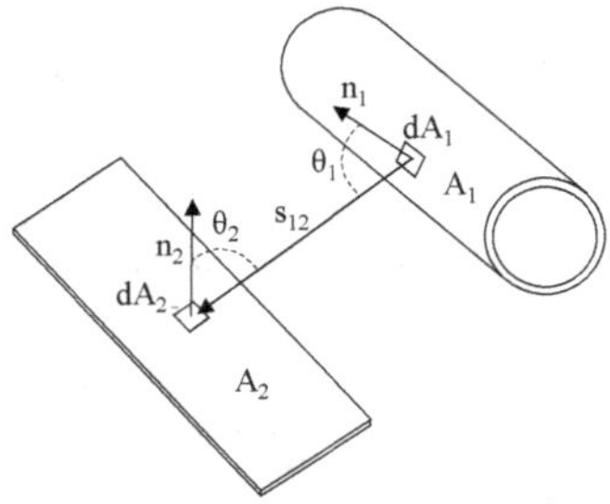

Fig. 2: Two finite-sized surfaces and their orientation and position in empty space [5]

Influence of the pre-cooling time in TemperBox®

Within the scope of this work, a test component in form of a hat profile is partially press-hardened at a trial facility of Mercedes-Benz AG and examined with regard to the semi-finished product temperature, hardness, tensile strength and elongation at break. For the investigations, the aluminised, cold-rolled and micro-alloyed 22MnB5 AS150 with a sheet thickness of $s_0 = 1.5$ mm is used. Table 1 summarises the chemical composition of 22MnB5 AS150.

Table 1: Chemical composition of 22MnB5 AS150

Alloying elements	C [%]	Mn [%]	B [%]	Cr [%]	Si [%]	Al [%]	Ti [%]	N [%]	Mo [%]
22MnB5/1,5 mm	0.23	1.18	0.003	0.18	0.25	0.04	0.04	0.004	0.004

The blanks are first heated in the chamber furnace with a furnace temperature of $T_O = 950$ °C and a furnace holding time of $t_O = 350$ s. Then the austenitized blanks are transported from the furnace chamber to the TemperBox® by a furnace feeder in 4.5 s and positioned there on the support frame between the masks. During the following partial cooling, the blank is distanced by pins intergated in the mask to ensure the distance of approx. 2 mm between the blank and the masks as well as to avoid direct contact between the masks and blank. For the parameter variation, the pre-cooling times 15 s, 30 s, 45 s, 60 s, 75 s and 90 s are chosen with regard to the direct influence on the phase transformation and cooling rate as well as due to the significance for the resulting component properties.

The partially austenitized blank is placed on the lower tool half after a transfer time of $t_T = 11$ s and is formed after 15 s by tool contact on both sides. In the hardening phase, the blank is partially quenched with a pressing force of $F_P = 1250$ kN and a hold time of $t_Z = 25$ s. The forming and hardening process is then repeated. The hydraulic press Imposite 2500 from Dieffenbacher GmbH is used for the forming and hardening process. The tools for the hat profile are made of tool steel 1.1730, which favours diffusionless phase transformation from austenite to martensite as a result of heat conduction effects due to its high thermal conductivity coefficient of $\lambda = 50$ W/(m·K).

The finite element software LS-Dyna® R12 is used to simulate the partial austenitization in the TemperBox® and the press hardening process. By calculating the view factors in LS-Dyna®, the heat radiation exchange between different surface segments can be mapped. Blankenhorn et al. [6] revised and used this calculation method to map the radiation exchange of a B-pillar segment in a paint curing oven. For the calculation of the thermal radiation exchange in the TemperBox® the

Materials Research Forum LLC
https://doi.org/10.21741/9781644902417-56

keyword *BOUNDARY_RADIATION_ENCLOSURE is used. For this purpose, the TemperBox® is modelled as a closed furnace chamber consisting of housing, heating coils, masks and blank (Fig. 3).

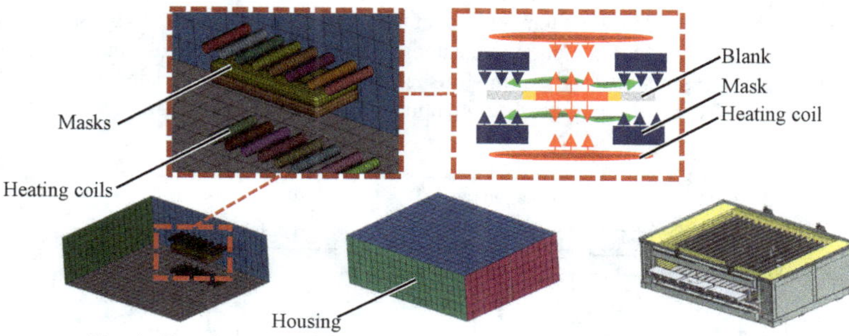

Fig. 3: Simulation model for the calculation of partial austenitization

The models for imaging the housing and the heating coils are constructed with shell elements whose unidirectional element normals point in the direction of the acting bodies. Since the heating coils affect the masks and the blank from below as well as from above, the blank and the masks are constructed with solid elements to avoid the restriction of the unidirectional radiation exchange. The thermal boundary conditions for the calculation of the radiation exchange are shown in Table 2. These were summarized from time-temperature measurements and literature values.

Table 2: Boundary conditions for the calculation of radiation exchange

Boundary conditions	Heating coil	Mask	Housing	Blank
Emission coefficient	-	0.2 [7]	-	0.75 [7]
Temperature	1633.15 K	Measurement of the time-temperature curve	323.15 K	Simulation

After the simulation of the partial austenitization, the history variables at the elements and the temperatures at the nodes of the blank are converted from volume elements to shell elements using the software Envyo®. This allows the forming and holding phases to be calculated more efficiently. In the forming and hardening phase, the process parameters such as hold time and pressing force are taken from the experimental investigations, whereby a constant Coulomb' friction coefficient of $\mu = 0.45$ is assumed. Another central aspect in the hardening phase is the heat transfer coefficient, which is integrated into the simulation as a function of the applied surface pressure based on the investigations by Svec [8]. The material model used for the 22MnB5 AS150 blank is *MAT_244, which calculates the phase transformation based on the work of Åkerström [9] and the Vickers hardness using the mathematical approach of Maynier [10].

Materials Research Proceedings **25** (2023) 455-462
https://doi.org/10.21741/9781644902417-56

To measure the temperature of the semi-finished product, type K thermocouples are welded on the surface of the blank in the heated and cooled area. From the measurement results shown in Fig. 4 it is evident that an increase in the pre-cooling time has an influence on the course of the semi-finished product temperature in the heated and in the cooled area. A comparison between the curves of the semi-finished product temperature at t_C = 15 s, t_C = 30 s and t_C = 45 s shows that the temperature in the heated and cooled area changes continuously. An increase in the pre-cooling time from 15 s to 45 s results in the semi-finished product temperature rising from 850 °C to 906 °C in the heated area after partial austenitization and decreasing from 742 °C to 633 °C in the cooled area.

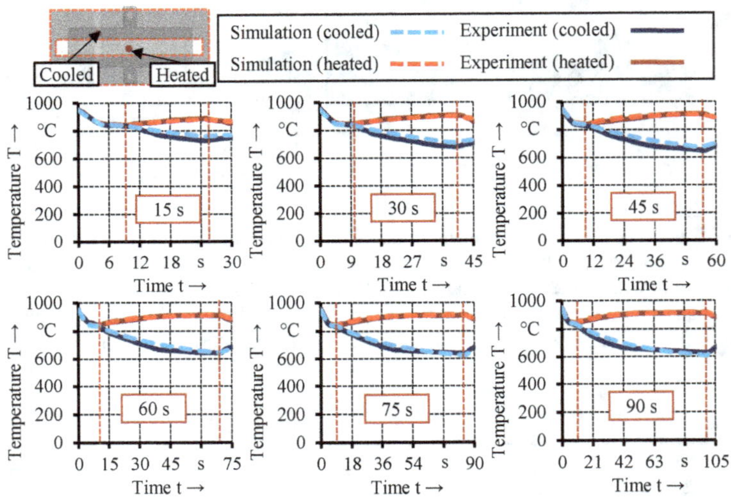

Fig. 4: Experimentally determined and simulatively calculated time-temperature curves at different pre-cooling times

Vickers hardness is an important mechanical parameter for the analysis of press-hardened components, that allows to evaluate the component properties of the semi-finished product and the resulting phase transformation. For this reason, hardness tests according to Vickers are carried out in cross-section with the semi-automatic hardness tester DuraScan 70 G5 from Struers GmbH. The results of the averaged hardness measurements for the different pre-cooling times are shown in Fig. 5. To determine the differences between experiment and simulation, the numerically calculated hardness curves and the experimentally determined hardness curves are compared. The diagrams show that the hardness curves can be calculated with good agreement. Analogous to the experimental results, no drop in hardness can be detected in the covered area from a pre-cooling time of t_C = 45 s onwards.

A significant characteristic of partially press-hardened components is the transition zone between hard and soft areas, which has an impact on their performance. The results of the hardness curves show that the pre-cooling time has an influence on the transition zone. With increasing pre-cooling time, homogeneous soft and hard areas are formed, which are characterised by narrower transition zones. Numerically, this effect can be determined by calculating the view factors and thermal interaction between the heating coils, the masks and the blank in a closed furnace chamber.

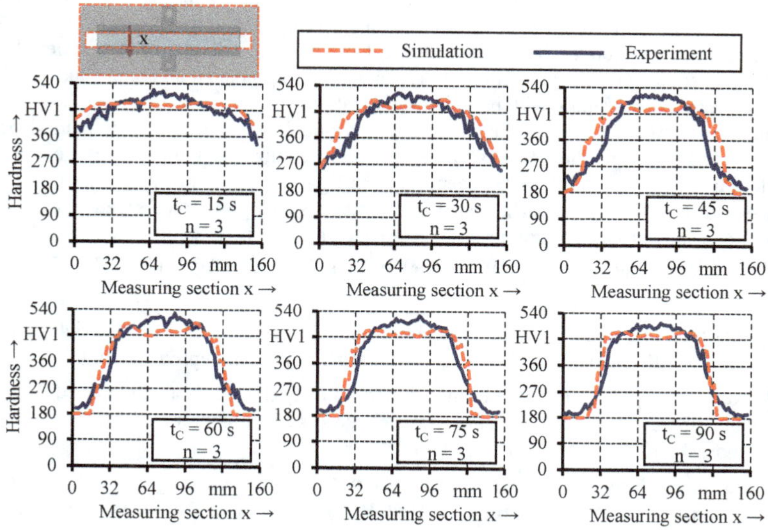

Fig. 5: Simulatively calculated and experimentally determined hardness curves

In addition to the hardness tests, tensile tests are carried out on a Z100 universal testing machine from Zwick Roll AG. The resulting mechanical properties of tensile strength and elongation at break are shown in the Fig. 6 for different pre-cooling times in the cooled and heated area.

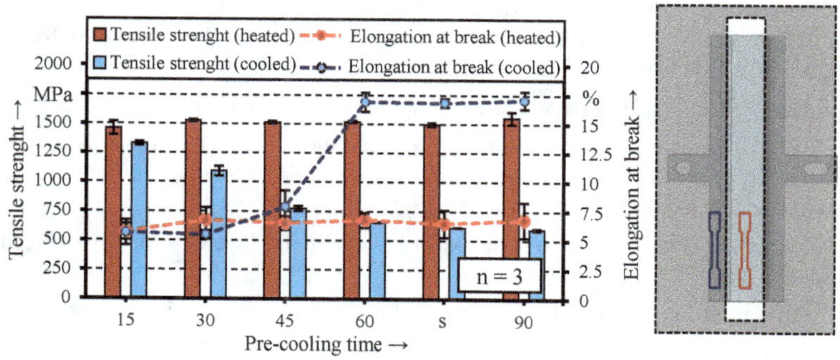

Fig. 6: Tensile strength and elongation at break in the cooled and heated area

Analysis of the diagram shows that the tensile strength in the cooled area decreases with increasing pre-cooling time up to 60 s. At the same time, the elongation at break increases significantly starting from the pre-cooling time of 60 s. The results with cooling times of 30 s and 45 s lead to a premature localisation of the strain due to the inhomogeneity of the strength. This effect is confirmed by the higher standard deviation in the respective mean tensile strengths and elongations at break. In the heated area, an effect of the pre-cooling time on the tensile strength and the elongation at break is not discernible.

Summary

In summary, the present work has shown that the partial pre-cooling time in the TemperBox® has an influence on the phase transformation and the resulting mechanical properties. Furthermore, an approach to simulate the partial austenitization with the TemperBox® was presented and validated. Overall, the combination of the view factor calculation and the use of the material model *MAT_244 is suitable for the prediction of the semi-finished product temperature and the Vickers hardness after press hardening with partial austenitization.

Future work will focus on extending the simulation model by measuring temperature-dependent emission coefficients at the blank surface and representing the temperature- and pressure-dependent tribological conditions between the forming tools and the heated blank. In addition, process-specific influencing variables such as the mask spacing, the recess geometry, the sheet thickness and the partial pre-cooling time are investigated both on semi-finished products with lower or higher strength classes and on new component concepts such as patchwork blanks.

References

[1] H. Karbasian and A. E. Tekkaya, A review on hot stamping, Journal of Materials Processing Technology 210 (2010) pp. 2103-2118. https://doi.org/10.1016/j.jmatprotec.2010.07.019

[2] M. Merklein, M. Wieland, M. Lechner, S. Bruschi, A. Ghiotti, Hot stamping of boron steel sheets with tailored properties, Journal of Materials Processing Technology 228 (2016) pp. 11-24. https://doi.org/10.1016/j.jmatprotec.2015.09.023

[3] H. Lehmann, New developments in furnaces for press-hardening. In: 5th International Conference on Hot Sheet Metal Forming of High Performance Steel (2015), pp. 331-341.

[4] H. D. Baehr, K., Stephan, Heat and Mass Transfer, Springer-Verlag Berlin Heidelberg, Berlin, Heidelberg, 2006.

[5] C. Balaji, Essentials of Radiation Heat Transfer, Springer eBook Collection. Springer International Publishing; Imprint Springer, Cham, 2021

[6] G. Blankenhorn, B., Gyesi, S., Malcolm, I., Gandkota, R., Grimes, F-.,H., Rouet, Validation of a Thermal Radiation Problem using *BOUNDARY_RADIATION_ENCLOSURE. In: 12th European LS-DYNA Conference (2019), Koblenz.

[7] Springer-Verlag GmbH: VDI-Wärmeatlas, Springer Berlin Heidelberg, Heidelberg, 2013.

[8] T. Svec, Untersuchungen zur Herstellung von funktionsoptimierten Bauteilen im partiellen Presshärtprozess mittels lokal unterschiedlich temperierter Werkzeuge, Friedrich-Alexander-Universität Erlangen-Nürnberg, Dissertation, 2015.

[9] P. Åkerström, Modelling and simulation of hot stamping. Lulea University of Technology, Dissertation, 2006.

[10] P. Maynier, B. Jungmann, J. Dollet, Creusot-loire system for the prediction of the mechanical properties of low alloy steel products, D.V. Doane, J.S. Kirkaldy (Eds.), Hardenability Concepts with Applications to Steel, (1978), p. 518.

Sheet Metal 2023
Materials Research Forum LLC
Materials Research Proceedings **25** (2023) 463-470
https://doi.org/10.21741/9781644902417-57

Investigation on the inter-ply friction when deforming magnesium-based fibre metal laminates at elevated temperature

Zheng Liu[1,a *], Enrico Simonetto[1,b], Andrea Ghiotti[1,c], Stefania Bruschi[1,d]

[1]Department of Industrial Engineering, University of Padova, Via Venezia 1, 35131 Padova, Italy

[a]zheng.liu@studenti.unipd.it, [b]enrico.simonetto.1@unipd.it, [c]andrea.ghiotti@unipd.it, [d]stefania.bruschi@unipd.it

Keywords: Friction, Hot Stamping, Fibre Metal Laminates

Abstract. Inter-ply friction plays a dominant role in inducing defects (e.g., wrinkling) during the hot stamping process of fibre metal laminates (FMLs). In particular, the metal/prepreg inter-ply friction is highly affected by the process parameters at elevated temperature as the molten matrix is significantly sensitive to possible changes in these parameters. In this paper, the metal/prepreg inter-ply friction was experimentally investigated at increasing relative sliding displacement and varying normal pressure. To do that, pull-through tests with a stop-start control strategy were conducted at elevated temperature. The obtained results showed the transition of the lubrication mode given by the Stribeck theory from a hydrodynamic to a mixed one as the relative sliding displacement rose at whatever normal pressure level. An increase in the inter-ply friction coefficient was found as well.

Introduction

Fibre metal laminates (FMLs), a form of hybrid laminated material system made up of thin metal sheets laminated with fibre-reinforced polymer (FRP) layers, take advantage of both the material categories, as they overcome the FRP reduced toughness, and, at the same, enhance the mechanical resistance of monolithic metal sheets [1]. Aluminium alloy-based FMLs, such as glass-reinforced aluminium laminates (GLARE), aramid-reinforced aluminium laminates (ARALL), and carbon-reinforced aluminium laminates (CARALL), were introduced in the aerospace sector to reduce the overall weight and improve the impact and fatigue responses of structural parts [2]. Recently, driven by the increasing demand for high strength-to-density parts, magnesium alloy-based FMLs have been introduced, with also the chance of widening the FMLs applications thanks to the effective electromagnetic interference shielding offered when magnesium alloy sheets are used [3].

For the composite core, there has been a rise in the adoption of continuous carbon fibre reinforced thermosetting polymers (CFRPs) for structural applications. Nevertheless, since thermoplastic polymers guarantee reduced cycle time, recyclability, and better compatibility with the currently available stamping lines [4], they are becoming an alternative to thermosetting ones as matrices of the composite core. However, thermoplastic polymers have a significantly higher melting viscosity than thermosetting ones, bringing various issues during the forming process that may affect the part final performances. FMLs are usually formed at elevated temperature to take advantage of the metal sheet higher formability. The deformation mechanisms taking place during stamping at elevated temperature mainly include intra-ply behaviours such as in-plane shear, out-plane bending, tension, transverse compression, and inter-ply behaviours consisting of ply/tool and ply/ply slippage [5]. In particular, the inter-ply friction is relevant to both the interface between two plies of the composite and the interface between the metal sheet and the composite. The inter-ply friction may have a significant impact on the onset of defects, such as wrinkling, during stamping, and can lead to undesired changes in the yarns, which, in turn, can lead to fibre in-plane undulations or in-plane waviness in the formed parts. Numerous researchers have investigated how

Content from this work may be used under the terms of the Creative Commons Attribution 3.0 license. Any further distribution of this work must maintain attribution to the author(s) and the title of the work, journal citation and DOI. Published under license by Materials Research Forum LLC.

process parameters such as the normal pressure, sliding velocity, and temperature may affect the tool/ prepreg inter-ply friction in case of hot stamping FRPs within the hydrodynamic lubrication range, and the obtained results were consistent with the Stribeck theory, which states that the kinetic friction coefficient can be linearly fitted as a function of the Hersey number, which is determined by the normal pressure (p), the viscosity of the liquid(η) and the relative sliding velocity(v), as shown in Fig. 1 from the work of Gorczyca-Cole et al. [6]. Chow [7] performed experiments to analyse the inter-ply friction of commingled glass polypropylene four-harness satin-weave fabric between the binder ring and the die: the mixed lubrication mode was found in this research, and an inter-ply friction analytical model was introduced accordingly to incorporate the weighted effects of the Coulomb and hydrodynamic friction models at varying process parameters. To measure the kinetic friction, Sachs et al. [8] compared and reviewed various friction test setups in terms of their measured coefficients of friction using a standardization procedure.

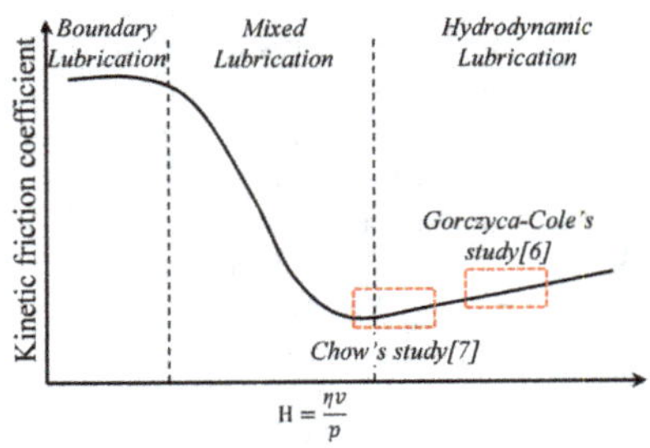

Fig. 1 Stribeck curve: kinetic friction coefficient versus Hersey number.

During the forming process carried out at temperature above the melting one of the thermoplastic polymer matrix of the composite, the thickness of the resin between the metal and composite layers varies as a consequence of the gradient of the normal pressure applied on the metal layers and the metal/composite inter-ply slippage. From the tribological viewpoint, the distance between two solid surfaces separated by a thin liquid film significantly affects the solid/liquid interfacial shear stress and further influences their inter-ply friction [9]. In this framework, in order to gain a comprehensive understanding of the mechanisms in the hot stamping process of magnesium alloy-based FMLs, the effect of the normal pressure on the inter-ply friction was investigated, as well as the coupled effect of the normal pressure and relative sliding displacement was studied.

Materials and Methods

Materials. The FMLs investigated in this study are made of glass fibre-reinforced PA6 prepregs (Tepex® 102-RG600 (2)/47% Type B, thickness 1 mm) and AZ31B magnesium alloy sheets (thickness 0.5 mm) arranged as described below. The mechanical characteristics of the magnesium alloy sheets and prepregs in the as-delivered condition were evaluated by carrying out standard tensile tests (ISO-6892) on a 50 kN MTS™ 322 hydraulic dynamometer, as shown in Table 1. The main characteristics of the prepregs are presented in Table 2.

Table 1 Mechanical properties of the AZ31B sheets and prepregs.

AZ31B sheets		Prepregs	
Elastic Modulus	45 GPa	Tensile modulus	18 GPa
Yield Strength	158±2 MPa	Tensile strength	380±5 MPa
Tensile Strength	248±4 MPa	Strain at break	2.3%
Shear Modulus	16.7 GPa	Flexural Modulus	16 GPa
Poisson's Ratio	0.35	Flexural Strength	300 ±5 MPa

Table 2 Main characteristics of the prepregs [10].

Layup	Value		Unit
	Longitudinal	Transversal	
Fibre	E-Glass		
Weaving style	Twill 2/2		
Area weight (dry fabric)	600		g/m^2
Yarn count	1200		tex
Yarn density	2.5	2.5	1/cm
Weight rate	50	50	%
Fibre content	47		Vol-%
Thickness per layer	0.5		mm
Laminate density	1.8		g/cm^3

Stop-start pull-through tests. According to the ASTM standard D3528 [11] for evaluating the tensile and shear strengths of adhesives for bonding metals, the testing apparatus shown in Fig. 2 (a) and (b) was designed and constructed for the measurement of the metal/prepreg inter-ply friction coefficient. The adopted cross-shaped specimen is made of two prepreg layers along the horizontal direction and one metal sheet between them along the vertical direction (see Fig. 3 for the sketch and dimensions of the specimen). The specimens were tested in a plate-plate configuration with a constant overlapping area of 20 x 20 mm^2, see Fig. 3. The tests were performed on the MTS™ 322 hydraulic dynamometer equipped with the MTS™ 651 environmental chamber to heat up the specimen to the testing temperature. To achieve a uniform temperature distribution, the set-up was heated up for 1 hour before assembling the specimen. The metal sheet was gripped to the lower clamp of the testing machine so that the pulling velocity could be controlled. Various normal forces F were adopted by controlling the stroke of the compression spring shown in Fig. 2. Therefore, the normal pressure p can be calculated from Eq. 1, where A is the overlapping area shown in Fig. 3. The homogeneous normal pressure distribution is assumed in this research. With the normal pressure applied on this area, the concave and convex structures of the fibres will be easily flattened at elevated temperatures, which contributes to a neglectable nonuniform normal pressure distribution.

$$p = \frac{F}{A}. \tag{1}$$

To assess the potential effect of the metal/prepregs relative sliding displacement on the inter-ply friction, the total stroke of the magnesium alloy sheet along the pulling direction, namely 25 mm, was uniformly divided into 5 stages as indicated in Fig. 3. A stop-start control strategy was applied in each stage to obtain the inter-ply friction coefficient. Pulling-through of the AZ31B sheet was regulated in a distance-controlled mode within each stage, so that the test was conducted at a given sliding velocity, meanwhile, the pulling force was monitored as a function of the distance. The test was stopped for one minute at the end of each stage to relax and relieve the stress, and then restarted for the next stage.

The pull-through tests were conducted at normal pressure of 0.5 MPa, 2.5 MPa, and 5 MPa, which covers the range of normal pressure in the thermoforming process of typical hat-shaped FMLs with the same material system [12]. To investigate the effect of normal pressure on the inter-ply friction, other influence factors were kept constant with the pulling velocity of the AZ31B sheet fixed at 0.5 mm/s as well as the testing temperature at 235°C, which is above the PA6 melting temperature, namely 220°C. For each testing condition, five tests were performed to assure the repeatability of the results.

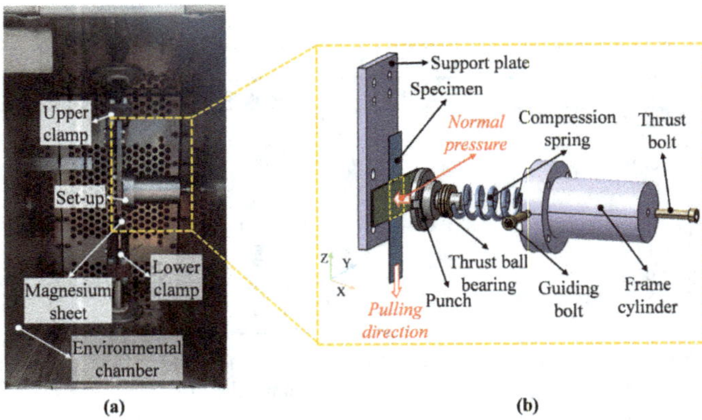

Fig. 2 Set-up for the stop-start pull-through tests: (a) assembly; (b) scheme.

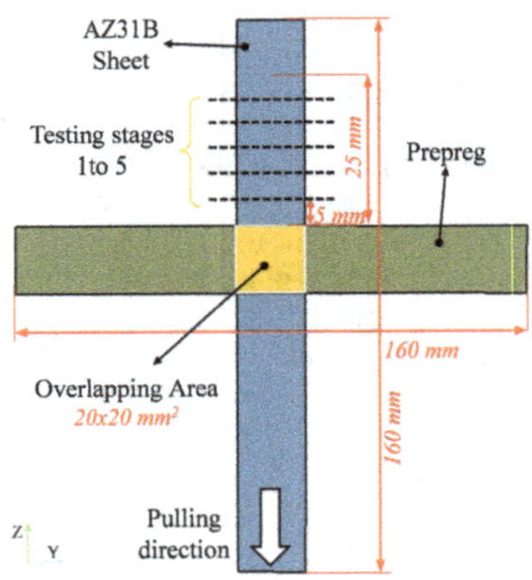

Fig. 3 Sketch and dimensions of the specimen used in the stop-start pull-through tests.

Test data elaboration. Fig. 4 illustrates the friction force developing between two surfaces separated by a liquid film. When the liquid film is thick enough, the friction force shows a typical liquid-like characteristic: it rises first at increasing relative sliding displacement, and then it reaches a stable value, named F_k, the kinetic friction force. As the liquid film thickness decreases, a hard-wall film occurs as a result of the increase in the liquid viscosity [13]. Therefore, a solid-like shear response is observed, characterized by the static friction force Fs.

The pulling force F measured during the stop-start pull-through tests, namely the friction force, can be used to calculate the inter-ply shear stress τ using Eq. 2.

$$\tau = \frac{F}{2A} \tag{2}$$

Furthermore, the static shear stress τ_S and kinetic shear stress τ_K can be obtained from Eq. 3 and Eq. 4, respectively.

$$\tau_S = \frac{F_S}{2A} \tag{3}$$

$$\tau_K = \frac{F_K}{2A} \tag{4}$$

Therefore, the static and kinetic friction coefficients can be calculated according to Eq. 5 and Eq. 6, respectively.

$$u_S = \frac{\tau_S}{p} \quad (5)$$

$$u_K = \frac{\tau_K}{p} \quad (6)$$

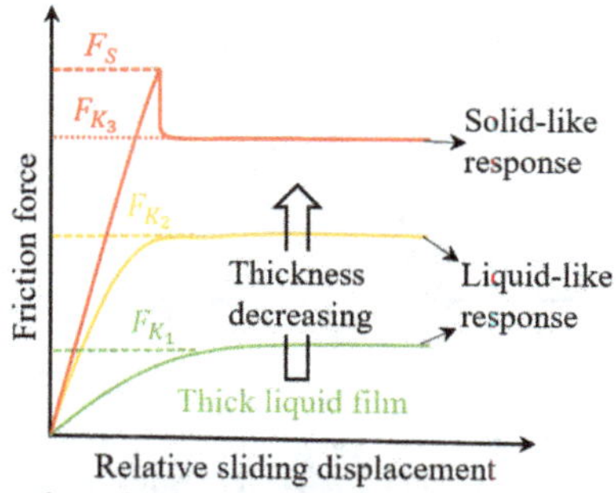

Fig. 4 Friction force between two surfaces separated by a liquid film.

Results and discussion

Influence of the relative sliding displacement on the inter-ply friction. Fig shows the inter-ply shear stress τ at different stages when applying 2.5 MPa as normal pressure p. It can be seen that the evolution of the shear stress is significantly different as the stage increases. A gradual increase in the shear stress followed by a steady state was found in case of the stages 1 and 2: this is indicative of a liquid-like inter-ply shear characteristic, which makes negligible the difference between the static and dynamic friction coefficients as shown in Fig. 5 (b) at these two stages. This can be further proved by the morphology of the residues on the surface of the tested magnesium alloy sheets (see Fig. 6). As most of the residues detected on the AZ31 sheets at stage 1 were of the polymer matrix, the inter-ply shear occurred at the interface between the metal and the polymer matrix. At the initial two stages, the molten polymer matrix actually worked as a sort of liquid lubricant at the metal/prepreg interface. As a result, the prepreg and the metal sheet were separated by a liquid film, which identifies the hydrodynamic mode of lubrication. From stages 1 to 2, more molten polymer matrix was removed from the connecting area due to the longer relative sliding displacement, and, as a consequence, the thickness of the polymer matrix becomes thinner. At stage 3, a mixture of polymer matrix and entangled fibres is visible on the tested magnesium alloy sheets as shown in Fig. 6 (b). This means that the mixed mode of the metal/polymer matrix and metal/fibre inter-ply friction makes the metal/prepreg inter-ply friction fall into the mixed lubrication range. A different shear stress response is also brought by the transition of the lubrication mode: the shear stress curve relevant to stage 3 in Fig. 5 (a), after reaching the peak point, decreases until a steady state level is reached. This curve clearly shows a solid-like inter-ply shear response. From stage 3 to stage 5, there is a growing divergence between the static and dynamic friction coefficients as shown in Fig. 5 (b). At stage 5, the difference reaches its maximum value, confirmed by the mixture of fibre and polymer matrix-dominate residues on the tested AZ31 sheets (see Fig. 6 (c)). What's more, the metal/fibre inter-ply friction coefficient, which can be interpreted as the Coulomb friction coefficient, is higher than the one at the metal/ polymer matrix interface, so the inter-ply friction coefficient increases at increasing fibre volume fraction. This explains why, as the stage moves from 3 to 5, both the inter-ply friction and shear stress rise. From stages 1 to 5, it can be seen that as the stage rose, the stroke reaching the shear stress peak declined. This can be explained by the fact that as the amount of molten polymer matrix in the contacting area reduces, the content of fibre grows, increasing the viscosity of the interface layer and further raising the system shear modulus (the ratio of the peak shear stress to the corresponding shear strain) [14].

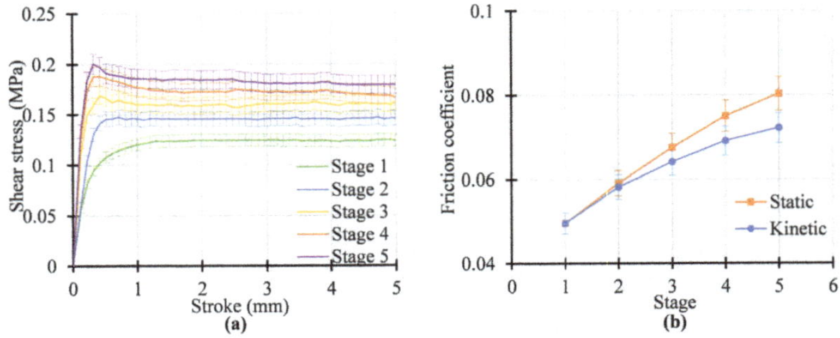

Fig. 5 Pull-through test results at different stages and 2.5 MPa normal pressure: (a) shear stress, and (b) friction coefficient.

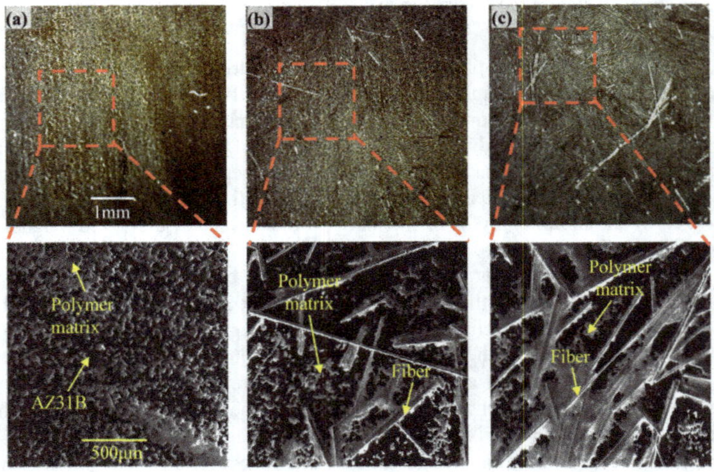

Fig. 6 Morphology of the residues on the tested AZ31 sheets at different stages:(a) stage 1, (b) stage 3, and (c) stage 5.

Influence of the normal pressure on the inter-ply friction. Fig. 7 shows the pull-through test results at varying normal pressure. As expected, from Fig. 7 (a), the higher the normal pressure (from 0.5 MPa to 5 MPa) the higher the shear stress (from $\tau^{p0.5}$ to τ^{p5}) at all the stages. However, the effect of increasing normal pressure on the increase in the inter-ply shear stress was less than linearly proportional. At stage 1, the liquid-like inter-ply friction at normal pressure from 0.5 MPa ($\mu^{p0.5}$) to 2.5 MPa ($\mu^{p2.5}$) made the inter-ply friction located in the hydrodynamic lubrication region. The net result was a decrease in the coefficient of friction at increasing normal pressure in this region, which is consistent with the results presented in [13]. This can be explained by the fact that the fibre layer was still characterized by a relatively rough surface when low values of the normal pressure were applied. On the contrary, the surface asperities were flattened at increasing normal pressure, which contributed to a decrease in the inter-ply friction coefficient. When the applied normal pressure was 5 MPa, the surface of the fibre layer was further flattened, however, the mixed lubrication mode due to the direct contact between the metal sheet and fibre layer, as shown in Fig. 8 (a), increased the friction coefficient. As a result, the inter-ply friction at 5 MPa normal pressure was a little higher than that at 2.5 MPa. Moreover, the inter-ply shear stress as a function of the relative sliding displacement at varying normal pressure shows different tendencies. It can be seen that a solid-like shear response was found at stage 1 when the FML was subjected to a 5 MPa normal pressure, leading to fibre residues on the tested magnesium alloy sheet as shown in Fig. 8 (a). When the normal pressure was decreased to 2.5 MPa, the transition from a liquid-like response to a solid-like one was found at stage 3. For a normal pressure of 0.5 MPa, a slight gap between the static and kinetic shear stress between the metal and fibre layers was found at the last two stages. It can be seen that the highest normal pressure contributed to accelerate the transition from a liquid-like inter-ply friction response to a solid-like one. This can be explained by the fact that the higher the normal pressure the more the molten polymer matrix squeezed out from the contacting area, leading to a decrease in its thickness between the metal and the fibre layers at the initial stage. On the other hand, as the relative sliding displacement increased, the higher shear stress caused by the higher normal pressure aggravated the loss of resin. This phenomenon contributed to the contact between the fibre and metal layers. What's more, with the increasing contact area between the metal and fibre layers, a wider gap between the static and kinetic shear stresses was found at the highest normal pressure.

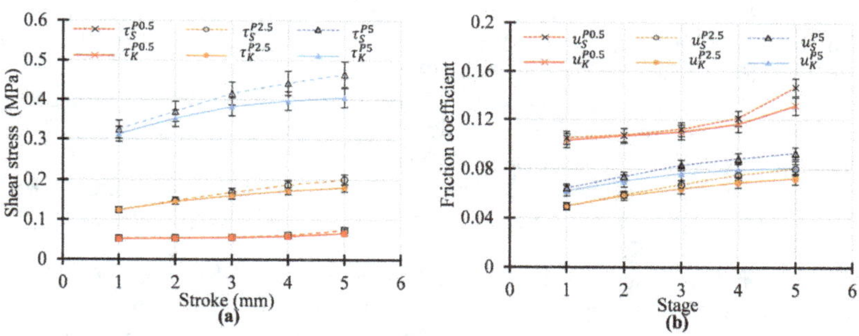

Fig. 7 Pull-through test results at varying normal pressure: (a) shear stress, and (b) friction coefficient.

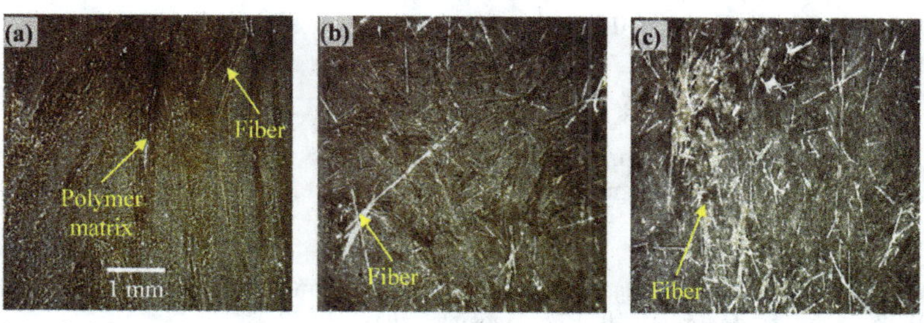

Fig. 8 Morphology of the residues on the tested AZ31 sheets at 5 MPa normal pressure: (a) stage 1, (b) stage 3, and (c) stage 5.

Summary

In this study, a comprehensive series of experiments were conducted to gain a better understanding of the effect of the normal pressure on the inter-ply friction when testing magnesium alloy-based

FMLs at elevated temperature. The main findings of the present research work can be summarized as follows. The metal/prepreg inter-ply friction at elevated temperature was found to be sensitive to the relative sliding displacement regardless of the applied normal pressure. As the relative sliding displacement rose, a transition from hydrodynamic lubrication mode to a mixed one was found, together with an increase of the inter-ply friction. Moreover, the increase in the normal pressure promoted the transition, further impacting the inter-ply friction.

References

[1] G. Lange, A. Carradò, H. Palkowski, Tailored sandwich structures in the focus of research, Mater Manuf Processes. 24 (2009):1150–4. https://doi.org/10.1080/10426910902978977

[2] X. Zhang, Q. Ma, Y. Dai, et al, Effects of surface treatments and bonding types on the interfacial behavior of fiber metal laminate based on magnesium alloy, Appl Surf Sci. 427 (2018) 897–906. https://doi.org/10.1016/j.apsusc.2017.09.024

[3] G. Lawcock, L. Ye, Y.W. Mai, C.Sun, The effect of adhesive bonding between aluminum and composite prepreg on the mechanical properties of carbon-fiber-reinforced metal laminates, Compos Sci Technol.57 (1997) 35–45. https://doi.org/10.1016/S0266-3538(96)00107-8

[4] H.M. El-Dessouky, CA. Lawrence, Ultra-lightweight carbon fibre/thermoplastic composite material using spread tow technology, Compos B Eng. 50 (2013) 91–97. https://doi.org/10.1016/j.compositesb.2013.01.026

[5] H. Chen, S. Li, J. Wang, A. Ding, A focused review on the thermo-stamping process and simulation, Composites Part B 224 (2021) 109196. https://doi.org/10.1016/j.compositesb.2021.109196

[6] J. L. Gorczyca-Cole, J. A. Sherwood, J.Chen A friction model for thermostamping commingled glass–polypropylene woven fabrics. Compos. Part A Appl. Sci. Manuf. 38 (2007) 393https://doi.org/10.1016/j.compositesa.2006.03.006

[7] S. Chow, Frictional interaction between blank holder and fabric in stamping of woven thermoplastic composites, Master thesis, Department of Mechanical Engineering, University of Massachusetts Lowell, 2002.

[8] U. Sachsa, R. Akkermanab, K. Fetfatsidisc, E. Vidal-Salléd, J.Schumachere, G. Ziegmanne, S. Allaouif, G. Hivetf, B. Marong, K. Vancloosterh, S. V.Lomov, Characterization of the dynamic friction of woven fabrics: Experimental methods and benchmark results, Compos. Part A Appl. Sci. Manuf. 67 (2014) 289–298. https://doi.org/10.1016/j.compositesa.2014.08.026

[9] S. Yamada, Nanotribology of Symmetric and Asymmetric Liquid Lubricants, Symmetry. 2 (2010) 320-345. https://doi.org/10.3390/sym2010320

[10] Information on www.tepx.com.

[11] ASTM D1894, Standard test method for static and kinetic coefficients of friction of plastic film and sheeting. Annu. B. ASTM Standard International.

[12] Ghiotti, A., Bruschi, S., Kain, M., Lizzul, L., Simonetto, E., et al., Simultaneous bonding and forming of Mg fibre metal laminates at high temperature. J Manuf Process; 72:105-114. https://doi.org/10.1016/j.jmapro.2021.10.017

[13] L. Wang, P. Xu, X. Peng, K. Zhao, R. Wie, Characterization of inter-ply slipping behaviors in hot diaphragm preforming: Experiments and modelling, Compos. Part A Appl. Sci. Manuf. 121 (2019) 28–35. https://doi.org/10.1016/j.compositesa.2019.03.012

[14] S. Erland, T.J. Dodwell, R. Butler, Characterization of inter-ply shear in uncured carbon fibre prepreg, Compos. Part A Appl. Sci. Manuf .77 (2015) 210–218. https://doi.org/10.1016/j.compositesa.2015.07.008

Keyword Index

About the Editors

Prof. Marion Merklein is full professor at the Friedrich-Alexander-Universität Erlangen-Nürnberg (FAU) and head of the Institute of Manufacturing Technology (LFT). After studying Material Science, she received her PhD in Manufacturing Engineering in 2001. In 2006 she got her postdoctoral lecture qualification and the venia legendi for Manufacturing Technology. Since April 1st, 2008, she has been head of LFT.

Prof. Merklein was spokesperson of the research unit FOR552 on hot stamping and is spokesperson of the CRC 73 on sheet-nulk metal forming. She is involved in the CRC814 on additive manufacturing and the CRC 285 on mechanical joining. In 2013 she received the prestigious Gottfried-Wilhelm-Leibniz price for her outstanding research activities. Prof. Marion Merklein is member of the international academy on production technology CIRP, member of the oldest German national academy Leopoldina, of Berlin-Brandenburgische Akademie der Wissenschaften (BBAW) and acatech. She is engaged as senator with the German Research Foundation (DFG). Beside other honors she received three calls from other national and international universities. Prof. Marion Merklein is acting as associate editor of the international Journal on Materials Processing Technology and the CIRP Journal of Manufacturing Science and Technology as well as editor of the international journal Production Engineering.

Prof. Hinnerk Hagenah is professor at the Friedrich-Alexander-Universität Erlangen-Nürnberg (FAU) at the Institute of Manufacturing Technology (LFT). He received his diploma in Computer Science from the FAU in 1995. In 2002 he received his PhD in Manufacturing Engineering at the LFT under the guidance of Prof. Geiger. The title of the thesis was "Simulation-based estimation of the expected accuracy in sheet metal bending". In 2010 he got his postdoctoral lecturer qualification focusing on "Knowledge-based methods in the design of manufacturing processes" and the venia legendi for Manufacturing Technology. Since 2019 he is professor at the LFT.

He has been managing director of the Collaborative Research Center 396 (Robust,shortened process sequences for lightweight sheet parts) and the Transregional Collaborative Research Center 73 (Manufacturing of Complex Functional Components with Variants by Using a New Sheet Metal Forming Process - Sheet-Bulk Metal Forming). His research and teaching interests are in the field of knowledge based approaches in manufacturing engineering, sheet metal processing and process optimization (6σ). Currently Prof. Hagenah is managing editor of the international journal Production Engineering and Secretary of ESAFORM (European Scientific Association for Material Forming).

Prof. Joost R. Duflou holds master degrees in Architectural and Electro-mechanical Engineering and a PhD in Engineering Sciences from KU Leuven, Belgium. After a number of years of industrial experience in different international companies, he has been a faculty member at the Mechanical Engineering Department of KU Leuven since 1997. He became a tenured Full Professor in 2012. His principal research activities are situated in the field of design support methods and methodologies, with special attention for Ecodesign and Life Cycle Engineering, and Sustainable Manufacturing. As chairholder of the LVD Chair on Sheet Metal Processing, he also leads a research group focussing on sheet metal oriented manufacturing processes and systems and he supervises the KU Leuven Division on Manufacturing Processes and Systems. He is a CIRP Fellow and has published over 400 international publications. As chair and board member of several spin-off companies and professional associations he contributes to research valorisation and dissemination. More detailed information can be obtained via http://www.kuleuven.be/wieiswie/nl/person/00016263.

Prof. Livan Fratini graduated with honours in Mechanical Engineering in 1993 and took his PhD in Production Engineering at University of Palermo in Italy in 1997. Currently he is full professor of Manufacturing at the University of Palermo. He has been Deputy Rector up to Oct. 2021; he has been formerly Vice-Rector for Research and Third Mission since 2018.

The scientific activity has been mainly focused on the development of numerical models for the analysis of the plastic behavior of metals during forming processes and to the development of experimental tests aimed to the validation of the numerical results with particular reference to the sheet metal stamping operations. Most recently he has been working on friction stir welding and on solid bonding processes. He has been coordinating or involved in several research projects both at a national and international level.

Prof. Fratini is co-author of over 350 papers on international scientific journals or on the proceedings of national and international conferences (ORCID: 0000-0002-1059-2850).

Prof. Fratini was awarded the F.W. Taylor Medal by the CIRP (2007). He has been Vice-President of AITEM (Italian Association of Manufacturing) in the period 2013-2017; he is Fellow of CIRP (The International Academy for Production Engineering). Currently he is President of ESAFORM (European Scientific Association for Material Forming) since 2020; he is President Elect of Namri/SME for 2021-22.

Paulo Martins is professor of manufacturing at Instituto Superior Técnico, University of Lisbon, Portugal. He obtained a PhD in mechanical engineering from Instituto Superior Técnico in 1991, the habilitation in mechanical engineering from Instituto Superior Técnico in 1999 and the degree of doctor technices honoris causa from the Technical University of Denmark in 2018.

He is fellow of the International Academy for Production Engineering (CIRP), member of the Standing Advisory Board of ICTP – The International Conference on Technology of Plasticity and received the (JSTP) Japan Society for Technology of Plasticity International Prize for R&D in Precision Forging in 2021.

He was president of the scientific board of Instituto Superior Técnico from 2009 to 2012 and president of the faculty council of Instituto Superior Técnico from 2017 to 2020. He is currently a member of the general council of the University of Lisbon.

His research interests are focused on metal forming, joining by forming and hybrid wire-arc additive manufacturing technologies and he is co-author of seven books, several international patents and more than 400 papers in international journals and conferences.

After graduating in mechanical engineering from the University of Paderborn in August 1994, **Prof. Dr.-Ing. Gerson Meschut** was employed as a research assistant at the Laboratory for Materials and Joining Technology (LWF) under the supervision of Prof. Dr.-Ing. Ortwin Hahn. He completed his doctoral thesis on "Investigations into the thermal conductivity of bonded joints, taking into account bond-specific boundary conditions" in 1998. At the beginning of 2000, he moved to Group Research at Volkswagen AG in Wolfsburg. On November 1, 2003, he was appointed to the management of Volkswagen AG, where he was responsible for optimizing existing joining systems and developing and establishing new ones for innovative lightweight construction concepts in the Volkswagen Group. At the beginning of 2005, he was appointed managing director of Wilhelm Böllhoff GmbH & Co. KG in Bielefeld. There he was responsible for worldwide research and development, application technology and patent management of the Böllhoff Group. On 01.09.2011 he was appointed as successor of Prof. Hahn at the University of Paderborn. A significant research focus of the LWF is on the development of joining techniques suitable for materials, especially for hybrid materials and mixed construction methods. Another competence of the LWF is the development of methods for experimental and numerical process simulation and stress analysis or lifetime prediction of joined hybrid structures. Through close cooperation with renowned industrial partners and research institutes in numerous bilateral or publicly funded projects, the LWF has developed into a recognized institution with nationally and internationally recognized expertise in the field of joining technology for lightweight construction. Prof. Meschut is the spokesperson of the Collaborative Research Center/Transregio TRR285.

Prof. Fabrizio Micari is professor at the University of Palermo (Department of Engineering).

He graduated in Mechanical Engineering from the University of Palermo with laude on July 1986. In 1999 he became full professor of Manufacturing Tecnology at the University of Calabria and in 2002 he moved to the University of Palermo.

He has been Head of the Department of Production and Management Engineering from 2008 to 2010, Dean of the Faculty of Engineering from 2010 to 2015 and finally Rector of the University of Palermo from 2015 to 2021.

He has been President of AITeM (Italian Association of Manufacturing Technology) from 2005 to 2009, President of the Italian Association of Mechanical Engineering from 2008 to 2010 and President of the Board of the Deans of the Faculty of Engineering of the Italian Universities from 2013 to 2015.

He is Fellow of CIRP (International Institution for Production Engineering Research), where he has been Technical Secretary of the CIRP Commission on Forming from 1996 to 2001, of ESAFORM and ASME. He has been Visiting Professor at the Laboratoire de Mecanique des Systemes ed des Procedes (LMSP) of the Ecole Nationale Supérieure d'Arts et Métiers (ENSAM) in Paris in 2003.

His research and teaching interests are in the field of modelling and optimization of metal forming and machining operations and in the field of R&D management.

www.ingramcontent.com/pod-product-compliance
Lightning Source LLC
LaVergne TN
LVHW052215200926
848481LV00001B/116
9781644902400